BUILDING TECHNOLOGY: Mechanical & Electrical Systems

BUILDING TECHNOLOGY: Mechanical & Electrical Systems

WILLIAM J. McGUINNESS
Professor of Architecture Emeritus,
    Pratt Institute
Partner, McGuinness and Duncan, Engineers

BENJAMIN STEIN
Consulting Engineer

JOHN WILEY & SONS
New York • Santa Barbara • London • Sydney • Toronto

Library of Congress Cataloging in Publication Data:

McGuinness, William J
  Building technology.

  Includes bibliographical references and index.
   1. Buildings—Mechanical equipment.
2. Buildings & Electric equipment. 3. Environmental engineering (Buildings) I. Stein, Benjamin, joint author. II. Title.

TH6010 M24   696   76-14961
ISBN 0-471-58433-9

Printed in the United States of America

10 9 8 7 6 5 4

Preface

In today's world the need for technologists is great and is constantly increasing. As new cities spring up in the far corners of every continent, our dependence on automation grows.

But today is also a period for special thoughtfulness and care. Only now do we realize how wasteful we have been with the world's limited supply of energy and how badly we have treated nature. Replanning and change are essential. The ways of the past are no longer adequate. In the future we must plan and build more intelligently.

This book is for students who will be doing this planning and building. Some will be learning technology in a college. Others, already engaged in technical work or just entering this field, will find it useful. In this sense, we are *all* students.

The advice of many educators has influenced this writing. To present useful material within the limits of a few hundred pages is not an easy task. For this reason the book's discussion is limited to the basic principles and systems of the following:

Heating, cooling, ventilation
Plumbing
Electricity
Lighting

This decision is based on the need for thoroughness. Ahead lies the challenge of large institutional buildings and high-rise structures. The technologist will advance to solve these problems as his or her experience increases.

In general, the systems discussed follow recent trends. Since major changes are to be expected, no effort is made to pinpoint "best" systems. Instead, we stress the need for an orderly approach, a knowledge of equipment and technical vocabulary, design steps and clear drawings.

It is hoped that this book will introduce many students to the fascinating subject of building equipment.

William J. McGuinness
Benjamin Stein

Acknowl-
edgments

We are deeply indebted to all those who have given us aid and advice.

The following educators were most generous in their responses to our inquiry about courses and educational needs at their respective colleges and technical institutes:

Professor L. C. Beauman
Jackson Area Vo-Tech.
Jackson, Minnesota

Professor Roy E. Berger, Chairman
Architectural Drafting
Des Moines Area Community College
Ankeny, Iowa

Professor Daryl Blanchard, AIA
University of Toledo—UCATC
Toledo, Ohio

Professor James B. Boggs, AIA
Chairman, Department of Architectural and
    Drafting Technology
Corpus Christi, Texas

Professor M. Copeland
Westard Community College
Fort Smith, Arkansas

Professor Thomas Dalton
State University of New York
Canton ATC
Canton, New York

Professor Robert W Duncan, P.E.
Mercer County College
Trenton, New Jersey

Professor James L. Flynn
Dutchess Community College
Poughkeepsie, New York

Professor Coswell E. Gerrald, AIA
Guilford Technical Institute
Jamestown, North Carolina

Professor Robert E. Kerns
Henry Ford Community College
Dearborn, Michigan

Professor L. E. Lippincott
Del Mar College
Corpus Christi, Texas

Professor Raymond Loer
Eau Claire Technical Institute
Eau Claire, Wisconsin

Professor Paul J. Lougeay
Chairman, Dept. of Interior Design
Southern Illinois University
Carbondale, Illinois

Professor Geoffrey B. Lynch
Architectural Department
Miami Dade Community College
South Campus
Miami, Florida

Professor Chester Orvold
Department of Architecture Technology
Southern Technical Institute
Marietta, Georgia

Professor Charles Paquette
Los Angeles Trade and Technical Institute
Los Angeles, California

Professor Gordon Phillips, AIA
Architectural Technology
City College of San Francisco
San Francisco, California

Professor R. J. Reinholt, Chairman
Architectural Technology
Schoolcraft College
Livonia, Michigan

Professor Dean R. Roll
Architectural Drafting
Southeast Community College
Milford, Nebraska

Professor Nick Scarlatis
Purdue University
Calumet Campus
Hammond, Indiana

Professor Robert Sebek, I/C
St. Petersburg Junior College
Clearwater, Florida

Professor Robert L. Simonds
Chairman, Construction Department
Orange Coast Community College
Costa Mesa, California

Professor H. William Succop
Dept. of Building Construction and Architectural Technology
Miami Dade Junior College
Miami, Florida

Professor Gene E. Trotter, AIA
Supervisor Architectural Technology
School of Technical Careers
Southern Illinois University
Carbondale, Illinois

Professor R. Troy
Architectural Technology
El Centro College
Dallas, Texas

Professor Ray O. Webb
Architectural Technology
Alta Loma, California

Professor Joseph Yohanan
Architectural Technology
William Rainey Harper College
Palatine, Illinois

Mr. James E. Ellison, AIA, Administrator of the Department of Education and Research of the American Institute of Architects, provided us with information about courses in architectural technology. (The Institute, prior to 1972, had established recommended curriculums that have now been adopted as guides by many colleges.)

Professor Donald S. Duncan of the Schools of Engineering and Architecture at Pratt Institute, in monitoring our work, assisted us greatly.

Budd Mogensen, AIA, Architect and Planner, and Samuel Scheiner, AIA, of Scheiner & Swit, Architects, contributed plans of buildings for which systems could be designed. Both hold professorships at New York Institute of Technology, Division of Architecture and Arts, Olindo Grossi, Dean.

By the kindness of Frank J. Versagi, Editor of *Air Conditioning and Refrigeration News*, our readers benefit by an introduction to the metric system through his excellent article "Metrication is Coming: Prepare for the Transition."

Several groups helped us to bridge the gap between textbook and the realism of the construction job. They include:

Professional societies
Trade associations
Code authorities
Manufacturers

We are grateful to Lila Stein and Mrs. Marvin H. Cohen for preparing the manuscript, and to Jonathan and Emily Stein for assistance with the proofs.

William J. McGuinness
Benjamin Stein

Symbols and Abbreviations

a-c	alternating current
AFF	above finished floor
AHAM	Association of Home Appliance Manufacturers
AIA	American Institute of Architects
amp	ampere (s)
ASHRAE	American Society of Heating, Refrigerating and Air Conditioning Engineers
AWG	American Wire Gage
Btu	British thermal unit(s)
Btuh	British thermal units per hour
$°C$	temperature, Celsius
$C°$	temperature, difference
c/b	circuit breaker
cps	cycles per second
d-c	direct current
DD	degree days
Δt	temperature difference
DL	developed length
DWV	drainage waste and vent
F	fuse
$°F$	temperature, Fahrenheit
$F°$	temperature difference
fc	footcandle
fpm	feet per minute
ft	foot, feet
gpd	gallons per day
gpm	gallons per minute
hz	hertz
I	symbol for current
I=B=R	Institute of Boiler and Radiator Manufacturers
in.	inch
in/wg	inches, water gauge
k	heat conductivity
kva	kilovolt-ampere
kw	kilowatt (s)
kwh	kilowatt-hour
lm	lumen
lpw	lumens per watt

m	meter	R	symbol for resistance
m²	square meter	R 19, R 7, etc.	thermal resistances
Mbh	thousands of British thermal units per hour	sq ft	square foot, feet
		sq in.	square inch, inches
MRT	mean radiant temperature	TEL	total equivalent length
NEMA	National Electric Manufacturers Association	U	transmission factor
		UG	underground
NEC	National Electrical Code	v	volt
No.	number	V	symbol for voltage
o/c	overcurrent	va	volt-ampere
OH	overhead	w	watt
%	percent	w/ft	watts per foot
pf	power factor	w/sq ft	watts per square foot
psi	pounds per square inch	Z	symbol for impedance

Contents

1 Basics of Heating, Cooling and Ventilation — 1

1.1	Indoor Comfort	2
1.2	Heat Loss	2
1.3	Evolvement of Modern Climate Control	7
1.4	Energy Sources for Heating	8
1.5	Heat Gain	14
1.6	Cooling Methods, Decentralized	16
1.7	Centralized Air Conditioning	18
1.8	Ventilation	20
	Problems	21
	Additional Reading	22

2 Heating by Hot Water and Electricity — 23

2.1	Hot Water Heating	24
2.2	Boiler and Controls	24
2.3	System Components	24
2.4	Design of a Hot Water Heating System	40
2.5	Electric Heating	47
2.6	Electric Heating Units	50
2.7	Electric Heating Design	54
2.8	The Use of Steam in Large Buildings	58
	Problems	61
	Additional Reading	61

3 Residential Heating — 63

3.1	A House to be Heated	64
3.2	The House	64
3.3	The Structure	64
3.4	Form and Geometry	64
3.5	Space Study	65
3.6	The Boiler Room	68
3.7	The Hardware	68

3.8 Design of a Hot Water Heating
System 69
3.9 Electric Heating Design 75
3.10 Critique 77

4 Nonresidential Heating 83
4.1 The Building 84
4.2 Institutional Convectors 84
4.3 Pumps, Systems and Friction 88
4.4 Design 90
4.5 Locate and Size Convectors 90
4.6 Circuits, Zones and Controls 97
4.7 Two-Pipe Systems 97
4.8 Tubing Expansion or Restraint 97
4.9 Boiler and Circulation 102
4.10 Summary of Data, Design Steps
and Results, Table 4.5 102
4.11 Ventilation 102
Problems 103
Additional Reading 104

5 Heating–Cooling, Air Systems 105
5.1 Air for Thermal Transfer 106
5.2 Air Systems 107
5.3 Central Equipment 110
5.4 Air Ducts 110
5.5 Registers and Grills 121
5.6 Design of a System 130
Problems 136
Additional Reading 136

6 Residential Heating and Cooling 137
6.1 Adapting to the Structure 138
6.2 The Heat Pump 138
6.3 Planning and Designing the
System 140
Problems 173
Additional Reading 173

7 Nonresidential Heating–Cooling 175
7.1 Incremental Units 176
7.2 Selection of a Type 176
7.3 Installation and Operation of
the EA Model 176
7.4 Unit Control 180
7.5 Design of a System 180
7.6 System Performance 181

7.7 Rule of Thumb for Cooling 181
Problems 186
Additional Reading 186

8 Principles of Plumbing 187
8.1 Water Services 188
8.2 Domestic Hot Water Heating 190
8.3 Sanitary Drainage 195
8.4 Materials for Drainage, Waste
and Vent (DWV) 196
8.5 Plumbing Fixtures 207
8.6 Private Sewage Treatment 207
8.7 Storm Drainage 210
Problems 218
Additional Reading 218

9 Residential Plumbing 219
9.1 Codes 220
9.2 Water Service 221
9.3 Domestic Hot Water Heating 227
9.4 Water Distribution 227
9.5 Sanitary Drainage 234
9.6 Plumbing Fixtures 237
9.7 Private Sewage Treatment 237
9.8 Storm Drainage and Disposal 244
Problems 249
Additional Reading 249

10 Nonresidential Plumbing 251
10.1 Planning and Layout 251
10.2 Water Service 252
10.3 Domestic Hot Water Heater 255
10.4 Sanitary Drainage 255
10.5 Plumbing Fixtures 255
10.6 Storm Drainage 255
10.7 Site Drawings 262
10.8 Space for Roughing 265
Problems 266
Additional Reading 266

11 Introduction to Electricity 267
11.1 Electric Energy 267
11.2 Batteries 268
11.3 Electric Power Generation 270
11.4 Voltage 271
11.5 Current 272
11.6 Resistance 273
11.7 Ohm's Law 273
11.8 Series Circuits 274
11.9 Parallel Circuits 275
11.10 Power and Energy 279

11.11 Energy Calculation 282
11.12 Circuit Voltage and Voltage Drop 283
11.13 Ampacity 284
11.14 Measurement in Electricity 285
11.15 Power and Energy Measurement 286
11.16 General 290
11.17 A-C and D-C; Similarities and Differences 290
11.18 Voltage Levels and Transformation 292
11.19 Voltage Systems 293
11.20 Single Phase and 3-Phase 293
Problems 296
Additional Reading 297

12 Branch Circuits 299
12.1 National Electrical Code 300
12.2 Drawing Presentation 300
12.3 Branch Circuits 302
12.4 Branch Circuit Wiring Methods 302
12.5 Wire and Cable 309
12.6 Connectors 312
12.7 Conduit 314
12.8 Other Raceways 316
12.9 Outlets 332
12.10 Receptacles and Other Wiring Devices 339
Problems 346
Additional Reading 347

13 Building Electric Circuits 349
13.1 Circuit Protection 350
13.2 Fuses and Circuit Breakers 353
13.3 Grounding and Ground Fault Protection 358
13.4 Panelboards 365
13.5 Procedure in Wiring Planning 368
13.6 The Architectural–Electrical Plan 370
13.7 Residential Electrical Criteria 370
13.8 Equipment and Device Layout 370
13.9 Circuitry Guidelines 371
13.10 Drawing Circuitry 376
13.11 Circuitry of the Basic Plan House 385
13.12 Basic Plan House—Electric Heat 387
Problems 390
Additional Reading 391

14 Lighting Fundamentals 393
14.1 Reflection of Light 394
14.2 Light Transmission 394
14.3 Light and Vision 396
14.4 Quantity of Light 396
14.5 Brightness and Brightness Ratios 398
14.6 Contrast 398
14.7 Glare 399
14.8 Diffuseness 402
14.9 Color 402
14.10 Footcandle Measurement 402
14.11 Reflectance Measurements 402
14.12 Artificial Light Sources 402
14.13 Incandescent Lamps 402
14.14 Quartz Lamps 406
14.15 Fluorescent Lamps—General Characteristics 406
14.16 Fluorescent Lamp Types 416
14.17 The HID Source 419
14.18 The Mercury Lamp 420
14.19 The Metal Halide Lamps 425
14.20 Sodium Lamps 425
14.21 Control of Light 425
14.22 Lampholders 428
14.23 Reflectors and Shields 428
14.24 Diffusers 435
14.25 Lighting Systems 435
14.26 Lighting Methods 439
14.27 Lighting Uniformity 439
14.28 Fixture Efficiency and Coefficient of Utilization 439
14.29 Metric Lighting Units 442
14.30 Illumination Calculations 442
14.31 Lighting Calculation Estimates 445
14.32 Conclusion 447
Problems 447
Additional Reading 448

15 Residential Electric Work 449
15.1 General 450
15.2 Living Areas 450
15.3 Kitchen and Dining Areas 463
15.4 Sleeping and Related Areas 467
15.5 Circulation, Storage, Utility and Washing Areas 470
15.6 Exterior Electrical Work 479
15.7 Circuitry 479
15.8 Load Calculation 480
15.9 Climate Control System 481
15.10 Electric Service Equipment 485
15.11 Electric Service—General 486
15.12 Electric Service—Overhead 486

15.13 Electric Service—Underground   486
15.14 Electric Service—Metering   492
15.15 Signal Equipment   492
       Problems   507
       Additional Reading   507

16 Nonresidential Electric Work   509

16.1   General   510
16.2   Electric Service   510
16.3   Emergency Electric Service   516
16.4   Building Main Electric Service
       Equipment   517
16.5   Electric Power Distribution   519
16.6   Motors and Motor Control   526
16.7   Guidelines for Layout and
       Circuitry   531
16.8   Classroom Layout   533
16.9   Typical Commercial Building   536
16.10 Special Topics   543
16.11 Conclusion   548
       Problems   549
       Additional Reading   549

Appendix Contents   551
Glossary   584
Index   591

BUILDING TECHNOLOGY: Mechanical & Electrical Systems

1. Basics of Heating, Cooling and Ventilation

Technical work flourishes under an orderly approach. Throughout this book this approach is followed. In all areas, notably in the major divisions of air conditioning, plumbing, electricity and lighting, step-by-step progress is the goal. Accordingly, the reader will be aware that the path toward achievement in any technical specialty comprises four stages.

1. Comprehension of basic principles.
2. Knowledge of the equipment (hardware) available.
3. The *design*, in which the technologist assists.
4. Preparation of the documents by which the system is built. In this the technologist is an important figure, utilizing the expertise that he has acquired in steps 1, 2 and 3.

The study of this chapter on basic principles (No. 1, above) will enable you to:

1. Convert Fahrenheit temperatures to the Celsius scale.
2. Calculate heat losses from both masonry and frame construction.
3. Select an outdoor "design temperature" to use in heat loss calculations.
4. Estimate the reduction of heat losses by the use of insulation.
5. Know the purpose of all the parts of a complete air conditioning system.
6. Draw a central plant to provide heat for a building.
7. Understand and identify the parts of a cooling plant using compressive refrigeration.
8. Distinguish between a "through the wall" cooling unit and one using a remote compressor–condenser.
9. Select an air furnace type for upflow, downflow or horizontal-flow of air.
10. Choose a location for a cooling coil to convert a warm air furnace to an all-year air conditioning plant.

1.1. Indoor Comfort

People living or working in fully enclosed parts of houses or buildings will expect to be provided with comfortable conditions. Although the requirements can vary greatly, depending on occupancy, outdoor climate, the season and geographic location, a complete system of indoor comfort control demands many desirable characteristics. These *could* include:

(a) Maintaining a uniformly warm indoor temperature in cold weather.

(b) A uniformly cool indoor temperature during hot weather.

(c) Adding humidity to the indoor air in winter and reducing it in summer.

(d) Providing, in winter (by insulation), warm interior wall surfaces that will not have a chilling effect on nearby occupants.

(e) Recirculating interior air and filtering out airborne dust.

(f) Controlling the speed of the recirculated air, fast enough to provide freshness and slow enough to avoid drafts.

(g) Exhausting odor-laden air from rooms such as kitchens.

(h) Introducing appropriately tempered (warm or cool) makeup fresh air from outdoors to reduce indoor stuffiness and to replace the air exhausted as mentioned in (g) above.

In densely occupied buildings of the commercial or institutional type, all of the foregoing amenities may be provided. Obviously, however, in some structures of lesser demand, such as houses, it is not always considered essential to include all of the items listed in (a) to (h), which comprise a classic and complete air conditioning facility. Perhaps the most basic and necessary item is heating. Cooling can often be optional but, from Florida to Montana, heating can be essential in normally expected weather conditions.

It is customary to heat many indoor areas to about 75° F whenever the outdoor temperature drops below 65° F (Figure 1.1). Table 1.1 shows that nowhere in the United States can heating (much or little) be *entirely* avoided. Notice that Table 1.1 also includes Celsius temperatures comparative to those of the Fahrenheit scale. At present, most professional technical societies continue to use the conventional English units, which include the Fahrenheit temperature scale. These standards are also the ones used throughout this book. Many foresee, however, a switch in the United States to metric units. Therefore technologists should also familiarize themselves with the metric measurements (see Appendix).

In summary, then, for comfort under the usual circumstances, indoor temperatures in winter should be raised to 75° F and, in summer, they should be reduced to 75° F. When arrangements permit the regulation of the relative humidity, it should be kept within the range of 30 to 50%.

Table 1.1 Winter Outdoor Design Temperatures[a]

Degrees Fahrenheit		Degrees Celsius
40	Miami Beach, Florida	4
30	Southern Texas	-2
20	San Diego, California	-7
10	Arkansas	-13
0	New York City	-18
-10	Kansas	-24
-20	Northern Maine	-29
-30	North Dakota	-35
-40	Regina, Saskatchewan	-41
-50	Montana	-46

[a]A random selection of suggested outdoor temperatures for heating design. Each represents a critically low outdoor temperature during the winter. It sometimes occurs that in many locations the *extreme* low may be 5 to 15 F° less than the ones shown here, but only on *very* infrequent occasions. In New York City temperatures of -14° F have been recorded.

1.2. Heat Loss

The temperature to be maintained in houses and buildings can, of course, vary to conform to the needs and activities of the occupants. A school playroom at 60° F might be comfortable for healthy and active children while a residential temperature of 80° F might be more suitable for the aging and infirm. Other considerations such as the present (and possibly continuing) energy shortage may, in the future, result in much lower standards to be supplemented by heavier clothing. For the present, 75° F may be considered a standard for the purposes of discussion and is used in Example 1.1.

An enclosure maintained at 75° F will continually lose heat to outdoor air that is at lower temperatures. Table 1.1 indicates that, in the United States, under critical winter conditions, the outdoor "design" temperatures in all states are less than the usual indoor temperatures of 75° F. It is seen that this "thermal pressure" or temperature difference,

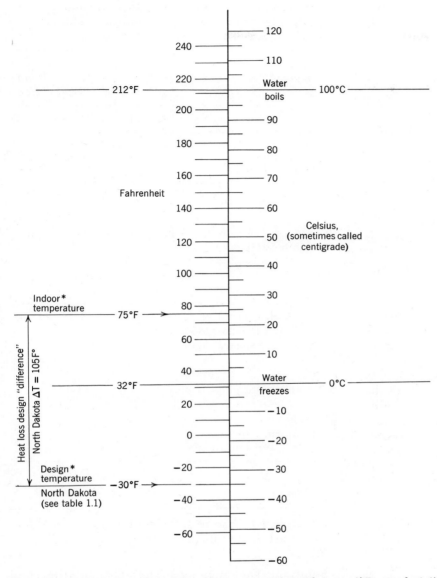

Figure 1.1 Conversion, Fahrenheit degrees to Celsius degrees (*Example 1.1).

delta T, (Δt) from 75° F to the usual winter low temperature is 35 F °* in Florida and 125 F ° in Montana. Thus in a house of equal size and equivalent construction in Montana the heat loss will be at a rate of about 3½ times that of the Florida house.

The first step in planning is to design the house for minimal heat loss so that boilers, convectors and fuel use can therefore all be reduced. Next it is essential to calculate the rate of heat loss per hour so that there will be a basis for the design of a heating system.

Those who will be on the staffs of architects, consulting engineers or mechanical contractors may well find that their tasks could include the making of such calculations. If so, they will find that the brief introduction to these processes given here will be useful. Complete resourcefulness in the making of such evaluations will, however, require the use of the much more complete reference material of the American Society of Heating, Refrigerating and Air Conditioning Engineers (ASHRAE). For advanced work, the student is referred to that material.

*Temperatures are stated as ° F. Temperature *differences* are stated as F °.

The common unit of heat *quantity* is the British thermal unit (Btu) and the *rate* of heat flow is expressed as British thermal units per hour (Btuh). It is important that they be distinguished (*quantity* in distinction to *rate*) and that the proper designation be used. You should recall that one Btu is the quantity of heat required to raise one pound of water one Fahrenheit degree.

The unit (rate) of heat flow through walls, roofs and glass is expressed as British thermal units per hour (Btuh) for one square foot of surface for one Fahrenheit degree difference in temperature between the air on the two sides, and is based on a wind velocity of 15 miles per hour. It is known as the U coefficient of transmission. Thus the rate of heat passing through a wall, for instance, of given area would be the product of the area in square feet, the U coefficient and the actual temperature difference (Δt) between the room and the outdoors:

$$\text{area} \times \text{U} \times \Delta t = \text{Btuh}$$

At wind velocities of 15 miles per hour or more, the air in an exterior room will change one or more times per hour. This introduces large volumes of cold air that must be heated to room temperature. Correspondingly, indoor warm air, previously heated to 75° F is lost on the downwind side of the structure. In Example 1.1, there are two operable windows and one fixed glass window. Since air leakage around fixed glass is negligible, it is considered that the room has only two walls with windows whose sash cracks admit outdoor air. This puts the room in category 3 of Table 1.3 and suggests 1½ air changes per hour.

Table 1.2 Revised U coefficients Resulting from Adding Glass Fiber Insulation—(k[a] value = 0.27) to Uninsulated Walls and Roofs

U Coefficient When Uninsulated	Improved U Coefficient When Insulation is Added in the Thicknesses Shown			
	2 in.	3 in.	4 in.	6 in.
0.40	0.101	0.074	0.057	0.040
0.35	0.097	0.072	0.056	0.040
0.30	0.093	0.069	0.055	0.039
0.28	0.091	0.068	0.054	0.039
0.26	0.089	0.066	0.054	0.038
0.24	0.087	0.065	0.053	0.038
0.22	0.084	0.064	0.052	0.037
0.20	0.081	0.062	0.050	0.037

[a]k value is expressed as the Btuh passing through one sq ft of the material, 1 in. thick when the difference in the temperatures between the surfaces is 1 F°.

Table 1.3 Infiltration. Air Changes Normally Taking Place under Average Conditions Not Including Air Provided for Ventilation

Room Type	Number Changes Taking Place per Hour
1. Rooms with no windows or exterior doors	½
2. Rooms with windows or exterior doors on one side	1
3. Rooms with windows or exterior doors on two sides	1½
4. Rooms with windows or exterior doors on three sides	2
5. Entrance Halls	2

Table 1.4 U Coefficients of Transmission, Glass in Exterior Walls

	U Coefficient, Btuh/(sq ft) (1 F°)
Single glass	1.13
Insulating glass, double	
Air space ³/₁₆ in.	0.69
¼ in.	0.65
½ in.	0.58
Insulating glass, triple	
Air spaces ¼ in.	0.47
½ in.	0.36
Storm Windows	
1 to 4 in. air space	0.56

Example 1.1 (Figure 1.2) Find the hourly heat loss from the living room of a residence in North Dakota, Figure 1.2.

Solution: The reference material, calculations and results are found in:

Figure 1.3 U coefficients of walls and roof. Also the edge factor which is expressed as Btuh loss per linear foot of exterior foundation wall perimeter.

Table 1.2 Adjustment of uninsulated U coefficients.

Table 1.3 Number of air changes per hour due to air infiltration.

Table 1.4 U coefficients for glass and windows.

Table 1.5 Calculations and results.

Note. For large rates of heat flow from rooms or houses, the totals are often expressed as Mbh meaning thousands of British thermal units per hour. Thus the room loss of Example 1.1 (15,617 Btuh) is abbreviated to 15.6 Mbh. The accuracy of more than

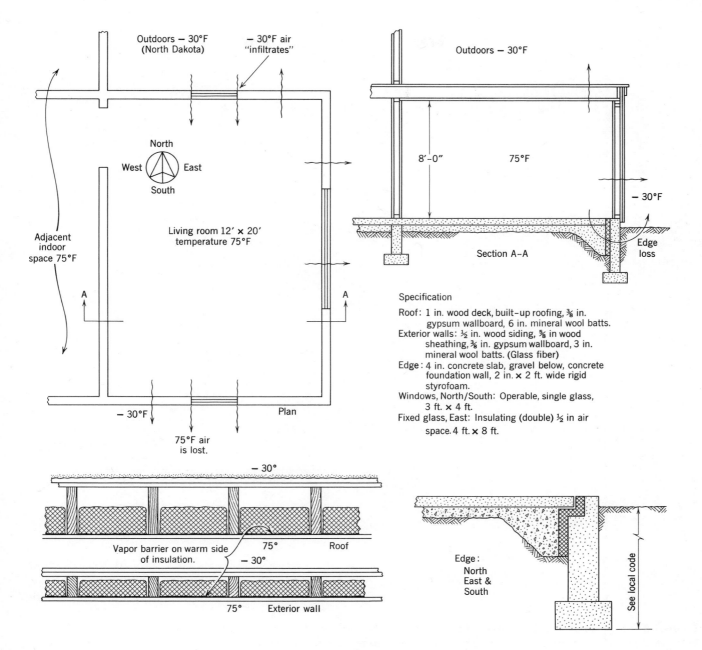

Figure 1.2 (Example 1.1) Sketch and specifications—a living room in North Dakota for heat loss calculations.

the first three significant digits is unnecessary. Indeed, it is not unusual or incorrect for several different technical persons to vary by as much as 5 or 10% in calculating the loss rate from complicated or unusual construction.

The examples in Figure 1.3 on the subject of U coefficients are, of course, few in number and are only a beginning in the subject. For practice, however, within the limited scope of this material, several applicable problems are given at the end of this chapter.

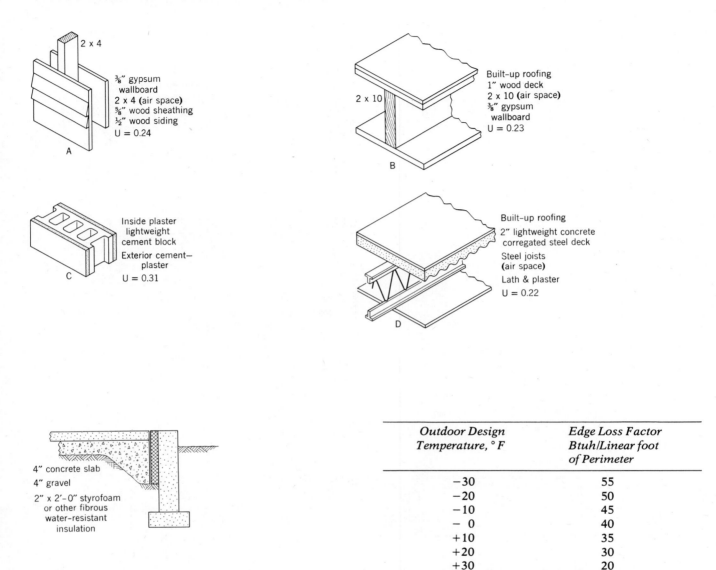

Outdoor Design Temperature, °F	Edge Loss Factor Btuh/Linear foot of Perimeter
−30	55
−20	50
−10	45
− 0	40
+10	35
+20	30
+30	20

Edge loss factors +40 and higher—omit insulation

Figure 1.3 A few basic typical examples of U coefficients. They are expressed in Btu per hour per square foot of surface for one Fahrenheit degree difference in temperature between the air on the two sides and based on an outside wind velocity of 15 miles per hour. Note that the edge loss *factor* is related to the outdoor temperature and is for one foot of *perimeter*. It is *not* a U coefficient.

Table 1.5 (Example 1.1) Calculations for the Hourly Heat Loss from a Living Room in a Residence in North Dakota under Design Conditions.

QUANTITIES		
Roof	12 ft × 20 ft	240 sq ft
Windows, N/S	2 × 3 ft × 4 ft	24 sq ft
Fixed glass, east	4 ft × 8 ft	32 sq ft
Walls, gross area N/E/S	(12 ft + 20 ft + 12 ft) × 8 ft =	
		352 sq ft
Less windows and glass		−56
Walls, net area		296 sq ft
Edge, N/E/S	12 ft + 20 ft + 12 ft	44 lin ft
Room volume	8 ft × 12 ft × 20 ft	1920 cu ft
Air change, cubic feet per hour	1920 × 1½ (Table 1.3)	
		= 2880 cu ft

U COEFFICIENTS		
	Uninsulated (Fig. 1.3)	*Insulated* (Table 1.2)
Roof	0.23 (6 in. batts)	0.037
Walls	0.24 (3 in. batts)	0.065
Windows	(Table 1.4)	1.13
Fixed glass	(Table 1.4)	0.58

EDGE FACTOR

For −30° F outdoor temperature (Figure 1.3) 55 Btuh/lin ft

TEMPERATURE DIFFERENCE, INSIDE TO OUT (TABLE 1.1)

$$\Delta t = 75 - (-30) = 105 \text{ F}°$$

HEAT LOSS CALCULATIONS

Transmission

	Area × U × Δt =	Btuh
Roof	240 × 0.037 × 105 =	939
Walls	296 × 0.065 × 105 =	2020
Windows	24 × 1.13 × 105 =	2847
Fixed Glass	32 × 0.58 × 105 =	1948

Edge Loss

44 lin ft × 55 Btuh/lin ft = 2420

Infiltration

2880 cu ft/hr × 0.018[a] × 105 = 5443

15,617[b]

[a] a factor—weight of air per cubic feet × specific heat of air, Btu per pound.

0.075 × 0.24 = 0.018 Btu per cubic feet

[b] Sometimes written as 15.6 Mbh, (thousands of Btu per hour).

1.3. Evolvement of Modern Climate Control

Having considered in the preceding sections the subjects of human comfort and heat loss from enclosures, we must now think ahead to heating systems. Similarly, when cooling is to be provided, cooling systems must be developed or combined in heating/cooling installations. Of prime interest to the architectural draftsman or technologist is not the design of these systems but, rather, an understanding of them and the ability to get them built to fulfill the intention of the designer. It would, however, be a great mistake after one has learned the details of *presently* popular climate control methods to assume that they would be applicable during a good many years of technical practice. Something always occurs to change the entire aspect of any technical discipline. Accordingly, it is appropriate for the scholar and the young practitioner to review briefly the history of what has happened, to study well what is now current and to prepare within a few years to accept something entirely different. At the present time, energy problems predominate. Gas and oil are in short supply, electricity is a popular choice (for a while), coal is making a comeback and nuclear power may be more available in 10 years. Meanwhile enterprising thinkers are trapping the energy from the sun and the wind and are considering the possibility of reducing comfort standards. To avoid any inclination toward complacency, let us look back (and ahead).

Circa

1870 Petroleum was known but not widely used. An eight-room house was heated by eight open coal fires. A pretty bad method (See Figure 1.4.)

1900 Coal-fired boilers supplied steam to one-pipe systems or hot water, circulated by gravity, to cast-iron radiators. Coal-fired furnaces warmed air that circulated by gravity. All very bulky, dirty and a lot of work.

1902 Dr. Willis Carrier developed the compressive refrigeration cycle for cooling, but several decades would pass before it was much used.

1935 Boilers fired by oil or gas were used to produce steam. Hot water systems similarly fired now used forced circulation. Warm air systems had blowers to circulate the air. This method was more compact, automatically controlled and efficient.

1945 In this year occurred the demise of the steam/cast-iron radiator heating system. Hot water radiant heating made its appearance and cooling started to become popular.

1955 Hot water radiant heating now declined in favor because it was expensive and hard to balance. Electric heating and air systems

Figure 1.4 Brooklyn, New York in the 1800s. Party wall of a recently demolished "brownstone" residence. Three flights up with the coal and three down with the ashes. Eight chimney flues signaled the early beginning of air pollution. There was a small circle of warmth around each fire while the flues carried away most of the warmed air. To fill this void, outdoor cold air was thus drawn in around un-weather-stripped window sash edges, accelerating the infiltration and chilling the backs of occupants facing the fires. It took a hundred years but designers found several better methods.

Sociologists may find interest in the small fireplaces in the servants' sitting room, ground floor front, and in the servants' bedroom, top floor front. Adequate cooking space appears, however, in the large kitchen hearth, ground floor rear.

were now much in use. Boilers became smaller and more efficient. Individual room control and multi-zone operation gained prominence.

1976   This illustration (see Figure 1.5.) shows a very commonly chosen comfort system. Yet it is only one method. The field is wide open for great originality and suitable adaptation

to special building types. Other systems are illustrated in this book.

2000   ?

1.4. Energy Sources for Heating

The so-called "fossil" fuels, coal, oil and gas, have long been standard for the firing of boilers and furnaces. The term boiler is generally applied to installations that produce hot water or steam. Furnace applies to units that produce warm air. Figure 1.6 gives some information about the usual arrangements of several heating installations. Electricity from a central utility plant has, in many instances, overtaken the fossil fuels in popularity and economy after a long period during which it had been much more expensive to operate except in a few locations. Previously, in a stable market, it had been possible to comment on the comparative cost of operation and the suitability of each of these sources of energy. Shortages and rapid changes in prices have caused the situation to become wide open. The reader and his associates will have to determine at the time of design and construction which system will be the most appropriate. Electricity does have several advantages. In addition to the elimination of fuel storage, air for combustion and provision for the elimination of the products of combustion, the furnace or boiler itself, can often be eliminated. Electricity can be supplied directly to baseboard heaters or other heating elements. For the warming of air, electric resistance elements can be located directly in the duct, effecting a more direct approach to the heating process.

Although Figure 1.6 shows the uses of the several energy sources available to produce *hot water* for heating, warm air can similarly be made available with any of the same energy sources by the use of a furnace instead of a boiler. Another comment is necessary in connection with Figure 1.6. For simplicity and easy comparison, the installations are shown in a *basement*. For many years this was the standard location for such equipment. Now, however, one finds heat-producing devices on concrete slabs, when no basement is planned, in attics and, in some large buildings, on upper stories. The use of gravity, which many years ago was the reason for a low elevation from which warm air or warm water could *rise*, has given way to power distribution that permits great freedom in planning the system.

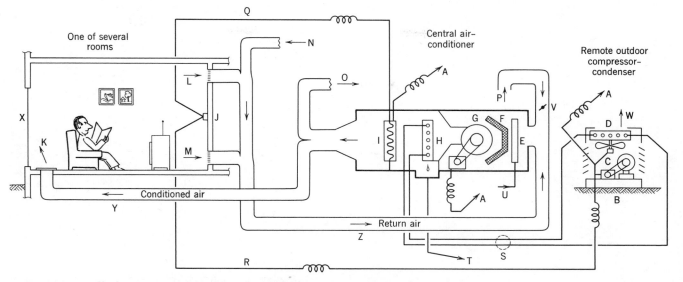

Figure 1.5 An all-electric complete central air conditioning system. It affords heating, cooling, humidity control, dust filtration and the introduction of fresh air. This kind of diagram is known as a "schematic functional drawing" and would not be part of contract documents for building. It is not to scale and the conditioning equipment is much enlarged to make the processes clear.

A	Wiring to electrical power panel	*N*	Return air from other rooms
B	Compressor, compresses the freon gas	*O*	Supply air to other rooms
C	Blower, passes outdoor air over the condenser	*P*	Outdoor fresh air, reduces odors
D	Condenser, liquifies the freon gas	*Q*	Electric signal, controls resistance heating unit
E	Humidifier, winter only	*R*	Electric signal, controls compressor/condenser for cooling
F	Filter, must be accessible for cleaning		
G	Blower, circulates all air	*S*	Refrigerant tubing, to and from the evaporator coil, *H*
H	Evaporator coil, freon evaporates (liquid to gas) absorbing heat from the air	*T*	Plumbing drain, carries away the room moisture that condenses on the cold evaporator coil
I	Electric resistance heating element		
J	Thermostat (summer–winter) best located on interior wall and near the return grills	*U*	Water supply to humidifier
		V	Damper, regulates rate of fresh air intake
K	Supply register, with deflecting vanes and volume damper	*W*	Hot air discharge from condenser, sometimes referred to as "heat rejection"
L/M	High and low return grills for reconditioning. *L* draws back warm air that accumulates there in summer. *M* returns cool air that settles there in winter.	*X*	Glass, double (insulating) when budget permits
		Y	Insulated supply duct
		Z	Return duct

The use of coal, so common early in this century, involves, of course, considerable labor. The suggestion that it now be used becuase it is in greater potential supply may run into difficulty because of the labor involved and also because smoke is not so gentle on the environment.

Nuclear energy may, in forthcoming decades, be an important source of power. Intensive study is now being given to decentralized sources such as wind, energy from the sun and geothermal energy that may utilize the power of the great natural heat under the earth's crust.

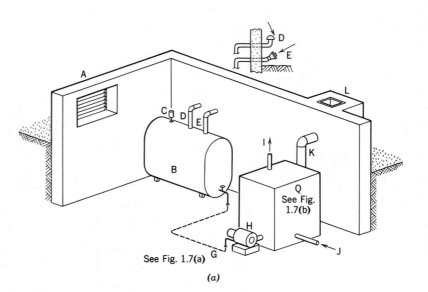

(a)

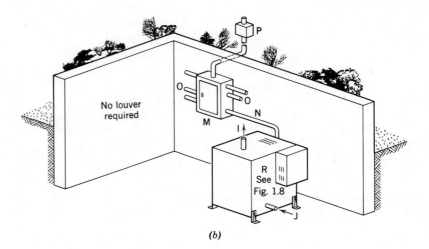

(b)

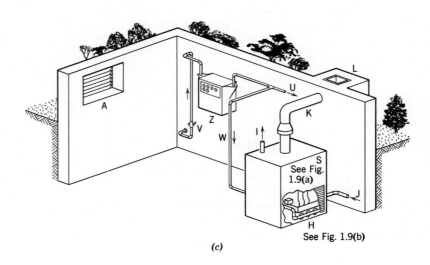

(c)

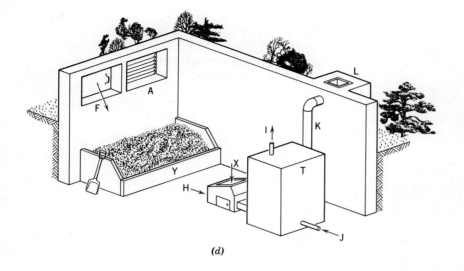

(d)

Figure 1.6 Heat sources (see also Figures 1.7, 1.8, 1.9). *(a)* Oil. *(b)* Electricity. *(c)* Gas. *(d)* Coal. Gas eliminates fuel storage. Electricity eliminates fuel storage, flue and chimney, and the need for combustion air.

A   Louver admits air to support combustion
B   Oil tank, 275 gal. Larger tanks are usually placed
    outdoors and below ground.
C   Oil gauge
D   Vent pipe relieves tank air
E   Oil fill pipe
F   Access for coal delivery
G   Oil supply to oil burner
H   Air to support combustion
I   Hot water to heating system
J   Hot water return to boiler
K   Smoke flue
L   Chimney flue, terra cotta lined
M   Electric panelboard
N   Electric circuit to boiler
O   Other electric circuits
P   Electric meter on exterior wall
Q   Oil-fired hot water boiler (see Figure 1.7a, b)
R   Electric hot water boiler (see Figure 1.8)
S   Gas-fired hot water boiler (see Figure 1.9a, b)
T   Coal-fired hot water boiler
U   Gas to other equipment
V   Gas service entry (below ground) and master shutoff
W   Gas to the burning unit
X   Coal hopper filled by shovel. Then feeds
    automatically by power to boiler fire.
Y   Coal storage
Z   Gas meter

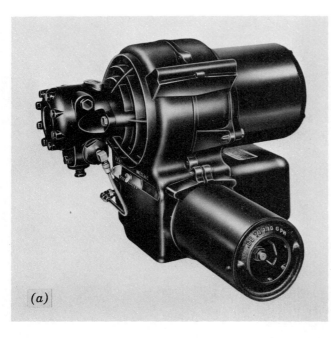

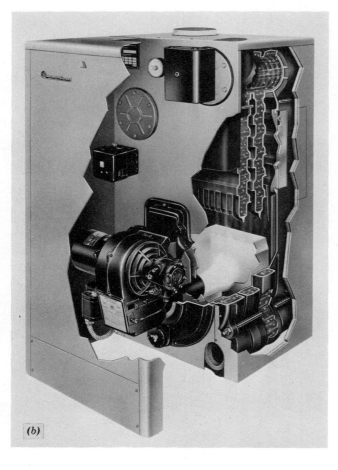

Figure 1.7 Oil-fired hot water heating boiler. *(a)* Oil burner unit. At left, oil pump and air ports to blower. Upper middle, blower. At right electric motor. Forward, projecting into boiler, nozzle, ignition electrodes and secondary air. The capacity, 0.65 to 1.10 gal of oil per hour represents a range of about 70 to 120 Mbh. *(b)* Boiler with unit in place. Boiler is of the cast-iron sectional type. (Courtesy of Burnham Corporation)

Figure 1.8 (Opposite page) Electric hot water boiler. Ratings 12 to 315 kw (kilowatts) (41 to 1092 Mbh). One watt equals 3.41 Btuh. Unlike oil, gas and coal, fossil fuels that are usually burned in a boiler to produce hot water or in a furnace to produce warm air, electricity may be used to heat the boiler or furnace *or* be routed directly to resistance elements in heating units such as electric heating baseboards. (Courtesy of Chromalox, Edwin L. Wiegand, Division of Emerson Electric Company)

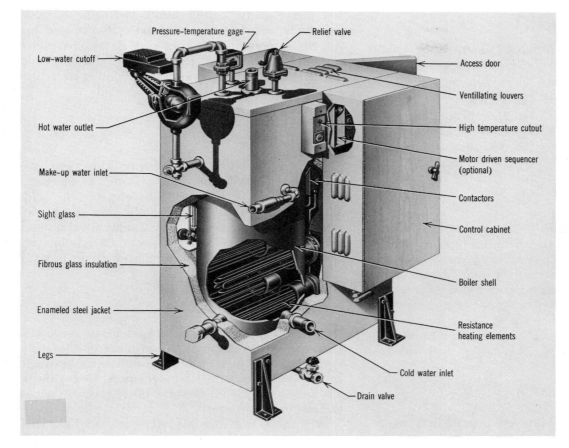

Pressure-temperature gage — Relief valve

Low-water cutoff

Access door

Ventillating louvers

Hot water outlet

High temperature cutout

Make-up water inlet

Motor driven sequencer (optional)

Contactors

Sight glass

Control cabinet

Fibrous glass insulation

Enameled steel jacket

Boiler shell

Resistance heating elements

Legs

Cold water inlet

Drain valve

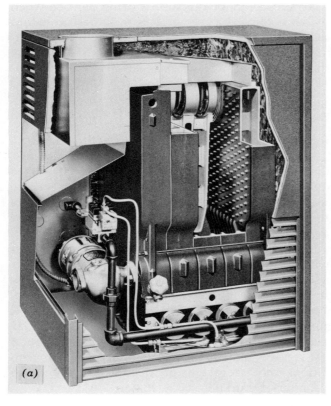

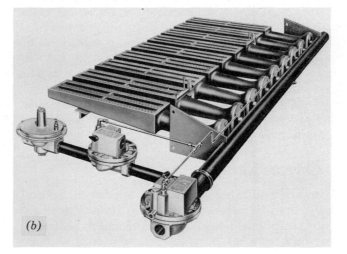

Figure 1.9 *(a)* Gas-fired hot water heating boiler. Gas-burning elements are seen below the cast-iron sections. (Courtesy of Weil McLain) *(b)* Gas burning unit. (Courtesy of Burnham Corporation)

(a)

(b)

1.5. Heat Gain

Unlike heat *loss* (see Section 1.2), summer heat *gain* is complex. It involves not less than 14 major considerations. They include:

(a) Time of day	**(i)** Lights
(b) Orientation	**(j)** Heat producing
(c) Latitude	equipment
(d) Solar gain (glass)	**(k)** Sensible/latent gain
(e) Thermal lag	**(l)** Daily "range"
(f) Shading	**(m)** Undersizing
(g) Ventilation	**(n)** Swing
(h) Occupancy	

Following the completion of the architect's preliminary design of any house or building, it is usually the responsibility of the consulting engineer or the air conditioning contractor to evaluate the heat gain in Btuh or tons of refrigeration. One ton equals 12,000 Btuh (12 Mbh). The "ton" is a carry-over term from the days when ice was used for cooling. It represents the amount of heat absorbed by one ton of ice melting to water in 24 hrs. Btuh and Mbh are the preferred terms.

a. Time of Day

Referring again to heat *loss*, the (usually unspoken) assumption is that the calculation applies to the hours of darkness. Obviously the heating plant must be capable of warming the house at night without the assistance of the sun. Conversely, heat *gain* in summer is primarily expected to occur during the hottest hours of the day. Strangely enough, it may actually occur at 10 A.M. or 4 P.M. instead of at about noon when the sun is highest and most intense. The principal reason for this could be that the major areas of glass (a most effective heat transmitter) could, if placed on the east, dominate the heat gain at 10 A.M. If the design located most of the glass facing West, then 4 P.M. could show the greatest heat gain. For a very complex structure, it would not be unusual for the engineer to calculate the gain at assumed 2-hr intervals from 8 A.M. to 6 P.M. The maximum gain might determine the size of a *central* plant. The hourly figures could be used to establish the optimal ratings of *decentralized* packaged equipment.

b. Orientation

A proposed building with a completed architectural design can be rotated slightly to reduce the heat gain. More sensibly, it can be *designed* to reduce it. One fact that has been long known is now being stressed by energy conservationists. In long, narrow buildings, the gain can be drastically reduced by arranging to point its narrow ends East and West to present minimum area to the orbit of the sun; and, in the Northern Hemisphere, by planning to locate glass on the South and to have it shaded.

c. Latitude

The heating effect of the sun on buildings is obviously less at locations remote from the equator. Also, the angle of the sun is different, and this will affect the arrangements for shading the glass and other surfaces of the structure. Tables of values are available in the ASHRAE Guide and Data Books for both sun effect and shading angles for all latitudes.

d. Solar Gain Through Glass

We refer again to heat *loss* through glass. We found that the transmission is the product of the glass area, the U coefficient and the difference between indoor and outdoor air temperatures. Heat *gain* is the same for north glass or any other glass not penetrated by the sun's rays.

When glass is in direct sunlight, another heat gain item is added to the above. It is called "solar gain" (warming effect of the sun). During periods when indoor cooling is required, it is most desirable that solar gain be minimized. This can be done by locating glass where the sun's rays do not strike it, or by shading it exteriorly.

e. Thermal Lag

In *winter*, heat transmission *out* is assumed to be a "steady state" condition, which indeed it *is* during a long winter night or a dull cloudy day (at a fixed outdoor temperature). In *summer*, however, when a wall or roof is subjected in the morning to the heat of the sun, its transmission of heat *into* the structure increases and occurs at its maximum rate *later*, a phenomenon that is known as thermal lag. It varies with the weight (mass) of the construction that is affected. For instance, assume that the solar energy striking a flat roof is maximum at noon. A *lightweight* frame roof may, when subjected to intense sun heat at noon, deliver to the interior its maximum rate of heat, not at noon, but at about 3 P.M. A heavy masonry slab roof under the same conditions may not give its greatest heat delivery into the interior until about 8 or 10 P.M.

f. Shading

Glass, of course, is the villain in the picture. Unfortunately, for the past several decades it had been very popular. No one can deny that the architectural effect is pleasant. Now, when fuel reserves in addition to dollars are at stake, it is essential to moderate the heat that one is willing to accept and later bail out at great expense in both energy and money. Three kinds of shading are possible.

1. *Exteriorly:* overhangs, baffles, trees, and awnings.
2. *Interiorly:* shades, drapes and venetian blinds.
3. *Intrinsically:* reflective glass and heat absorbing glass.

In the tropics where the sun is distinctly overhead, roofs are especially vulnerable. There, it is not uncommon to provide a "roof shade" (a second roof or "sun-intercepter") some distance above the principal roof with a fully ventilated air space between.

g. Ventilation

It is usual to introduce some fresh outdoor air to enclosed spaces to reduce odors resulting from human occupancy. Much energy is required for this operation. The air must be cooled from about 95° F or more to about 75° F. Also it is necessary to reduce the moisture that this outdoor air carries with it. For every pound of water comprising such moisture, roughly 1000 Btu are required for the condensing process. Conservationists are recommending drastic reductions in ventilation rates.

h. Occupancy

Humans in sedentary or mildly active situations each give off about 400 to 600 Btuh. In *winter*, densely occupied spaces often do not need to be fully heated since people take over a large part of the task. Therefore, the heating system can run at reduced capacity. Still, it must be designed to keep the place warm until humans arrive to do the job. Thus in heat *loss* calculations, the effect of their contribution is neglected. Not so in summer. A place of assembly such as an auditorium deep within a building, a location where the sun effect has no influence, can have as its major heat gain item the heat given off by its occupants.

i. Lights

Every watt of electricity is equivalent to 3.41 Btuh. Office lighting, for instance, puts a great amount of energy in the form of unwanted heat into the space it serves. True, some of this heat can be collected and transferred to other parts of the building if it happens to be needed there. Yet, lighting intensities above 150 footcandles (fc) are usually sufficient to heat an office space in winter. In summer, they could pose a problem.

j. Heat-producing Equipment

Kitchens and great computer banks produce a lot of heat. From kitchens the heated air is often thrown away because of the difficulty of reducing the odors that would be spread by recirculation. Computer installations often need an air cooling system of their own even when the heat gain load is not much burdened by personnel, sun or lighting.

k. Sensible/Latent Gain

Very often one third or more of the refrigeration energy supplied to condition a space is not for the purpose of reducing the air temperature but to condense the moisture in the air so that the relative humidity in the occupied space will not be excessive. Moisture that is given off by people through breathing and perspiration represents, in warm weather, a large part of the 400 to 600 Btuh attributable to each person. Ventilation by outdoor air also contributes unwanted moisture that must be removed. Thus the heat gain load is usually expressed as:

1. Sensible heat, requiring energy to reduce the air temperature.
2. Latent heat, requiring energy to condense moisture.

l. Daily Temperature Range

In some geographic locations that are at high elevation, the temperature drop during the night can be as much as 30 F°. This is known as a "high" daily temperature range. In regions where the temperature is moderated by the thermal stability of surrounding waters, the drop might be about 15 F°. This would be considered an area of "low" daily temperature range. Hence, there are three divisions in the United States: high, medium, and low. Under certain weather conditions, some cooling might be needed all through the day and night in all of these locations. But the low range locations where the temperature did not drop much during the night would demand the greatest cooling capacity.

m. Undersizing

One of the errors in early cooling practice was to be generous in choosing cooling units of larger capacity than were required. The result was blasts of cooling with long periods of shutdown which permitted humidity to accumulate to the discomfort of occupants. Thus in more recent practice cooling units, especially in residences, are undersized by a small amount that assures reasonably continuous operation.

n. Swing

Recommended practice in residences is to design for 24-hr operation in hot weather. This, together with the undersizing usually observed and the resulting almost continuous running of the system, is the cause of an indoor temperature rise that is known as "swing." It occurs for an hour or so in the hottest part of the afternoon. It is considered acceptable because an advantage of continuous action that offsets this air temperature rise is the fact that the walls and other room surfaces have been kept continuously cool and have a lower "mean radiant temperature" (MRT). Thus occupants have a feeling of surrounding coolness and, therefore, do not mind the slight temporary increase in indoor air temperature, which might be from the optimum of 75° F to about 78° F.

1.6. Cooling Methods, Decentralized

It is not difficult to understand how to make *heat*. It is only necessary to set fire to a combustible material (such as wood, coal, gas or oil) give it enough air to support the combustion, and to provide a means of getting rid of the resulting gases. But how does one make "cool"? The problem, of course, is to maintain, indoors, a little island of coolness (75° F or less) when the temperature there *could* approach the outdoor design temperature of 95 or 100° F. The answer is that heat has to be *pumped* out. The pumping system is shown in Figure 1.10. It is known as the refrigeration cycle. Heat cannot be destroyed. It just has to be moved to another place. While summer heat is constantly penetrating in through walls, roofs and glass, it must be continuously moved back out. This is one reason why, in crowded cities with many air-conditioned buildings, it is much hotter outdoors than it would be under *normal* summer conditions. This is not yet our problem, but at some future time it may be if we are to recognize the opinions of ecologists. It is not likely that a technologist in the fields of architecture or construction will be called on to design air conditioning equipment, but he will surely draw, specify and install it. It is therefore essential that the principle of cooling be understood.

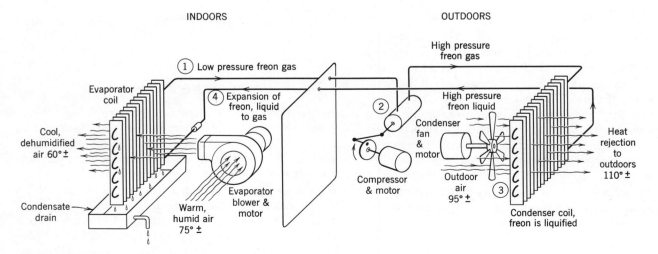

Figure 1.10 Principle of cooling by means of the compressive refrigeration cycle using Freon gas as the refrigerant. The components in this diagram are drawn to illustrate a principle and have no resemblance to the actual parts of a manufactured product. See Section 1.6 of this chapter for a discussion of the thermal interchanges that take place.

In the conventional refrigeration cycle the heat transfer is by *change of state*. It will be recalled from studies of physics that many chemical compounds can exist in three states: gas, liquid and solid. Water is a good example. At normal temperatures, it is a liquid—water. When heat is drawn *out* of liquid water at 32° F, it becomes ice, still at 32° F. This is known as latent heat. When at normal temperatures, ice reverts back to water, it draws back into its mass the same amount of latent heat as had been extracted when it froze. But water is not our refrigerant. It *was*, however, in times long forgotten when truckloads of ice were delivered each day to melt and to cool the air to be circulated in theaters.

Now we have a new and less bulky refrigerant, Freon. Unlike liquid water, Freon is a *gas* at normal temperatures. As we stepped water down to ice by extracting heat, so we can step Freon down from a gas to a liquid by extracting heat from *it*. But Freon does not want to be a liquid at normal temperatures and pressures. In reverting from a liquid to a gas, it draws back the latent heat that had been taken out of it. The heat thus acquired is drawn from whatever is near it, namely the air to be circulated in the occupied (conditioned) space.

Let us trace the process in Figure 1.10. At the upper left, in step 1, low pressure Freon gas is drawn into the compressor. In step 2, the compressor puts it under high pressure and hence, makes it quite ready to be liquified if heat is extracted from it. Step 3, in passing through a very efficient heat transfer finned coil (similar to an automobile radiator), heat is extracted from the Freon by nice cool 95° F outdoor air, condensing it to a liquid. High and low temperatures are relative, of course, and the 95° "cool" air gets heated up to 110° F or higher and constitutes the "heat rejection" of the cycle. Passing now to step 4, the liquid Freon experiences a reduction in pressure as it passes through an expansion valve. In expanding (evaporating) to a gas, it absorbs back its latent heat from whatever surrounds it, which happens to be, by design, the circulated room air. The medium for this action is another heat transfer coil that is very similar to the condenser coil. And so the temperature reduction of the room air is concurrent with the increase in temperature of the outdoor air that is used to condense and liquify the Freon. Note another important process. The temperature of the surfaces of the evaporator coil is below the "dew point" of the humid air passing through. The water vapor (humidity/moisture) is condensed. Thus the room receives cool *dry* air that it promptly warms up and humidifies again by the sensible and latent heat gains within the space. The air then returns for reconditioning.

Now consider Figure 1.11, which begins to look a little more like a real through-wall unit. To the basic cycle a few things have been added. A filter has been shown as well as an indication of the method of disposing of the condensed moisture. Provision for the adding of outdoor fresh air is also indicated. Think of the outdoor side of the conditioner as a manufacturing plant that produces liquid Freon under pressure, ready to expand to a gas and perform its cooling and dehumidifying functions.

These through-wall conditioners are often referred to as "room air conditioners," since they are seldom very effective for *several* rooms, especially when doors are closed. For this reason their capacity ratings are not very high. See Table 1.6

Table 1.6 Cooling Capacity and Electrical Characteristics of Through-Wall Models of One of 56 Manufacturers Listed by AHAM in a Recent Directory.

Section 2[a] Through-wall Models

The certified room air conditioner models listed in Section 2 are of special interest to architects, contractors, builders, consulting engineers, those associated with building design, construction, or maintenance, and others making such installations. All of the models listed in this section are for through-wall application. However, some of these through-wall models may also be used for window installation. Models with heating capacity are also shown in Section 3[a].

Model Number	AHAM-Certified Cooling Data				Energy Efficiency Ratio
	Volts	BTU/hr	Amperes	Watts	
AIRTEMP					
Airtemp Div, Chrysler Corp.					
Dayton, Ohio 45401					
A07-10H-NA	115	6,500	7.5	860	7.6
A09-20H-NA	115	9,000	12.0	1,350	6.7
A10-80H-NA	208	10,000	9.8	1,820	5.5
A10-30H-NA	230	10,000	8.8	1,780	5.6
R12-8EG-NR	208	12,000	10.3	1,925	6.2
HEATING	208	12,000	19.0	3,900	
A12-80H-NA	208	12,000	10.3	1,925	6.2
A12-30H-NA	230	12,000	9.7	2,000	6.0
A14-80H-NA	208	13,500	12.3	2,355	5.7
A14-30H-NA	230	13,500	11.7	2,470	5.5

Source. Reprinted by permission from *1974 Directory of Certified Room Air Conditioners*, Edition 3, June 1974, AHAM–Association of Home Appliance Manufacturers

[a]The section numbers referred to are those of the Association.

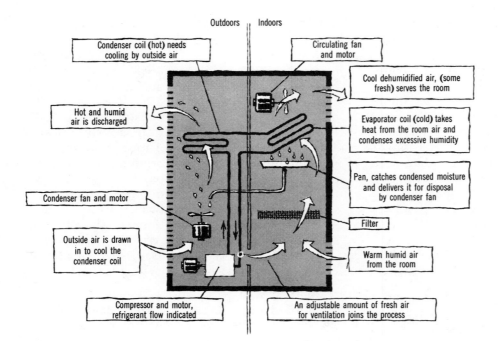

Outdoors | Indoors

Figure 1.11 Schematic diagram of a through-the-wall "room" air conditioning unit. Sometimes referred to as a package unit or a decentralized unit. The compressor-condenser *must* be outdoors. In this compact and complete refrigeration cycle assembly, the usual capacities are about ½ to 1 ton (6 to 12 Mbh). See also Figure 1.10. Table 1.6 lists the capacities and electrical characteristics of the products of one manufacturer.

1.7. Centralized Air Conditioning

It is evident from Figure 1.10 that the cooling cycle can be divided into two parts. One is the compressor plus condenser and condenser fan. This serves an outdoor function. The other is the evaporator coil plus its fan or blower. This serves an indoor function. Now there is a choice. The two functions can be joined as we have seen in a compact decentralized through-wall unit. This unit has a barrier to separate the air circuits and to partially exclude from the room the noise of the compressor. The other choice, which constitutes a *central system*, is to separate the two parts of the refrigeration cycle (see Figure 1.12). An example of the location of an evaporator coil is shown in Figure 1.13. The coil is at the end of the

airstream. Its blower is at the bottom of this upflow air furnace (or air handler as it is also known). In winter when the evaporator coil is not used, the blower still has the function of moving the air that is warmed by the gas-fired heating elements. There is, of course, a complete separation of the room air and the other air-gas circuit of combustion air and flue gas.

Air furnaces and air-handling units are terms that are now almost synonymous. A quite complete installation is shown in Figure 1.5. In very large installations in high rise buildings the several functions are combined in a custom designed and specially built air handling unit. Typical warm air furnace types, some with adjunct cooling, humidification (for winter), and the like can be seen in Figure 1.14.

Since this chapter deals primarily with heating/cooling basics, the subject of ducts is considered in a later discussion.

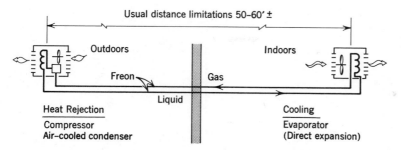

Usual distance limitations 50–60' ±

Outdoors   Indoors

Freon   Gas

Liquid

Heat Rejection   Cooling

Compressor   Evaporator
Air–cooled condenser   (Direct expansion)

Figure 1.12 In central air conditioning systems the refrigeration cycle equipment is split. The compressor/condenser is placed outdoors for heat rejection, and the evaporator is found indoors as part of the heating/cooling/air-handling assembly. It is seen at the top of the upflow unit in Figure 1.13. Airflow is effected by the blower at the bottom of that unit.

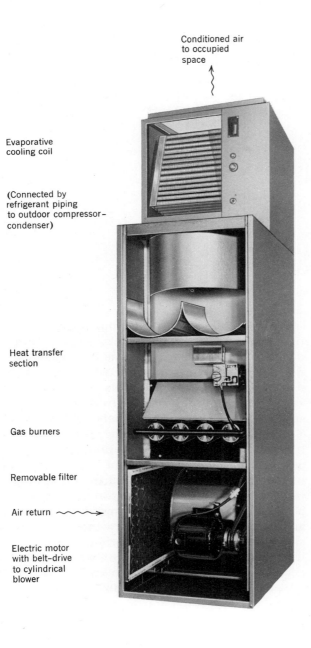

Conditioned air
to occupied
space

Evaporative
cooling coil

(Connected by
refrigerant piping
to outdoor compressor–
condenser)

Heat transfer
section

Gas burners

Removable filter

Air return

Electric motor
with belt-drive
to cylindrical
blower

Figure 1.13 Small central cooling/heating air-handling unit. Not shown in this view are the smoke connection and a condensate drain from the cooling coil. If humidification were to be added, it could be at the lower left in the return airstream just ahead of the filter. Flexible, (canvas-asbestos) connections to supply and return ducts can minimize vibration that might be carried to the duct system. Heating can be provided by gas, as shown, by oil, or by an electric resistance element. This figure illustrates an upflow type. Downflow or horizontal flow equipment is also obtainable. See Figures 1.5 and 1.14a. (Courtesy of American Furnace Division, The Singer Company)

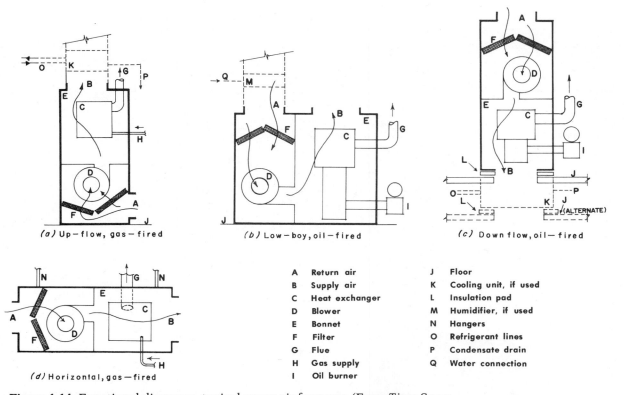

A	Return air	J	Floor
B	Supply air	K	Cooling unit, if used
C	Heat exchanger	L	Insulation pad
D	Blower	M	Humidifier, if used
E	Bonnet	N	Hangers
F	Filter	O	Refrigerant lines
G	Flue	P	Condensate drain
H	Gas supply	Q	Water connection
I	Oil burner		

Figure 1.14 Functional diagrams, typical warm air furnaces. (From *Time-Saver Standards for Architectural Design Data*, Fifth Edition by John Hancock Callender, Copyright © 1974, McGraw-Hill. Used with permission of McGraw-Hill Book Company.)

1.8. Ventilation

Houses that are kept tightly closed in winter become stuffy. Odors build up. The increase in the use of summer cooling has made it essential that houses so cooled be also kept sealed during *that* season. Only in mild, pleasant days in spring and fall is it possible to take full advantage of generous open-window ventilation. Yet it is important to admit a reasonable amount of fresh outdoor air during the months when indoor climate control requires that windows be kept closed.

Table 1.3 shows that a certain amount of natural ventilation normally occurs during breezy times when the air in a room experiences a complete change once or twice each hour. Although the effects of such infiltrating air must go into the thermal accounting that sets up the required output of the heating or cooling system, this air is usually insufficient to reduce indoor odors.

Figure 1.5 depicts a system that offers more positive ventilation. If an air system is used, fresh outdoor air can be drawn in on the suction side of the blower and can be controlled in quantity by a damper (V in Figure 1.5). This process obviously adds air to the air drawn back from the conditioned space and puts the room under a slight pressure. Several advantages accrue. The correct amount of fresh air can be added, and the pressure in the room acts to reduce the accidental outdoor air that comes in by infiltration. Finally, the fresh ventilation air passes through the central conditioning unit and is, accordingly, cooled and dehumidified in summer and heated and humidified in winter. In both seasons it is filtered to remove airborne dirt. It is evident that such an arrangement is far better than accepting, in the vicinity of windows, dirty infiltrating air of the wrong temperature which is also too humid in summer and too dry in winter.

When fresh air is introduced, some *indoor* air usually needs to be exhausted. Concentrations of humidity occur in kitchens and baths due to cooking, bathing and showering. This humidity, together with odors that originate in these rooms, must be removed. Small exhaust fans are used for this purpose. It is well to keep the total of the ratings of the several fans, expressed in cubic feet per minute (cfm), less than the rate of ventilation air admitted through the conditioner. Otherwise, a suction may occur in the room. One must not forget that the combustion process of the heating boiler or furnace requires its own air supply which must not be affected by pressures or suctions caused by the ventilation system. It is becoming more and more customary in modern houses of tight construction to specify exhaust fans in kitchens, baths and laundries and *not* to depend on any windows in these rooms for needed ventilation.

In public buildings, such as schools and theaters, the dense occupancy results in body odors. Codes usually require specific ventilation rates stated in cubic feet per minute of fresh air per person and dependent on the volume of the space occupied. For schools, this rate has often been of the order of about 10 to 40 cfm per person. Because of considerations of energy conservation, a reduction of these rates is now being considered.

Problems

1.1. Assume that, in the future, ASHRAE and other professional societies adopt metric standards. Express the following in Celsius degrees:

  (a) The usual outdoor winter design temperature in North Dakota.
  (b) The commonly selected indoor temperature.
  (c) The temperature difference, Delta T.

1.2. Calculate the hourly heat loss of a small manufacturing plant in New York City built in accordance with the following specifications:

Plan dimensions:	50 ft × 100 ft
Height:	12 ft
Walls:	Type C, Figure 1.3
Roof:	Type D, Figure 1.3
Windows:	Single glass on *each* of the 100 ft walls, 120 sq ft.
Windows:	Single glass on *each* of the 50 ft walls, 50 sq ft
All doors:	Total 260 sq ft U coefficient = 0.49
Edge insulation:	As shown in Figure 1.3
Infiltration:	Assume three air changes per hour.

1.3. Calculate the hourly heat loss in the plant described in Problem 1.2, but with the following modifications.
Roof: add 2 in. of glass fiber insulation
Walls: add 3 in. of glass fiber insulation
All windows: insulating glass, double, ½-in. air space. All other conditions unchanged.

1.4. In Problems 1.2 and 1.3 calculations were made to establish the hourly heat loss under design conditions for a small manufacturing plant. Based on the uninsulated value (Problem 1.2), calculate the percentage of heat saving that would be achieved by insulating the plant as in Problem 1.3.

1.5. Inspect and sketch the heating plant in your classroom building, dormitory or home. Discuss the following:
(a) Fuel and fuel storage.
(b) Provision for combustion air.
(c) Flue and chimney.
(d) Space required for the plant.
Do *not* include controls, ducts, piping or the heating elements in the occupied space.

1.6. An uninsulated masonry wall has a U coefficient of 0.35. What are the improved U coefficients if glass fiber insulation (k = 0.27) is added in the following thicknesses?
(a) 2 in.   (b) 3 in.   (c) 4 in.   (d) 6 in.

1.7. Wood frame roof construction A is uninsulated. Its U coefficient is 0.22. Wood frame roof B is of the same construction but has 6 in. of glass fiber insulation added. Which of the following statements is correct?

Roof A will lose:

(a) 3.7 times as much heat as roof B.

(b) 5.9 times as much heat as roof B.

(c) 8.5 times as much heat as roof B.

Show calculations.

1.8. A room A has windows on one side only. Room B has windows on three sides. How many times as much heat will room B lose by infiltration as room A?

1.9. How many times as much heat will be lost through a single pane of glass as through triple-pane insulating glass with ½-in. air spaces?

Additional Reading

Design with Climate, Victor Olgay, Princeton University Press. For advanced thinking, based on high level research, this volume offers the architect and designer the best guidance in planning for human comfort under widely varying climatic conditions.

Design of Insulated Buildings, Tyler Stewart Rogers, F. W. Dodge Corporation. Far ahead of its time in energy conservation, Tyler Rogers' book is an excellent resource in insulation techniques.

Handbook of Fundamentals, ASHRAE GUIDE AND DATA BOOKS, 1972, The society of Heating, Refrigerating and Air Conditioning Engineers. ASHRAE publications are for the professional engineer and are a bit above the needs of the beginning technologist who should, however, be familiar with them.

Directory of Certified Room Air Conditioners, AHAM, Association of Home Appliance Manufacturers. This directory is published quarterly and lists available package cooling units including through-wall types varying in capacity from 5000 to 36,000 Btuh.

Air Conditioning with Packaged Equipment, Carrier Air Conditioning Company. Before attempting to absorb the highly developed material of the ASHRAE books, the beginner in air conditioning would do well to work with this much simpler approach.

2. Heating by Hot Water and Electricity

The work of the architect and engineer includes choosing a heating system, determining the fuel to be used, fitting the system to the house or building and sizing the parts. In such work, the technical person is part of the design team. Depending on his experience, there need be no limit to his duties subject to check by the professional to whom he reports.

Recent technical advances have resulted in: greater use of one-pipe hot water systems in preference to two-pipe systems, better baseboard heaters (both hot water and electric) and recessed (below floor) heating units for use below exterior glass walls. Let us now study some currently popular heating methods.

After completing this chapter on heating, you will be prepared to:

1. Identify all of the piping and controls of a hot water heating boiler.
2. Allow for the expansion of water in the system as water temperature increases.
3. Order a "package" type boiler of required output and with its own circulator and controls.
4. Design a series perimeter-loop hot water heating system.
5. Allow for expansion of metal in long runs of copper tubing.
6. Lay out a multiple-zone system of one-pipe circuits.
7. Select finned-tube baseboard heating units based on water temperature and room heat loss.
8. Select proper *types* of electric heaters: baseboard, floor inserts or downflow.
9. Design an electric heating system.
10. Choose proper electric baseboard units based on system voltage and space heat loss.

2.1. Hot Water Heating

Water is a very efficient medium for transmitting heat to the various parts of a building. Steam, its forerunner for many decades, had been, in houses, bulky, essentially intermittent in performance, slow in response, and the father of that most unsightly item the cast-iron radiator. Automation and flexibility are the keynote features of "forced" hot water (hydronic) systems that universally employ an aquastat. The aquastat constantly assures a boiler full of heated water that stands ready to be circulated (pumped) to the room-heating elements on call from a thermostat.

The typical system is a closed, self-sufficient plant with the need for only minimal adjustments other than regulation of the thermostat. Indeed, *that* control need not be touched if it is of the night setback type, and if the occupant is satisfied with a fixed choice of room conditions. Makeup water is added automatically when (infrequently) it is needed. A compression tank adapts to the water volume as it changes with changing temperatures. Entrained air in the water can be vented out automatically at high points where it collects, which prevents "air binding" that can obstruct effective circulation. A drain valve at the low point in the return line permits draining of the system. This will prevent freezing and subsequent bursting of pipes, boiler and other parts in an unoccupied and unheated house. Other low points or trapped sections of piping should be similarly provided with drains. The piping system, usually of copper tubing, may be run level instead of sloped for drainage. Accessible fittings can be provided in the piping, through which residual water can be blown out.

The boiler may be placed higher than the heating elements, such as baseboards or convectors. This is termed a downfeed system. It may also be lower than the heating units. This is called an upfeed system. An adjunct feature can be the heating of "domestic hot water" for use in baths, laundries and kitchens. A tap water coil can be immersed in the boiler water for heat transfer to the domestic water. This, of course, makes it necessary to select a boiler of greater capacity than one that is used for house-heating only.

2.2. Boiler and Controls

As we explain in the preceding chapter, any of the three fossil fuels may be used. They are ignited and are burned in a combustion chamber, within which there is a water-filled assembled group of cast-iron sections or a steel chamber. In an electrical boiler, watertight electric resistance elements extend directly into the boiler water. From this body of hot water, which may be considered as a heat reservoir, the hydronic and electrical controls take over as shown in the accompanying illustrations. (See Figures 2.1 to 2.5).

2.3 System Components

In sections 2.1 and 2.2, and in Figure 2.1 to 2.5 we described and illustrated the basic heating plant and its accessories for hot water heating. Now we must consider the tubing and other components by which heat is delivered to the several rooms, based on heat loss calculations like those given in Chapter 1. They comprise the following:

(a) Tubing circuit types   (e) Expansion fittings
(b) Tubing and fittings   (f) Special tees
(c) Convectors   (g) Thermostat
(d) Air vents

a. Tubing Circuit Types

The series perimeter loop is very suitable for small houses. The temperature drop is not very great and, therefore, all baseboards, including Nos. 1 and 4 of Figure 2.6, can be selected for the same average water temperature. Adjustment of the heat output in each room is possible only by use of the damper in the baseboard. No valves are possible, since any closure would affect the entire loop.

Somewhat better control is obtained by the use of the one-pipe method, which permits the shutoff of any convector by closing its valve. Water at proper temperature still flows past to heat the more distant convectors. The branch runouts to convectors make the installation a bit more expensive. Multiple circuits of this method are shown in Figure 2.7.

The two-pipe reverse return is the best choice for very long runs in large buildings. More tubing is needed for this method than for either the loop of one-pipe system. Each unit receives water at nearly full boiler temperature, the cooler water being carried away by the return tube.

The two-pipe direct return should be avoided. It illustrates a possible error in planning. The "short circuit" through the nearest convector steals water that should flow through the rest of the sytem.

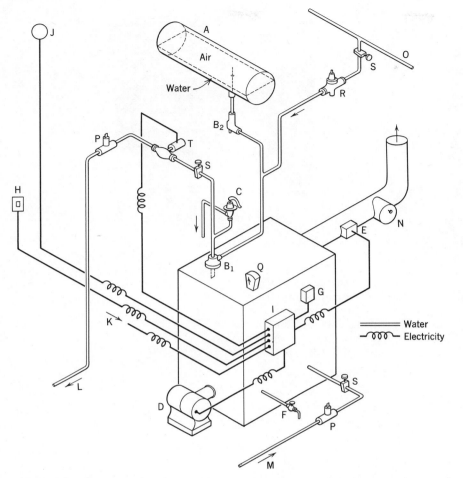

Figure 2.1 Hydronic and electrical controls; an oil-fired boiler for heating by hot water.

A   *Compression Tank.* Accommodates the expansion of the water in the system.

B   *Air Control Fittings.* Vent out unwanted air in the boiler and maintain the level in the compression tank.

C   *Pressure Relief Valve.* Usually set for 30 psi. Initial cold pressure about 12 psi. Relieves excessive system pressure.

D   *Oil Burner.* Responds to aquastat or thermostat.

E   *Stack Temperature Control.* Senses stack temperature and stops oil injection if ignition has not occurred.

F   *Drain Valve.* At low point in the water system.

G   *Aquastat.* Maintains temperature of boiler water by starting the oil burner when temperature of water drops below the aquastat's setting. Sometimes set at about 180° F.

H   *Remote Switch.* At a safe distance from the boiler so that the plant can be turned off in case of trouble during which the boiler cannot be approached.

I   *Junction Box and Relays.* General control center.

J   *Thermostat.* When the room temperature drops below its setting, it turns on both the oil fire and the circulating pump.

K   *Electrical Power Source.* Operates from a separate individual circuit at the power panel.

L   *Hot Water Supply.* Copper tubing to convectors or baseboards.

M   *Hot Water Return.* Copper tubing from convectors or baseboards.

N   *Draft Adjuster.* Regulates the draft (combustion air) over the fire.

O   *House Cold Water Main.* From which water is fed automatically into boiler.

P   *Flow Control Valves.* Prevent casual flow of water by gravity when the circulator is not running.

Q   *Temperature/Pressure Gauge.* Indicates water temperature and pressure. Sometimes supplemented by immersion thermometers in supply and return mains.

R   *Pressure Reducing Valve.* Admits water into the system when the pressure there drops below about 12 psi. Has a built-in check valve to prevent backflow of boiler water into the water main.

S   *Shutoff Valves.* Normally open. Can be closed to isolate the system and permit servicing of components.

T   *Circulator.* Centrifugal circulating pump that moves the water through the tubing and heating elements.

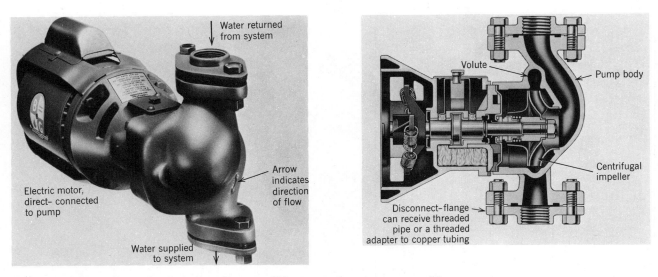

Figure 2.2 Typical circulating pump for a small hot water heating system. Motors are usually of fractional horsepower. The sectional view at the right shows the pumping action. Water is drawn in through the upper opening and flows through the pump body to the receiving opening of the centrifugal impeller element. Fast rotation throws the water into the volute under pressure and from there it goes to the lower opening. See Figure 2.1(t). Flow may be horizontal or vertical. (Courtesy of ITT Bell & Gossett)

Figure 2.3 Details, hydronic accessories and controls. (Courtesy of ITT Bell & Gossett) (Figure continued on following page.)

Pressure reducing valve. See Figure 2.1, R. Admits water to the system automatically as needed but closes against flow of system water back into the house water system if the pressure there should drop below 12 psi.

Flow control valve. See Figure 2.1, P. Usually placed near the boiler on both supply and return mains. The valves open when the circulator is activated but close to prevent gravity flow when the circulator stops.

Bottom view of compression tank. See Figure 2.1, *A.* Proper adjustment of the water level and the air chamber above it gives assurance that the pressure relief valve need not operate except in an emergency.

Tank air-control fitting. See Figure 2.1, B_2. Small amounts of air, which may be drawn into the suction side of the system or entrained in the entering make-up water admitted through the pressure reducing valve, find its way to the top of the tank. Water level can be controlled. See Figure 2.4.

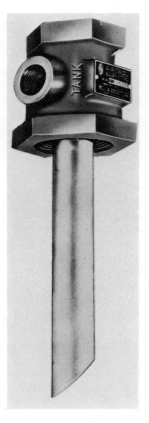

Boiler air-control fitting or "dip-tube." See Figure 2.1, B_1. Air collecting at the top of the boiler rises into the tank through the side connection of the fitting and cannot be drawn up with the water, which enters from the bottom of the tube.

Pressure relief valve. Excessive pressure is relieved through the side opening. It should operate only in an emergency and should not leak out water during normal changes of volume of the water in the system. See Figure 2.1, *C.*

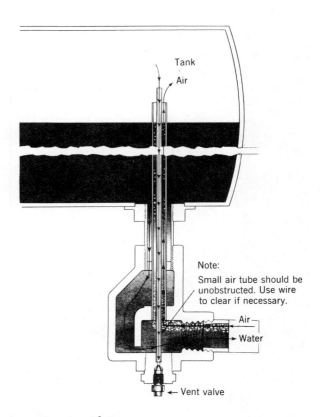

Figure 2.4 Air control. As illustrated, tap water added to the system carries air with it. Collected in the tank, it affords a cushion for the expansion of the system water as it increases in temperature. Very small amounts of air may still find their way into the system piping and need to be relieved by vents at high points where air commonly collects. (Courtesy of ITT Bell & Gossett)

Figure 2.5 A "package type" oil-fired boiler for a hot water heating system (opposite page), the controls for which are illustrated in Figure 2.1.

In addition to the manufacturer's excellent explanatory details surrounding the central sectional view, some additional features may be identified. Their specific locations often vary in different units.

• At the top is a pressure relief valve that will operate only in emergency situations. A pipe connected to it will carry overflow water to an open plumbing drain.

• The smoke outlet is at the top and will connect to the iron smoke breeching that leads to the flue.

• A circulating pump (mid-height on the front) is an integral part of the package. The system return-pipe connects to the upper flange of the pump. The L-shaped return pipe (black) delivers the system water to the water-filled cast-iron boiler sections. Note the drain faucet at the lowest point in the water system.

• Apparent are the electrical connections to the motors of the oil-firing unit and the circulator. Notice the circular pressure-temperature gauge below the aquastat, which is to the left of the black disc.

Jacket extension to cover oil burner and controls is available as optional equipment. Front panel is removable.

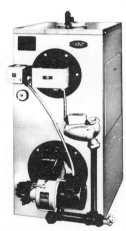

Insulated metal jacket is finished in hammer tone Blue. Panels on each side lift out for "thorough-cleaning" of boiler flues.

Extra large tankless heater for fresh hot water at all times.

Rugged cast iron boiler sections designed for top efficiency in operation.

Stainless steel combustion chamber

• For efficiency of heat transfer the walls of the cast iron sections are extended into baffles to enlarge the surface area that separates the fire chamber from the system water that flows through the sections.

• Insulation in the jacket walls minimizes energy loss and assures a cool boiler room.

• Not part of the space-heating equipment but a useful adjunct item is the tankless heater (upper right detail), and also shown immersed in the boiler-jacket water. It provides *domestic* hot water to baths, kitchens, and so forth. Cold water from the water main under normal pressure is piped to one of the two connections in front of the black disc. When a faucet is opened in a plumbing fixture, water flows through the cylinder coil with extended fins, is heated to a suitable temperature and flows out of the other connection and into the house *hot-water* main.

Note. The primary source of heat can be oil as shown here or, as indicated in Figure 1.6, page 10, it could be electricity, gas or coal (Courtesy of The H. B. Smith Company, Inc.)

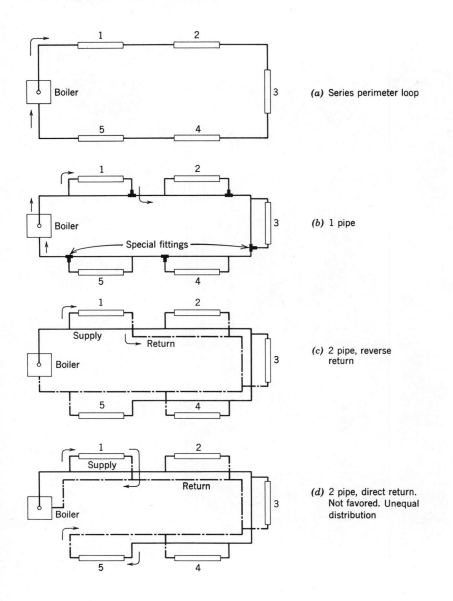

(a) Series perimeter loop

(b) 1 pipe

(c) 2 pipe, reverse return

(d) 2 pipe, direct return. Not favored. Unequal distribution

Figure 2.6 Plan views (*a* to *d*) showing principles of tubing arrangements for water distribution to heating elements (baseboard convectors shown here) in hot water heating systems. Scheme (*e*) is a *two-circuit* series perimeter loop system. (Courtesy of Dunham-Bush, Inc.)

Note. For Example 2.1 of this chapter, type (*a*) is a good choice.

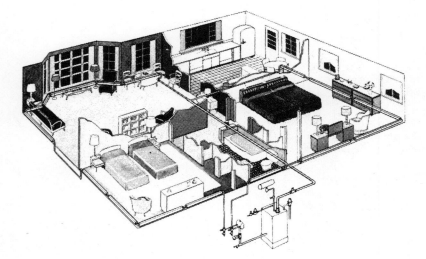

(*e*) Two-circuit one-pipe series perimeter loop. (Courtesy of Dunham-Bush, Inc.)

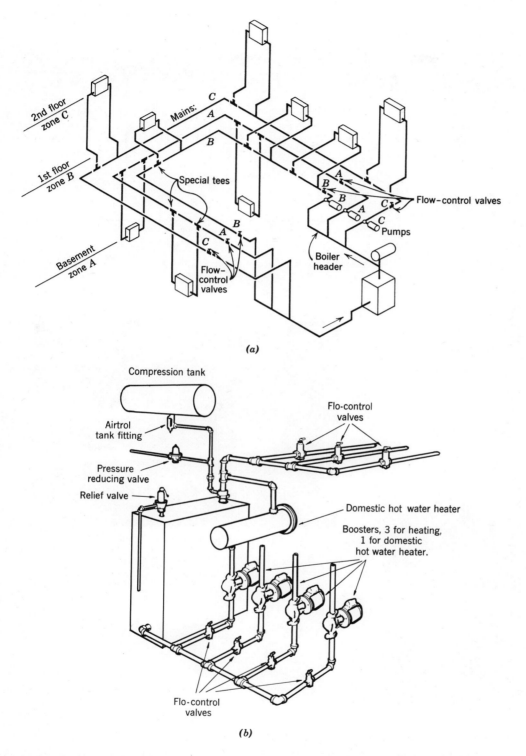

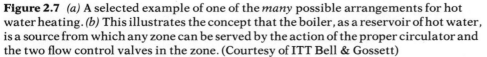

Figure 2.7 *(a)* A selected example of one of the *many* possible arrangements for hot water heating. *(b)* This illustrates the concept that the boiler, as a reservoir of hot water, is a source from which any zone can be served by the action of the proper circulator and the two flow control valves in the zone. (Courtesy of ITT Bell & Gossett)

Filler metals for brazing

AWS Classification	Principal Elements, percent					
	Silver	Phosphorus	Zinc	Cadmium*	Tin	Copper
BCuP–2		7.00–7.50				Balance
BCuP–3	4.75–5.25	5.75–6.25				Balance
BCuP–4	5.75–6.25	7.00–7.50				Balance
BCuP–5	14.5–15.5	4.75–5.25				Balance
BAg–1*	44–46		14–18	23–25*		14–16
BAg–2*	34–36		19–23	17–19*		25–27
BAg–5	44–46		23–27			29–31
BAg–7	55–57		15–19		4.5–5.5	21–23

*WARNING: BAg1 and BAg2 contain cadmium. Heating when brazing can produce highly toxic fumes. Avoid breathing fumes. Use adequate ventilation.

(a)

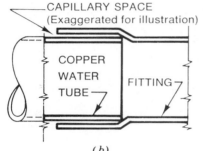
CAPILLARY SPACE
(Exaggerated for illustration)
COPPER WATER TUBE
FITTING

(b)

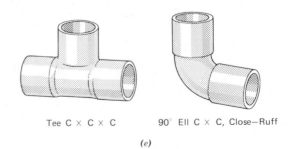
Feeding brazing alloy to horizontal joint

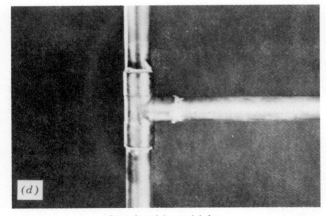

(d)

Completed brazed joint

Tee C × C × C   90° Ell C × C, Close—Ruff

(e)

Figure 2.8 Copper tubing and its fittings. (Courtesy of Copper Development Association, Inc.)
(a) It is obvious that for brazed, sometimes called "soldered", fittings, ordinary plumbers solder is *not* suitable. Quality specifications must be followed.
(b) Brazing alloy fills the capillary space between water tube and fitting.
(c) The fifth of eight steps in preparing and completing the connection between tube and fitting.
(d) Final condition following the removal of residue.
(e) Two of the many fittings available for conventional and special uses.

b. Tubing and Fittings

Specifications and details for all copper-tubing types (Figure 2.8) are much too lengthy to be described here. The principal kinds are identified as K, L and M. They vary as to wall thickness, ductility (for bending) and resistance to pressure. It is sufficient for our purpose to say that type M is recommended for straight runs in heating systems like those we will describe. Bronze fittings and joining methods appear in Figure 2.8.

c. Convectors

The Hydronics Institute issues, periodically, ratings of the Btuh output of radiation (including finned-tube baseboards), convectors, commercial radiation and heating boilers. These booklets should be in the reference file of technologists, who thus can make certain a wide choice of equipment from a number of manufacturers. (See Figures 2.9 to 2.11.)

For the purpose of this chapter, only one style of baseboard convector is used. Reference to its approved ratings can be found in Item 7 of Figure 2.10. That item refers to "I=B=R ratings." This is the identification of the Institute of Boiler and Radiator Manufacturers whose symbol and scope are now in the jurisdiction of the Hydronics Institute.

d. Air Vents

For proper water flow all tubing must be completely water-filled. Air that may collect in high, trapped sections of tubing may interfere with water circulation. Such air must be vented out. See Figure 2.12.

e. Expansion Fittings

The cylindrical cover surrounding the compressible (and expandable) copper element prevents a lateral "hernia" when that element is compressed by the motion of the lengthening tubing. See Figure 2.13.

f. Special Tees

The use of special tees is essential in one-pipe systems (Figures 2.14 and 2.15). Without them, the water would take the easier path along the supply main with very little water diverted to the supply branch leading to the convector.

g. Thermostat

Some boiler "packages" include a thermostat (Table 2.4). Since the boiler to be chosen in solving Example 2.1 includes such an item (T822), it will be used in the solution of that example.

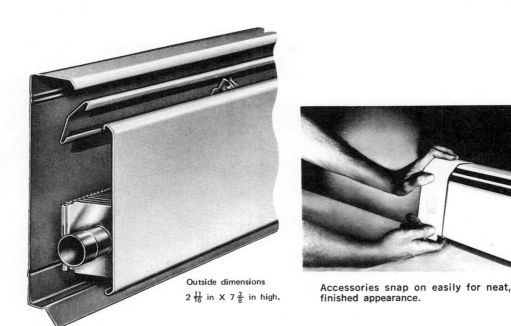

Outside dimensions
$2\frac{11}{16}$ in X $7\frac{3}{8}$ in high.

Accessories snap on easily for neat, finished appearance.

Figure 2.9 (a) Typical section through baseboard heating element. Manufacturers designation is "Model R Heatrim non-ferrous residential baseboard." (b) Snap-on element can conceal a shutoff valve or, as in the case of Example 2.1, an air vent. Note adjustable damper that is part of the enclosure. (Courtesy of Burnham Corporation)

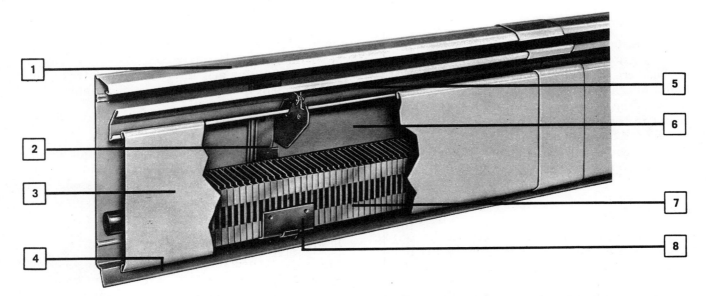

1—enclosure top and back panel—is of one-piece steel construction for strength and durability. Smooth surface won't collect dust and lint. Lower portion is extended ¾" out from wall to cover any openings between wall and finished flooring and to let panel stand unaided while being fastened to wall.

2—snap-in brackets—are die-formed of one-piece heavy gauge steel to support front panel and element where required; can be located at any point along back panel.

3—enclosure front panel—is of flat, dent-resisting steel; firmly snaps onto brackets either way (there is no top or bottom edge) and is easily removed for cleaning when necessary.

4—air intake space—allows unrestricted circulation of air for maximum heating efficiency and permits easy cleaning underneath element.

5—pivot-mounted damper—allows positive fingertip control for easy opening and closing from any point along its length; can be set in any position to regulate heat flow. Damper is standard equipment... no extra charge.

6—enclosure interior—gray prime coat provides background so that brackets and element blend in.

7—heating elements—two high-output elements are available, each with heat-reflecting interlocked aluminum fins (vented for maximum performance) bonded to seamless copper tubing for quick response, high I=B=R approved ratings.

8—element slide cradle—of lead-coated metal (terneplate) for lubricant slickness; snaps onto element and rides in bracket for noiseless linear movement of element.

two high-capacity elements

E-500 (½-in. nominal copper tubing)   E-750 (¾-in. nominal copper tubing)

element slide cradle

support bracket

Interlocked box-fins... 2⅛ inches square... tubes have one end expanded for easy coupling with unexpanded end of next element.

Figure 2.10 Details of heating baseboard and finned-tube elements. Note that for Example 2.1, E-750 (¾-in. nominal tubing) is chosen. (Courtesy of Burnham Corporation)

Table 2.1 Manufacturers Data For Hot Water Baseboard Convectors (Courtesy of Burnham Corporation)

approved I=B=R water ratings■

model R-750 Heatrim—capacities Btu/hr per lineal foot

number of lineal feet ●	water flow rate 500 lbs/hr average water temperature °F						water flow rate 2000 lbs/hr average water temperature °F					
	170°	180°	190°	200°	210°	220°	170°	180°	190°	200°	210°	220°
1	500	560	630	690	760	820	530	590	670	730	800	870
2	1000	1120	1260	1380	1520	1640	1060	1180	1340	1460	1600	1740
3	1500	1680	1890	2070	2280	2460	1590	1770	2010	2190	2400	2610
4	2000	2240	2520	2760	3040	3280	2120	2360	2680	2920	3200	3480
5	2500	2800	3150	3450	3800	4100	2650	2950	3350	3650	4000	4350
6	3000	3360	3780	4140	4560	4920	3180	3540	4020	4380	4800	5220
7	3500	3920	4410	4830	5320	5740	3710	4130	4690	5110	5600	6090
8	4000	4480	5040	5520	6080	6560	4240	4720	5360	5840	6400	6960
9	4500	5040	5670	6210	6840	7380	4770	5310	6030	6570	7200	7830
10	5000	5600	6300	6900	7600	8200	5300	5900	6700	7300	8000	8700
11	5500	6160	6930	7590	8360	9020	5830	6490	7370	8030	8800	9570
12	6000	6720	7560	8280	9120	9840	6360	7080	8040	8760	9600	10440
13	6500	7280	8190	8970	9880	10660	6890	7670	8710	9490	10400	11310
14	7000	7840	8820	9660	10640	11480	7420	8260	9380	10220	11200	12180
15	7500	8400	9450	10350	11400	12300	7950	8850	10050	10950	12000	13050
16	8000	8960	10080	11040	12160	13120	8480	9440	10720	11680	12800	13920
17	8500	9520	10710	11730	12920	13940	9010	10030	11390	12410	13600	14790
18	9000	10080	11340	12420	13680	14760	9540	10620	12060	13140	14400	15660
19	9500	10640	11970	13110	14440	15580	10070	11210	12730	13870	15200	16530
20	10000	11200	12600	13800	15200	16400	10600	11800	13400	14600	16000	17400
21	10500	11760	13230	14490	15960	17220	11130	12390	14070	15330	16800	18270
22	11000	12320	13860	15180	16720	18040	11660	12980	14740	16060	17600	19140
23	11500	12880	14490	15870	17480	18860	12190	13570	15410	16790	18400	20010
24	12000	13440	15120	16560	18240	19680	12720	14160	16080	17520	19200	20880
25	12500	14000	15750	17250	19000	20500	13250	14750	16750	18250	20000	21750
26	13000	14560	16380	17940	19760	21320	13780	15340	17420	18980	20800	22620
27	13500	15120	17010	18630	20520	22140	14310	15930	18090	19710	21600	23490
28	14000	15680	17640	19320	21280	22960	14840	16520	18760	20440	22400	24360
29	14500	16240	18270	20010	22040	22780	15370	17110	19430	21170	23200	25230
30	15000	16800	18900	20700	22800	24600	15900	17700	20100	21900	24000	26100

■ Approved I=B=R water ratings shown above for American-Standard Heatrim Panels (with Model E-750 element) are based on a water flow of 500 pounds per hour with a pressure drop of 0.047 inches of water per lineal foot and a water flow of 2000 pounds per hour with a pressure drop of 0.530 inches of water per lineal foot. As allowed by the Institute of Boiler and Radiator Manufacturers (I=B=R) Testing and Rating Code for Baseboard Type of Radiation, 15% is added to water heat capacity.

The use of I=B=R ratings at water flow rates of 2000 pounds per hour is limited to installation where the water flow rate through the baseboard unit is equal to or greater than 2000 pounds per hour. Where the water flow rate through the baseboard is not known, the I=B=R rating at the standard water flow rate of 500 pounds per hour must be used.

● These ratings are based on active (finned) Heatrim lengths. Difference between active length and total length of the standard Heatrim heating elements is 2¹⁵⁄₃₂ inches. Elements are unpainted.

Non-ferrous fins on Model E-750 elements measure 2⅛ x 2⅛ x 0.008 inches, spaced 52 fins per foot.

use these numbers for ordering

description	E-750 (¾″) element only	R-750 (¾″) element and enclosure assembled	model R enclosure only	
			assembled (individual package)	unassembled (bulk package—3 per box)
	ordering number	ordering number	ordering number	ordering number
2-ft. length	—	R-750-A2	—	—
3-ft. length	E-750-3	R-750-A3	R-3-A	R-3-B
4-ft. length	E-750-4	R-750-A4	R-4-A	R-4-B
5-ft. length	E-750-5	R-750-A5	R-5-A	R-5-B
6-ft. length	E-750-6	R-750-A6	R-6-A	R-6-B
7-ft. length	E-750-7	R-750-A7	R-7-A	R-7-B
8-ft. length	E-750-8	R-750-A8	R-8-A	R-8-B
10-ft. length	—	—	R-10-A	R-10-B
12-ft. length	—	—	R-12-A	R-12-B

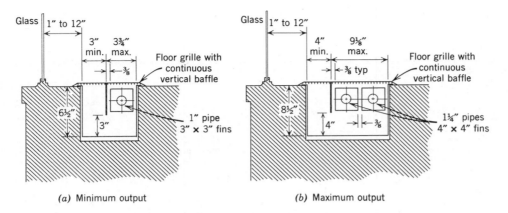

(a) Minimum output　　　　　(b) Maximum output

Figure 2.11 Underfloor convectors for hydronic systems. Essential when glass extends to the floor. Ratings shown below are based on 0° F outdoor temperature and 70° F room temperature and only for *glass* walls, not conventional walls. Average temperature of downdraft, 47° F. Average air velocity, 60 ft/min inside glass temperature 25° F. Sections (a) and (b) are for specific selected outputs. Consult manufacturer for all controlling conditions and intermediate ratings.

Output rating at a (reasonable) selected water temperature of 210° F
(a) Minimum 930 Btuh/lin ft.
(b) Maximum 2430 Btuh/lin ft.

Note. Architectural treatment of hatched area must conform to structure and provide insulation. See also Figures 3.7 and 3.8, page 70. (Courtesy of Dunham Bush, Inc.)

average water temperature °F.	maximum length of straight run (feet)
220	26
210	28
200	30
190	33
180	35
170	39
160	42
150	47

(a)

Figure 2.13 Expansion. (a) Observe that for the higher water temperatures (200° F and over) 26 to 30 ft are as far as *straight* tubing can be run without causing it to be overstressed or tend to buckle. When such straight runs occur, the expansion can be accommodated by the expansion unit seen in (b). The bellows or accordion-type element moves to avoid the stress that would be caused in the rigid tubing.

Note. In Example 2.1 the *linear* runs are, indeed, greater than 30 ft. However, they are not *straight;* the straightness is broken by one or several offsets of tubing down into the basement. A slight rotation of the fittings absorbs the movement. [(a) Courtesy of Copper Development Association; (b) Courtesy of Dunham Bush, Inc.]

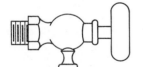

(a) Manual, for convectors

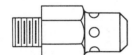

(b) Automatic, for convectors

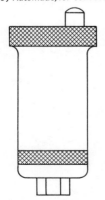

(c) Automatic, for piping

Figure 2.12 Air-vent valves. See also Figure 3.9b, page 71.

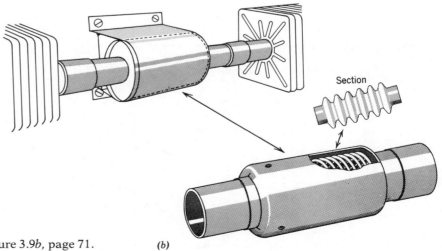

(b)

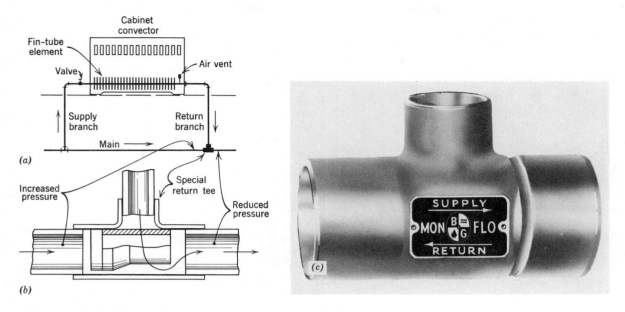

Figure 2.14 (a) Special fitting for one-pipe systems. Venturi-jet-type tee used here on the return branch connection to the main. It induces flow through the convector by retarding the flow to force water into the supply branch and producing a jet to reduce the pressure in the main following the return branch. (b) Illustrative sectional view to describe the function. No scale. Check with manufacturer as to dimensional details of the tee fitting. (c) Manufacturer's photo. (Courtesy of ITT Bell & Gossett)

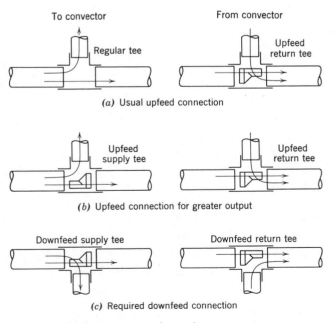

Figure 2.15 Venturi tees in branch convector connections in one-pipe systems. Their application can be noted in Figures 2.6 and 2.7.

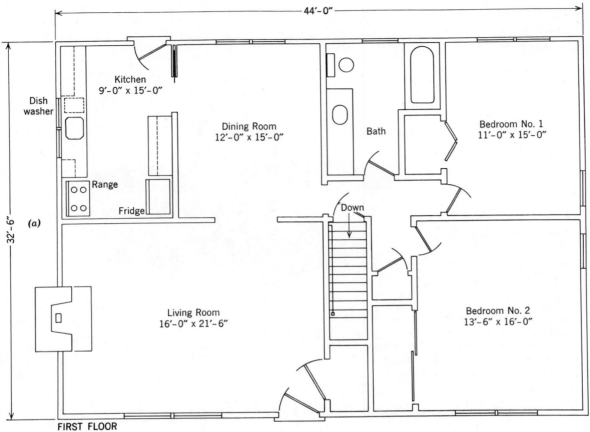

FIRST FLOOR

(a)

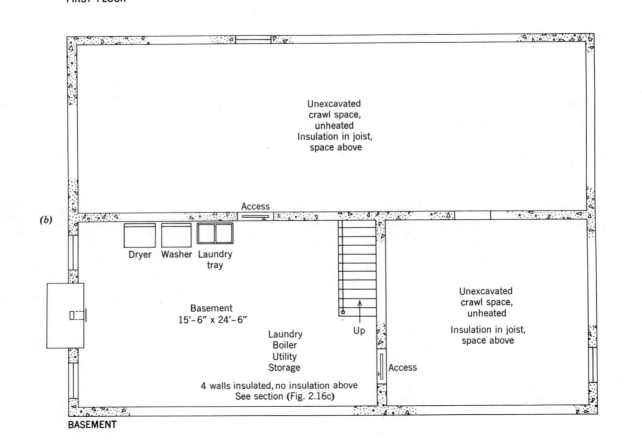

BASEMENT

(b)

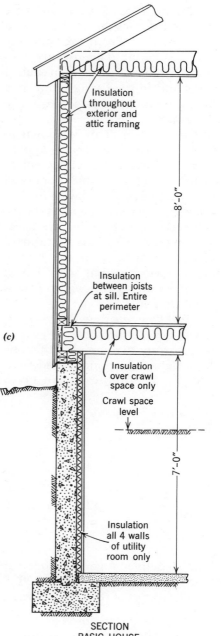

(c)

SECTION
BASIC HOUSE

Specification, Significant Items Only

Superstructure	Wood frame
Foundation	8-in. poured concrete
Windows	All insulating glass, double ½-in. air space
Insulation	k value = 0.27 for both batts and rigid insulation

6-in. mineral wool batts $R = 22.2$[a]
In ceiling joists between first floor and attic

In floor joists over crawl space
Behind wood sill, entire perimeter

3-in. mineral wool batts $R = 11.2$[a]
In stud space, all exterior frame walls

2-in. rigid insulation $R = 7.4$[a]
Four walls of utility space in basement

NONE In floor joists between living room and utility room below

Vapor barriers	Polyethylene sheet on *warm* side of *all* insulation
Ventilation	Vents to crawl space and attic Open in summer/closed in winter

[a]See page 54 regarding these resistance R values.

Heating Design Data: New York area

Indoor temperature, winter	75° F
Outdoor temperature, winter	0° F

Heat Losses

Living Room	9,300
Kitchen	4,800
Dining Room	4,300
Bath	2,900
Bedroom 1	5,200
Bedroom 2	7,500
Basement	4,900
Total	38,900 Btuh

Figure 2.16 *(a)* (Examples 2.1 and 2.2) Basic Plan, first floor. *(b)* Basic Plan, basement. *(c)* Section and selected specifications.

2.4. Design of a Hot Water Heating System

Example 2.1 Now that you have gained some knowledge of the products available for the composition of an effective hot water heating system, it is desirable to begin at once a specific design. For that purpose this example is presented. It is for a very simple but well insulated, house (Basic Plan) in the vicinity of New York City.

Figures 2.16 *a, b,* and *c* set the requirements for the design. The designer assembles some reference material. It is given here in selected items shown in Figures 2.17 to 2.19 and in Tables 2.1 to 2.4. The approach and procedure of any designer would surely affect the sequence of his personal decisions. However, the following sequence is used in *our* solution.

(a) Select a system type.
(b) Select an average water temperature.
(c) Choose a convector type.
(d) Size the convectors for the several rooms.
(e) Make a sketch of the system.
(f) Establish the required flow in gpm.
(g) Check the capacity of the circulator.
(h) Reconcile circulator "head" with that of the system.
(i) Select a boiler.
(j) Locate the thermostat.
(k) List the lengths of baseboard enclosures.
(l) Check the water temperature drop in the system.

Solution: Details follow. See summary in Figure 2.20 *a, b,* and *c* and in Table 2.5.

(a) *Select a system type.* For this house a perimeter loop system is the best and most economical.
(b) *Select an average water temperature.* Even though water boils at 212° F at atmospheric pressure, higher temperatures can be used because the system is under pressure. However, a 200° F average temperature is a good choice. At this relatively high temperature the convectors can be much smaller than, for instance, a 150° F temperature. Smaller convectors obviously reduce the cost of installation.
(c) *Choose a convector type.* From Table 2.1, model R-750 is chosen. Its fins are on ¾-in. tubing, and the prescribed water circulation rate to achieve the listed outputs is 2000 lb of water per hour.
(d) *Size the convectors for the several rooms.* Using

the heat losses in Figure 2.16*c*, select from Table 2.1 (200° F column) the lengths of the finned sections of all baseboards. From the ordering list in that table, the available lengths are chosen. The results are listed in Table 2.5. In each of the two bedrooms and in the living room, two sections of finned tube must be joined to make up the proper length.

(e) *Make a sketch of the system.* This is shown in Figure 2.20*c*. So as not to have the mains appear in the rooms at points where there is no enclosure, the mains drop into the basement or crawl space where they must be provided with drains. In the crawl spaces, they must have insulative wrapping. Tentatively, 1-inch mains are used, subject to reconciling the frictional resistance of the system with the capacity of the pump. Hence, we have a series system with 1-inch mains reducing to ¾ in. as the water passes through the finned elements.

(f) *Establish the required flow rate in gallons per minute.* In converting *pounds* per *hour* to *gallons* per *minute,* it is important to remember that one gallon of water weighs 8.33 pounds and that, obviously, there are 60 minutes in an hour. Therefore:

$$\text{pounds per hour} = \text{gallons per minute} \times 60 \times 8.33$$

Transposing

$$\text{gallons per minute} = \frac{\text{pounds per hour}}{60 \times 8.33}$$

Substituting

$$\text{gallons per minute} = \frac{2000}{60 \times 8.33} = 4 \text{ gallons per minute}$$

The circulator must deliver 4 gal or more per minute of water at average temperature 200° F to achieve the listed ratings of the convectors.

(g) *Check the capacity of the circulator.* The boiler that will be chosen (item i) for the solution to Example 2.1 provides a 1¼-in. circulator piped to the boiler return. This will be used if its capacity suits the system. Figure 2.18 shows that a 1¼-in. circulator will deliver the required 4 gpm at a "head" of 6 ft.

(h) *Reconcile circulator "head" with that of the system.* Pressure loss in pounds per square inch (psi) at any given rate of flow through any given tube size can be expressed also as "feet of head" by the equation

$$\text{feet of head} = \frac{\text{pounds per square inch}}{0.433}$$

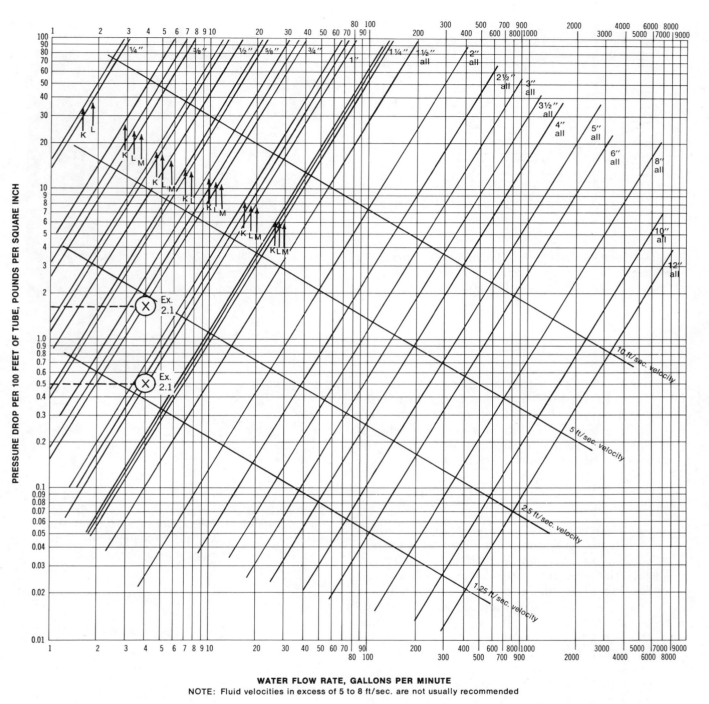

WATER FLOW RATE, GALLONS PER MINUTE
NOTE: Fluid velocities in excess of 5 to 8 ft/sec. are not usually recommended

Figure 2.17 Pressure loss and velocity relationships for water flowing in copper tube.
(Courtesy of Copper Development Association, Inc.)

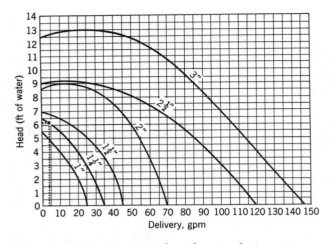

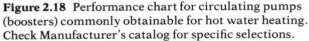

Figure 2.18 Performance chart for circulating pumps (boosters) commonly obtainable for hot water heating. Check Manufacturer's catalog for specific selections.

To find the pounds per square inch lost in the system, we are first concerned about tube sizes and the lengths of each tube size. Referring to Fig. 2.20c, we may scale from the drawings the lengths of 1-in. and of ¾-in. tubing. These lengths are as follows:

1 in.	130 ft
¾-in.	48 ft

These are known as "developed lengths," which means the total linear distance that a drop of water passes through in making the entire circuit. However, pressure is lost when water passes through fittings such as ells or the reducers from 1 to ¾ in. Tables such as Table 2.3 can be useful in converting fittings into feet of pipe through which the pressure loss would be the same as through the fitting. However, it is easier and sometimes sufficiently accurate to add 50% to the *developed* length (DL) to arrive at a "total equivalent length" (TEL), an imaginary longer length of tubing that causes the same pressure loss as the DL plus the effect of the fittings. Thus the TEL is

1 in.	130 × 1.5	=	195 ft TEL
¾ in.	48 × 1.5	=	72 ft TEL

Table 2.2 gives the pressure loss per 100 ft. of tubing of the several sizes for various flow rates in gallons per minute. At 4 gpm the unit loss per 100 ft for 1-in. tubing is 0.5 and for ¾ in. is 1.8.

Figure 2.19 GHB2 series hot water boiler. See Table 2.4 for ratings and other data. Note symbols indicating conformance to AGA and I=B=R standards. (Courtesy of The H. B. Smith Company, Inc.)

Thus the total pressure loss is

1 in.	1.95 × 0.5	=	0.975
¾ in.	0.72 × 1.8	=	1.296
			2.271 psi

Now convert to feet of head

$$\frac{2.271}{0.433} = 5.2 \text{ ft of head}$$

Since 5.2 is less than the 6 ft that the circulator *can* deliver, it will pump a little more than 4 gpm, which will provide even more than the 2000 lb/hr required to assure the output of the convectors. Observe that Figure 2.17 verifies these unit pressure losses and reveals the water velocities at about 1.50 and 2.25 ft per second (fps) for the two tube sizes. These are both below the limiting maximum of about 5 to 8 fps set in

Table 2.2 Pressure Loss Due to Friction in Type M Copper Tube Based on Figure 2.17 (Courtesy of Copper Development Association, Inc.)

Flow, gpm	Pressure Loss per 100 Feet of Tube, psi											
	Standard Type M Tube Size—inches											
	$3/8$	$1/2$	$3/4$	1	$1 1/4$	$1 1/2$	2	$2 1/2$	3	4	5	6
1	2.5	0.8	0.2									
2	8.5	2.8	0.5	0.2								
3	17.3	5.7	1.0	0.3	0.1							
4	28.6	9.4	1.8	0.5	0.2							
5	42.2	13.8	2.6	0.7	0.3	0.1						
10		**46.6**	8.6	2.5	0.9	0.4	0.1					
15			17.6	5.0	1.9	0.9	0.2					
20			**29.1**	8.4	3.2	1.4	0.4	0.1				
25				12.3	4.7	2.1	0.6	0.2				
30				**17.0**	6.5	2.9	0.8	0.3	0.1			
35					8.5	3.8	1.0	0.4	0.2			
40					11.0	4.9	1.3	0.5	0.2			
45					**13.6**	6.1	1.6	0.6	0.2			
50						7.3	2.0	0.7	0.3			
60						10.2	2.7	1.0	0.4			
70						**13.5**	3.6	1.2	0.5	0.1		
80							4.6	1.6	0.7	0.2		
90							5.7	2.0	0.9	0.2		
100							**7.5**	2.7	1.0	0.3	0.1	
200								**8.5**	3.6	1.0	0.3	0.1
300									**8.0**	2.0	0.7	0.3
400										**3.3**	1.2	0.5
500											1.7	0.7
750											**3.6**	1.5
1000												**2.5**

NOTE: Numbers in bold face correspond to flow velocities of just over 10 ft per sec.

Table 2.3 Allowance for Friction Loss in Valves and Fittings Expressed as Equivalent Length of Tube (Courtesy of Copper Development Association, Inc.)

Fitting Size, inches	Equivalent Length of Tube, feet						
	Standard Ells		90° Tee		Coupling	Gate Valve	Globe Valve
	90°	45°	Side Branch	Straight Run			
$3/8$	0.5	0.3	0.75	0.15	0.15	0.1	4
$1/2$	1	0.6	1.5	0.3	0.3	0.2	7.5
$3/4$	1.25	0.75	2	0.4	0.4	0.25	10
1	1.5	1.0	2.5	0.45	0.45	0.3	12.5
$1 1/4$	2	1.2	3	0.6	0.6	0.4	18
$1 1/2$	2.5	1.5	3.5	0.8	0.8	0.5	23
2	3.5	2	5	1	1	0.7	28
$2 1/2$	4	2.5	6	1.3	1.3	0.8	33
3	5	3	7.5	1.5	1.5	1	40
$3 1/2$	6	3.5	9	1.8	1.8	1.2	50
4	7	4	10.5	2	2	1.4	63
5	9	5	13	2.5	2.5	1.7	70
6	10	6	15	3	3	2	84

NOTE: Allowances are for streamlined soldered fittings and recessed threaded fittings. For threaded fittings, double the allowances shown in the table.

Table 2.4 Manufacturer's Data for GHB2 Boilers (See Figure 2.19) (Courtesy of The H. B. Smith Company, Inc.)

GHB2 SERIES COMPLETELY PACKAGED GAS $\frac{\text{STEAM}}{\text{WATER}}$ BOILERS

FACTORY ASSEMBLED AND WIRED • 60,000 TO 240,000 BTU GROSS OUTPUT
FOR USE WITH NATURAL GAS • DRY BASE CONSTRUCTION • NO TANKLESS

STANDARD EQUIPMENT
Shipped with wirebound Skid Bottom Crate with Polyethylene Cover to Protect Boiler from moisture and dust.

Assembled Boiler with Extended Jacket
L8148 Combination High Limit and Circulator Relay Control
Combination Pressure-Temperature Gauge
B & G 1¼" Circulator—Wired and Piped to Boiler
Boiler Drain Valve
Gas Burners and Manifold Installed
Gas Controls—furnished as a combination unit which includes:

> Automatic Gas Valve (V8146), 24V. with Pilot Filter
> Pressure Regulator
> 100% Pilot Shut Off and Main Gas Valve
> Pilot Burner and Thermocouple

In Separate Carton Packed in Crate
> Relief Valve
> T822 Thermostat

In Separate Carton NOT Packed in Crate
> Draft Diverter

WATER

WATER BOILER MODEL NO.	NATURAL GAS			APPROX. SHIPPING WEIGHT
	A.G.A. INPUT MBH	A.G.A. OUTPUT MBH	NET I-B-R RATING MBH	
GHB2-W-3	75.0	60.0	52.2	370
GHB2-W-4	112.5	90.0	78.3	465
GHB2-W-5	150.0	120.0	104.3	540
GHB2-W-6	187.5	150.0	130.4	615
GHB2-W-7	225.0	180.0	156.5	690
GHB2-W-8	262.5	210.0	182.6	765
GHB2-W-9	300.0	240.0	208.7	840

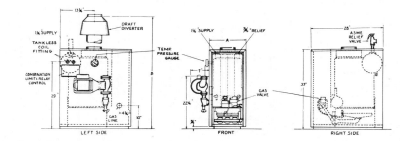

UNIT ARRANGED FOR INTERMITTENT CIRCULATION OPERATION

STANDARD EQUIPMENT
Shipped with wirebound Skid Bottom Crate with Polyethylene Cover to protect Boiler from moisture and dust.

Assembled Boiler with Extended Jacket
> 467B Low Water Cutoff—mounted
> PA404 Pressure Limit Control
> Water Gauge Glass
> Boiler Drain Valve
> Steam Pressure Gauge
> Gas Burners and Manifold Installed

Gas Controls—furnished as a combination unit which includes:
> Automatic Gas Valve (V8146), 24V. with Pilot Filter
> Pressure Regulator
> 100% Pilot Shut Off and Main Gas Valve
> Pilot Burner and Thermocouple

In separate Carton Packed in Crate
> Safety Valve

In Separate Carton NOT Packed in Crate
> Draft Diverter

STEAM

STEAM BOILER MODEL NO.	NATURAL GAS				APPROX. SHIPPING WEIGHT
	A.G.A. INPUT MBH	A.G.A. OUTPUT MBH	NET I-B-R RATING MBH	NET I-B-R SQ. FT. STEAM	
GHB2-S-3	75	60	45	188	345
GHB2-S-4	105	84	63	263	445
GHB2-S-5	140	112	84	350	520
GHB2-S-6	175	140	105	438	595
GHB2-S-7	210	168	126	525	670
GHB2-S-8	245	196	147	613	745
GHB2-S-9	280	224	168	700	820

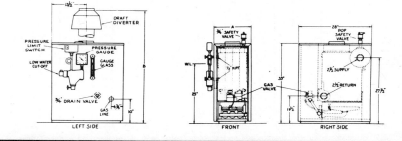

Table 2.5 (Ex 2.1) Design Hot Water Heating, Basic Plan[a, b]

Engineer's Calculations		Design Results					General Data and Results
Space	Heat Loss (Btuh)	Baseboard Convector Number	Baseboards, See General Data for Type——>			Enclosure Length (Feet)	
			Length (Feet)	Output, (Btuh) (Table 2.0)	Ordering Catalog No. (Table 2.1)		
Living room	9300	1	6	4380	E-750-6	17 ft 6 in.	Average Water Temperature: 200° F
		2	7	5110	E-750-7		Baseboards: R-750 Heatrim 2000 lb of water per hour, ¾ in. tube (Table 2.1)
Kitchen	4800	3	4	2920	R-750-A4	4 ft 0 in.	
		4	2	1460	R-750-A2	2 ft 0 in.	
Dining room	4300	5	6	4380	E-750-6	12 ft 0 in.	Circulator: B & G 1¼ in., (Table 2.4)
Bath	2900	6	4	2920	E-750-4	6 ft 0 in.	Thermostat: T822 (Table 2.4)
Bedroom No. 2	7500	7	8	5840	E-750-8	11 ft 0 in.	Boiler: GHB2-W-3, Net I=B=R Rating 52.2 Mbh (Table 2.4)
Bedroom No. 1	5200	8	6	4380	E-750-6	13 ft 6 in.	Convector output: 35 Mbh
		9	5	3650	E-750-5		
Basement	4900	NONE[c]					
Total	38,900		Total	35,040			
Total, first floor only	34,100		Specified 35.0 Mbh				
		Required 34.1 Mbh					

[a]See Figure 2.20 *a*, *b*, and *c*.
[b]*Temperatures* are degrees Fahrenheit (° F). Temperature *differences* are Fahrenheit degrees (F °).
[c]The basement is warmed by heat from the boiler.

the note below Figure 2.17. Excessively high velocities can cause a system to be noisy.

(i) *Select a boiler.* (Table 2.4) An H. B. Smith GHB2-W delivers (net rating) 52.2 Mbh, which is more than is required by our 38,900 Btuh house (Table 2.4). The excess is enough to keep the basement warm.

(j) *Locate the thermostat.* This is correctly positioned as shown in Figure 2.20*a* in an accessible spot and away from cold exterior walls, direct sunlight and warm areas near the convectors.

(k) *List the lengths of the enclosures.* These are recorded in Table 2.5 and conform in some cases to the room dimension along the exterior wall.

(l) *Check the water temperature drop through the system.* The heat balance formula is

$$Btuh = \text{gallons per minute} \times 8.33 \times 60 \times 1^* \times TD$$
$$TD = \text{temperature drop}$$

Transposing

$$TD = \frac{Btuh}{\text{gpm} \times 8.33 \times 60 \times 1}$$

Substituting

$$TD = \frac{38,900}{4 \times 8.33 \times 60 \times 1} = 19.4 \text{ F}°$$

This is OK. The usual drop in this kind of system is about 20 F°. Hence, the water will be delivered at 210° F and will return to the boiler at 190° F. Average temperature is 200° F.

*Specific heat of water.

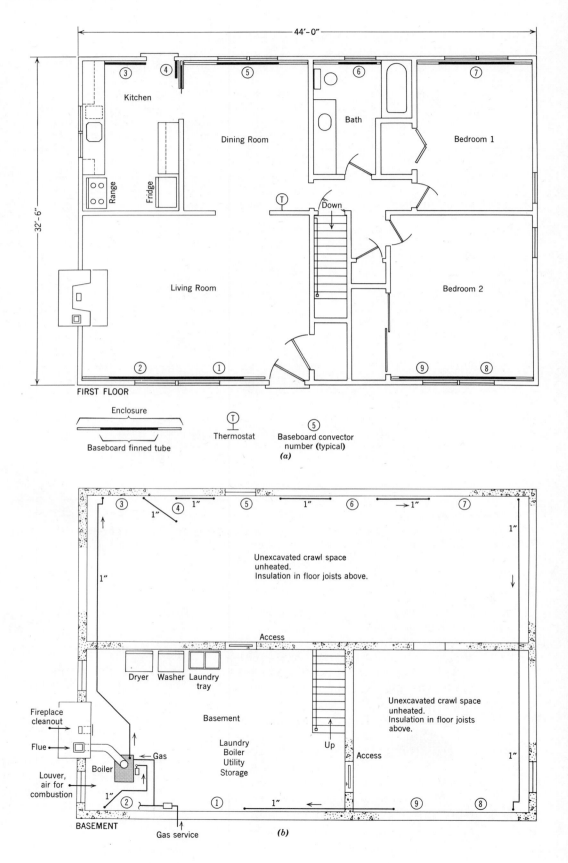

FIRST FLOOR

Enclosure

Baseboard finned tube

Thermostat

Baseboard convector
number (typical)

(a)

BASEMENT

Gas service

(b)

Figure 2.20 (Example 2.1) Solution and layout, hot water heating. *(a)* First floor plan. *(b)* Basement plan. All 1-in. tubing in unheated crawl space to have insulated covering. Elsewhere uncovered. *(c)* Schematic isometric. See Figure 2.1 for boiler controls, omitted here. See Table 2.5.

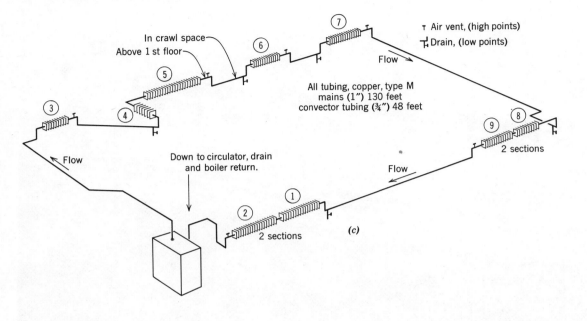

2.5. Electric Heating

When economic considerations are favorable, heating by electricity is found to have many advantages. The responsibility for this clean and adaptable energy source is the problem of the local power station. There, at the utility company's generating plant, fossil fuel, often oil or coal, is used to produce steam. Turbines powered by the steam move the great wire coils through the generator's magnetic field to produce electricity that is then transmitted to the user's building where it is converted to heat.

The first and obvious advantage is the elimination of the need for combustion at the building site. The chimney, fuel storage, air to support combustion and the resulting undesirable flue gases are problems delegated to the electric company. When direct electric resistance heating is used (baseboards or wall units), the need for a boiler or furnace is also avoided. The system becomes a very compact one. For this reason the *installation* cost of a simple electric heating system is usually a great deal less than for a hot water (hydronic) or a warm air installation.

A designer's study of the overall expense of installing *and using* electric heating is not so simple. Low first cost (for installation) must be considered together with cost of periodic maintenance and repairs and, most important, the electric bills over the years. Repairs are not, in general, much different than for hot water or air heating. The cost of electric energy, however, deserves special study. For several years prior to this writing, electric energy proved to be relatively cheap in the face of shortages and the rising cost of oil and gas. During that period, a great increase occurred in the choice for electric systems. At the moment, *electric* rates are rising. Some consumers are suffering from the increased cost of this energy, and a few are considering a switch to other heating methods. The market is, at least for now, unstable. In future decisions it would appear wise to consider the geographic location of the project and the availability and cost of electricity as opposed to other energy sources.

Some *operational* advantages are to be found in electric heating. Individual thermostats in each heating unit or in each room are easily arranged and afford great flexibility. Energy conservation may be achieved by the fact that heat may be turned down or off in unused rooms. When turned on again, the response is rapid and room comfort is speedily restored. Comfort is served because the temperature in each room can be adjusted to the pleasure of the occupant by the use of the room thermostat.

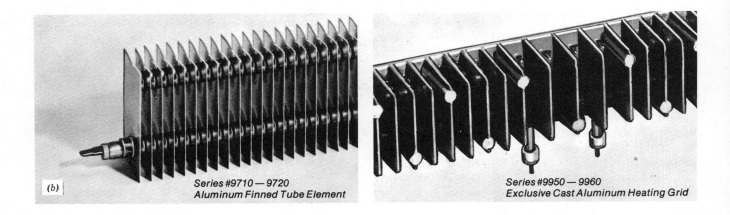

Series #9710 — 9720
Aluminum Finned Tube Element

Series #9950 — 9960
Exclusive Cast Aluminum Heating Grid

Figure 2.21 *(a)* Baseboard heating unit in place. *(b)* Metal heating elements through which the resistance wiring runs. Cast-aluminum grid retains its warmth somewhat longer than the aluminum finned-tube element. *(c)* Cutaway view of a heater employing a cast-aluminum grid. End sections with removable covers may be used for wiring, convenience outlets or for built-in thermostat to control the baseboard. *(d)* Specifications. (Courtesy of The Singer Company, Climate Control Division)

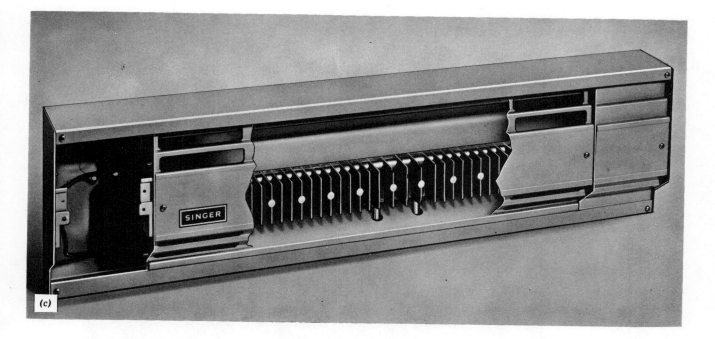

(c)

SPECIFICATIONS

Dual Voltage

Watts Per Foot	Watts	Length, Inches (1)	Btuh	Catalog Number			
				120v	208v	240v	277v
CAST ALUMINUM GRID 187	500	40	1707	9950	9955-A	9955-E	9950-M
	750	56	2560	9950-C	9955-B	9955-F	9950-N
	1000	72	3415	9950-F	9955-C	9955-G	9950-P
	1500	104	5120	9950-J	9955-D	9955-H	9950-Q
250	650	40	2219	9960	9960-B	9955-A	9955-E
	1000	56	3415	9960-C	9960-E	9955-B	9955-F
	1300	72	4439	9960-F	9960-H	9955-C	9955-G
	2000	104	6830		9960-L	9955-D	9955-H
ALUMINUM FINNED TUBE ELEMENT 187	375	28	1280	9710	9715	9715-A	9710-A
	565	40	1930	9710-B	9715-B	9715-C	9710-C
	750	52	2560	9710-D	9715-D	9715-E	9710-E
	940	64	3210	9710-F	9715-F	9715-G	9710-G
	1125	76	3840	9710-H	9715-H	9715-J	9710-J
	1500	100	5120		9715-K	9715-L	9710-L
	1875	124	6400		9715-M	9715-N	9710-N
250	500	28	1707	9720	9720-A	9715	9715-A
	750	40	2560	9720-B	9720-C	9715-B	9715-C
	1000	52	3415	9720-D	9720-E	9715-D	9715-E
	1250	64	4265	9720-F	9720-G	9715-F	9715-G
	1500	76	5120	9720-H	9720-J	9715-H	9715-J
	2000	100	6830		9720-L	9715-K	9715-L
	2500	124	8530		9720-N	9715-M	9715-N

(d)

(1) 7-1/2" high, 3-1/16" deep. Finish: Mist Gray

2.6. Electric Heating Units

There are a great many more electric heating methods than can be described in the limited space of this book. Two have already been shown. One is in Figure 1.5(I) (page 9). It consists of a heating element in an airstream. The other (Figure 1.8) (page 13) places the heating coil in a water boiler. Respectively, they are adjunct parts to a warm air heating system and to a hot water heating system. Other methods described here that can be considered more "direct" are:

3 and 4 KW Models
21" high, 16¼" wide,
extend 1½" from wall

1.5 and 2 KW Models
16¼" high, 13" wide,
extend 1⁵⁄₁₆" from wall

specifications and ordering information

CATALOG NO.	MODEL NO.	CONTROL	WATTS	VOLTS	BTUH	CFM
5990	W-15	Manual	1500	120	5122	65
5990-A	W-15	Manual	1500	240	5122	65
5990-F	W-20	Manual	2000	240	6830	65
5990-H	W-30	Manual	3000	240	10245	200
5990-R	W-40	Manual	4000	240	13660	210
5990-S	WA-15	Automatic	1500	120	5122	65
5990-T	WA-15	Automatic	1500	240	5122	65
5990-M	WA-20	Automatic	2000	240	6830	65
5990-N	WA-30	Automatic	3000	240	10245	200
5990-W	WA-40	Automatic	4000	240	13660	210

Figure 2.22 Electric downflow wall heaters with automatic control. Both fan and heating element start up whenever the room temperature drops below the selected thermostat temperature-setting. (Courtesy of The Singer Company, Climate Control Division)

(a) Electric baseboard heaters
(b) Electric wall heaters
(c) Electric floor inserts
(d) Electric radiant cable
(e) Electric bathroom heaters

And one of many control items:

(f) Wall mounted thermostat

a. Electric Baseboard

Unlike *hot water* baseboards, which afford finned elements of specific length, and enclosures which, for appearance, may be extended wall-to-wall, electric baseboards are manufactured integrally with their enclosures. They are available in lengths such as those shown in Figure 2.21*d*.

(b)

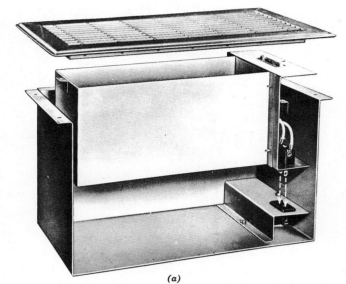

(a)

SPECIFICATIONS

Cat. No.	Model No.	Width	Length	Depth	Volts	Watts
8180	FI-4	6¹⁄₁₆″	14⅛″	8¼″	120	400
8180-A	FI-4	6¹⁄₁₆″	14⅛″	8¼″	240	400
8181	FI-8	6¹⁄₁₆″	30⅛″	8¼″	120	800
8181-A	FI-8	6¹⁄₁₆″	30⅛″	8¼″	240	800

Floor Opening: 6¼″ x 14¼″ for Model FI-4, 6¼″ x 30¼″ for Model FI-8
Heater furnished with grille in gray enamel, which can be refinished to match floor if desired.

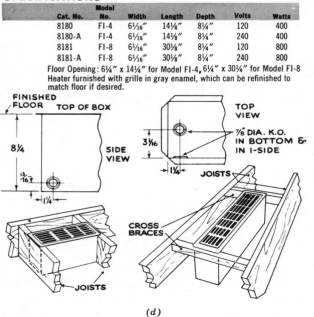

(d)

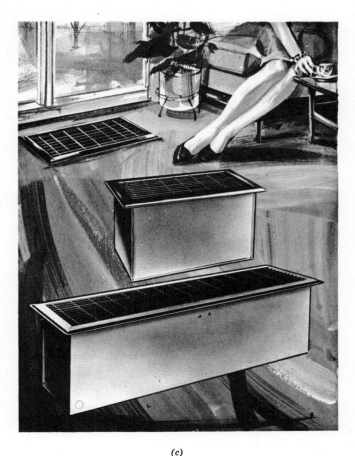

(c)

Figure 2.23 Electric floor insert heaters. *(a)* Sectional view of the heater. *(b)* Electric heating element, a removeable part of the assembly. *(c)* Models Fl 4 and Fl 8 and the application of Fl 4 in a house. *(d)* Specifications and framing. Output in Btuh can be obtained by multiplying watts by 3.41 (Courtesy of The Singer Company, Climate Control Division)

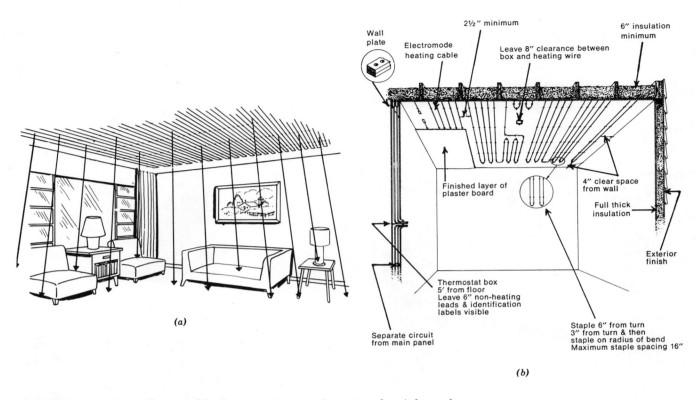

Figure 2.24 Heating with no visible elements. Outputs (not given here) depend on cable selection and spacing. *(a)* Application sketch. In the finished work the cables do not show but are covered by finished layer of plaster board as shown in *(b)*. (Courtesy of The Singer Company, Climate Control Division)

b. Electric Wall Heaters

More compact and presumably a bit faster to respond, these units employ a fan to accelerate the delivery of the heat (Figure 2.22). Not quite as silent as baseboards, they are often chosen for special conditions, such as in the kitchen of Example 2.2.

c. Electric Floor Inserts

These find their greatest use below floor-length glass doors or glass-to-the-floor. See Figure 2.23. Yet they may be used in other locations where, except for the grill, they are "out of sight."

d. Electric Radiant Cables

These, of course, are completely "out of sight." See Figure 2.24. Among their other qualities, they cannot be tampered with unless one knocks a hole in the ceiling. Sometimes they are used in apartment houses when economics are favorable.

e. Electric Bathroom Heaters

These (Figure 2.25) provide instant comfort in the bathrooms of electrically heated houses *or* in houses where a nonelectric system is not controllable room by room.

f. Thermostats

Although not illustrated here, thermostatic control can be located in the end of baseboards (see Figure 2.21*c*) and is an integral part of *automatic* wall heaters. See Figure 2.22. When control of one or several electric units is desired from a convenient, centralized location in the room, a thermostat like that shown in Figure 2.26 can be chosen. It is often located on the wall next to the door that gives entrance to the room and is placed higher than the light switch. See the solution to Example 2.2.

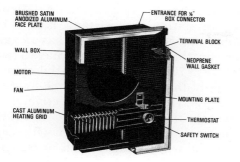

Automatic   **Manual**

surface mounting

The surface wall box is built to accommodate the complete heater for use where installation must be made directly to concrete, brick, stone or other surfaces into which it is impossible or impractical to cut. This box is furnished separately and should be ordered as follows: Catalog No. 8754. 16½" high, 13¼" wide, 4¼" deep.

recessed mounting

All models are designed to fit within the standard 16" stud centers and wall thickness, with only the decorative aluminum frame and front panel extending not more than 1⁵⁄₁₆" beyond the wall surface. A service entrance for ½" conduit fitting is conveniently located at the top right corner of the wall box.

semi-recessed mounting

This wall box is for installations where the heater may be recessed up to two inches but not to its full depth. The complete heater including regular wall box inserts into the semi-recessed box for a neat, simplified installation. This box is furnished separately and should be ordered as follows: Catalog No. 8754-A. 16½" high, 13¼" wide, 2" deep.

Specifications

Catalog Number	Model	Control	Watts	Volts	BTUH	Ship. Wgt. Lbs.
8750	WJ-12*	Manual	1250	120	4269	13
8751	WJA-12*	Automatic	1250	120	4269	13
8751-B	WJA-12*	Automatic	750	120	2550	13

Finish: Brushed Satin Anodized Aluminum
Wall Box Opening: 14⁵⁄₁₆" high, 11¼" wide, 3¾" deep
Front Plate: 16¼" high, 13" wide. Extends 1⁵⁄₁₆" from wall

*Not equipped with fan delay switch.

Figure 2.25 Electric bathroom heaters. Often used in bathrooms and other small rooms with wall space too limited to accommodate baseboard heaters. In houses with hot water heat, they are sometimes used to supplement the central water system to give a little warmth in bathrooms when the hot water system is not operating. See Figure 3.10a, page 72. (Courtesy of The Singer Company, Climate Control Division)

Figure 2.26 One type of wall-mounted thermostat suitable for controlling electric heating units. Bottom knob sets the temperature desired and the upper reading indicates the actual temperature in the space (Courtesy of Mears Controls, Inc.)

2.7. Electric Heating Design

At an earlier time before the operating cost of electric heating was competitive with that of fossil fuels, proponents of the use of electrical energy established new and upgraded standards of insulation. By this means, the annual cost of electric energy for heating usually compared favorably with the (then) cost of less expensive fuel that was used in houses with lesser insulative values. At that time many utility companies would not consider supplying electricity for heating unless the new higher standards were met. Now, with the prospect of curtailed fuel supplies and rising costs of *all* energy sources, the higher standards of insulation are often being chosen for *all* construction regardless of the proposed fuel.

For this reason, Examples 2.1 and 2.2 (Hot Water Heating and Electric Heating, respectively) are based on the basic plan detailed in Figure 2.16c, where very adequate insulation is specified.

The Electric Energy Association (EEA) has set certain criteria, expressed as R (for resistance) values, which are followed and sometimes exceeded in Figure 2.16. See Table 2.6a. Before we compare the R values of our Basic Plan with the suggested EEA values, the basis for R values should be known. The k value of insulation previously mentioned is a *conductivity* unit. It is the number of Btuh that will pass through 1 sq ft of material 1-in. thick when 1 F° exists between the two faces. The inverse or reciprocal of the unit of conductivity could be termed resistivity. So, R = $1/k$. This, of course, is for 1 in. of thickness. Resistance, of course, increases directly with the increasing thickness of the material. Hence, for x inches of insulation, the resistance is $1/k \times x$ or x/k. The insulation used in Figure 2.16c has a k value of 0.27. Thus 1-in. thickness of it would have a resistance of $1/0.27 = 3.70$. The resistances, therefore, of the 2, 3 and 6-in. thick insulation of the house in Figure 2.16 are 7.4, 11.1, and 22.2, respectively.

Now let us make the comparison. The NEMA values are for "normal" frame construction with nominal 4-in. studs and with siding, shingle or masonry veneer exterior.

	Suggested Values (EEA) Table 2.6	Actual Values (Figure 2.16)
Ceilings	R 19	R 22.2
Opaque walls	R 11	R 11.1
Floors over vented crawl space	R 11	R 22.2
Basement walls (masonry)	R 7	R 7.4

It is evident that the house (Figure 2.16) is well insulated, and this fact, together with "insulating double glass," puts it in a category where minimal heat loss will be the result.

Another measure may be taken to check the effect of insulation on heat loss. Subject to specific conditions as shown in Table 2.6a, for electrically heated houses in the New York area (about 5000 degree days) (see Table 2.6b) the heat loss, expressed in watts, should not exceed 8.2 w/sq ft of floor area of the space to be heated. Table 2.7 shows the required wattage, including that for the basement, to be 11400. The heated area including the basement utility room is 1885 sq ft.

To compare with the amount of 8.2w, we have 11,400/1885 = 6.04, well below the 8.2 w/sq ft maximum suggested by EEA.

Example 2.2 Design an electrical heating system for the Basic Plan as shown in Figure 2.16a, b and c.

Solution: Although many different kinds of electrical heating methods could be used, this solution is confined to the use of resistance baseboard heaters and one wall heater (in the kitchen). The design makes use entirely of Figures 2.21 and 2.22 from which heating units are selected to provide heat at rates to balance the heat loss in all rooms as shown in Figure 2.16c. The use of 240 v items in preference to those of 120 v reduces the size of the required electrical conductors. Thermostats for control of each room are shown at appropriate locations. The kitchen thermostat is integral with the wall heater. Summary of the design is shown in Table 2.7 and Figure 2.27.

Table 2.6 Two Recommendations Selected from "The All Weather Comfort Standard" (Reprinted by permission of the Electric Energy Association)

Recommendation a: Recommended Thermal Performance Values
The maximum heat loss limits shown in Recommendation b will generally be met by utilizing the thermal performance values shown below for the opaque sections of the structure.

	Type of Exterior Wall Construction	
	Frame: Nominal 4-in. Studs (With Siding, Shingle or Masonry Veneer Exterior)	Masonry: Lightweight Block or Brick Cavity
Ceilings:		
U-Value of Section	0.05	0.04
Insulation Resistance[a]	R-19	R-22
Opaque walls:		
U-Value of Section	0.07	0.11
Insulation Resistance[a]	R-11	R-7
Floors over vented crawl spaces:		
U-Value of Section	0.07	0.05
Insulation Resistance[a]	R-11	R-19
Floors over unheated basements:		
U-Value of Section	0.07	0.07
Insulation Resistance[a]	R-11	R-11
Floors and walls between separately heated units:		
U-Value of Section	0.07	0.07
Insulation Resistance[a]	R-11	R-11
Concrete slabs on grade:		
Insulation Resistance[a]	R-5	R-5

[a]Insulation R values shown refer to the *resistance of the insulation ONLY.*

Recommendation b: Maximum Winter Heat Loss Limits for Electrically Heated Homes
When based on an infiltration rate of three-quarters air change per hour, the total calculated heat loss shall not exceed the values shown below. These values are expressed in watts (or Btuh) per square foot of floor area of the space to be heated, measured to the outside of exterior walls.

	Maximum Heat Loss Values[b]	
Degree Days[c]	W/Sq. Ft.	Btuh/Sq. Ft.
Over 8000	10.0	34
7001 to 8000	9.4	32
6001 to 7000	8.8	30
4501 to 6000	8.2	28
3001 to 4500	7.6	26
2000 to 3000	7.0	24

[b]Typical means of attaining these values are shown in recommendation a. It is recognized that these limits may sometimes be exceeded in the installation of electric heating in existing homes.
[c]The approximate annual degree days in various parts of the United States will be found in Figure 2.28.

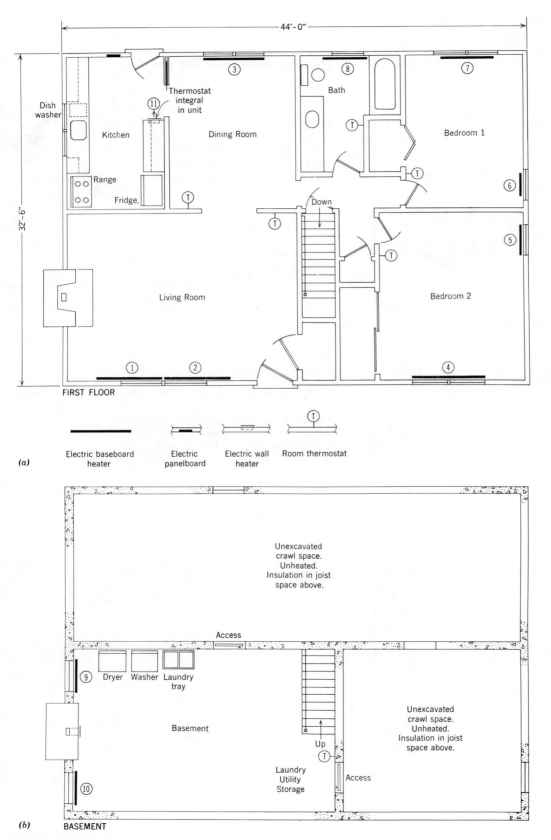

Figure 2.27 (Example 2.2) Layout and identification of parts, electric heating, basic plan, (a) first floor, (b) basement plan. For results of design see Table 2.7.

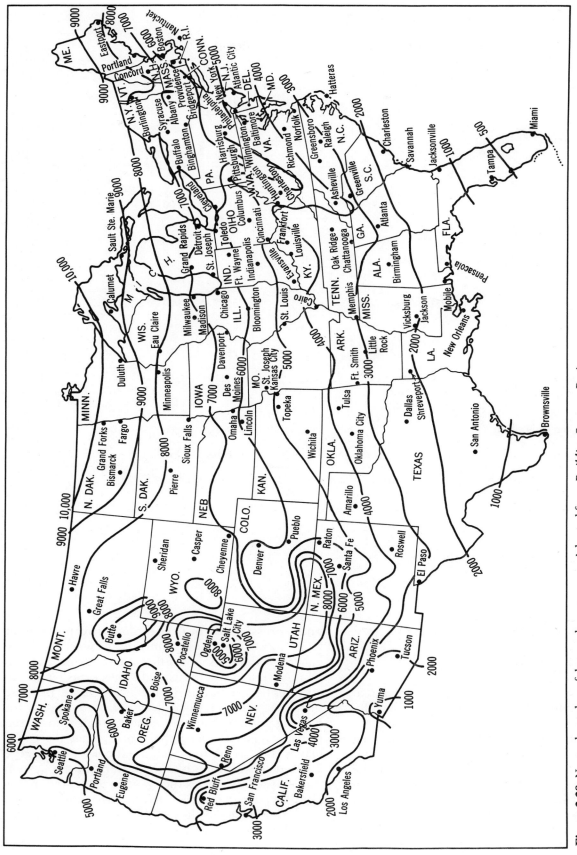

Figure 2.28 Normal number of degree days per year. Adapted from *Building Systems Design*.

Table 2.7 (Example 2.2) Summary of Design, Electric Heating, Basic Plan[a, b]

Engineer's Calculations				Design Results					
Space	Heat Loss Rate		Heater Type Volts	Heater Number	Output		Length (Inches)	Catalog Number	
	Btuh	Watts			Btuh	Watts			
Living room	9300	2720		1	5120	1500	76	9715H	
				2	5120	1500	76	9715H	
Dining room	4300	1250	Aluminum finned-tube 240	3	4265	1250	64	9715F	
Bedroom No. 2	7500	2200		4	5120	1500	76	9715H	
				5	2560	750	40	9715B	
Bedroom No. 1	5200	1520	Baseboard (Figure 2.21)	6	1707	500	28	9715	
				7	3415	1000	52	9715D	
Bath	2900	850	250 w/lin ft	8	3415	1000	52	9715D	
Basement	4900	1430		9	2560	750	40	9715B	
				10	2560	750	40	9715B	
Kitchen	4800	1420	Downflow 240 wall heater (Figure 2.22)	11	5122	1500	Size 16¼ in. × 13 in. × 1⁵⁄₁₆ in.	5990T Model WA-15	
Totals	38,900	11,390		Totals	40,964	12,000			
	Required 11.4 kw				Specified 12.0 kw				

Temperatures are degrees Fahrenheit (° F). Temperature *Differences* are Fahrenheit Degrees (F °).

[a]OUT=0° F, IN=75° F, ΔT=75 F °.

[b]See Figure 2.27 *a* and *b*.

2.8. The Use of Steam in Large Buildings

As we stated in Section 2.1, hot water is now much preferred to steam for heating. This is especially true in *houses*. In large buildings, *steam* is often the principal medium for heat transfer. It is generated in large boilers, often of the type shown in Figure 2.29. Sometimes a number of boilers supply the steam to large pipes known as "headers" from which it is distributed to its destinations. Such an installation could produce many *millions* of Btuh.

Often the reason for this scheme is that steam under pressure can be used for many purposes other than heating. These include power to drive generators and other equipment, cooking in commercial kitchens, industrial "processes" and instrument sterilization in hospitals. Even in these instances, steam is very often *not* used directly for heating. Instead, the heat of the steam is transferred in a converter to circulated water. After the water leaves the "steam to water" heat converter (Figure 2.30), it flows through a hot water heating system that closely resembles the systems presented in this chapter. In principle, the steam boiler-plus-converter takes the place of the hot water boiler that is shown in Figures 2.5 and 2.19.

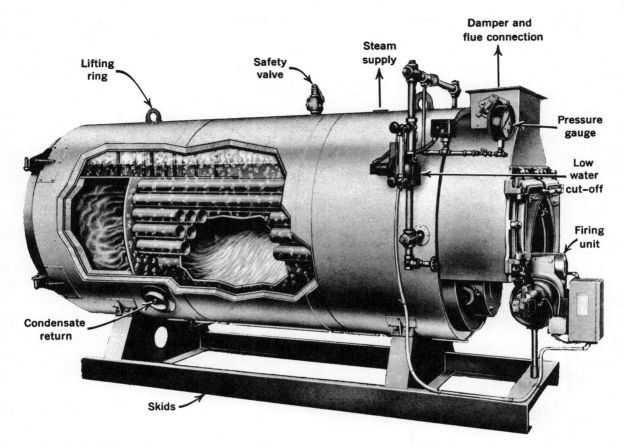

Figure 2.29 Package Type Fire-tube steam boiler. Capacity 600,000 to 3,000,000 Btuh, adaptable for oil, gas, or both. Complete with controls and all fittings.

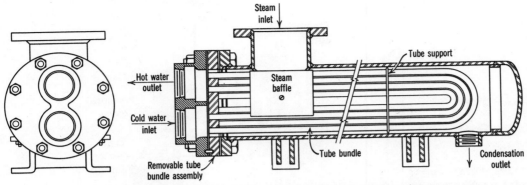

(a) Section illustrating the principle of heat transfer from steam to water.

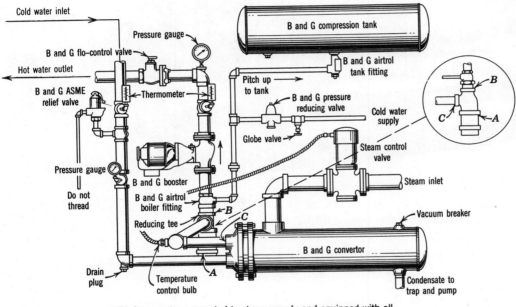

(b) A convertor connected to steam supply and equipped with all devices necessary for a complete hot water heating system.

Figure 2.30 Conversion. Transferring heat from steam to hot water. (Courtesy of ITT Bell & Gossett)

Problems

2.1. Refer to Example 1.1 (page 4) and select hot water baseboards, R-750 Heatrim, flow rate 2000 lb of water per hour, average water temperature 170° F for the room. Inside dimensions 12 × 20 ft. Make a sketch indicating the length of baseboard finned elements and enclosures.

2.2. As an alternate to Problem 2.1, use the correct number of electric baseboards, cast-aluminum grid, 187 w/ft, 120 v, each 72 in. long for the room. Make a sketch and show the location of the units. By what percentage does your installation exceed in output the demands of the room on the basis of its heat loss?

2.3. A hot water heating system has a developed length of 200 ft of 1-in. tubing. In this circuit there are soldered fittings, also 1 in. as follows: 60 standard 90° ells, 10 standard 45° ells and 4 gate valves. Calculate the total equivalent length of the system.

2.4. How many gallons per minute (gpm) will a 1½-in. circulator deliver against a head of 6½ ft?

2.5. If chosen as an alternate to electric baseboards, how many floor insert electric heaters (Catalog No. 8181A) would be needed for the living room of Example 2.2?

2.6. The conditions proposed in Problems 2.1 and 2.2 above might not lead to the most economical installations. How would you improve them?

2.7. The total equivalent length of a water circuit is 300 ft. The total pressure to be lost in friction is 9 psi. The flow is 20 gpm. Select a type M copper tube size.

2.8. Refer to Figure 2.20*b* and make an isometric drawing of the corner of the utility room showing boiler (Figure 2.19 and Table 2.4) and all piping, controls, smoke pipe, the circulator, and the like, but not including tankless coil or domestic hot water items.

2.9. Make a measured and dimensioned scale drawing of kitchen heater No. 11 (Figure 2.27*a*) in place showing a front view and a cross-sectional view. Assume that the heater cabinet is 6 in. deep. Refer to Figure 2.22. Describe in writing the operation of the heater.

Additional Reading

Baseboard Ratings, I=B=R Ratings for Finned-Tube (Commercial) Baseboard.

SBI Ratings for Steel Boilers, The Hydronics Institute. Products of many manufacturers can be found in these two publications.

Copper Tube Handbook, Copper Development Association. Standards for design, fabrication and installation.

Electric Heating and Cooling Handbook, Electric Energy Association. Contains sections such as Electric Resistance Comfort Heating Economic Analysis, and the like.

Heating and Cooling of Houses: Hot Water Heating/Electric Heating Time Saver Standards, 5th Edition, McGraw Hill. This section of a standard reference for architects includes typical design examples for hot water and electric heating.

Hydronic Specialties Catalog, Bulletin A 50, ITT Bell & Gossett. Complete and detailed information about hydronic controls.

Systems, ASHRAE Guide and Data Books, 1973, Chapter 13. A very complete treatment of the subject of steam heating for large buildings.

3. Residential Heating

In Chapter 2, we are given heat losses for a Basic Plan. Based on these losses, we design and draw up both hot water and electric heating systems. This is a relatively simple problem. However, large houses, especially those of custom design, are more complex.

The heating engineer must primarily make the system effective for comfort. Other demands are that the system be as inconspicuous as possible and that, if possible, temperature controls be provided in each room.

The study of this chapter will enable you to:

1. Adapt a heating system to a complex structure.
2. Design a one-pipe hot water heating system.
3. Specify venturi tees, proper valves and suitable air vents for each heating element.
4. Connect 2 one-pipe circuits to one boiler.
5. Detail a fuel oil storage tank.
6. Design an electric heating system using mainly floor-recessed drop-in heating units.
7. Distinguish between the output rating of hot water baseboard according to water flow rate in gallons per minute.
8. Detail a structural floor recess in masonry or frame construction to receive a recessed heating unit.
9. Prepare an orderly tabulation of heat output as against heating requirements for each space.
10. Properly locate balancing valves to adjust water flow in each of two circuits.

3.1. A House To Be Heated

The Merker residence (shown in Figures 3.1 to 3.5) presents a typical problem of climate control that must be solved by the architect and his mechanical engineer. The house is in construction at the time of this writing. With the gracious permission of the owner and his architect, the structure will be a clinical framework for our creative studies. Architect and planner, Budd Mogensen, AIA, designed this house for the Sherman Merker family.

As in all well-coordinated projects, the scheme for interior climate control was considered as a basic element of the general design. It was selected and developed along with the architectural plans. It is not good practice to delay the heating design until after the architectural design is complete.

Of course, the architectural design *is* complete. So, for the purpose of *study only*, we will learn how two systems, hot water heating and electric heating,

adapt to this kind of construction. As in any task, we cannot proceed without knowledge. An understanding of the structure is therefore essential.

3.2. The House

Located on the north shore of Long Island and occupying a large plot, this house looks out over the waters of Long Island Sound. All of the principal rooms face the view. Nestled into a hill, the house presents a two-story facade to the West. As we can see in Figure 3.3, the (uphill) east elevation resembles that of a one-story house. Two skylight dormers reach up to trap the morning sun, lighting the entry foyer and the master bedroom. Conventional windows provide east light for the living room and master bath. The upstairs guest bath accepts the sun through a plastic roof bubble but is otherwise windowless.

On the drawings, the elevations are identified directly as the points of the compass. Actually, they are 45 deg away from these directions. See the north arrows in Figure 3.4. Thus the front elevation faces Southwest rather than directly West. In our discussion, we will call it West.

3.3. The Structure

Figure 3.5 shows footings and east wall below grade to be of poured concrete. This east wall turns the corner at both ends to extend partially on the south and north elevations. See Figure 3.4. The construction photograph (Figure 3.1), indicates wood frame construction on the west facade. The entire upper story is of wood frame construction. Wood studs, joists and rafters make up the structural frame. A few steel beams carry long spans. Otherwise the house is wall bearing, using stud walls.

Throughout there is heavy thermal insulation that is equivalent to the EEA recommendations in Table 2.6, page 55. Windows and doors are weather-stripped. All glass, fixed and movable, is of the double (insulating) type.

Figure 3.1 The Merker residence, Sands Point, New York. Construction phase. Directional guide to site. West elevation. Terraces face Long Island Sound.

3.4. Form and Geometry

The construction photograph (Figure 3.1) shows clearly that the house is divided into three sections.

Figure 3.2 Partial views. *(a)* Master bedroom and study, garages below. *(b)* Kitchen and dining room. Two bedrooms below. *(c)* Living room. Family room below. *(d)* Living room interior, looking south.

The divisions are evident in both floor plans (Figure 3.4). Views a, b and c of Figure 3.2 emphasize the left, center and right-hand sections of this three-part scheme.

Independent of this three-section arrangement, the upper and lower floors are each well planned for their respective uses. The upper or principal living unit affords access to all rooms from a central foyer. The lower story, intended for visiting family and guests, places all of *its* rooms conveniently around the central *hall*.

3.5. Space Study

One may have to search for areas suitable for boilers, air ducts or other equipment. There is no basement or crawl space. The garage is of conventional *width*, but it does have a generous 27 ft depth. This,

however, would be only minimally adequate to suit long cars, possible boat storage, or a workbench and a few garden tools. The outdoor storage shed is for terrace furniture.

Let us now look at the *upstairs*. The section in Figure 3.5 and the interior view (Figure 3.2*d*) tell us that there is no attic above the living room. However, there *is* a very small wedge-shaped passage above the glass doors. It might be used for tubing or air ducts.

There is, in addition, a somewhat larger attic. See Figure 3.5. It, too, is wedge shaped, and is about 18 ft wide and 7 ft high. It extends over the northern two thirds of the upper story.

We have said that heating equipment may be located at high points in a structure. In selecting a location for the boiler room of a hot water heating system, *this* attic would not be suitable. It would be a poor decision to place a heavy boiler above habitable rooms in this light wood structure. Lightweight

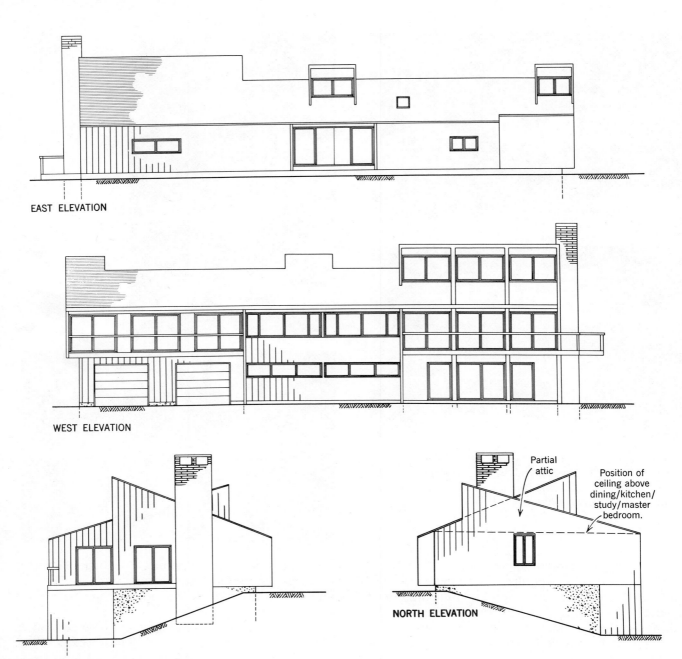

EAST ELEVATION

WEST ELEVATION

SOUTH ELEVATION

Partial attic

Position of ceiling above dining/kitchen/ study/master bedroom.

NORTH ELEVATION

Figure 3.3 Elevations, the Merker residence. See north arrows in Figure 3.4 for exact orientation.

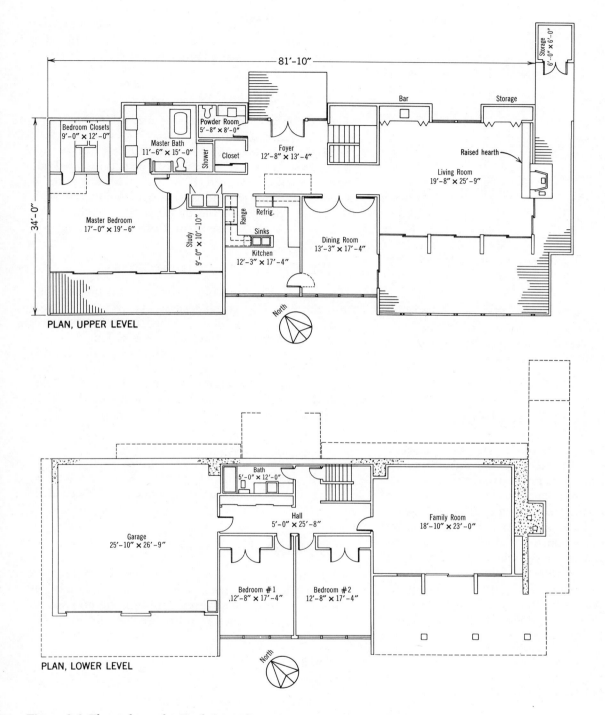

Figure 3.4 Floor plans, the Merker residence. Dimensions are approximate.

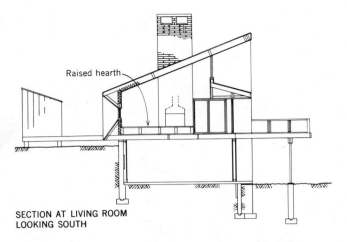

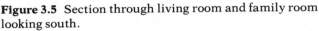

Figure 3.5 Section through living room and family room looking south.

air handling equipment (see Figure 1.5, page 9) might be used in such a space. Use of the attic, therefore, is considered in a later chapter. At present, we are planning a *hot water* system.

3.6. The Boiler Room

It is apparent that we need a boiler room in which the boiler can stand on a concrete slab. Since the architect had granted us permission (for study purposes) to modify his design, we discuss this need with him. His recommendation is that we use the south one third of the family room. This involves eliminating the glass on the end of this new room. Because the assigned space is larger than required for a boiler room, a beach shower room is created at the front. Compare plans in Figure 3.4 with Figures 3.10*b* and 3.12. A flue for the boiler can be provided in the masonry of the chimney. The family room is reduced in size as indicated.

3.7. The Hardware

Before designing the hot water heating system, let us check the equipment that we may plan to use. The engineer will usually make his choice of items and specific products. He will then write a tight specification. At times, he will insist on an item that he calls for. At other times, he will accept an "equal" product if it is of comparable quality and performs its function correctly. In this book, we illustrate various products. In a few instances, we show the items of more than one manufacturer. You can make a comparison of ratings and catalog style.

a. Tubing and Fittings

Type M copper tubing with soldered or "sweat" type bronze fittings (Figure 2.8, page 32) will be used. Tubing must be supported at suitable intervals by hangers or fasteners. In general, the tubing need not be wrapped or covered, However, where it runs in exposed or cold locations, it must be covered with insulative pipe covering or must be placed on the warm side of the house insulation.

b. Baseboards and Convectors

Dunham-Bush, Inc., has products, output ratings and details concerning *recessed* convector (finned-tube) strips that will be located in trenches or recessed enclosures below floor levels. This recessed detail is necessary where exterior glass extends down to the floor. (We will use it in the living room, study, master bedroom and family room.) Baseboard elements and their enclosures of the same manufacturer will be specified. See Figures 3.6, 3.7 and 3.8.

c. Expansion Fittings

Where straight runs of tubing exceed prescribed limits, these fittings will be used. See Figure 2.13, page 36.

d. Special Venturi Tees

A one-pipe system in preference to a perimeter loop will require these tees on the return "branch" or runout, to force the water through the convector. See Figure 2.14, page 37, and Figure 3.11.

e. Boiler

The Smith-Pac 12 will be a suitable boiler for this house. See Figure 3.13, Table 3.3.

f. Circulators and Boiler Controls

Circulators, compression tank, pressure-reducing valve and other items not part of the "package" equipment of Smith-Pac 12 can be of ITT Bell & Gossett manufacture. See Figure 2.2, page 26, and Figure 2.3, page 26.

HOT WATER RATINGS IN BTU PER LINEAL FOOT ACTIVE LENGTH

TYPE E-10	FLOW RATE		AVERAGE WATER TEMPERATURE																		
	GPM	Ft. per second	240°	235°	230°	225°	220°	215°	210°	205°	200°	195°	190°	185°	180°	175°	170°	165°	160°	155°	150°
	4	2.36	1070	1040	990	960	930	890	860	810	780	740	710	670	630	600	560	530	490	450	410
	1	.59	1010	980	940	910	880	840	810	770	740	700	670	630	600	570	530	500	460	430	390

NOTE 1. Ratings at water flow rate of 4 g.p.m. (2000 lb/hr.) should be used for E-10 Baseboard, Series-Connected Systems, where the water flow rate through the baseboard unit is equal to or greater than 4 g.p.m. For other piping arrangements resulting in lower water flow rate, or where the water flow rate is not known, the rating at the std. water flow rate of 1 g.p.m., (500 lb/hr.) must be used.

NOTE 2. All above ratings include 15% heating effect * and are based on active length (finned length). Total length equals active length plus 2-1/2".
NOTE 3. Standard Heating Element: 3/4" Nom. Copper Tube. Fins 2-1/4" x 2-1/2" x .012 unpainted alum. 59 fins per foot.
NOTE 4. Friction drop per lineal foot 3/4" Nom.—4 g.p.m., 525 mil-inches; 1 g.p.m., 47 mil-inches.

Figure 3.6 Webster Baseboard (hydronic heating units). (Courtesy of Dunham-Bush, Inc.) (See note 2 of Table 3.1 about heating effect.)

g. Water and Air Controls

The branch tubing circuits through the convectors may be shut off by a valve (Figure 3.9a). Air is relieved from the branch by a "water valve," which is also known as an air vent valve (Figure 3.9b).

h. Thermostat and Aquastat

If the thermostat and aquastat are not part of the boiler package, they are separately specified to control the circulator and oil burner.

i. Fuel Storage System

A steel storage tank is used. If it is underground, it should be asphalt-coated to retard corrosion. For tank and its accessories, see Figure 3.12.

The items listed above do not form a complete specification. They are part of it, however. A complete "spec" always accompanies the working drawings. Both the drawings *and* the specification together with the general contract are legally binding on the heating contractor *and* general contractor.

3.8. Design of a Hot Water Heating System

Example 3.1 Design a hot water heating system for the Merker residence just described (refer to Tables

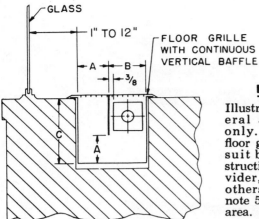

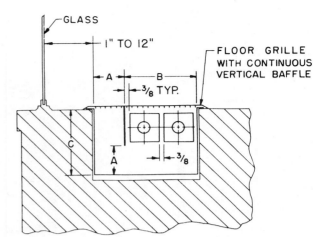

NOTE

Illustrations for general arrangement only. Trench and floor grille design to suit building construction. Grille, divider, supports by others. See rating note 5 for grille free area.

SCHEDULE OF DIMENSIONS

HEATING ELEMENT			TRENCH DIMENSIONS		
SIZE	SYMBOL	ROWS WIDE	A (Min.)	B (Max.)	C*
1″ —3x3—50	3HS	1	3″	3¾″	6½″
1″ —3x3—50	3HS	2	3″	7⅛″	6½″
1¼″—4x4—50	4HS	1	4″	4¾″	8½″
1¼″—4x4—50	4HS	2	4″	9⅛″	8½″

*To under-side of louvers in floor grille.

APPLICATION:

Installation of heating element below the floor, as illustrated, is applicable with modern construction where glass walls or doors are installed at, or close to, the floor line.

The ratings indicated below are only for glass walls and will not hold true for other wall materials unless the material used will provide the conditions of down-draft temperature and velocity indicated by Note 2 under rating table.

DESCRIPTION:

The heating element is Webster non-ferrous Walvector heating element.

The trench, preferably under the General Contract, must be made air-tight and provisions made for cleanout and drainage. If the area surrounding the trench is subject to freezing temp. the trench should be insulated.

The vertical baffle can be attached to floor grille or be part of trench construction.

The heating contractor should support and install the heating elements.

STEAM AND HOT WATER RATINGS—BTU/LF/HR*

HEATING ELEMENT			STEAM RATINGS		HOT WATER RATINGS AVERAGE WATER TEMPERATURE & FACTOR										HEATING EFFECT FACTOR %
SIZE	SYMBOL	ROWS WIDE	SQ. FT. EDR PER LF	BTU/hr PER LF	240 1.25	230 1.14	220 1.05	210 .95	200 .86	190 .78	180 .69	170 .61	160 .53	Installed Height	
1″ —3x3—50	3HS	1	4.08	980	1230	1120	1030	930	840	770	680	600	520	as shown	15
1″ —3x3—50	3HS	2	7.42	1780	2220	2030	1870	1690	1530	1390	1230	1080	940	as shown	15
1¼″—4x4—50	4HS	1	5.92	1420	1770	1620	1490	1350	1220	1110	980	860	750	as shown	15
1¼″—4x4—50	4HS	2	10.66	2560	3200	2920	2690	2430	2200	1990	1760	1560	1350	as shown	15

*1. Ratings based on glass windows or doors going to floor where cold glass creates a natural down-draft.

2. Ratings based on 0° outdoor temp., 70°F room temp., 47° Average temp. of down-draft, and 60 ft. per minute average velocity, and 25° inside glass temp.

3. Ratings will decrease (1) with increase in outdoor temp. (2) with increase in down-draft temp. (3) with decrease in velocity.

4. Trench containing heating elements must be air-tight.

5. Floor grille to have at least ⅔ free area.

Figure 3.7 Underfloor hydronic heating elements. Recessed in trench or in joist construction of self-supporting floors. See Figure 3.8. (Courtesy of Dunham-Bush, Inc.) (See notes 2 and 3 of Table 3.1 concerning heating effect factor.)

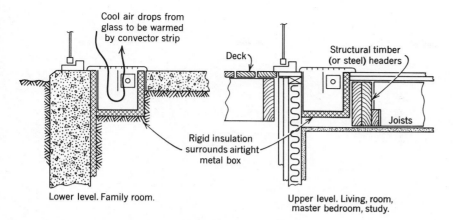

Cool air drops from glass to be warmed by convector strip

Deck

Structural timber (or steel) headers

Joists

Rigid insulation surrounds airtight metal box

Lower level. Family room.

Upper level. Living, room, master bedroom, study.

Figure 3.8 Recessed finned-tube convector strips. Insulative details and structural adaptation suggested by the engineer to the architect. See Figure 3.7. *Electric* floor insert heaters are similar. See Figure 2.23, (page 51).

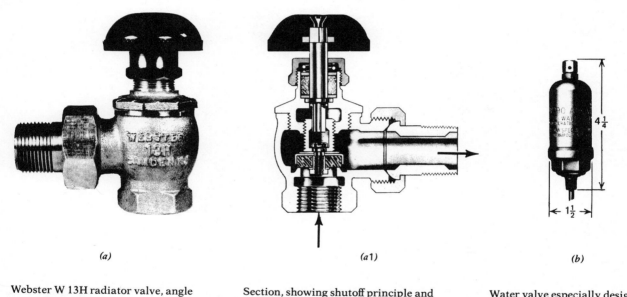

(a)

(a1)

(b)

Webster W 13H radiator valve, angle model ¾-in. pipe size.

Section, showing shutoff principle and direction of water flow *when valve is open.*

Water valve especially designed for removing air from convectors, baseboard, and wall radiation. Safety drain connection at the top for discharging moisture entrained in the vented air. Fitting and ferrule for ³/₁₆-in. OD tubing. Telescopic Siphon Tube.
Size conn.: ⅛-in. straight shank.
Max. oper. press.: 30 PSI.

Figure 3.9 *(a)* Hot water shutoff valve. This style, illustrating the *principle* of valve closure, has *threaded* connections. For use with copper tubing, a valve with *soldered* connection to the tubing would be used. See Figure 2.8, (page 32). *(b)* Convector air-vent valve (Hoffman Specialty ITT). For other types, see Figure 2.12, (page 36).

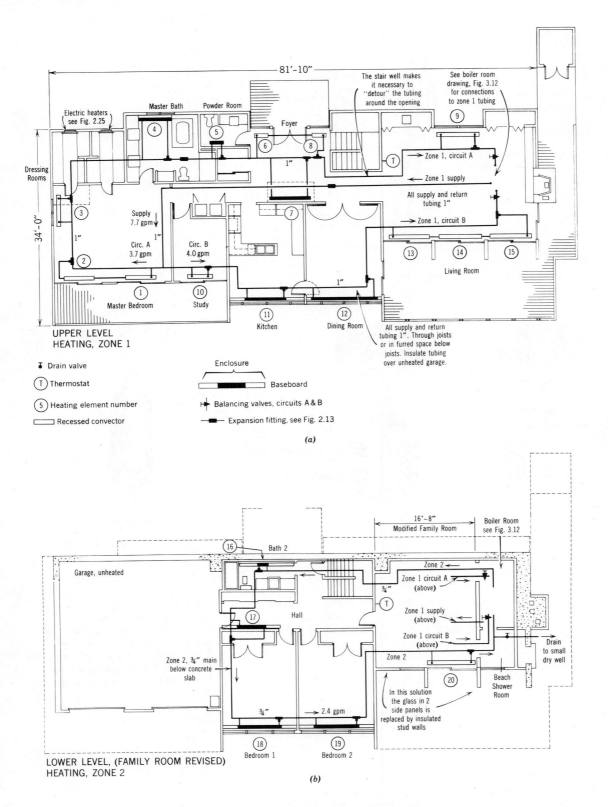

Figure 3.10 Layout of the solution to Example 3.1. Design of a hot water heating system for the Merker residence. See Tables 3.1 and 3.2 and Figure 3.12. For further information on expansion fitting, see Figure 2.13, page 36.

3.1 and 3.2). The hourly heat losses are listed in the left column of Table 3.2 and total 97,500 Btuh.

Solution: Following generally the applicable suggested steps, we begin our design by doing the following steps. A summary of the design is seen in Figure 3.10 to 3.12; Tables 3.1 to 3.3.

(a) *Selecting a system type.* A one-pipe system.
(b) *Determining average water temperature.* 200° F.
(c) *Choosing convector type.* Dunham Bush E-10 baseboard (Figure 3.6), recessed convectors and 1¼ in.–4 × 4–50, 4 HS, 1 row (Figure 3.7).
(d) *Sizing the convectors.* Divide the room heat loss figures in the first column of Table 3.2 by 740 Btuh/lin ft. (baseboards, see Figure 3.6) or by 1060 Btuh/lin ft. (recessed convectors, see Figure 3.7). Lay out and number heater locations (Figure 3.10) and list their outputs (Table 3.1). Complete the third and fourth columns of Table 3.2. Add outputs to assure the fulfilling of room requirements.
(e) *Laying out tubing circuits.* Figure 3.10 shows a reasonable system. Since the heat loss in the *upper* level is very large, the tubing is divided into two circuits. At the end of each circuit there is a balancing valve. These can be used to equalize the hot water flow if any unbalance should occur.
(f) *Establishing the required flow in gallons per minute.* We state again the formula that ties the heat loss in the space to the gallons per minute to be circulated.

$$\text{Btuh} = \text{gallons per minute} \times 8.33 \times 60 \times 1 \times \text{TD}$$

This time, instead of pumping enough water to assure 2000 lb/hr (4 gpm), we assume a 20 F° temperature drop. Then, we find the gallon per minute rates through the three circuits (Zone 1A, Zone 1B and Zone 2).

Transposing for gallons per minute and using a TD of 20 F°, we have

$$\text{gallons per minute} = \frac{\text{Btuh (in the circuit)}}{8.33 \times 60 \times 1 \times 20}$$

$$= \frac{\text{Btuh}}{9996 \text{ (say, 10,000)}}$$

and, hence, the *standard* formula when designing for a TD of 20 F° is

$$\text{gallons per minute} = \frac{\text{Btuh}}{10,000}$$

Applying this to the two circuits of Zone 1 and their supply circuit, and to Zone 2, we have:

$$\text{\textit{Gallons per Minute}}$$

Zone 1, Circuit A $\dfrac{37,100}{10,000} = 3.7$

Zone 1, Circuit B $\dfrac{39,700}{10,000} = 4.0$

Zone 1, supply $\dfrac{76,800}{10,000} = 7.7$ (sum of A and B)

Zone 2 $\dfrac{24,100}{10,000} = 2.4$

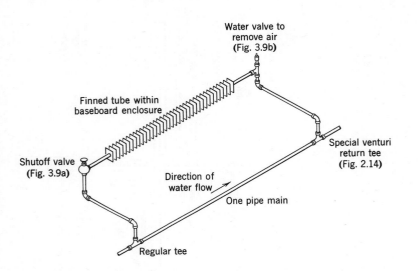

Figure 3.11 Typical supply and return "runouts." Suitable for one-pipe systems such as those in Figure 3.10. Locations of air vent, shutoff valve and special tee are indicated. For further information on special venturi return tee, see Figure 2.14, page 37.

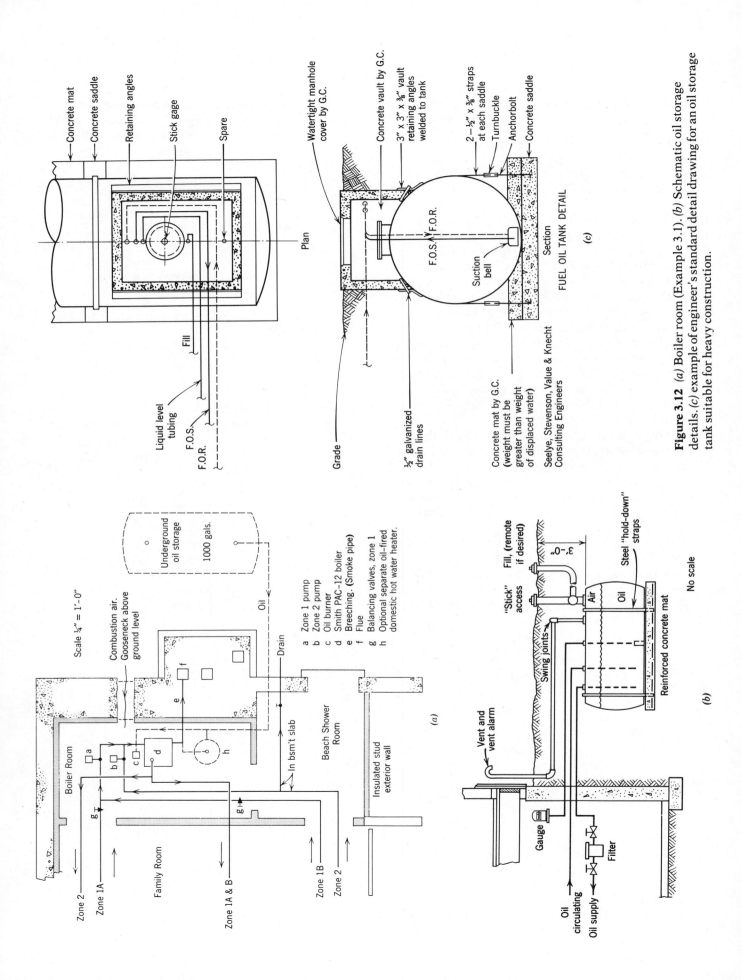

Figure 3.12 (a) Boiler room (Example 3.1). (b) Schematic oil storage details. (c) example of engineer's standard detail drawing for an oil storage tank suitable for heavy construction.

Plan labels:
Concrete mat
Concrete saddle
Retaining angles
Stick gage
Spare
Watertight manhole cover by G.C.
Plan
Liquid level tubing
F.O.S.
F.O.R.
Fill

Section labels:
Concrete vault by G.C.
3″ x 3″ x ⅜″ vault retaining angles welded to tank
2 – ½″ x ⅜″ straps at each saddle
Turnbuckle
Anchorbolt
Concrete saddle
F.O.R.
F.O.S.
Suction bell
Grade
½″ galvanized drain lines
Concrete mat by G.C. (weight must be greater than weight of displaced water)
Section
FUEL OIL TANK DETAIL
Seelye, Stevenson, Value & Knecht Consulting Engineers
(c)

Boiler room (a) labels:
Scale ¼″ = 1′-0″
Combustion air. Gooseneck above ground level
Underground oil storage
1000 gals.
Oil
Drain
a Zone 1 pump
b Zone 2 pump
c Oil burner
d Smith PAC-12 boiler
e Breeching. (Smoke pipe)
f Flue
g Balancing valves, zone 1
h Optional separate oil-fired domestic hot water heater.
Boiler Room
In bsm't slab
Beach Shower Room
Insulated stud exterior wall
Family Room
Zone 2
Zone 1A
Zone 1A & B
Zone 1B
Zone 2
(a)

Schematic (b) labels:
"Stick" access
Fill, (remote if desired)
3′-0″
Vent and vent alarm
Air
Oil
Swing joints
Steel "hold-down" straps
Reinforced concrete mat
Gauge
Filter
Oil circulating
Oil supply
No scale
(b)

Figure 3.13 (Example 3.1) The Smith-Pac 12 oil-fired boiler. Selected as suitable for a hot water heating system. Components may be identified from similar illustration in Figure 2.5, page 29. For ratings of the Smith-Pac 12, see Table 3.3. (Courtesy of The H. B. Smith Company, Inc.)

(g) *Establishing the required flow of the circulators*

Zone 1 circulator 7.7 gpm (see Figure 3.10a)
Zone 2 circulator 2.4 gpm

(h) *Finding the frictional resistance of both zones.*

TOTAL EQUIVALENT LENGTH

	Feet
Zone 1 supply	117
Zone 1, Circuit A	170
Zone 1, Circuit B	114
Zone 2	200

FRICTIONAL RESISTANCE, ZONE 1, DATA

Item	Supply	Circuit A	Circuit B
TEL (ft)	117	170	114
Flow (gpm)	7.7	3.7	4.0
Tube size (in.)	1	1	1
Friction (psi/100 ft)	1.6	0.40	0.50
			(Figure 2.17, page 41)

FRICTIONAL RESISTANCE, ZONE 1, CALCULATIONS

	(TEL÷100)		(psi/100 ft)		(psi in the run)
Supply	1.17	×	1.5	=	1.76
Circuit A	1.70	×	0.40	=	0.68
Circuit B	1.14	×	0.50	=	0.58

In Figure 3.10a by trial and error, Circuits A and B are arranged to present *approximately* equal friction. We see that 0.68/0.58 = 1.17. They are equal to within about 17%. This is close enough. They can be balanced more precisely by means of the balancing valves. Now, add the friction of flow in the supply to that of the more resistant circuit: 1.76 + 0.68 = 2.44 psi.

$$Head, Zone\ 1 \quad \frac{2.44}{0.433} = 5.6 \text{ ft pumping head}$$

FRICTIONAL RESISTANCE, ZONE 2, DATA

TEL (ft)	200
Flow (gpm)	2.4
Tube size (in.)	¾
Friction (psi/100 ft)	0.70 (Figure 2.17, page 41)

FRICTIONAL RESISTANCE, ZONE 2, CALCULATIONS

(TEL÷100)		(psi/100 ft)		(psi in the zone)
2.0	×	0.70	=	1.40

$$Head, Zone\ 2 \quad \frac{1.40}{0.433} = 3.23 \text{ ft}$$

CIRCULATOR SELECTION

Zone 1 7.7 gpm at 5.6 ft 1¼-in. circulator, ¹/₁₂ hp
Zone 2 2.4 gpm at 3.2 ft 1-in. circulator, ¹/₁₂ hp

Observe that the 1¼-in. circulator furnished with the boiler package can be relocated to serve Zone 1. A 1-in. circulator is ordered for Zone 2.

(i) *Selecting a boiler.* A Smith-Pac boiler FD12-W-4 (see Table 3.3) is selected with a net rating of 109,600 Btuh. This is adequate to carry the heat loss of 97,500 Btuh. If the 3.25 gpm of *domestic* hot water listed in Table 3.3 is insufficient for the *plumbing* system, a separate oil-fired domestic hot water heater may be used.

(j) *Determining location of thermostats.*

LOCATIONS

Zone 1	Living Room
Zone 2	Family Room

3.9. Electric Heating Design

Example 3.2 Design a system of electric heating for the Merker residence (voltages: 120/240).

Table 3.1 (Example 3.1) Hot Water Ratings of Finned Sections of Baseboards and of Recessed Convectors Used in the Solution of Example 3.1.

Heating Unit (See Figure 3.10a and b)	Baseboard 740 Btuh/Lin ft (Type E-10) (See Figure 3.6)		Recessed Heating Element 1060 Btuh/Lin ft (See Note 3 below and Figures 3.7 and 3.8)		Summary Output of Units by Zones	
	(Length, ft)	(Output, Btuh)	(Length, ft)	(Output, Btuh)	(Btuh)	
1			5	5,300	5,300	
2			5	5,300	5,300	
3			5	5,300	5,300	Zone 1
4	4	2,900			2,900	Circuit
5	2	1,500			1,500	A
6			2	2,100	2,100	37,100
7	7	5,200			5,200	Btuh
8			2	2,100	2,100	
9			7	7,400	7,400	
10			4	4,200	4,200	Zone 1
11	7	5,200			5,200	Circuit
12	11	8,100			8,100	B
13			7	7,400	7,400	39,700
14			7	7,400	7,400	Btuh
15			7	7,400	7,400	
16	2	1,500			1,500	
17	5	3,700			3,700	Zone 2
18	7	5,200			5,200	24,100
19	7	5,200			5,200	Btuh
20			8	8,500	8,500	
					100,900	

Notes: (1) Average Water Temperature 200° F. (2) Baseboard rating of 740 Btuh/lin ft *includes* 15% heating (pickup) factor. See Note 2 of Figure 3.6. (3) To achieve fast heating (pickup) a 15% longer recessed element length must be used. The rating for 4HS, 1 row at 200° F is 1220 Btuh/lin ft. Thus, the length to be used must be selected on the basis of $1220/1.15 = 1060$ Btuh/lin ft. See Figure 3.7. (4) It will be noted in Figure 3.6 that the hot water ratings for E-10 type baseboards are about 5% greater if 4 gpm are circulated instead of 1 gpm. Since, in our design, only one of the three circuits exceeds 4 gpm flow, baseboards are selected at 1 gpm ratings.

Solution: The "hardware" for this solution will be found in:

Figure 2.21   Baseboard heaters, page 48
Figure 2.22   Downflow wall heaters, page 50
Figure 2.23   Floor insert heaters, page 51
Figure 2.25   Bathroom heaters, page 53

In designing an electric heating system, the hourly heat losses are, of course, expressed in watts instead of in Btuh. The standard conversion is 1 w = 3.41 Btuh. Beginning with Table 3.4, the losses are listed, room by room, in watts. The family room is calculated for a greater heat loss than it was for hot water heating. The reason is that there need not be a boiler room. The family room is built as

shown in the original drawings, one third larger than in the hot water scheme of Example 3.1.

Recessed-type convectors must be used in the living room, master bedroom and family room. They would also be used in the study except for inappropriate ratings. The room requires 1100 w. Although one FL-4 plus one FL-8 would total 1200 w, their lengths do not agree (14⅛ in. and 30⅛ in.). Hence, a downflow heater is suggested at 1500 w. The 750 w "bathroom"-type heater in the bedroom closets are for occasional use. All other heaters are baseboard aluminum finned-grid elements at 250 w/ft. Unlike hot water baseboard, the enclosures cannot be extended wall to wall. They are related to the size of the finned-grid and are integral with it. All heaters

Table 3.2 (Example 3.1) Calculated Hourly Heat Losses under Design Conditions (70° F[a] Indoors and 0° F Outdoors). Heating Elements Selected.

Space	Engineer's Heat Loss Calculations (Btuh)	Heating Unit Numbers (Figure 3.10 a,b)				Heating Units Output (Btuh)	
Living room	28,600	9	13	14	15	29,600	
Dining room	8,000	12				8,100	
Kitchen	5,100	11				5,200	
Study	3,900	10				4,200	
Master bedroom	15,000	1	2	3		15,900	
Master bath	2,800	4				2,900	
Powder room	1,500	5				1,500	Upper level
Foyer	9,400	6	7	8		9,400	76,800
Family room	7,900[b]	20				8,500	
Bedroom No. 1	5,200	18				5,200	
Bedroom No. 2	4,900	19				5,200	
Bath	1,500	16				1,500	Lower level
Hall	3,700	17				3,700	24,100
	97,500					100,900 (3% more than 97,500)	

[a]In Chapters 1 and 2 we speak of 75° F as a "standard" value for indoor temperatures in residences. For the Merker residence (Example 3.1) 70° F was chosen in the interest of conserving energy.

[b]This is the value when the room is reduced in size as shown in Figure 3.10*b*. Otherwise it is 12,700 Btuh when built as originally planned (Figure 3.4).

use 240 v circuits except in the bedroom closets where the "bathroom" elements use 120 v. The higher (240 v) voltage in the other circuits permits greater flexibility in circuiting and reduces the number of circuits.

The process of design is similar to that for hot water heating. The elements are selected for the proper output in each room. Sometimes it is possible to use an element of exactly the required wattage. For instance, this is true in the kitchen (required: 1500 w; element rating: 1500 w). In other locations the output may sometimes be a little less than required. The living room needs 8400 w and has 8000. This difference of about 5% is quite acceptable. It is well, however, that the entire house-heating wattage installed be the same or slightly larger than the entire heat loss calls for. The electric and hot water solutions exceed the requirements by 1% and 3% respectively. For a summary of Example 3.2 (electrical design) see Table 3.4 and Figure 3.14.

3.10. Critique

Although the preparation of the contract documents is the special interest of the technologist, he may also aid in other decision making. A most important decision is which heating method to use. The preceding sections of this chapter tell *how to* design, draw and specify. There is no intention to imply that either of the two (hot water or electric), methods is distinctly the better choice.

In a comparison of systems, a few comments can be made. The electric system is more compact. Its arteries (light wiring) can be easily concealed in the structure. Tubing is not so easy to run. Electric methods of this type require no boiler room. Control in each room is somewhat easier. Fuel storage is not needed and smoke nuisance is avoided. The annual cost of oil or gas versus electric energy could be a deciding factor.

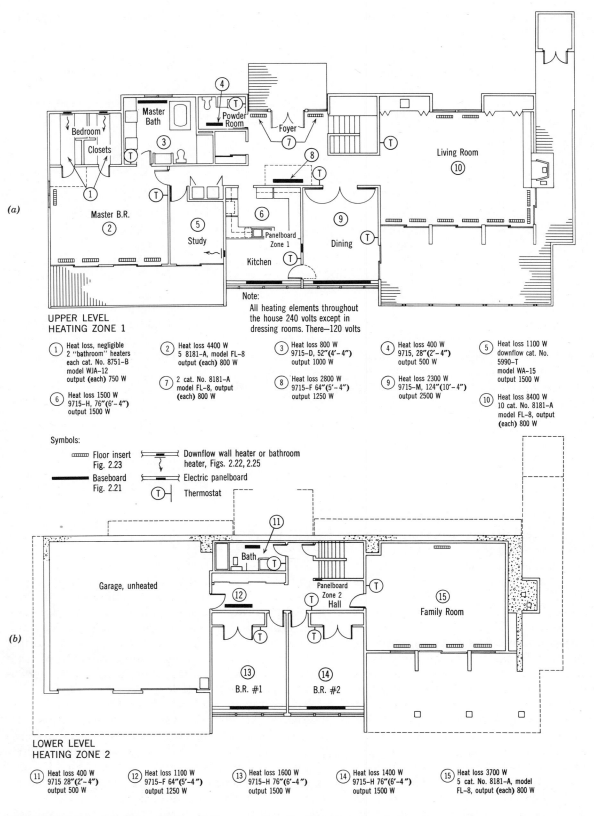

(a)

UPPER LEVEL
HEATING ZONE 1

Note:
All heating elements throughout
the house 240 volts except in
dressing rooms. There—120 volts

① Heat loss, negligible
2 "bathroom" heaters
each cat. No. 8751–B
model WJA–12
output (each) 750 W

② Heat loss 4400 W
5 8181–A, model FL–8
output (each) 800 W

③ Heat loss 800 W
9715–D, 52"(4′–4")
output 1000 W

④ Heat loss 400 W
9715, 28"(2′–4")
output 500 W

⑤ Heat loss 1100 W
downflow cat. No.
5990–T
model WA–15
output 1500 W

⑥ Heat loss 1500 W
9715–H, 76"(6′–4")
output 1500 W

⑦ 2 cat. No. 8181–A
model FL–8, output
(each) 800 W

⑧ Heat loss 2800 W
9715–F 64"(5′–4")
output 1250 W

⑨ Heat loss 2300 W
9715–M, 124"(10′–4")
output 2500 W

⑩ Heat loss 8400 W
10 cat. No. 8181–A
model FL–8, output
(each) 800 W

Symbols:

▭▭▭ Floor insert
Fig. 2.23

▬▬▬ Baseboard
Fig. 2.21

Ⓣ⊣ Thermostat

⌇⌇ Downflow wall heater or bathroom
heater, Figs. 2.22, 2.25

⌇⌇ Electric panelboard

(b)

LOWER LEVEL
HEATING ZONE 2

⑪ Heat loss 400 W
9715 28"(2′–4")
output 500 W

⑫ Heat loss 1100 W
9715–F 64"(5′–4")
output 1250 W

⑬ Heat loss 1600 W
9715–H 76"(6′–4")
output 1500 W

⑭ Heat loss 1400 W
9715–H 76"(6′–4")
output 1500 W

⑮ Heat loss 3700 W
5 cat. No. 8181–A, model
FL–8, output (each) 800 W

Figure 3.14 Layout of the solution to Example 3.2. Design of an electric heating system
for the Merker residence. See Table 3.4.

Table 3.3 Ratings and Specifications for the Smith-Pac 12 Boiler Shown in Figure 3.13.

BOILER NUMBER	I=B=R RATINGS B.T.U. PER HR.		I=B=R BURNER CAPACITY (GALLONS PER HOUR)	CHIMNEY*	
(I=B=R Reg. U. S. Pat. Off.)	GROSS RATING	NET RATING		SIZE (INCHES)	HEIGHT (FEET)
FD12-W-3	86,000	74,800	.80	8 x 8	20
FD12-W-4	126,000	109,600	1.15	8 x 8	21
FD12-W-5	166,000	144,300	1.50	8 x 8	22
FD12-W-6	206,000	179,100	1.85	8 x 8	23
FD12-W-7	246,000	213,900	2.20	8 x 12	23

The Net I=B=R Water Ratings shown are based on an allowance of 1.15.
The manufacturer should be consulted before selecting a boiler for installations having unusual piping and pickup requirements such as intermittent system operation, extensive piping systems, etc.
For forced hot water heating systems where the boiler and all the piping are within the area to be heated, the boiler may be selected on the basis of its Gross Output.
All boilers hydrostatically tested — A.S.M.E. standard.
Maximum allowable working pressure 40 psi water.
* Chimney heights shown are for natural draft operation. Sealed construction of the boiler, and the burner design, allow operation with only a 7″ vent to outdoors.

Dimensions (inches)

OVERALL LENGTH "A"	SUPPLY TAPPING	RETURN FLANGE	FLUE OUTLET
23¾	1¼	1¼	7
27⅞	1¼	1¼	7
32	1¼	1¼	7
36⅛	1¼	1¼	7
40	1¼	1¼	7

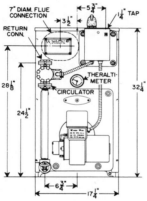 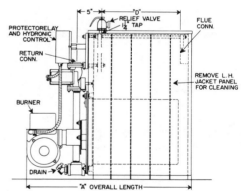

Tankless Heater Schedule

NUMBER OF SECTIONS IN BOILER	CAPACITY Gals, per min. W/Boiler Water 200°F
3	3.00
4	3.25
5	3.75
6	4.50
7	5.00

All connections to tankless heater are ½″ N.P.T. female

Standard Equipment

- Flame retention-type oil burner.
- Insulated metal jacket.
- Circulator and return piping with provision for additional zones.
- Preformed combustion chamber.
- Combination Protectorelay with hydronic control and circulator relay.
- ASME relief valve (30 psi).
- Theraltimeter.
- Brass drain cock.
- Low-voltage thermostat.
- Heater cover plate and gasket (units without tankless heater only).
- Wiring harness.
- Barometric draft control.

Optional Equipment

- Tankless heater (see schedule)

Source. The H. B. Smith Company, Inc.

Table 3.4 (Example 3.2) Summary of Design, Electric Heating System.

Space	Engineer's Heat Loss Calculations (watts)	Output of Units Selected (watts)	
Living room	8,400	8,000	
Dining room	2,300	2,500	
Kitchen	1,500	1,500	
Foyer	2,800	2,850	
Powder room	400	500	
Master bath	800	1,000	
Master bedroom[a]	4,400	4,000	Upper
Study	1,100	1,500	level
Family room	3,700	4,000	21,850
Bedroom no. 1	1,600	1,500	
Bedroom no. 2	1,400	1,500	Lower
Bath	400	500	level
Hall	1,100	1,250	8,750
	29,900	30,600	(30.6 kw)
	(30 kw)	1% more than 29,900	

[a]The small heaters in the upper level bedroom closets, although not based on appreciable heat loss, add 1.5 kw to upper level power demand and to total demand.

Problems

3.1. Using Dunham Bush Type E-10 baseboard at 4 gpm flow rate and 170° F average water temperature, select the active finned length of heating baseboard for the exterior wall of a living room. The room has a heat loss of 10 Mbh and a 20 ft long exterior wall. The baseboard must include the usual 15% heating effect.

3.2. If in Problem 3.1, 220° F average water temperature were used, what active length would be needed if all *other* conditions were the same?

3.3. If, under the conditions of Problem 3.1, the flow rate is changed from 4 to 1 gpm, by what percentage is the baseboard rating in Btuh per foot decreased?

3.4. (a) Using Singer Company electric baseboard heaters, 250 w/ft, 240 v, aluminum, finned-grid element, select units and give catalog numbers for the living room of Problem 3.1.
(b) Would they fit on the 20-ft-long exterior wall?

3.5. A large residence has an hourly heat loss under design conditions of 176,000 Btuh. On the basis of "net rating," select a Smith-Pac 12 boiler for this house.

3.6. For the boiler you have selected in problem 3.5, what are its ratings for the following?
(a) Gross rating
(b) Burner capacity, gallons of oil per hour.
(c) Chimney size.
(d) Chimney height.

3.7. Make an isometric sketch of the boiler (Problem 3.5) and its burner (no other controls) and add the overall dimensions.

3.8. Sometimes space in a boiler room is limited. What is the overall *length* of Smith-Pac boilers?
(a) FD 12-W-3.
(b) FD 12-W-7.

3.9. As part of the standard equipment of Smith-Pac 12 boilers:
(a) What is the setting of the relief valve?
(b) Is the metal jacket insulated?
(c) What size tap receives the relief valve?

(d) Is a drain provided?
(e) Is a circulator included in the "package"?

3.10. Make a scaled and dimensioned cross-sectional drawing of a recessed heater 1¼-in.–4 × 4–50, two rows wide using trench dimensions of Figure 3.7 and set in a *wood frame* floor, joist 2 in. × 10 in. Construction is similar to the right-hand illustration in Figure 3.8. Use clearance dimensions called for in Figure 3.7.

Additional Reading

Magazines of the Profession: *Heating, Piping and Air Conditioning, ASHRAE Journal, Actual Specifying Engineer, Building Systems Design.* These publications reflect the most recent developments in technical fields. They are for engineers.

Architectural Magazines: *Progressive Architecture* and *Architectural Record.* Technical articles in these two journals are most useful.

Architectural Graphic Standards, American Institute of Architects and C. G. Ramsey and H. R. Sleeper, Sixth Edition, John Wiley. This standard reference for architects has many graphical illustrations of heating equipment.

4. Nonresidential Heating

Each new project adds to our experience in the design and layout of heating systems. We acquire some new ideas in the example that is chosen for this chapter. They include:

Use of cabinet-type convectors.
Design of a two-pipe hot water system.
Embedment of long tube-runs in concrete slabs.
A different approach to selecting a circulator.
Exhaust ventilation of interior rooms.

After studying this chapter on the subject of heating in industrial buildings, you will be able to:

1. Select a type of wall convector suitable for a hot water heating system.
2. Determine the need for wall-recessed convectors in small rooms.
3. Distinguish between pressure needed to lift water to a height and pressure needed to overcome friction in tubing.
4. Lay out a boiler room.
5. Design a two-pipe, reverse-return hot water heating system.
6. Select a gas-fired boiler for the system.
7. Locate thermostats for the several zones.
8. Select circulators for the zones.
9. Make a drawing of the entire system.
10. Provide exhaust ventilation from interior rooms.

4.1. The Building

Again, through the courtesy and permission of its architect, we have a building with which to experiment. Designed and built for light manufacturing in the clothing industry, it is of a type also suitable for other industrial uses.

Samuel Scheiner of Scheiner and Swit, Architects, has provided us with information and details of the building. They appear in the first five illustrations of this chapter. See Figures 4.1 to 4.5. The plan has several principal components:

Administrative wing
Work area
Storage area
Shipping and Receiving
Parking field

Industrial buildings have greater headroom than residences. In the administration and work areas the ceiling clearances are 9 ft. In the storage area the clearance is 12 ft to the underside of the roof trusses.

Exterior walls are of 8-in. concrete block or are of 4 in. of brick backed with 4-in. concrete block. A one-level concrete floor slab is convenient for possible use of rolling carriers for materials and merchandise. Windows are of commercial or architectural steel sash, single-glazed. Perimeter insulation reduces heat loss at slab edges. Between the concrete plank of the roof and the built-up roofing, there are 3 in. of rigid insulation.

4.2. Institutional Convectors

The heating system to be developed for this building will be forced hot water. Two kinds of heating units are possible choices. High-output commercial baseboard or cabinet-type convectors could be used. The latter will be chosen. Seven styles are shown in Figure 4.8. Details of one of them (FH-freestanding) are shown in Figure 4.6. Water enters through one end of the finned-tube heating element and leaves at the other end. The one shown has *three* tubes running through the aluminum fins. The number of tubes and the depth of the unit relate to its output. Depth and number of tubes are as follows:

Depth	Number of Tubes
4 in.	2
6 in.	3
8 in.	4
10 in.	5

Below and in front of the heating element is an inlet opening for air. As the air passes over the finned-tube unit, it is warmed and creates an upward convection-flow. It then leaves through the upper outlet grill. An adjustable damper controls this flow, regulating the heat output of the unit. Details of the damper and of the piping and air-releif vent are in Figure 4.7.

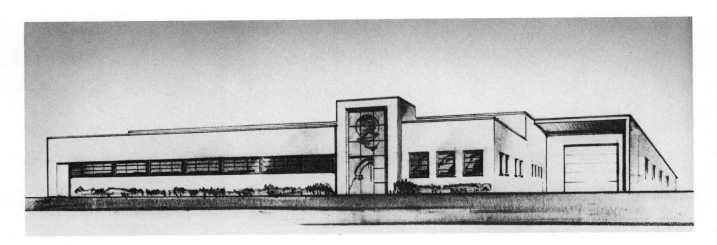

Figure 4.1 Architect's rendering, building for light industry. North elevation. (Courtesy of Scheiner and Swit, Architects, 1926 Oakland Ave. Wantagh, N.Y.)

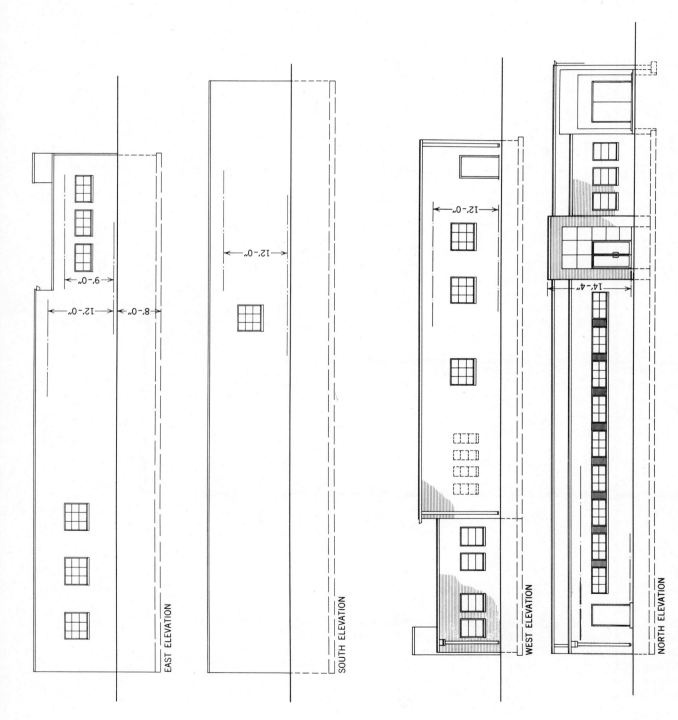

Figure 4.2 Industrial building. Elevations.

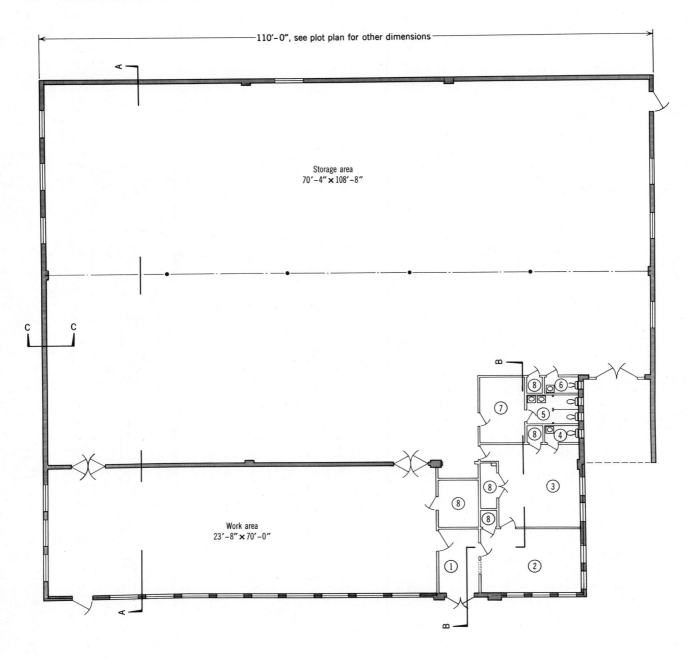

Figure 4.3 Building for light industry. Space use, administrative wing.

1. Lobby   7 ft 0 in. by 12 ft 0 in.
2. Administrative office   12 ft 0 in. by 18 ft 0 in.
3. Private office   14 ft 6 in. by 15 ft 0 in.
4. Private toilet   3 ft 0 in. by 6 ft 8 in.
5. Women's toilet   6 ft 0 in. by 9 ft 6 in.
6. Men's toilet   3 ft 0 in. by 6 ft 8 in.
7. Women's rest room   8 ft 0 in. by 12 ft 0 in.
8. Closets and storage

For sections *A-A*, *B-B* and *C-C*, see Figure 4.4.

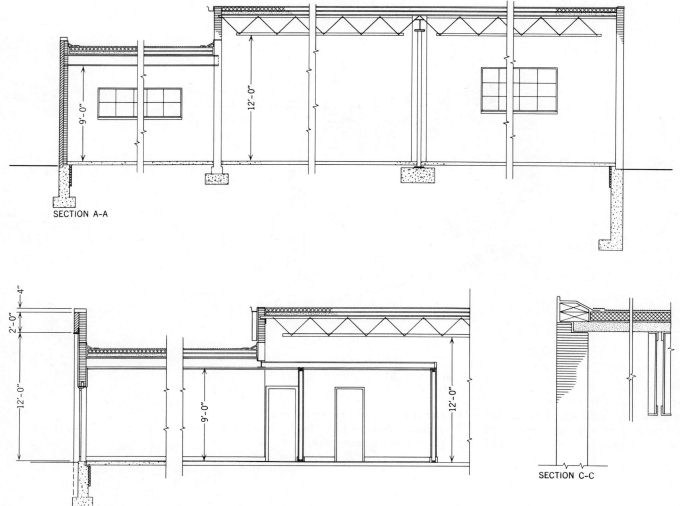

Figure 4.4 Sections. For locations, see Figure 4.3.

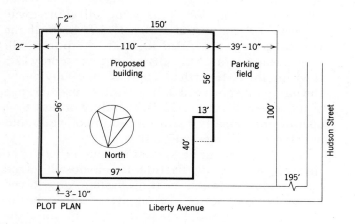

Figure 4.5 Industrial building. Plot plan.

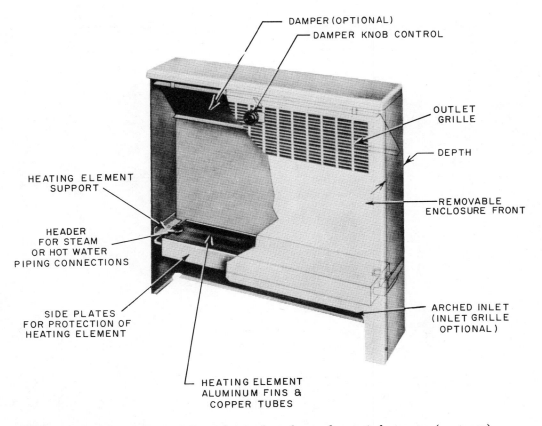

DAMPER (OPTIONAL)
DAMPER KNOB CONTROL
OUTLET GRILLE
DEPTH
REMOVABLE ENCLOSURE FRONT
HEATING ELEMENT SUPPORT
HEADER FOR STEAM OR HOT WATER PIPING CONNECTIONS
SIDE PLATES FOR PROTECTION OF HEATING ELEMENT
ARCHED INLET (INLET GRILLE OPTIONAL)
HEATING ELEMENT ALUMINUM FINS & COPPER TUBES

Figure 4.6 A heating convector. One of a number of types for use in hot water (or steam) heating. See Figure 4.8 for other types; See Figures 4.9 and 4.10 for the two models to be used in Example 4.1. (Courtesy of Standard Fin-Pipe Radiator Corporation).

Cabinet convectors are a somewhat better choice than baseboard for this design. The reason is that they provide a concentration of heat at windows where the heat loss is greatest. Two types will be used in our design. They are

SH   (Figure 4.9)
FH-CM   (Figure 4.10)

Reasons for their use are given in the caption below each illustration.

4.3. Pumps, Systems and Friction

Before we begin designing the system, let us discuss circulating pumps in a little more detail.

Pumps that raise water are rated by the gallons per minute that they can deliver to various heights. Figure 4.11a is a typical performance curve. A 1-in. PR pump can lift 20 gpm to a height (head) of 10.5 ft. If it must lift the water to 16 ft, it will deliver only 10 gpm. A height of 17.5 ft can be achieved but no water will flow. The impeller of the pump will churn away to sustain a column of water 17.5 ft high. If the vertical pipe is extended to 20 ft, the pump will continue to operate but will raise the water only 17.5 ft within the 20 ft vertical length of pipe. Obviously, the pump must be used only within the useful range of its power. Figure 4.11b is, therefore, a task that operates principally against a column of water and not significantly against friction.

Now we introduce a system in which there is no lifting at all but *only* friction. It is a closed, forced circulation hot water heating system, *any* forced circulation hot water heating system. Sketches of Figure 4.11c, *d* and *e* suggest three. Because the system is a closed loop rather than a single vertical

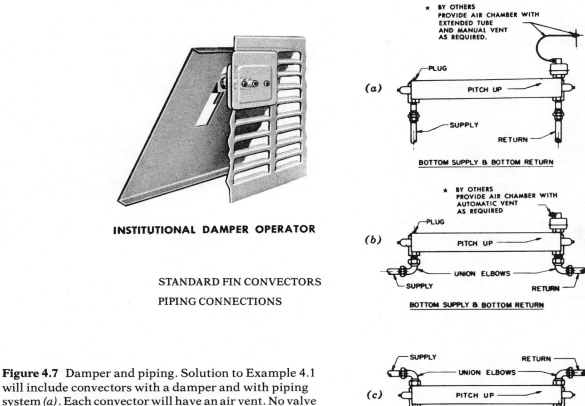

INSTITUTIONAL DAMPER OPERATOR

STANDARD FIN CONVECTORS

PIPING CONNECTIONS

Figure 4.7 Damper and piping. Solution to Example 4.1 will include convectors with a damper and with piping system *(a)*. Each convector will have an air vent. No valve is necessary at each convector. Instead, the flow of warm *air* can be regulated by the damper. (Courtesy of Standard Fin-Pipe Radiator Corporation)

pipe, the water in the two vertical sections of the system balance. The *only* purpose of the pump is to move the water through the circuits. The tubing resists this effort. Two things can occur:

1. For the same flow rate, smaller tubes cause greater friction.
2. For a fixed tube size, greater flow rate causes greater friction.

We have established "feet of head" in Figure 4.11*b*. One foot of head is the pressure resulting from a column of water one foot high. Now a new unit is introduced. It is the "milinch" of water. This is the pressure caused by a column of water one one-thousandth of an inch high. Therefore, a foot of head equals 12 × 1000 = 12000 milinches.

The milinch is a very small unit. Its use in heating systems is to find the total head in friction of the system. We must now translate the flow rate in gallons per minute to the heat delivery rate in Btuh. We

know that a commonly used water temperature drop is 20 F°. We also know the formula for the relation of gallons per minute and Btuh for this drop. It is:

$$\text{gallons per minute} = \frac{\text{Btuh*}}{10,000}$$

If we know the Btuh to be carried in a tube for any selected "milinch per foot" pressure drop, its size can be found in Table 4.4. For instance, if a tube must deliver 118 Mbh at a 300 milinches per foot pressure drop and 20 F° temperature drop, its size must be 1¼ in. If all the tubing in the system is selected for a fixed, chosen pressure drop, then the total "head" of the system can be found. If a system has a total equivalent length of 240 ft and is designed for 300 milinches per foot drop, then its total frictional resistance head is 240 × 300 = 72,000 mil-

*(See Chapter 3 for explanation of the factor 10,000.)

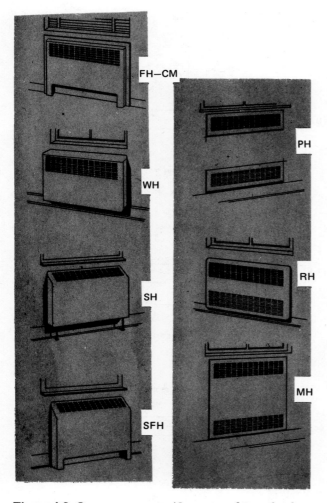

Figure 4.8 Convector types. (Courtesy of Standard Fin-Pipe Radiator Corporation)

inches. Expressed in *feet* of head, this is

$$\frac{72,000}{12,000} = 6 \text{ ft of water.}$$

Or we can just look it up in Section A of Table 4.4. Follow the 300 milinches column up to a TEL of 240 and then, to the extreme left, find 6 ft. With this head and the gallons per minute of the system, a pump may be selected from Figure 4.15.

High milinch drops (400, 500) permit small tubing and require high capacity pumps. Low milinch drops (150, 200) require larger tubing. Pumps can then be of low capacity.

4.4. Design

Example 4.1 Design a forced hot water heating installation for the industrial building of Figures 4.1 to 4.5. Heat losses are found in Table 4.1.

Solution:
(a) Locate and size the convectors.
(b) Arrange the circuits, zones and controls.
(c) Size the tubing for one- and two-pipe systems.
(d) Select the circulators and boiler.
(e) Summarize data of design steps and results.

4.5. Locate and Size Convectors

The selected locations of convectors are shown in Figures 4.12 and 4.13. A larger scale is used for the plan of the administrative wing (Figure 4.13). An average water temperature of 200° F is chosen. The convectors are selected from Table 4.2, which is based on a 20 F° temperature drop. Now let us select convectors for one space, the storage area. Eleven convector locations are planned (Figure 4.12). Table 4.1 says that the engineer requires 196.6 Mbh to maintain 70° F in the area.

$$\frac{196.6}{11} = 17.8 \text{ Mbh}$$

Each convector must produce this heat using 200° F water and with a drop of 20 F° in the system. Type SH has been chosen. A height of 32 in. is O.K. It will fit easily within the window sill height of 4 ft 6 in. Now we must pick a unit from the lower half of Table 4.2 where SH types are rated. In the 200° F column for 32-in.-high units, we find that a 10-in.-deep, 48-in.-long convector will produce 17.6 Mbh. This is close enough to the output required.

Notice that in the toilet rooms where the heat losses are very small, 20-in.-high convectors are used. To save space, they are only 4 in. deep and are recessed 2 in. into the wall.

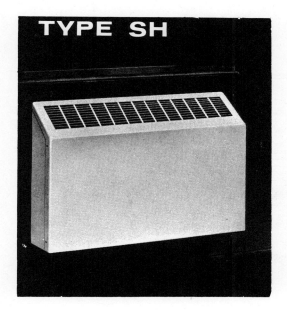

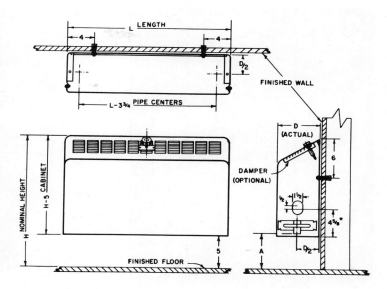

DIMENSIONS							
D (Nominal)	D (Actual)	L (Length)	H (Height) NOMINAL	Cabinet Height	A (Max.)	A (Min.)
4	4¼	20-24-28-32-36-40-44-48-56-64	20-24-32-38	15-19-27-33	5¹³⁄₁₆	5⁵⁄₁₆
6	6¼	20-24-28-32-36-40-44-48-56-64	20-24-32-38	15-19-27-33	5¹³⁄₁₆	5⁵⁄₁₆
8	8¼	32-36-40-48-56-64	20-24-32-38	15-19-27-33	5¹³⁄₁₆	5⁵⁄₁₆
10	10¼	36-40-48-56-64	20-24-32-38	15-19-27-33	5¹³⁄₁₆	5⁵⁄₁₆
All dimensions are in inches and are subject to slight variations.							

* Knockouts are optional—furnished only on request.

TYPE SH This model is a fully exposed wall hung unit with outlet grille located in sloping top. Enclosure wraps around unit and fastens to sides with screws. Air inlet is through open bottom of unit. Slope of top is 30°.

Dampers and Air Vents available at extra cost. Two 13/16" Dia. Holes provided in back of enclosure for attaching to wall, bolts not included. Mansory wall mounting shown above. For frame wall, provide stud—headers opposite mounting holes. Header tappings—¾" top and bottom on 4, 6 and 8" depths; 1" top and bottom on 10" depth.

Figure 4.9 Type SH convector. This wall-hung model is good in industrial spaces. The floor can be mopped below the unit. Also, employees cannot place objects on top of the cabinet as they might on other styles. (Courtesy of Standard Fin-Pipe Radiator Corporation)

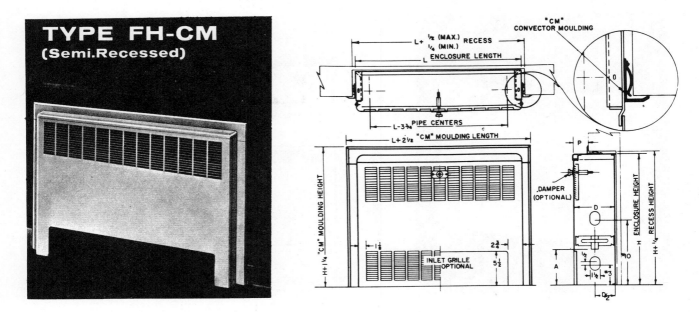

DIMENSIONS

D (Nominal)	D (Actual)	L (Length)	H (Height)	A (Max.)	A (Min.)
4	4¼	20-24-28-32-36-40-44-48-56-64	20-24-32-38	5¹³⁄₁₆	5⁵⁄₁₆
6	6¼	20-24-28-32-36-40-44-48-56-64	20-24-32-38	5¹³⁄₁₆	5⁵⁄₁₆
8	8¼	32-36-40-48-56-64	20-24-32-38	5¹³⁄₁₆	5⁵⁄₁₆
10	10¼	36-40-48-56-64	20-24-32-38	5¹³⁄₁₆	5⁵⁄₁₆

P-1¼" minimum.   All dimensions are in inches and are subject to slight variations.

* Knockouts are optional—furnished only on request.

TYPE FH SEMI-RECESSED Cabinet design is same as FH model, furnished with three-piece metal molding to seal space between enclosure and recess in wall. Enclosure projects a minimum of 1¼" from wall. Complete unit includes enclosure, front panel with outlet grille and arched inlet opening, heating element, and three-piece CM metal molding trim. Front panel is easily removed for cleaning or access to heating element.

Dampers, Inlet Grille and Air Vents available at extra cost. "CM" Molding available for each size of FH Convector—no cutting required. Finished in baked—on grey prime. Molding trim easily inserted between convector and recess opening.

Holes provided in ½" flange at bottom of enclosure for fastening cabinet to floor. Header tappings—¾" top and bottom on 4, 6 and 8" depths; 1" top and bottom on 10" depth.

Figure 4.10 Type FH-CM convector. In the three toilet rooms of our industrial building, this style is chosen. Space in these rooms is very limited. By recessing the cabinet into the wall, only 2 in. of its 4-in. depth projects into the room. (Courtesy of Standard Fin-Pipe Radiator Corporation)

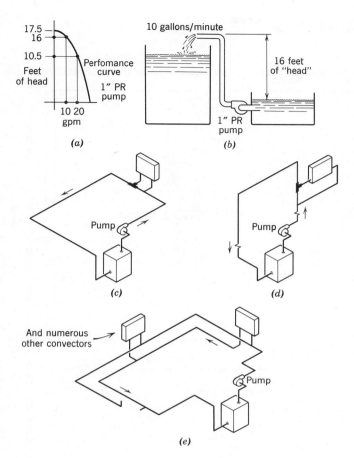

Figure 4.11 Pumps, systems and friction. *(a)* Typical performance curve. *(b)* Illustration of one pumping task (lifting). *(c)* One-story horizontal pumping circuit. *(d)* Multistory pumping circuit. *(e)* System to be considered in two zones of Example 4.1.

Note. Pumps in *heating* systems *(c)*, *(d)* and *(e)* do *not* lift water. The water in each of the vertical legs of any one of these systems is of exactly the same height. The legs balance. The pump (circulator) operates ONLY to circulate the water against the frictional resistance of the water flowing through the tubing. For further explanation see Section 4.3.

Table 4.1 (Example 4.1) Results of Engineer's Calculations of hourly heat losses for an Industrial Building[a]

	Mbh
Work area	52.6
Storage area	196.6
Administration	14.4
Private office	8.5
Men's toilet	1.5
Women's toilet	3.2
Private toilet	1.5
Lobby	4.6
Women's rest room	(Negligible)
	282.9 Mbh

[a]These are to be used as a basis for the design of a heating system as called for in Example 4.1

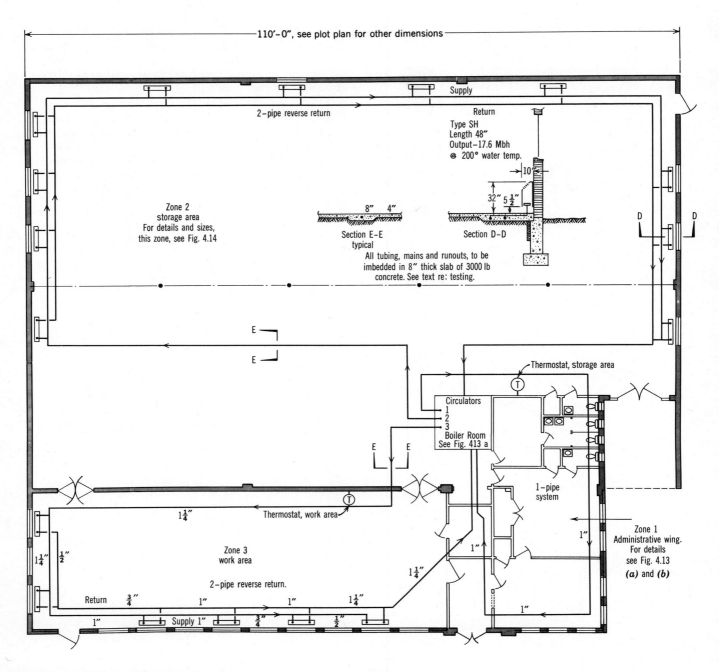

110'-0", see plot plan for other dimensions

Supply

2-pipe reverse return

Return
Type SH
Length 48"
Output–17.6 Mbh
@ 200° water temp.

10'

32" 5½"

8" 4"

Zone 2
storage area
For details and sizes,
this zone, see Fig. 4.14

Section E-E
typical

Section D-D

D   D

All tubing, mains and runouts, to be
imbedded in 8" thick slab of 3000 lb
concrete. See text re: testing.

E

E

Thermostat, storage area

T

Circulators
1
2
3
Boiler Room
See Fig. 413 a

1-pipe
system

Zone 1
Administrative wing.
For details
see Fig. 4.13
(a) and (b)

E   E

T

Thermostat, work area

1¼"

Zone 3
work area

2-pipe reverse return.

1¼" ½"

1¼"

1"

1"

Return ¾"   1"   1"   1¼"

1"   Supply 1"   ¾"   ½"

1"

Figure 4.12 (Example 4.1) Layout of three-zone system for industrial building.
Enlarged details are shown in Figures 4.13 and 4.14 and results are given in Tables 4.3
and 4.5.

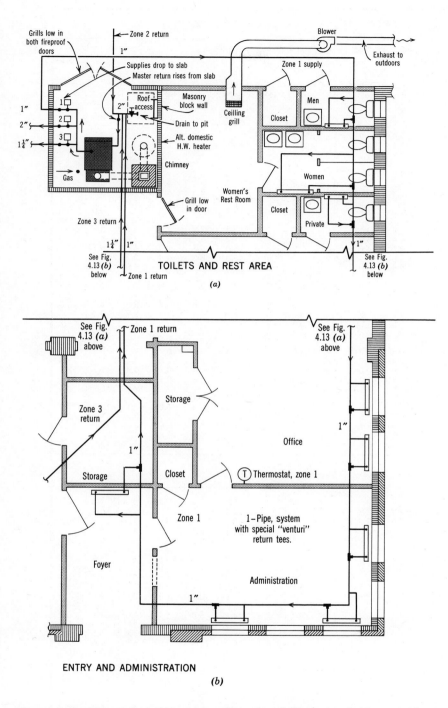

Figure 4.13 (Example 4.1) Layout and results of the design for heating the administrative wing, Zone 1.

Table 4.2 (Example 4.1) Tables for the Selection of Convectors

Hot Water Ratings—MBH—20° Temp. Drop (Av. W. T.)

FRONT OUTLET CONVECTORS—TYPES FH, RH, MH, WH, PH

Depth Inches	Length	20" HEIGHT 180°	190°	200°	215°	24" HEIGHT 180°	190°	200°	215°	32" HEIGHT 180°	190°	200°	215°	38" HEIGHT 180°	190°	200°	215°
4	20	1.5	1.7	1.9	2.2	1.7	1.9	2.2	2.6	1.9	2.2	2.4	2.8	2.0	2.3	2.5	2.9
	24	1.8	2.1	2.3	2.7	2.1	2.4	2.7	3.1	2.2	2.7	3.0	3.5	2.4	2.8	3.1	3.6
	28	2.2	2.5	2.8	3.3.	2.5	2.9	3.2	3.8	2.7	3.2	3.5	4.1	2.9	3.3	3.7	4.3
	32	2.5	2.9	3.3	3.8	2.9	3.3	3.7	4.4	3.2	3.7	4.1	4.8	3.3	3.8	4.3	5.0
	36	2.9	3.3	3.7	4.3	3.3	3.7	4.2	5.0	3.6	4.2	4.7	5.5	3.8	4.3	4.9	5.7
	40	3.2	3.7	4.1	4.9	3.7	4.2	4.7	5.6	4.1	4.7	5.3	6.2	4.2	4.9	5.5	6.4
	44	3.6	4.1	4.6	5.4	4.1	4.7	5.3	6.2	4.6	5.2	5.9	6.9	4.7	5.4	6.1	7.1
	48	3.9	4.5	5.0	5.9	4.5	5.2	5.7	6.8	5.0	5.7	6.4	7.5	5.2	5.9	6.6	7.8
	56	4.6	5.3	5.9	7.0	5.3	6.1	6.8	8.0	5.9	6.7	7.5	8.9	6.1	7.0	7.8	9.2
	64	5.3	6.1	6.8	8.0	6.1	7.0	7.8	9.2	6.7	7.7	8.7	10.2	7.0	8.1	9.0	10.6
6	20	2.1	2.4	2.7	3.1	2.4	2.8	3.1	3.6	2.6	3.0	3.4	3.9	2.7	3.1	3.5	4.1
	24	2.6	3.0	3.3	3.9	3.0	3.4	3.8	4.5	3.2	3.7	4.2	4.9	3.3	3.8	4.3	5.0
	28	3.1	3.5	4.0	4.6	3.6	4.1	4.6	5.4	3.9	4.5	5.0	5.8	4.0	4.6	5.1	6.0
	32	3.6	4.1	4.6	5.4	4.1	4.8	5.4	6.2	4.5	5.2	5.8	6.8	4.6	5.3	5.9	7.0
	36	4.1	4.7	5.2	6.1	4.7	5.4	6.0	7.1	5.1	5.9	6.6	7.7	5.3	6.0	6.8	7.9
	40	4.5	5.2	5.9	6.9	5.3	6.1	6.8	7.9	5.7	6.6	7.3	8.7	5.9	6.8	7.6	8.9
	44	5.1	5.8	6.5	7.6	5.8	6.7	7.5	8.8	6.3	7.3	8.1	9.6	6.5	7.5	8.4	9.9
	48	5.5	6.4	7.1	8.4	6.4	7.4	8.2	9.6	7.0	8.0	9.0	10.5	7.2	8.2	9.2	10.8
	56	6.5	7.5	8.4	9.7	7.5	8.7	9.7	11.4	8.2	9.3	10.6	12.4	8.5	9.7	10.9	12.7
	64	7.5	8.7	9.7	11.3	8.7	10.0	11.2	13.1	9.5	10.9	12.2	14.3	9.8	11.2	12.5	14.7
8	32	4.4	5.1	5.7	6.7	5.0	5.7	6.4	7.5	5.7	6.6	7.3	8.6	5.9	6.7	7.5	8.8
	36	5.0	5.8	6.5	7.6	5.7	6.5	7.3	8.6	6.5	7.5	8.4	9.8	6.7	7.7	8.6	10.0
	40	5.7	6.5	7.3	8.5	6.4	7.3	8.2	9.6	7.3	8.4	9.4	11.0	7.5	8.6	9.6	11.3
	48	6.9	7.9	8.9	10.4	7.7	8.9	10.0	11.7	8.9	10.2	11.5	13.4	9.1	10.4	11.7	13.7
	56	8.1	9.4	10.4	12.2	9.1	10.5	11.7	13.7	10.5	12.0	13.5	15.8	10.7	12.3	13.7	16.2
	64	9.3	10.8	12.0	14.7	10.5	12.1	13.5	15.9	12.1	13.9	15.6	18.2	12.3	14.2	15.9	18.6
10	36	6.4	7.3	8.2	9.6	6.9	7.9	8.8	10.3	7.7	8.8	9.9	11.6	7.9	9.1	10.2	11.9
	40	7.2	8.2	9.2	10.8	7.7	8.9	9.9	11.6	8.7	10.0	11.1	13.1	8.9	10.2	11.5	13.4
	48	8.7	10.0	11.4	13.1	9.4	10.8	12.1	14.2	10.5	12.1	13.5	15.9	10.8	12.5	13.9	16.3
	56	10.4	11.8	13.4	15.5	11.1	12.7	14.2	16.7	12.4	14.3	16.0	18.7	12.8	14.7	16.4	19.2
	64	11.8	13.6	15.2	17.8	12.8	14.7	16.4	19.2	14.3	16.5	18.4	21.6	14.7	16.9	18.9	22.2

SLOPE TOP CONVECTOR—TYPE SH, SFH

Depth Inches	Length	20" HEIGHT 180°	190°	200°	215°	24" HEIGHT 180°	190°	200°	215°	32" HEIGHT 180°	190°	200°	215°	38" HEIGHT 180°	190°	200°	215°
4	20	1.6	1.8	2.0	2.3	1.7	1.9	2.2	2.5	2.0	2.3	2.6	3.0	2.1	2.3	2.6	3.1
	24	1.9	2.2	2.5	2.9	2.1	2.4	2.7	3.1	2.4	2.8	3.1	3.7	2.5	2.9	3.2	3.8
	28	2.3	2.6	2.9	3.4	2.5	2.9	3.2	3.7	2.9	3.3	3.7	4.4	3.0	3.5	3.9	4.6
	32	2.7	3.1	3.4	4.0	2.9	3.3	3.7	4.4	3.4	3.9	4.4	5.1	3.5	4.1	4.5	5.3
	36	3.0	3.5	3.9	4.5	3.3	3.8	4.2	5.0	3.8	4.4	5.0	5.8	4.0	4.6	5.1	6.0
	40	3.4	3.9	4.3	5.1	3.7	4.2	4.7	5.6	4.3	5.0	5.6	6.5	4.5	5.2	5.8	6.7
	44	3.8	4.3	4.8	5.7	4.1	4.7	5.3	6.2	4.8	5.5	6.2	7.2	5.0	5.7	6.4	7.5
	48	4.1	4.7	5.3	6.2	4.5	5.2	5.8	6.8	5.3	6.0	6.8	7.9	5.5	6.3	7.0	8.2
	56	4.8	5.6	6.2	7.3	5.3	6.1	6.8	8.0	6.2	7.1	8.0	9.4	6.4	7.4	8.3	9.7
	64	5.6	6.4	7.2	8.4	6.1	7.0	7.8	9.2	7.1	8.2	9.2	10.8	7.4	8.5	9.5	11.2
6	20	2.7	3.1	3.4	4.0	2.9	3.3	3.7	4.3	3.3	3.8	4.2	5.0	3.4	4.0	4.4	5.2
	24	3.3	3.8	4.2	4.9	3.5	4.1	4.6	5.3	4.1	4.7	5.2	6.1	4.3	4.8	5.5	6.4
	28	3.9	4.5	5.1	5.9	4.2	4.9	5.4	6.4	4.9	5.6	6.3	7.4	5.1	5.9	6.5	7.7
	32	4.6	5.2	5.9	6.9	4.9	5.7	6.3	7.4	5.7	6.5	7.3	8.5	5.9	6.8	7.6	8.9
	36	5.2	6.0	6.7	7.8	5.6	6.4	7.2	8.4	6.4	7.4	8.3	9.7	6.7	7.7	8.6	10.1
	40	5.8	6.7	7.5	8.8	6.3	7.2	8.1	9.5	7.2	8.3	9.3	10.9	7.5	8.6	9.6	11.3
	44	6.4	7.4	8.3	9.7	7.0	8.0	9.0	10.5	8.0	9.2	10.3	12.1	8.4	9.6	10.8	12.6
	48	7.1	8.1	9.1	10.7	7.6	8.8	9.8	11.5	8.8	10.1	11.3	13.3	9.2	10.5	11.8	13.8
	56	8.3	9.6	10.7	12.6	9.0	10.4	11.6	13.6	10.4	11.9	13.3	15.6	10.8	12.4	13.9	16.3
	64	9.6	11.1	12.4	14.5	10.4	12.0	13.4	15.7	11.9	13.7	15.4	18.0	12.5	14.3	16.0	18.8
8	32	5.9	6.8	7.6	8.9	6.3	7.3	8.2	9.6	7.5	8.6	9.7	11.3	7.8	8.9	10.0	11.7
	36	6.7	7.7	8.6	10.1	7.2	8.3	9.3	10.9	8.5	9.8	11.0	12.9	8.8	10.1	11.3	13.3
	40	7.5	8.7	9.7	11.4	8.1	9.3	10.4	12.2	9.6	11.0	12.3	14.4	9.9	11.4	12.7	14.9
	48	9.2	10.6	11.8	13.8	9.9	11.3	12.7	14.8	11.7	13.4	15.0	17.6	12.0	13.8	15.5	18.1
	56	10.8	12.5	13.9	16.3	11.6	13.3	14.9	17.5	13.7	15.8	17.7	20.7	14.2	16.3	18.2	21.4
	64	12.5	14.3	16.0	18.8	13.4	15.4	17.3	20.2	15.9	18.3	20.4	23.9	16.4	18.8	21.1	24.7
10	36	8.1	9.3	10.4	12.2	9.0	10.4	11.6	13.6	10.0	11.5	12.9	15.1	10.3	11.8	13.2	15.5
	40	9.1	10.5	11.8	13.8	10.2	11.7	13.1	15.3	11.3	13.0	14.5	17.0	11.6	13.3	14.9	17.5
	48	11.1	12.8	14.3	16.8	12.4	14.2	15.9	18.6	13.7	15.8	17.6	20.7	14.1	16.2	18.1	21.2
	56	13.1	15.1	16.8	19.7	14.6	16.7	18.7	21.9	16.1	18.6	20.8	24.3	16.6	19.1	21.4	25.0
	64	15.1	17.3	19.4	22.7	16.8	19.3	21.6	25.3	18.6	21.4	23.9	28.0	19.1	22.0	24.6	28.8

Source. Standard Fin-Pipe Radiator Corporation. Circles relate to Example 4.1.

4.6. Circuits, Zones and Controls

For Zone 1, the administrative wing, a one-pipe circuit is chosen. For Zone 2, the storage area, a two-pipe, reverse-return method is adopted. For the longer tube-runs this scheme is more efficient. Zone 3, the work area, also uses the two-pipe method.

For the pumping rate in Zone 1, we consult Table 4.3. The convector group in this zone calls for 36,300 Btuh heating rate. The pumping rate for a 20 F° drop is, therefore,

$$\frac{36,300}{10,000} = 3.6 \text{ gpm}$$

In Section 4.3 of this chapter we explained Table 4.4. It is one of many shortcut design methods for engineers and technologists. It has four principal items:

1. Pressure drop.
2. Capacities of mains (Mbh for various tube sizes).
3. Total equivalent length of systems.
4. Head pressures in feet.

Pressure drops in the middle range, 200 to 400, are suitable for the systems of Example 4.1. We use 200 and 300 in our designs. *Very* high drops can produce high velocity and noisy flow. Low drops can result in larger (and, therefore, more expensive) tubing.

Therefore, returning to Zone 1, we find that, using a 300 milinch per foot drop, a 1-in. main will deliver 53 Mbh (Table 4.4, Section B). This is satisfactory for the 36.3 Mbh (Table 4.3) of this zone. The tubing circuit has a developed length (DL) of 124 ft and a total equivalent length (TEL) of $124 \times 1.5 = 186$ ft. In Section A of Table 4.4, we find that for 180 ft TEL (close enough), the pumping "head" is 4½ ft. Condition "A" (Figure 4.15) for these values (3.6 gpm and 4½ ft) shows that a 1-in. circulator is a good choice.

Thermostats control the three circulators that pump water through the zone circuits. The one for the "executive" area, Zone 1, is located in the private office of the manager. Thermostats in Zones 2 and 3 must each be, as shown, within the area served. Their locations should be *away* from heating units, drafts and direct sunlight. In these two *employee* areas, they should be of the locked type. Control can thus be under the charge of a foreman. This prevents confusion. An aquastat in the boiler can keep the water at 210° F. It can then be circulated at design temperature.

4.7. Two-Pipe Systems

Figure 4.14 is a "foldout" diagram of Zone 2. It makes the design easier. In this two-pipe arrangement, water at full boiler temperature supplies each convector. These runouts are at points B, C, D, to L at the last (No. 11) convector. The *return* main starts at point M and picks up the cooler water that flows out of Convector No. 1. As it continues toward the boiler, it picks up return water at points N, O, P, to U, V and W. Obviously the supply main can reduce in size as it supplies fewer and fewer convectors. Conversely, the return main increases in size until it carries *all* of the water that is circulated in the system.

Now, notice that water finding its way through tubing circuit A, B, M, X (Convector No. 1) passes through exactly the same distance as other water through circuit A, L, W, X (Convector No. 11). Hence, either route is the "developed length" (DL). Also each convector is served equally.

In this zone (2), we will design for 300 milinches per foot of total equivalent length (TEL). As we stated previously, the TEL is often considered to be DL × 1.5. In Zone 2, the DL (ABMX) or any other flow path is 320 ft. TEL is thus 480 ft. Table 4.4 shows us at once that the head is 12 ft.

The next step is to size the sections of the main. Based on their Mbh carrying capacity and 300 milinches per foot friction drop, the sizes can be selected at once from Table 4.4.

Actually the *return* main cannot be considered to *deliver* Mbh values. Yet, because Mbh and gallons per minute are directly proportional (gpm = Mbh/10,000), the returns can be sized by the Mbh values that they *pick up*. For instance, section MN *picks up* the flow after 17.6 Mbh is supplied to Convector No. 1. Thus, MN is 1 in., just as KL is 1 in., because it *delivers* 17.6 Mbh to Convector No. 11.

4.8. Tubing Expansion or Restraint

You will notice (Figure 4.12) that all of the mains and part of each runout are embedded in 8 in. of 3000 psi concrete. In Chapter 2, the expansion of tubing is discussed. In Figure 2.13 (page 36), limits were placed on straight runs of tubing allowed between expansion joints. The scheme was that the *ends* of long runs be anchored and that the expansion joint would take up the motion at the middle of

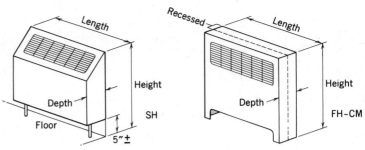

Table 4.3 (Example 4.1) Selection of Convectors[a]

Space	Heat loss (Mbh)	Number of Convectors	Type	Height (inches)	Length (inches)	Depth (inches)	Output of Each (Mbh)	Product	Total Output (Mbh)	
						Convectors selected, (Table 4.2)				
Work area	52.6	6	SH	32	40	6	9.3	6 × 9.3 = 55.8		Zone 3
Storage area	196.6	11	SH	32	48	10	17.6	11 × 17.6 = 193.6		Zone 2
Admin. office	14.4	3	SH	32	36	4	5.0	3 × 5.0 = 15.0		
Private office	8.5	2	SH	32	32	4	4.4	2 × 4.4 = 8.8		
Men's toilet	1.5	1	FH–CM	20	20	4[b]	1.9	1 × 1.9 = 1.9		
Women's toilet	3.2	1	FH–CM	20	36	4[b]	3.7	1 × 3.7 = 3.7	36.3	
Private toilet	1.5	1	FH–CM	20	20	4[b]	1.9	1 × 1.9 = 1.9	Zone 1	
Lobby	4.6	1	SH	32	36	4	5.0	1 × 5.0 = 5.0		
Women's rest room	(negligible)									

$$\frac{\text{Output}}{\text{Heat required}} = \frac{285.7}{282.9} = 1.009$$

282.9 Mbh   Less than 1%   285.7 Mbh

[a]Average water temperature, 200° F; temperature drop, 20 F°.
[b]Recessed 2 in.

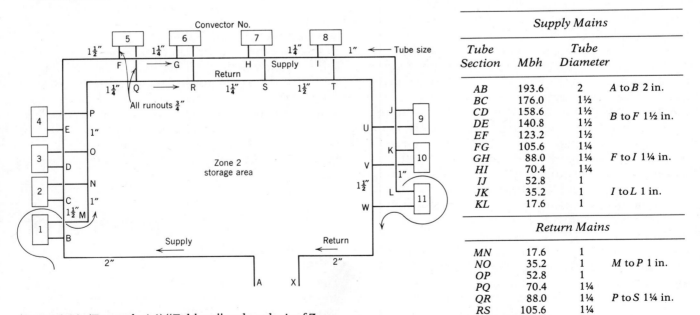

Supply Mains		
Tube Section	Mbh	Tube Diameter
AB	193.6	2 · · · A to B 2 in.
BC	176.0	1½
CD	158.6	1½ · · · B to F 1½ in.
DE	140.8	1½
EF	123.2	1½
FG	105.6	1¼
GH	88.0	1¼ · · · F to I 1¼ in.
HI	70.4	1¼
IJ	52.8	1
JK	35.2	1 · · · I to L 1 in.
KL	17.6	1

Return Mains		
MN	17.6	1
NO	35.2	1 · · · M to P 1 in.
OP	52.8	1
PQ	70.4	1¼
QR	88.0	1¼ · · · P to S 1¼ in.
RS	105.6	1¼
ST	123.2	1½
TU	140.8	1½ · · · S to W 1½ in.
UV	158.6	1½
VW	176.0	1½
WX	193.6	2 · · · W to X 2 in.

Note. Uniform Pressure Drop is 300 milinches per foot of tubing. See Table 4.4.

Figure 4.14 (Example 4.1) "Foldout" and analysis of Zone 2. Sizes of supply and return mains. Sizes are selected from Table 4.4. They are found by using the 300 milinch per foot column and the Mbh to be carried by the tube.

Average water temperature is 200° F; temperature drop is 20 F°, pump delivery is 19.3 gpm, 11 Convectors at 17.6 Mbh each, DL 320 ft, TEL 480 ft, head 12 ft, runouts ¾ in.

Table 4.4 Pipe Sizing Table for Mains, Forced Circulation Hot Water Systems

SECTION A

Zones, Example 4.1

		1 and 2						3

Booster Head Pressures (ft)

Total Equivalent Length of Pipe in Feet

	2	40	48	60	68	80	96	120	160	240
	2½	50	60	75	86	100	120	150	200	300
	3	60	72	90	103	120	144	180	240	360
	3½	70	84	105	120	140	168	210	280	420
	4	80	96	120	137	160	192	240	320	480
Zone 1	4½	90	108	135	154	[180] 1	216	[270] 3	360	540
Zone 3	5	100	120	150	171	200	240	[300]	400	600
	5½	110	132	165	188	220	264	330	440	660
	6	120	144	180	206	240	288	360	480	720
	6½	130	156	195	223	260	312	390	520	780
	7	140	168	210	240	280	336	420	560	840
	7½	150	180	225	257	300	360	450	600	900
	8	160	192	240	274	320	384	480	640	960
	8½	170	204	255	291	340	408	510	680	1020
	9	180	216	270	308	360	432	540	710	1080
	9½	190	228	285	325	380	456	570	760	1140
	10	200	240	300	342	400	480	600	800	1200
	10½	210	252	315	360	420	504	630	840	1260
	11	220	264	330	377	440	528	660	880	1320
	11½	230	276	345	394	460	552	690	920	1380
Zone 2	12	240	288	360	411	[480] 2	576	720	960	1440

SECTION B (Based on 20° Temperature Drop)

Main Capacities (in Thousands of Btuh)

Pipe Size (in.)		Pressure Drop in Pipe in Milinches per Foot								
		600	500	400	350	[300]	250	[200]	150	100
	½	19	18	16	15	13	12	10	9	7
	¾	41	37	33	30	28	26	23	20	15
Zone 1	1	80	71	64	59	[53] 1	48	42	37	31
Zone 3	1¼	170	160	140	130	118	102	[90] 3	78	63
	1½	260	240	210	185	175	156	140	121	94
Zone 2	2	500	450	410	360	[322] 2	294	261	227	182
	2½	810	750	670	610	551	523	460	385	310
	3	1600	1400	1300	1150	1000	900	800	680	550
	3½ᵃ	2300	2100	1850	1650	1500	1350	1190	1020	825
	4ᵃ	3200	2900	2600	2300	2100	1950	1700	1350	1140

[a]Trunk main capacities only. Fittings are not made larger than 3 in.

Note: The figures shown in these tables apply to both steel pipe and Type M copper tubing, as capacity differences are not sufficient to cause design errors. ITT Bell & Gossett

Added notes relate to Example 4.1

the run. This was illustrated in the upper story of the house in Figure 3.10a, page 72. The *lower* story of that house (Figure 3.10b) had shorter runs. Even in short runs, there is a small amount of motion. In that zone of Figure 3.10b the tubing was placed in the gravel fill *below* the concrete slab. Vertical runouts through the slab pass through oversize metal sleeves. This would avoid any strain on the runout or its connection to the main.

In Section D-D and E-E of Figure 4.12, the embedment of the tubing in 8 in. of concrete is shown. This *prevents* any expansion of the tubing. By restraining it, a small stress, uniformly distributed, is set up in the metal. It is well within the stress that the metal can take. No damage is possible.

It is most important that the bond between concrete and copper be *effective*. Concrete with an ultimate strength of 3000 psi is used. To prevent leaks, the tubing is tested at increased water pressure. This is done before the concrete is poured. If leaks are seen, correction is made. Then the concrete is poured. This careful routine is most important to preclude future leakage of the mains and to prevent motion of the copper.

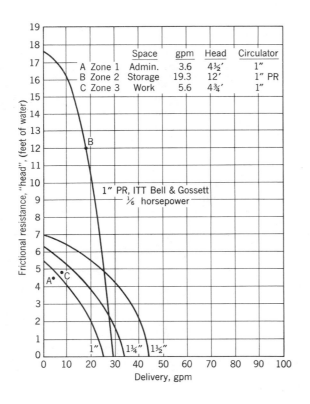

	Space	gpm	Head	Circulator
A Zone 1	Admin.	3.6	4½'	1"
B Zone 2	Storage	19.3	12'	1" PR
C Zone 3	Work	5.6	4¾'	1"

1" PR, ITT Bell & Gossett ⅙ horsepower

Check actual performance curves in manufacturers' catalogs and follow their engineering recommendations.

Figure 4.15 (Example 4.1) Circulators. Choice of pumps (boosters) suitable for the three zones of the hot water heating system of Example 4.1. Flow based on a 20 F° drop in water temperature. Head dependent on total equivalent length of tubing, heating demand and milinch per foot pressure drop.

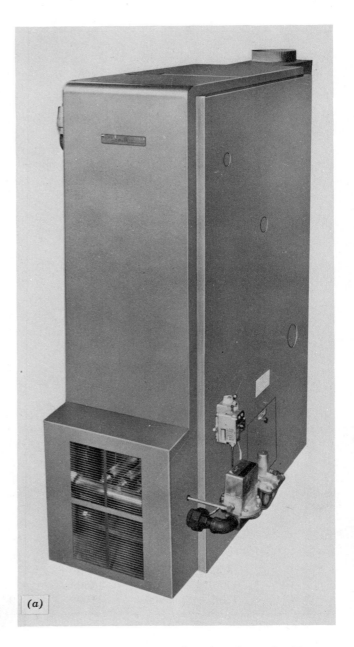

(a)

Figure 4.16 (a) Photograph of a (typical) gas-fired hot water heating boiler, G300-3W. (b) Dimensions and ratings of boilers in this series. In Example 4.1, a G 300-7W boiler is chosen. G gas-fired; 300, series number; 7, seven sections; W, water boiler (S if steam). (Courtesy of H. B. Smith Company, Inc.)

DIMENSIONS AND RATINGS

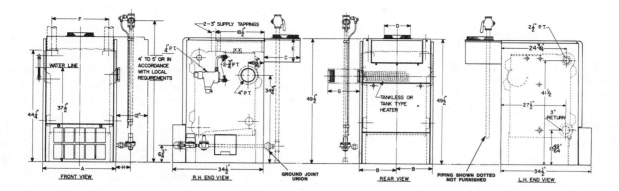

HYDROSTATICALLY TESTED — A.S.M.E. STANDARD
MAXIMUM ALLOWABLE WORKING PRESSURE — STEAM 15 LBS., WATER 40 LBS.

BOILER NUMBER	RATINGS NATURAL AND PROPANE GAS					DIMENSIONS (Inches)								CHIMNEY SIZE, INCHES X HGT. IN FT.	SIZE OF GAS SHUT-OFF NATURAL & PROPANE
	A. G. A. RATINGS		I=B=R NET RATINGS												
	INPUT BTUH	OUTPUT BTUH	STEAM		WATER	A	B	C	D	E	F	G	H		
			SQ. FT.	BTUH	BTUH										
1	2	3	4	5	6	7	8	9	10	11	12	13	14	15	16
G300-3*	150,000	120,000	375	90,000	104,400	16	8	11	6	8½	9⅝	24	5⅜	8x8x20'	¾ "
G300-4*	225,000	180,000	563	135,100	156,500	20	10	12	8	8½	13⅝	24	5⅜	8x8x20'	¾ "
G300-5*	300,000	240,000	750	180,000	208,700	24	12	12	10	8½	17¾	26	5⅜	8x12x20'	¾ "
G300-6*	375,000	300,000	938	225,100	260,900	28	14	14	10	8½	21¾	26	5⅜	8x12x20'	¾ "
G300-7*	450,000	360,000	1125	270,100	313,000	32	16	14	12	9	25⅝	34	5⅜	8x12x20'	¾ "
G300-8*	525,000	420,000	1313	315,100	365,200	36	18	16	12	9	29¾	38	5¾	12x12x20'	1 "
G300-9*	600,000	480,000	1500	360,100	417,400	40	20	16	14	9	33¾	42	5¾	12x12x20'	1 "
G300-10*	675,000	540,000	1688	405,100	469,600	44	22	18	14	9½	37⅞	46	5¾	12x16x20'	1 "
G300-11*	750,000	600,000	1875	450,100	521,700	48	24	18	14	9½	41⅞	46	5¾	12x16x20'	1 "
G300-12*	825,000	660,000	2063	495,100	573,900	52	26	18	14	9½	46	54	5¾	12x16x20'	1 "

*Insert S for Steam, W for Water.

The net I=B=R Steam Ratings shown are based on a piping and pickup allowance of 1.333.

The net I=B=R Water Ratings shown are based on an allowance of 1.15. The manufacturer should be consulted before selecting a boiler for installations having unusual piping and pickup requirements such as intermittent system operation, extensive piping systems, etc.

For forced hot water heating systems where the boiler and all the piping are within the area to be heated, the boiler may be selected on the basis of its gross output.

NOTE: For altitudes above 2,000 feet reduce ratings 4 percent for each additional 1,000 feet.

STANDARD EQUIPMENT — STEAM

● Main manual shut-off valve ● Manual pilot shut-off valve ● Pilot filter ● Gas pressure regulator ● Main gas valve ● Transformer ● Automatic safety pilot ● Low water cut-off ● Water gauge glass and fittings ● Pressure limit control and syphon ● Steam pressure gauge ● Drain cock with hose connection ● Pop safety valve, set at 15 lbs. ● Flush jacket ● Draft hood, horizontal type ● Flue brush

STANDARD EQUIPMENT — WATER

● Main manual shut-off valve ● Manual pilot shut-off valve ● Pilot filter ● Gas pressure regulator ● Main gas valve ● Automatic safety pilot ● Altitude gauge and thermometer ● High-limit Aquastat and circulator relay combination control ● Water pressure relief valve ● Drain cock with hose connection ● Flush jacket ● Draft hood, horizontal-type ● Flue brush

OPTIONAL EQUIPMENT

● Room thermostat   ● Tankless heater

Form No. 10M 10/74 CAT. NO. 2508

(b)

4.9. Boiler and Circulators

Boiler G300W7 is selected from Figure 4.16b. An 8 by 12 in. chimney 20 ft high is used. It will project 6 ft above the high roof of the storage area. This assures a good draft and high discharge of the flue gases. The boiler room has walls of masonry block and a fireproof ceiling. Air for ventilation and combustion is drawn in through grills in both of the fireproof doors.* The large volume of this building makes it reasonable to draw combustion air from the storage area rather than from outdoors. The fire precautions mentioned here are intended to suggest a more complete study of fire safety. Planning for safety should be based on recommendations in the literature of the National Fire Protection Association.

Circulators are chosen from Figure 4.15 on the basis of their required performance in gallons per minutes and feet of head as found in Table 4.4.

4.10. Summary of Data, Design Steps and Results, Table 4.5

It is usually desirable to summarize the results of a design. They form a basis for working drawings and specifications. A few comments follow about Table 4.5.

Column No. 3. The Mbh of convectors or other heating units is seldom much greater than the engineer's requirements. Yet it is well to remember that the system (boiler and tubing) must develop the output of the *installed heaters*.

Column No. 4. The factor 10,000 is *only* for a 20 F° temperature drop. Sometimes other "drop" values are used.

Column No. 5. Reviewing the idea of *developed length*. This length includes horizontal, vertical and other tube runs from boiler back to boiler or for the length of a partial circuit. It is as though the tubing system were unfolded, laid out in one straight line, and then measured.

Column No. 6. The factor 1.5 is only an estimate. It assumes that fittings, valves, and the like increase the frictional resistance of the developed length by 50%. In some detailed work the fittings are counted and evaluated in their "equivalent length of tube or pipe." See Table 2.3, page 43.

---

*Some codes require, for safety, a secondary exit leading directly to outdoors. This could affect the location and planning of the boiler room. The roof access and fixed ladder in the corner of boiler room (Figure 4.13a) would be desirable in this regard.

4.11. Ventilation

State labor laws usually require special facilities for women employees. Adjacent to toilets, a resting or lounge area must be provided. Ventilation is required for such a space, especially if it is fully interior with no windows to the outdoors.

Locker rooms, rest rooms and other spaces could be required by some codes to have about five air changes per hour. If this were applied to the women's restroom of our industrial building, the cubic feet per minute (cfm) air-rate of the exhaust fan would need to be specified. For five air changes per hour the cubic feet per minute of the fan (blower) would have to be (for this 8 by 12 by 9 ft high room):

$$\frac{8 \text{ ft} \times 12 \text{ ft} \times 9 \text{ ft} \times 5 \text{ (changes)}}{60 \text{ (min/hr)}} = 72 \text{ cfm}$$

The blower would also be rated to overcome the friction of the air being forced out through the duct. This is expressed in inches of water per 100 ft of duct.

Table 4.5 (Example 4.1) Data, Design Steps and Results

	Column Numbers for this Table			For Reference					
1	*2*	*3*	*4*	*5*	*6*	*7*	*8*	*9*	*10*
Zone	Space	Mbh for Design	gpm at 20 F° TD	Developed Length, Feet	Total Equivalent Length, (Feet)	Milinches per Foot	"Head" (Feet)	Maximum Size of Main, (Inches)	Pump Size
1	Admin. wing	36	3.6	124	186	200	4½	1	1 in.
2	Storage area	193	19.3	320	480	300	12	2	1 in. PR
3	Work area	56	5.6	190	285	200	4¾	1¼	1 in.
		Table 4.3 (last col.)	Btuh ÷ 10,000	Measured, Figure 4.12	TEL = DL × 1.50	See Section 4.3	Table 4.4	Table 4.4	Col. numbers 4 and 8 plus Figure 4.15

Boiler selection (Figure 4.16*b*) G300 W7 is chosen. Its I=B=R net output rating of 313,000 Btuh is adequate for the total convector demand of 285,700 Btuh. Chimney size 8 in by 12 in by 20 ft high.

Problems

4.1. Select four convectors, type SH. Average water temperature is 180° F, temperature drop is 20 F°, convector requirements are height 38 in., depth 6 in.

For these conditions select convectors of proper *length* to deliver:
(a) 4.0 Mbh.
(b) 10.6 Mbh.
(c) 6.7 Mbh.
(d) 3.0 Mbh.

4.2. How many gallons per minute will a 1-in. circulator deliver against a head of:
(a) 4 ft?
(b) 2 ft?

4.3. One-pipe system, 150 milinches/ft design, 360 ft TEL, 78 Mbh, 20 F° temperature drop, wanted the following:
(a) Tube size.
(b) Head.
(c) Gallons per minute.
(d) Circulator size.

4.4. What are the maximum allowable working pressures in an H. B. Smith Co. G300 Series boiler for: (a) Steam? (b) Water?
Note: psi is sometimes stated as "pounds."

4.5. For a G300-11 (11 sections) boiler what are the ratings in Btuh for:
(a) Input?
(b) Output?
(c) Net rating steam?
(d) Net rating, water?

4.6. At locations in altitudes greater than 2000 ft, are boiler ratings *increased* or *decreased* as suggested in H. B. Smith literature?

4.7. For this series of boilers would tankless domestic hot water heaters be available as optional equipment?

4.8. On what types of gas are the shutoff sizes of the 300 series boilers based?

4.9. On which boiler type (steam or water) will one find a high limit aquastat and circulator relay combination control?

4.10. Make a dimensional drawing to scale of the FH-CM 20 recessed convector in the private toilet (Figure 4.13*a*, Figure 4.10, and Table 4.3). Show convector in place in a 4-in. nominal stud partition in front view and cross section (no inlet grill).

Additional Reading

Life Safety Code, NFPA 101

Gas Appliances, and Gas Piping NFPA 54. These and other publications of the National Fire Protection Association apply to fire safety.

Uniform Building Code, Volume II (Mechanical Code), International Association of Plumbing and Mechanical Officials. Combustion and ventilation standards are included in this code.

Radiation, Convectors AIA 30-c-4 Catalog SFC 321, Standard Fin Pipe Radiator Corp. A very good reference for heating units in industrial work.

Systems Handbook and Product Directory, AS-HRAE, 1973. A most complete manual of system design for air conditioning, heating, industrial ventilation and refrigeration.

I=B=R Ratings for Cast-Iron Boilers, 1971, The Hydronics Institute. This and other publications of the Institute are useful in the design of heating systems.

5. Heating – Cooling, Air Systems

Now we tackle the problem of cooling a building as well as heating it. Air systems are often used for this purpose. Figure 1.10, page 16, introduces us to the refrigeration cycle that is the heart of cooling systems. Before applying heating-cooling to more complex structures, we "work out" again with our Basic Plan. In doing so, we learn about ducts and blowers and the actual use of refrigeration equipment.

After studying this chapter, you will be able to:

1. Calculate the flow rate of air in cubic feet per minute for a given heat loss or heat gain in a building.
2. Locate and size supply registers and return grills.
3. Select air furnaces, cooling coils and heating coils.
4. Size ducts.
5. Make a drawing of an air distribution system for several schoolrooms.
6. Specify splitter-dampers, duct-turns and opposed-blade dampers as required.
7. Call for the correct thickness of sheet metal for rectangular ducts of various cross-sectional dimensions.
8. Specify details for the construction of round ducts.
9. Plan for control of system air-velocities in schools, public buildings, theaters and industrial buildings.
10. Make heating-cooling drawings for a residence.

5.1. Air for Thermal Transfer

Either water or air can be used to deliver heat. Air, as we shall see, can also *remove* heat from warm rooms in summer. In very large buildings, *water* may also be used for cooling. That method is a two-stage scheme. Chilled water cools an airstream. The air is then passed through the rooms to cool them. So, in both cases, *air* is the final medium that regulates the temperature of the occupied space.

Let us confine our thinking to the requirements of our Basic Plan. If 70° F is chosen as a suitable indoor temperature in winter, air in the room should not drop below that temperature. Therefore, it is drawn back through a grill at 70° F and warmed in a furnace. It is then redelivered at a higher temperature. The temperature "rise" can be between 45 F° and 75 F°. Thus air enters the room at a temperature somewhere between 115° F and 145° F. This is considered a comfortable level for the entering air. See Figure 5.1.

It is important to know the necessary "flow rate" of air. This rate is used in the design. An example may help in explaining this.

Example 5.1 A house has an hourly heat loss in winter of 60,000 Btuh. If the temperature rise in the furnace is 55 F°,* find the required flow in cubic feet per minute.

*Remember that temperature differences are given by F° while *specific* temperatures are °F.

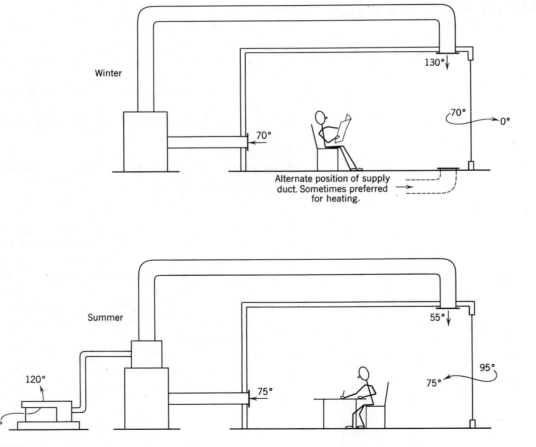

Figure 5.1 Airflow/heatflow. Air is a thermal "sponge." *Winter.* It soaks up the heat in the furnace and drops it in the room. It must balance the heat lost to outdoors by transmission. *Summer.* Air soaks up the heat that enters the room by transmission. In the cooling (evaporator) coil above the furnace, it loses the heat to the refrigerant. The condenser disposes of this heat the outoors. Temperatures are approximate.

Solution: We need to know two facts about the physical qualities of air. One is the approximate weight of air. This is considered to be about 0.075 lb/cu ft. The other item is its "specific heat." This is the amount of heat (Btu) necessary to raise one pound of air one Fahrenheit degree. It is approximately 0.24 Btu per pound of air for 1 deg. Note that *heat* flow is expressed in Btu per *hour* while *air* flow is expressed in cubic feet per *minute*. The fact that there are 60 min in an hour will enter our calculation. The heat equation develops as follows:

cubic feet per minute $\times$ 0.075 = pounds of air circulated per minute

cubic feet per minute $\times$ 0.075 $\times$ 60 = pounds of air circulated per hour

cubic feet per minute $\times$ 0.075 $\times$ 60 $\times$ 0.24 = Btuh to raise the air 1 F°

cubic feet per minute $\times$ 0.075 $\times$ 60 $\times$ 0.24 $\times$ 55 Btuh to raise the air 55 F°*

This must equal the hourly heat loss from the house:

cubic feet per minute $\times$ 0.075 $\times$ 0.24 $\times$ 60 $\times$ 55 = Btuh (loss)

The unknown in this equation is the cubic feet per minute.

Transposing, we have

$$\text{cubic feet per minute} = \frac{\text{Btuh}}{0.075 \times 0.24 \times 60 \times 55}$$

Substituting

$$\text{cubic feet per minute} = \frac{60,000}{0.075 \times 0.24 \times 60 \times 55}$$

$$\text{cubic feet per minute} = 1010 \quad \text{(say, 1000 cfm, } answer)$$

When summer cooling is planned, the indoor temperature should not rise above about 75° F. This 75° F air must be cooled before reentering the space. Fifty-five degrees is about as cold as the reentering air should be for comfort.

Example 5.2 A house is to be provided with 3 tons of cooling (36,000 Btuh). With a cooling temperature *drop* of 20 F°, how many cubic feet per minute must be circulated?

Solution:

$$\text{cubic feet per minute} = \frac{\text{Btuh}}{0.075 \times .24 \times 60 \times \Delta t}$$

*If *this* temperature rise is chosen.

Substituting

$$\text{cubic feet per minute} = \frac{36,000}{.075 \times .24 \times 60 \times 20} = 1666$$
$$\text{(say, 1700 cfm, answer)}$$

Because of the limited cooling drop (20 F°), the air that must be circulated for cooling is often greater than that required for heating.

5.2. Air Systems

Air introduced into each room must be returned to the furnace-evaporator for heating or cooling. See Figure 5.2. The best practice calls for supply registers *and* return grills in every room. When there are no doors in a room, the return grill can be located in a hall or other adjacent space. A way to avoid return ducts from rooms *with* closable doors is to place a grill in the lower part of the door or to cut 1 in. off the bottom of the door. Air comes out of the register under slight pressure and returns through the grill or the underdoor opening under slight suction. There must be free air passage between such locations. An exception usually occurs in bathrooms and kitchens. Because of moisture and odors, it is customary *not* to recirculate the air from these rooms. Instead, exhaust fans pass the air to the outdoors. An air intake on the suction side of the furnace brings in fresh air to replace the air thus exhausted. An intake of this kind is shown in Figure 5.2*b*. In very small houses, it is sometimes omitted. This is evident in Figure 5.2*a* where there is a single return directly at the furnace. Doors in such a house would require grills or undercutting to permit return of the *circulated* air. Certain parts of a house, notably near windows or fixed glass, are subject to the greatest heat loss or gain. The conditioned air (warm or cool) is best directed to these locations. When cooling is included, *return* grills are often placed at both high and low levels. This aids in returning cool air that drops to the floor in winter and in returning warm air that rises to the ceiling in summer. Long ago, when air was distributed and collected by gravity, special requirements applied to *supply registers*. They were placed low for heating because the warmed air *rose*. For cooling, they were placed high because cool air descended. Effective blowers have changed this. Air may be blown up or down across glass. The *up* direction is sometimes preferred for heating. Another idea is to have it fanned out in a

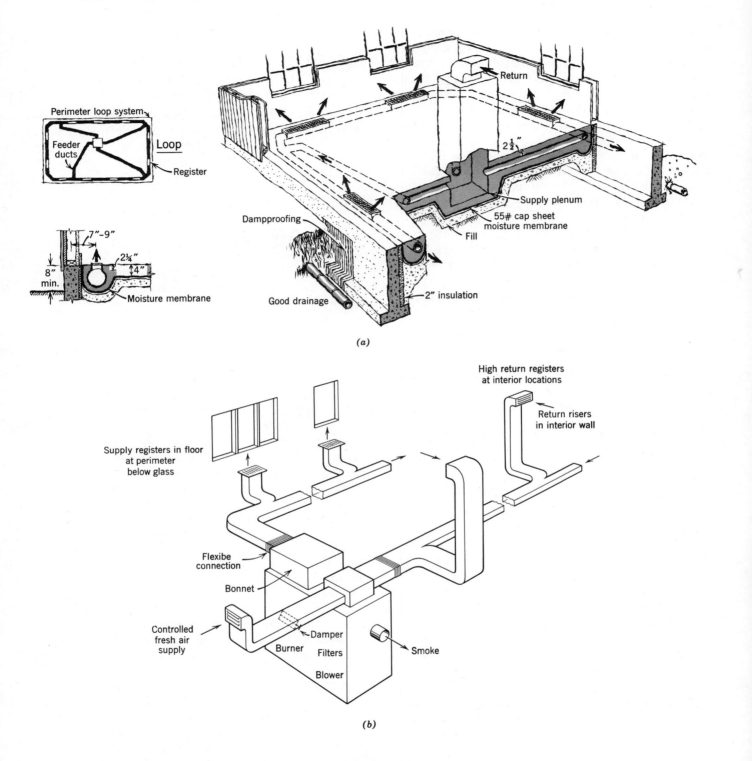

Figure 5.2 Warm air heating systems. Two popular methods of air distribution. With additional components and correct design, either method is also suitable for cooling. *(a)* Schematic view of house on concrete slab using perimeter loop system. A downflow furnace is used. *(b)* Conventional warm-air furnace and ducts. This system with supply registers in the floor under glass and high interior return-registers is suitable for heating *or* cooling. Furnace and ducts in basement of this basement-and-one-story house.

horizontal plane just below the ceiling. It is essential to avoid direct drafts on people.

Figure 5.2 shows only two of a great number of possibilities. The scheme in (a) is a good choice for houses on concrete slabs. The scheme in (b) is the more conventional method. It employs ducts for both supply and return sides of the system. The furnace shown is called by some manufacturers a "low-boy." Other types are shown and described in the next section.

Low air speeds in ducts and well-made ducts that do not rattle, help to assure a quiet system. Vibration that might be transmitted along the metal ducts is much reduced by flexible (canvas-asbestos) connections between the furnace and the ducts. The noise and motion of the blower and burner are thus reduced. Thermostats that start heating or cooling devices should be placed near return grills. They sense air that is too cool in winter or too warm in summer. Very often "continuous circulation" is planned. The blower runs 24 hr a day. This causes even distribution of the air; it thus avoids "dead spots" and stuffiness.* In such an arrangement, the thermostat does not control the blower but only the burner or the condenser-evaporator. Usually the fan switch next to the thermostat offers three choices:

(a) *Continuous circulation.* This is described above.
(b) *Automatic.* The blower *and* the heater (or cooler) come on together
(c) *Off.* This shuts down the entire system.

---

*In recent practice these *comfort* advantages are sometimes sacrificed in favor of energy savings. Thus (b), automatic, might be chosen. Then the blower would operate intermittently instead of continuously.

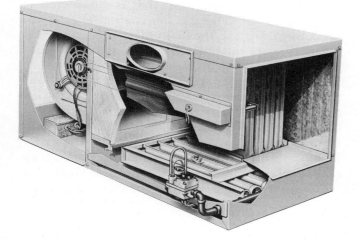

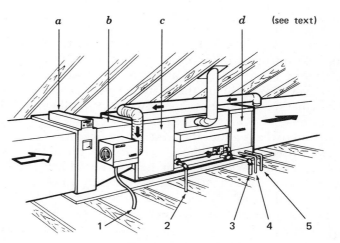

Figure 5.3 Forced-air gas furnaces, GS10 Series, horizontal stowaway. 75,000 to 140,000 Btuh input. Add-on cooling 1½ to 5 nominal tons (Courtesy of Lennox Industries, Inc.)

1. Tap water *in* to humidifier.
2. Gas *in* to furnace.
3. Drain. Condensate *out*.
4. Liquid refrigerant *in* to evaporator.
5. Expanded (gas) refrigerant *out* to condenser.

Typical Applications

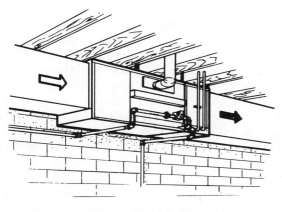

Basement installation
with cooling coil

Attic installation
with cooling coil, electronic air
cleaner and automatic humidifier

5.3. Central Equipment

Air furnaces for warm air heating together with their cooling units include a wide variety of equipment. Only a few items are shown here (Figures 5.3 to 5.8). They are selected from the catalogs of one manufacturer:

Figure Numbers	Item
HEATING	
5.3	Horizontal Stowaway, gas
5.4	Horizontal Stowaway, oil
5.5	Down-Flo, oil
5.6a	Up-Flo, electric
5.6b	Electric data
5.6c	Blower data/dimensions and specifications
COOLING	
5.7	Evaporator, Up-Flo
5.8a	Condensing unit
5.8b	Selector schedule/dimensions

Horizontal Stowaway units are most adaptable to fit into otherwise unused spaces in houses. The typical applications shown in Figures 5.3 and 5.4 include basement, crawl space and attic installations. The most completely equipped is the suggested attic installation (Figure 5.3). It includes in the order of airflow:

(a) Electronic air cleaner
(b) Automatic humidifier
(c) Gas-fired air furnace
(d) Cooling coil

With a well-designed duct and distribution system, this is most typical of a *complete* air conditioning or "climate control" job. It must be understood that, in any location or for any type (horizontal/upflow/downflow), all of these items may be included whether or not they are shown here in other illustrations.

It is good to think about the relative importance of the four units listed. Foremost, of course, is the air furnace. This is basic wherever heat is needed. If economy is important, it could be the *only* item. Perhaps second would be the choice to add the cooling coil (and outdoor condenser). Two other very desirable items are the electronic air cleaner and the automatic humidifier. The cleaner supplements the standard furnace filter for much greater efficiency. The humidifier adds to indoor comfort under winter

conditions. Note its need for air under pressure from the *supply* duct.

Five principal connections are shown in Figure 5.3. In addition, electricity is needed for power and for the controls of the furnace, humidifier and cleaner. Add to all of these a flue to carry away the products of combustion.

It is quite evident that this busy group of devices and connections cannot be *really* "stowed away" and forgotten. Adequate space for access, servicing, adjustment and parts-replacement must be planned.

Refer now to Figure 5.6a, b and c. Of the four furnaces shown in this book, this is the only one for which engineering data have been included. Even this is not as complete as the full information furnished by the manufacturer. The limitations of space do not allow us to reproduce the rest of these data, or *any* of the engineering facts about the *other* three furnaces. We have been more thorough in Figure 5.6, since it is the unit that is considered in Example 5.3, given later in this chapter.

Cooling units are shown in Figures 5.7 and 5.8a. Again, they are examples of only two selected items of a great number of available products. Please turn back to Figure 1.5, page 9. Now compare the functions of the condenser and evaporator with the *actual* equipment shown here:

Condenser   Figure 1.5(D) (page 9) to Figure 5.8a
Evaporator   Figure 1.5(H) to Figure 5.7

In selecting the right evaporator and condenser for a given cooling load, the *specifications* of Figure 5.7 and the *selector* table (Figure 5.8b) are used.

5.4. Air Ducts

Galvanized steel, aluminum and glass fiber are the materials most often used for ducts. Glass fiber ducts (Figure 5.9) are good for reducing sound. They also provide their own thermal insulation. This helps to maintain the air temperature (warm or cool) within the duct. Sheet metal ducts are *covered* with glass fiber. See Figures 5.11 and 5.12. For sound reduction, they are lined with similar material that is manufactured for this purpose. Construction details for sheet metal ducts are given in Tables 5.1 and 5.2. Air controls are shown in Figure 5.10. Detail sheets (Figure 5.10e and f) are typical of engineering drawings for large buildings. They are an important part of the contract documents. Such detail sheets are not always used for small residential projects.

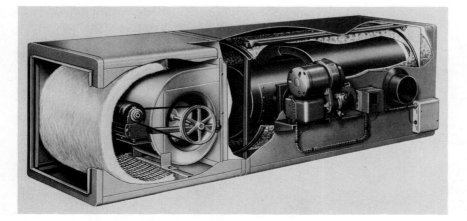

Typical Applications

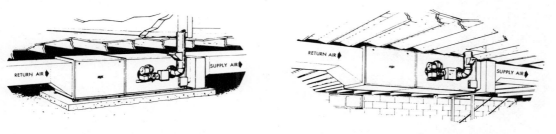

Crawl Space Horizontal Installation   Suspended Horizontal Installation

Figure 5.4 Forced Air Oil Furnaces, OS7 Series horizontal stowaway. 91,000 to 189,000 Btuh input. Add-on cooling, 2 to 5 nominal tons. (Courtesy of Lennox Industries, Inc.)

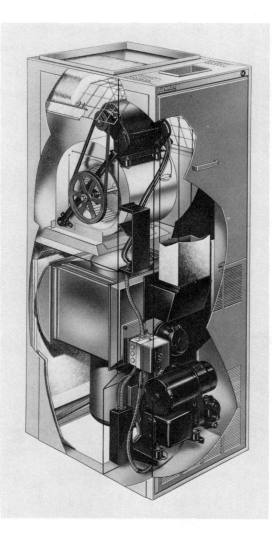

Figure 5.5 Forced-air oil furnaces, Down-Flo, 08R Series. 105,000 to 168,000 Btuh input. Add-on cooling 1½ to 5 nominal tons. See application in Figure 5.2a, page 108. (Courtesy of Lennox Industries, Inc.)

TYPICAL APPLICATIONS

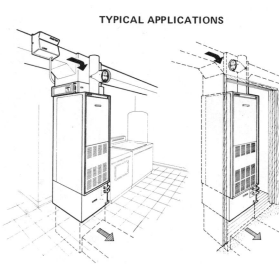

Utility Room Installation With Electronic Air Cleaner, Cooling Coil and Humidifier.

Closet Installation With Cooling Coil

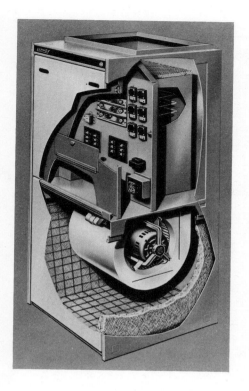

Figure 5.6 *(a)* Forced-air electric furnaces, EQ 10 Series, Up-Flo, 28,000 to 112,600 Btuh output. Add-on cooling 1½ to 5 nominal tons. *(b)* Electrical data (page 114) *(c)* Blower data and dimensional data (page 115). (Courtesy of Lennox Industries, Inc.)

Typical Applications

Basement Installation
With cooling coil, return air
cabinet and power humidifier

Basement Installation
With cooling coil, electronic air
cleaner and power humidifier

Closet Installation
With cooling coil and
electronic air cleaner.

(a)

ELECTRICAL DATA

Model No.	No. of Steps	No. of Elements 1 phase	Volts Input	KW Input	Btuh Output	Maximum Unit Amps 1 phase	*AWG Wire Size 1 phase	Time Delay Fuse Fusetron (Amps) 1 phase	Disconnect Rating Hp	Disconnect Rating Amps
E10Q2-371	2	2	208	8.3	28,300	41.6	6	60	10	----
			220	9.2	31,400	43.8	6	60	10	----
			230	10.1	34,400	45.8	4	60	10	----
			240	11.0	37,500	47.6	4	60	10	----
E10Q2-561	3	3	208	12.4	42,300	61.5	3	80	----	100
			220	13.8	47,100	64.8	2	90	----	100
			230	15.2	51,900	67.8	2	90	----	100
			240	16.5	56,300	70.5	2	90	----	100
E10Q3-561	3	3	208	12.4	42,300	63.3	3	80	----	100
			220	13.8	47,100	66.6	2	90	----	100
			230	15.2	51,900	69.6	2	90	----	100
			240	16.5	56,300	72.3	2	90	----	100
**E10Q3-751	4	4	208	16.6	56,600	43.4/39.8	6/6	60/50	10/7.5	----
			220	18.4	62,800	45.6/42.0	4/6	60/60	10/10	----
			230	20.2	68,800	47.6/44.0	4/4	60/60	10/10	----
			240	22.0	75,000	49.4/45.8	4/4	60/60	10/10	----
**E10Q5-751	4	4	208	16.6	56,600	46.7/39.8	4/6	60/50	10/7.5	----
			220	18.4	62,800	48.9/42.0	4/6	70/60	10/10	----
			230	20.2	68,800	50.9/44.0	4/4	70/60	10/10	----
			240	22.0	75,000	52.7/45.8	4/4	70/60	10/10	----
**E10Q5-941	5	5	208	20.7	70,600	59.7/46.7	3/4	80/60	----/10	100/----
			220	23.0	78,500	63.0/48.9	3/4	80/70	----/10	100/----
			230	25.3	86,300	66.0/50.9	2/4	90/70	----/10	100/----
			240	27.5	93,800	68.7/52.7	2/4	90/70	----/10	100/----
**E10Q5-1121	6	6	208	24.8	84,600	59.7/66.6	3/2	80/90	----	100/100
			220	27.6	94,200	63.0/69.9	3/2	80/90	----	100/100
			230	30.4	103,800	66.0/72.9	2/2	90/100	----	100/100
			240	33.0	112,600	68.7/75.6	2/2	90/100	----	100/100

*Up to 100' of run. Local codes take precedence.
**Units use two power supplies to keep within 100 amp disconnects. See field wiring diagrams.
NOTE—All units are equipped with internal fusing according to NEC.
NOTE—E10Q5-1123 three phase unit is available on special order only.

FIELD WIRING

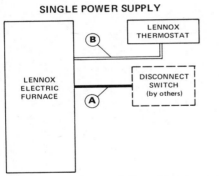

SINGLE POWER SUPPLY

DUAL POWER SUPPLY

A—Two wire power supply (not furnished). See Electrical Data for size.
B—Two wire low voltage (not furnished)—18 ga. minimum.
NOTE—All wiring must conform to NEC and local electrical codes.

BLOWER DATA

E10Q2-371 & E10Q2-561 BLOWER PERFORMANCE

External Static Pressure (in. wg)	Air Volume (cfm) @ Various Speeds High	Medium	Low
0	1265	945	660
.05	1245	940	665
.10	1220	935	675
.15	1195	930	680
.20	1170	925	680
.25	1145	920	680
.30	1115	910	675
.40	1055	880	670
.50	980	830	645
.60	890	760	605
.70	790	685	----

NOTE—All cfm data is measured external to the unit with the air filter in place.

(b)

BLOWER DATA

E10Q3-561 & E10Q3-751 BLOWER PERFORMANCE

External Static Pressure (in. wg)	Air Volume (cfm) @ Various Speeds			
	High	Med-High	Med-Low	Low
0	1490	1275	1040	880
.05	1470	1270	1040	880
.10	1455	1260	1040	880
.15	1435	1255	1040	880
.20	1415	1245	1035	875
.25	1395	1235	1030	875
.30	1375	1220	1025	870
.40	1335	1190	1005	855
.50	1285	1155	980	830
.60	1230	1105	945	800
.70	1175	1050	900	----
.80	1100	985	850	----
.90	1005	900	790	----
1.00	900	805	----	----

NOTE—All cfm data is measured external to the unit with the air filter in place.

E10Q5-751, E10Q5-941 & E10Q5-1121 BLOWER PERFORMANCE

External Static Pressure (in. wg)	Air Volume (cfm) @ Various Speeds				
	High	Med-High	Medium	Med-Low	Low
0	2680	2480	2080	1920	1600
.05	2640	2440	2050	1880	1580
.10	2610	2410	2020	1840	1560
.15	2570	2375	1990	1810	1530
.20	2540	2340	1960	1780	1500
.25	2500	2310	1935	1750	1475
.30	2465	2275	1905	1720	1450
.40	2390	2205	1850	1660	1290
.50	2310	2130	1790	1600	1330
.60	2220	2050	1720	1540	1280
.70	2140	1970	1640	1480	1220
.80	2040	1880	1570	1410	----
.90	1935	1790	1490	1340	----
1.00	1810	1680	1410	1270	----

NOTE—All cfm data is measured external to the unit with the air filter in place.

DIMENSIONS (in.)

Electric Furnace

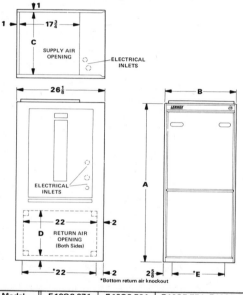

Return Air Cabinet

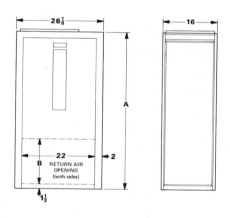

NOTE—Return air cabinet shipped knocked down and must be field assembled.

Model No.	E10Q2-371 E10Q2-561	E10Q3-561 E10Q3-751	E10Q5-751, E10Q5-941 E10Q5-1121
A	49	49	53
B	16-1/4	21-1/4	26-1/4
C	14	19	24
D	14	14	18
E	11	16	21

Model No.	RA10-16-49	RA10-16-53
A	49	53
B	14	18

SPECIFICATIONS

Model No.	E10Q2-371 E10Q2-561	E10Q3-561 E10Q3-751	E10Q5-751 E10Q5-941 E10Q5-1121
Blower wheel nominal diam. x width (in.)	9 x 7	10 x 7	12 x 12
Blower motor hp	1/4	1/3	3/4
Free filter area, sq. ft.	6.1	6.9	9.4
Filter cut size, (in.)	38 x 28 x 1	42 x 28 x 1	56 x 28 x 1
Tons of cooling that can be added	1-1/2, 2 or 2-1/2	2, 2-1/2 or 3	3-1/2, 4 or 5
Net weight (lbs.)	120	150	190
Number of packages in shipment	1	1	1
Return Air Cabinet — Model No.	RA10-16-49	RA10-16-49	RA10-16-53
Return Air Cabinet — Net weight (lbs.)	50	50	75

(c)

Figure 5.7 Cooling units. C4 Series, evaporators—Up-Flo 26,000 to 59,000 Btuh cooling capacity. (Courtesy of Lennox Industries, Inc.)

SPECIFICATIONS

Evaporator Coil Assembly		C4-41FF	C4-51FF	C4-65FF
*Nominal tons cooling capacity		3	4	5
Evaporator Coil	Net face area (sq. ft.)	3.44	4.01	4.58
	Tube diam. (in.)	3/8	3/8	3/8
	No. of rows	3	3	3
	Fins per inch	13	13	13
	Suction line connection (in. flare)	3/4	3/4	3/4
	Liquid line connection (in. flare)	1/2	1/2	1/2
	Condensate drain connection mpt (in.)	3/4	3/4	3/4
Refrigerant		R-22	R-22	R-22
Coil net weight (lbs.) 1-Pkg.		35	40	47
Evaporator Coil Cabinet		C4-41-00	C4-51-00	C4-65-00
Cabinet net weight (lbs.) 1-Pkg.		9	11	12
Expansion Valve Kit		BM-5985	BM-5986	BM-5987

*For actual Btuh capacity with different condensing units, refer to condensing unit engineering data sheets.

Typical Applications

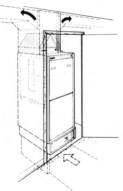

Closet Installation
With electric furnace and
electronic air cleaner

Basement Installation
With gas furnace, return air
cabinet and power humidifier

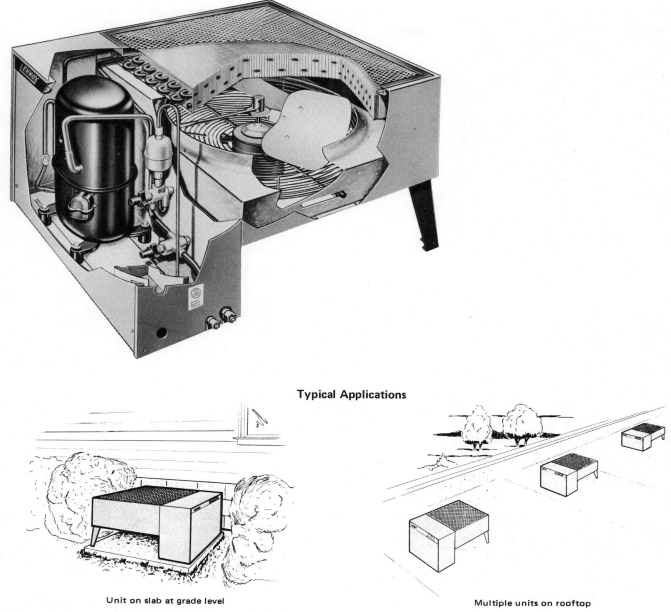

Typical Applications

Unit on slab at grade level

Multiple units on rooftop

NOTE—Specifications, ratings and dimensions subject to change without notice.

(a)

Figure 5.8 *(a)* Cooling units, HS8 Series condensing units (2 to 2½, 3 to 3½, 4 and 5 ton) 25,000 to 59,000 Btuh cooling capacity. Note the three essential elements—compressor, condenser and fan. *(b)* Selector schedule and dimensions (page 118). (Courtesy of Lennox Industries, Inc.)

SELECTOR

Lennox Condensing Unit Model No.	ARI Standard 210 Certified Ratings			Lennox Evaporator Unit		
	*Btuh Cooling Capacity	Total Unit Watts	Dehumidifying Capacity	Up-flo	Down-flo	Horizontal
HS8-261FF	25,000	3900	25%	**CB3-41VFF **CBH3-41VFF	**CB3-41VFF **CBH3-41VFF	**CB3-41VFF **CBH3-41VFF
	26,000	3500	24%	C4-41FF	CR4-41FF	C3-41VFF, CH3-41VFF CH1-41VFF
HS8-311FF	29,000	4300	29%	**CB3-41VFF **CBH3-41VFF	**CB3-41VFF **CBH3-41VFF	**CB3-41VFF **CBH3-41VFF
	30,000	3800	29%	C4-41FF	CR4-41FF	C3-41VFF, CH3-41VFF CH1-41VFF
HS8-411FF HS8-413VFF	36,000	4700	30%	-----	CR4-41FF	----
	†36,000	5400	30%	**CB3-41VFF **CBH3-41VFF	**CB3-41VFF **CBH3-41VFF	**CB3-41VFF **CBH3-41VFF
	37,000	4900	30%	C4-41FF	----	C3-41VFF, CH3-41VFF CH1-41VFF
	†37,000	5400	26%	**CB3-51V **CBH3-51V	**CB3-51V **CBH3-51V	**CB3-51V **CBH3-51V
	38,000	5000	26%	----	CR4-51FF	C3-51V, CH3-51V
	39,000	5000	25%	C4-51FF	----	----
†HS8-461FF †HS8-463VFF	38,000	5200	32%	----	CR4-41FF	----
	39,000	5800	32%	**CB3-41VFF **CBH3-41VFF	**CB3-41VFF **CBH3-41VFF	**CB3-41VFF **CBH3-41VFF
	40,000	5200	32%	C4-41FF	----	C3-41VFF, CH3-41VFF CH1-41VFF
	41,000	5800	27%	**CB3-51V **CBH3-51V	**CB3-51V **CBH3-51V	**CB3-51V **CBH3-51V
	42,000	5300	27%	C4-51FF	CR4-51FF	C3-51V, CH3-51V
†HS8-511FF †HS8-513V	46,000	6000	28%	C4-51FF	----	----
	47,000	6100	28%	C4-51FF	----	----
†HS8-511FF †HS8-513V	47,000	7100	28%	**CB3-51V **CBH3-51V	**CB3-51V **CBH3-51V	**CB3-51V **CBH3-51V
	49,000	6400	28%	----	CR4-51FF	C3-51V, CH3-51V
	51,000	6600	26%	C4-65FF	----	CH3-65V, C3-65V
		7100	26%	**CB3-65V **CBH3-65V	**CB3-65V **CBH3-65V	**CB3-65V **CBH3-65V
	52,000	6600	25%	----	CR4-65FF	LSH2-500V
HS8-651FF HS8-653V	55,000	8600	33%	**CB3-51V **CBH3-51V	**CB3-51V **CBH3-51V	**CB3-51V **CBH3-51V
	56,000	7800	33%	----	----	C3-51V, CH3-51V
		7700	33%	C4-65FF	----	----
	58,000	8900	27%	**CB3-65V **CBH3-65V	ſ**CB3-65V **CBH3-65V	**CB3-65V **CBH3-65V
	59,000	8100	26%	----	CR4-65FF	C3-65V, CH3-65V
		8000	27%	----	----	LSH2-500V

*Rated in accordance with ARI Standard 210; 450 cfm evaporator air volume per ton of cooling, 95F outdoor air temperature, 80db/67wb entering evaporator air and 25' of connecting refrigerant lines.
**Denotes blower powered evaporator. Wattage for blower motor is included in total unit watts listed.
†Derate 1,000 Btuh for 208 volt operation.

(b)

DIMENSIONS (in.)

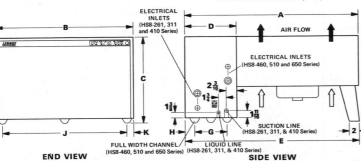

Model No.	A	B	C	D	E	F	G	H	J	K
HS8-261FF	35-11/16	28	18	10-5/8	35-11/16	28	6	2-1/16	24-1/2	1-3/4
HS8-311FF HS8-411FF, HS8-413VFF	39-11/16	28	18	10-5/8	39-11/16	28	6	2-1/16	24-1/2	1-3/4
HS8-461FF, HS8-463VFF HS8-511FF, HS8-513V HS8-651FF, HS8-653V	49-1/8	33	25-1/2	15-1/8	49-1/8	33	8-1/2	1-3/4	27	3

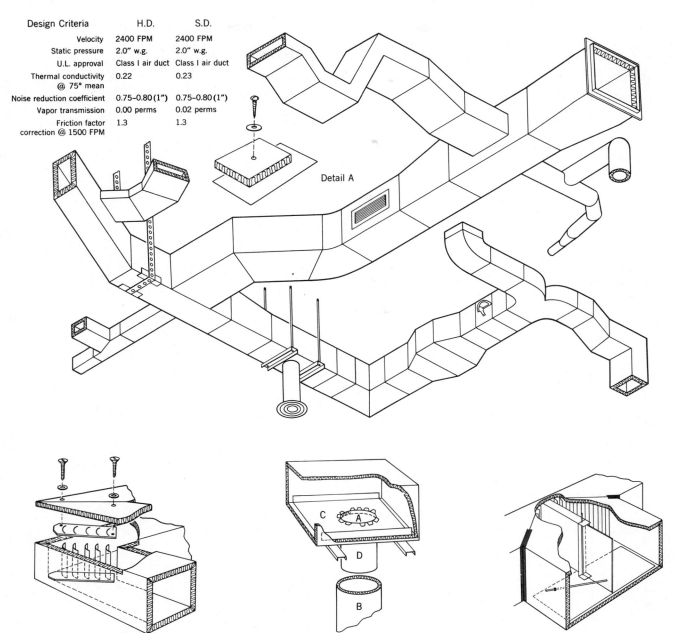

Design Criteria	H.D.	S.D.
Velocity	2400 FPM	2400 FPM
Static pressure	2.0" w.g.	2.0" w.g.
U.L. approval	Class I air duct	Class I air duct
Thermal conductivity @ 75° mean	0.22	0.23
Noise reduction coefficient	0.75–0.80 (1")	0.75–0.80 (1")
Vapor transmission	0.00 perms	0.02 perms
Friction factor correction @ 1500 FPM	1.3	1.3

Detail A

Figure 5.9 Fiberglass ducts. Design criteria and installation techniques are typified by this page selected from the many standards for fiberglass ducts. (1) *A 90-deg duct with turning vane.* A 4-ft section of duct is mitered at a 45-deg angle and one of the pieces is rotated 160 deg. A standard sheet metal turning vane is inserted between the two sections. The sheet metal turning vane is attached to the duct as per detail "A," the two pieces of the duct are now placed together using detail "A" to adhere the other one half of the duct to the sheet metal turning vane. Mastic and reinforcing fabric are applied along the mitered cut and over screws and washers as per standard instructions. (2) *Bottom diffuser connector.* A round hole is cut into the bottom of the duct. The diameter of this hole (A) is equivalent to the inside diameter of the round duct. (B) A sheet metal "shoe" the width of the duct interior (C) with round opening the same size and having 1-in. high sides is pre-

placed in the bottom of the duct. A sheet metal duct (D) with 1-in. metal dovetails, or tabs, is inserted through the duct and sheet metal "shoe." Slide round fiberglass duct (B) over the sheet metal duct. The diffuser is attached directly to the sheet metal duct with sheet metal screws. The fiberglass duct must be supported by hangers under the sheet metal shoe or by wires attached to the diffuser. (3) *Splitter-damper tee section.* The "tee" section is constructed with turning vanes. A standard splitter damper is inserted with the rod being attached to the same sheet metal bearing surfaces. The sheet metal bearing surfaces are attached to the duct as per detail "A." The adjusting rod passes through the duct and the handle of the splitter rod is attached to the wall of the duct by using a sheet metal bearing surface in back of this handle. All sheet metal is secured to the duct per detail "A." (Courtesy of Owens-Corning Fiberglass Corp,)

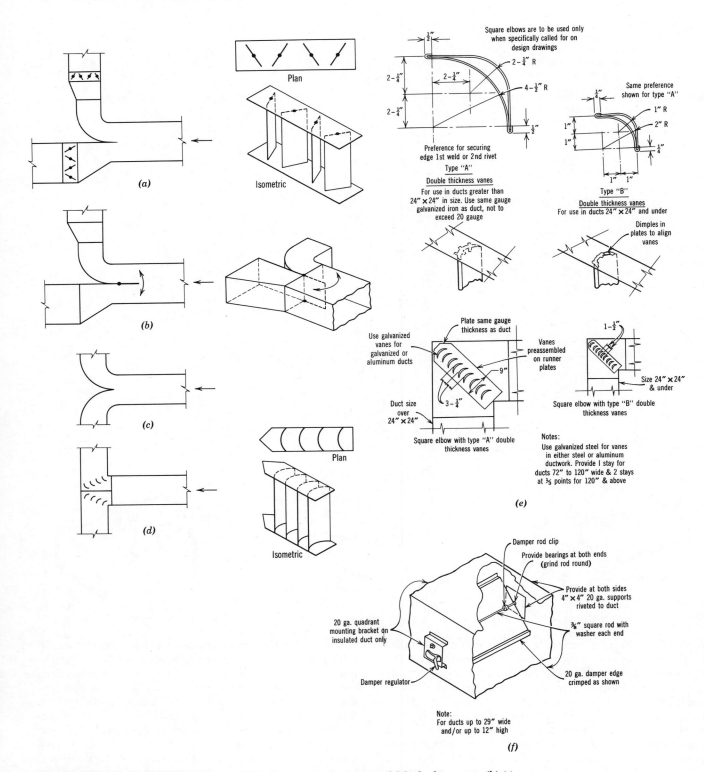

Figure 5.10 Air controls in ducts. *(a)* Air adjustment by opposed-blade dampers. *(b)* Air adjustment by splitter damper. *(c)* Conventional turns in ducts. *(d)* Right-angle turns with turning vanes (more compact). *(e)* and *(f)* Examples of developed details used by engineers in contract work in large buildings. ((e) and (f) Courtesy Seelye, Stevenson Value and Knecht, Consulting Engineers)

Controls in duct *systems* are found in Figures 5.9, 5.14 and 5.19*b*.

There are usually several different kinds of layouts for duct systems. Figure 5.19*b* is the type of layout that an engineer makes. Figure 5.14 is a *contractor's* shop drawing for installation of a warm air heating system in two classrooms of a school. See Figures 5.13 and 5.14. It shows exactly the size of each duct section as required to fit in the building. From this drawing, special drawings are often made to help the shop mechanic fabricate each part. In plans, the duct *width* is given first and the *depth* last.

5.5. Registers and Grills

Handling large quantities of air into and out of rooms usually requires careful planning. Direct drafts on people must be avoided. Registers are shown in Figures 5.16 and 5.17. The 40/41/41F series of floor registers spread the airstream into a flat fan shape. By diverting the air left and right in a flat pattern against the glass, drafts are avoided. The "sidewall" registers in Figure 5.17 have a similar action. The series 10V type has, in front, vertical vanes set at 30° to spread the air stream left and right. Behind these are adjustable horizontal blades. Other sidewall types have horizontal front vanes and vertical (movable) rear vanes. In each classroom of Figure 5.14, one sees two sidewall registers and two downflow registers of similar type. See Figure 5.15. Flow either *down* or *up* along the surface of the glass is satisfactory. Dampers (sometimes called "valves"), for volume regulation are usually located behind all registers.

Grills are, in general, for the return of air from the space. Figure 5.18 gives a choice of horizontal (100H) or vertical (100V) vanes. They are sometimes curved to reduce the sound of the air. Mainly, their purpose is to screen the opening of the duct and to prevent direct vision into it. In Figure 5.19, floor and sidewall registers are seen as well as 100H return registers.

Wrap and staple

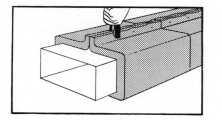

Wrap and hold with wire

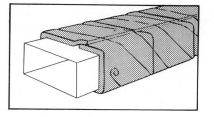

Wrap and adhere

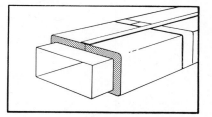

Figure 5.11 Duct insulation. Three of a number of effective ways of applying thermal insulation to the outside of sheet metal ducts. Duct *liner*, an acoustical product, may be used on the *inside* of ducts. It reduces fan noises and doubles as thermal insulation. Follow manufacturers instructions regarding all application procedures. (Courtesy of Gustin-Bacon Manufacturing Company)

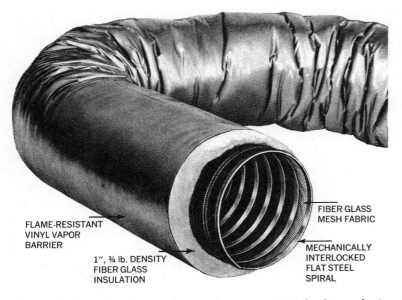

FLAME-RESISTANT
VINYL VAPOR
BARRIER

1″, ¾ lb. DENSITY
FIBER GLASS
INSULATION

FIBER GLASS
MESH FABRIC

MECHANICALLY
INTERLOCKED
FLAT STEEL
SPIRAL

Figure 5.12 Flexible ducts. This product, type 59K for low velocity systems, has an inner duct made of a fiberglass mesh fabric mechanically interlocked with its supporting steel spiral. With this type of construction:

Ducts can be handled, cut, spliced, formed into oval ends and terminated without raveling or separating.

Tighter terminations are possible. Drill screws, clamps and duct tape can be used—no special fittings required.

Type 59K offers in-line sound attenuation, dynamically tested.

Longer service life—mesh liner protects the insulation from air stream erosion. (Courtesy of the Wiremold Company)

Figure 5.13 South elevation of the Vincent Smith School in construction. Budd Mogensen, Architect and Planner. Four zones of warm air heating. Each zone serves two classrooms and part of the hall and common room. Central equipment and ducts are entirely overhead. See Figures 5.14 and 4.15a and b.

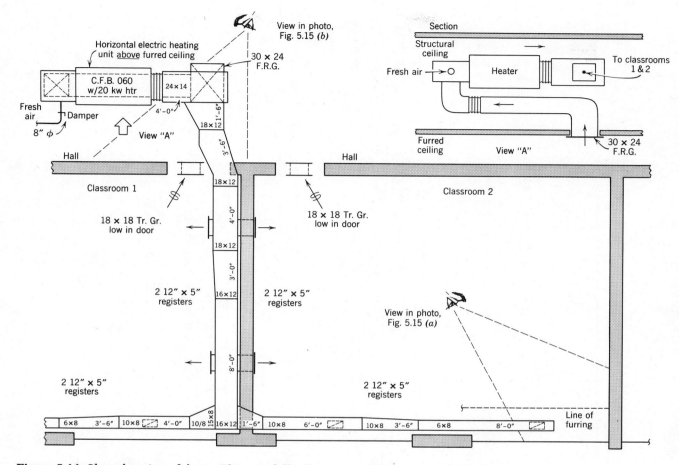

Figure 5.14 Shop drawing of ducts. This carefully dimensioned drawing made by the contractor is for the ductwork supplying two classrooms in the Vincent Smith School (Figure 5.13). (Courtesy of R. Ferina, Inc., Contractor)

Table 5.1 Recommended Construction for Rectangular Low Pressure Ducts

Dimension of Longest Side, Inches	Sheet Metal Gage (All Four Sides)[a]			Transverse Reinforcing[c]						
				Between Joints[d]				At Joints[e]		
					Flat S Slip / Drive Slip	Standing S Slip		Standing Seam Joint	Pocket Lock	Alternate Standing S Slip
	Steel Gage	Aluminum Alloy[b] Thickness, In.	Copper Oz Per Sq Ft	Minimum Reinforcing Angle Size and Maximum Longitudinal Spacing	Min. Gage	Min. Gage	Min. Angle Size	Min. Gage for Pocket Lock or Standing S Slip	Min. Angle Size	Min. Height, Inches
Up thru 12	26	0.020	16	None Required	26	24	None Required	24	None Required	1
13–18	24	0.025	24	None Required	24	24	None Required	24	None Required	1
19–30	24	0.025	24	$1 \times 1 \times \frac{1}{8}$ @60 in.	—	24	None Required	24	None Required	1
31–42	22	0.032	32	$1 \times 1 \times \frac{1}{8}$ @60 in.	—	22	None Required	22	None Required	1
43–48	22	0.032	32	$1\frac{1}{2} \times 1\frac{1}{2} \times \frac{1}{8}$ @60 in.[f]	—	22	$1\frac{1}{2} \times 1\frac{1}{2} \times \frac{1}{8}$	22	None Required	$1\frac{1}{2}$
49–54	22	0.032	32	$1\frac{1}{2} \times 1\frac{1}{2} \times \frac{1}{8}$ @48 in.	—	22	$1\frac{1}{2} \times 1\frac{1}{2} \times \frac{1}{8}$	22	None Required	$1\frac{1}{2}$
55–60	20	0.040	36	$1\frac{1}{2} \times 1\frac{1}{2} \times \frac{1}{8}$ @48 in.	—	22	$1\frac{1}{2} \times 1\frac{1}{2} \times \frac{1}{8}$	22	None Required	$1\frac{1}{2}$
61–84	20	0.040	36	$1\frac{1}{2} \times 1\frac{1}{2} \times \frac{1}{8}$ @24 in.	—	22	$1\frac{1}{2} \times 1\frac{1}{2} \times \frac{1}{8}$	22	$1\frac{1}{2} \times 1\frac{1}{2} \times \frac{1}{8}$	$1\frac{1}{2}$
85–96	18	0.050	48	$1\frac{1}{2} \times 1\frac{1}{2} \times \frac{3}{16}$ @24 in.	—	22	$1\frac{1}{2} \times 1\frac{1}{2} \times \frac{3}{16}$	22	$1\frac{1}{2} \times 1\frac{1}{2} \times \frac{3}{16}$	$1\frac{1}{2}$
97–120	18	0.050	48	$2 \times 2 \times \frac{1}{4}$ @24 in.	—	22	$2 \times 2 \times \frac{1}{4}$	22	$2 \times 2 \times \frac{1}{4}$	$1\frac{1}{2}$
121 and Over	18	0.050	48	$2 \times 2 \times \frac{1}{4}$ with tie rods @120 in. along angle	—	22	—	22	$2 \times 2 \times \frac{1}{4}$ with tie rods @120 in. along joint	$1\frac{1}{2}$

[a] Flat areas of duct over 18 in. wide shall be stiffened by crossbreaking unless duct will have non-conductive covering or sound absorbing lining.

[b] Suitable aluminum alloys are: Commercial Designation 3003 Temper H14 and Duct Sheet.

[c] Transverse reinforcing size is determined by dimension of side to which angle is applied. Angle sizes are based on mild steel. Reinforcing made in other shapes or of other materials must be of equivalent strength and rigidity.

[d] There is no restriction on the length of duct sections between joints. Ducts are normally made in sections of 4, 8, 10 or 12 ft in length. The longitudinal spacing of the transverse reinforcing between joints may necessarily be less than the spacing recommended in the table in order to conform to the selected length module.

[e] Other joint types of equivalent strength, rigidity and air tightness may be used.

[f] For aluminum or copper ducts 43 in. through 48 in. maximum-dimensions, the maximum longitudinal spacing of transverse reinforcing is 48 in.

Reprinted by permission from *Systems and Equipment*, of the American Society of Heating, Refrigerating and Air Conditioning Engineers.

Table 5.2 Recommended Construction for Round Ducts

Duct Diameter, Inches	Steel—Galv. Sheet Gage				Girth Reinforcing	Girth Joints[a] (Continuously Welded or as Below)	
	Low Pressure Ducts and Fittings	Medium and High-Pressure Ducts			Minimum Reinforcing Angle Size and Maximum Longitudinal Spacing	Low Pressure Ducts	Medium and High Pressure Ducts
		Spiral Lock Seam Duct	Longitudinal Seam Duct	Welded Fittings*			
Up thru 8	26	26	24	22	None required	Crimped and beaded joint	2 in. long slip joint
9–13	26	24	22	20	None required	Crimped and beaded joint	4 in. long slip joint
14–22	24	24	22	20	None required	Crimped and beaded joint	4 in. long slip joint
23–36	→	22	20	20	None required	—	4 in. long slip joint
37–50	→	20	20	18	$1\frac{1}{4} \times 1\frac{1}{4} \times \frac{1}{8}$@72 in.	—	$1\frac{1}{4} \times 1\frac{1}{4} \times \frac{1}{8}$ angle flanged joint
51–60	→	—	18	18	$1\frac{1}{4} \times 1\frac{1}{4} \times \frac{1}{8}$@72 in.	—	$1\frac{1}{4} \times 1\frac{1}{4} \times \frac{1}{8}$ angle flanged joint
61–84	→	—	16	16	$1\frac{1}{2} \times 1\frac{1}{2} \times \frac{1}{8}$@48 in.	—	$1\frac{1}{2} \times 1\frac{1}{2} \times \frac{1}{8}$ angle flanged joint

[a]Flanged joints may be considered as girth reinforcing.   → = use next recommended construction.

Reprinted by permission from *Systems and Equipment,* of the American Society of Heating, Refrigerating and Air Conditioning Engineers.

Figure 5.15 Construction photographs. Vincent Smith School. See Figures 5.13 and 5.14. *(a)* This picture shows the furring that surrounds the 6 × 8 in. duct in classroom 2. Shown also is the opening that will receive one of the four 12 × 5 in. supply registers that deliver warm air to the room. The air is drawn back through the 18-in. square grill in the door. *(b)* Ceiling opening that will receive the 30 by 24 in. flush return grill (F.R.G.). Suction from the unit blower draws the air back for reheating. Classroom 1 is seen through the door opening.

series 40

series 41

series 41 F

series 40, 41

NOTE: 6 x 10, 6 x 12, 6 x 14 available in Series 40 only

SIZE	FREE AREA SQ. IN.		3045 / 855 / 40	4565 / 1280 / 60	6090 / 1710 / 80	7610 / 2135 / 100	9515 / 2670 / 125	11415 / 3200 / 150	13320 / 3735 / 175	15220 / 4270 / 200	17125 / 4805 / 225	19025 / 5340 / 250	20930 / 5870 / 275	22830 / 6405 / 300	24735 / 6940 / 325	26635 / 7470 / 350	30440 / 8540 / 400	34245 / 9605 / 450	38050 / 10675 / 500	45660 / 12810 / 600
		Heating BTU/h	3045	4565	6090	7610	9515	11415	13320	15220	17125	19025	20930	22830	24735	26635	30440	34245	38050	45660
		Cooling BTU/h	855	1280	1710	2135	2670	3200	3735	4270	4805	5340	5870	6405	6940	7470	8540	9605	10675	12810
		C.F.M.	40	60	80	100	125	150	175	200	225	250	275	300	325	350	400	450	500	600
2¼" x 10"	19	T.P. Loss	.013	.022	.035	.045	.060	.092	.120	.150										
		Vert. Throw (ft.)	3.0	4.5	5.5	6.5	8.5	10.5	13.0	15.5										
		Vert. Spread (ft.)	6	8	10	12	15	18	23	26										
		Face Velocity	309	463	617	771	964	1159	1352	1545										
2¼" x 12"	21	T.P. Loss	.009	.015	.027	.037	.050	.080	.105	.134										
		Vert. Throw (ft.)	3	4	5	6	8	10	12	14										
		Vert. Spread (ft.)	6	8	10	11	14	17	22	25										
		Face Velocity	280	420	565	705	880	1050	1230	1400										
2¼" x 14"	24	T.P. Loss	.006	.010	.021	.031	.042	.070	.093	.121	.150									
		Vert. Throw (ft.)	3	4	4.5	5.5	8	9.5	11	12.5	14									
		Vert. Spread (ft.)	6	8	9	11	14	16	19	22	25									
		Face Velocity	245	365	490	610	760	915	1065	1220	1370									
4" x 10"	32	T.P. Loss		.008	.021	.026	.032	.045	.062	.084	.110	.134	.163							
		Vert. Throw (ft.)		3	4	5	7	8.5	10	11	12	13	14							
		Vert. Spread (ft.)		6	8	9	12	14	17	19	22	24	26							
		Face Velocity		265	355	445	555	665	775	890	1000	1120	1220							
4" x 12"	39	T.P. Loss		.004	.010	.016	.023	.033	.042	.058	.075	.089	.107	.128	.159					
		Vert. Throw (ft.)		2	3	4	6.5	8	9	10	11	12	13	14	15					
		Vert. Spread (ft.)		4	6	8	12	14	16	18	20	23	24	26	28					
		Face Velocity		220	295	370	460	555	645	735	830	925	1020	1110	1200					
4" x 14"	46	T.P. Loss			.006	.010	.016	.021	.028	.039	.051	.060	.080	.101	.124	.137	.167			
		Vert. Throw (ft.)			3	4	6	7	8	9	10	11	12	13	14	15	16			
		Vert. Spread (ft.)			6	8	11	12	14	16	18	20	22	24	26	28	30			
		Face Velocity			255	320	395	475	555	635	715	790	870	950	1025	1110	1270			
6" x 10" Series 40 Only	52	T.P. Loss				.009	.014	.019	.027	.035	.044	.054	.064	.078	.090	.104	.135	.171	.210	.314
		Vert. Throw (ft.)				4	5.5	6.5	7.5	8.5	10	11	12	13	14	15	17.5	19	21.5	26
		Vert. Spread (ft.)				8	9.5	11.5	13.5	15.5	17	19	21	23	25	27	31	34.5	38.5	46
		Face Velocity				278	348	417	487	556	626	695	765	834	904	973	1112	1251	1390	1668
6" x 12" Series 40 Only	59	T.P. Loss				.007	.011	.015	.021	.027	.035	.042	.050	.060	.071	.083	.107	.134	.165	.242
		Vert. Throw (ft.)				4	5	6	7	8	9	10.5	11.5	12.5	13.5	14	16.5	18.5	20.5	29
		Vert. Spread (ft.)				7	9	11	12.5	14.5	16	18	20	21.5	23.5	25	29	32.5	36	43
		Face Velocity				245	307	368	429	491	552	613	675	736	797	859	981	1103	1227	1472
6" x 14" Series 40 Only	66	T.P. Loss				.005	.009	.012	.017	.022	.028	.034	.041	.050	.058	.067	.088	.110	.132	.194
		Vert. Throw (ft.)				4	5	6	7	8	8.5	9.5	10.5	11.5	12	13	15	17	19	23
		Vert. Spread (ft.)				7	8.5	10	12	13.5	15	17	18.5	20.5	22	24	27	30.5	34	41
		Face Velocity				219	274	329	384	439	494	549	604	659	713	768	878	988	1098	1317

Figure 5.16 Floor registers. (Courtesy of Lima Register Company)

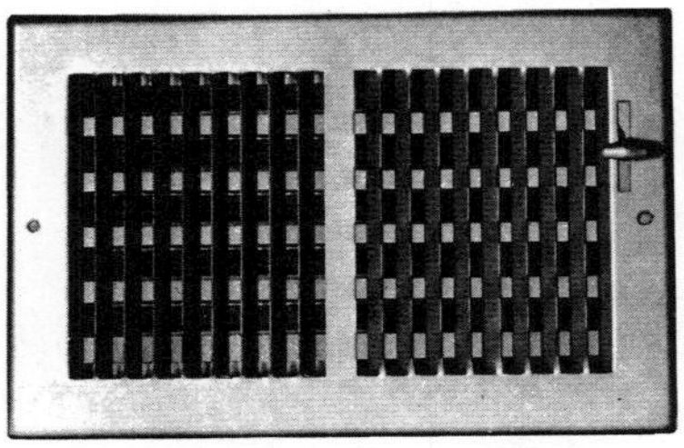

series 10V

series 10V

SIZE	FREE AREA SQ. IN.		3805	5710	7610	9515	11415	15220	19025	22830	26635	30440	34245	38050	45660	53270	60880	68490	76100	91320	106540
		Heating BTU/h	3805	5710	7610	9515	11415	15220	19025	22830	26635	30440	34245	38050	45660	53270	60880	68490	76100	91320	106540
		Cooling BTU/h	1070	1600	2135	2670	3200	4270	5340	6405	7470	8540	9605	10675	12810	14945	17080	19215	21350	25620	29890
		C.F.M.	50	75	100	125	150	200	250	300	350	400	450	500	600	700	800	900	1000	1200	1400
6″ x 4″	13	T.P. (In. W. G.)	.027	.059	.108	.168	.240														
		Throw (ft.)	6	9	12	15	18														
		Vel. (fpm)	550	830	1110	1390	1670														
8″ x 4″	18	T.P. (In. W. G.)	.014	.031	.054	.088	.120	.215													
		Throw (ft.)	5	8	10	13	15	20													
		Vel. (fpm)	393	590	787	983	1180	1574													
8″ x 5″ 10″ x 4″	24 24	T.P. (In. W. G.)		.018	.033	.052	.075	.130	.195												
		Throw (ft.)		7	9	11	13	18	22												
		Vel. (fpm)		450	600	750	900	1200	1500												
8″ x 6″ 12″ x 4″ 10″ x 5″	29 30 31	T.P. (In. W. G.)		.012	.021	.034	.049	.085	.125	.158											
		Throw (ft.)		6	8	10	12	16	20	24											
		Vel. (fpm)		360	480	600	720	960	1200	1440											
14″ x 4″	35	T.P. (In. W. G.)		.015	.025	.035	.062	.094	.130	.180											
		Throw (ft.)		7	9	11	15	19	22	26											
		Vel. (fpm)		411	514	617	823	1028	1234	1440											
12″ x 5″ 10″ x 6″	38 38	T.P. (In. W. G.)		.013	.021	.030	.052	.080	.110	.150	.195										
		Throw (ft.)		7	9	11	14	18	21	25	28										
		Vel. (fpm)		379	474	568	758	947	1137	1326	1516										
16″ x 4″ 8″ x 8″	40 40	T.P. (In. W. G.)		.012	.019	.027	.048	.073	.104	.140	.180										
		Throw (ft.)		7	9	10	14	17	21	24	28										
		Vel. (fpm)		360	450	540	720	900	1080	1260	1440										
14″ x 5″ 18″ x 4″ 12″ x 6″	45 45 46	T.P. (In. W. G.)			.015	.021	.038	.058	.082	.110	.142	.180									
		Throw (ft.)			8	10	13	16	20	23	26	29									
		Vel. (fpm)			400	480	640	800	960	1120	1280	1440									
20″ x 4″ 16″ x 5″ 10″ x 8″	51 52 52	T.P. (In. W. G.)				.017	.030	.045	.066	.088	.112	.142	.170								
		Throw (ft.)				9	12	15	18	21	25	28	31								
		Vel. (fpm)				423	565	706	847	988	1129	1270	1412								
14″ x 6″ 22″ x 4″	55 56	T.P. (In. W. G.)				.015	.026	.039	.056	.076	.100	.122	.150	.230							
		Throw (ft.)				9	12	15	18	21	24	27	29	35							
		Vel. (fpm)				393	524	655	786	916	1047	1178	1309	1571							
18″ x 5″	59	T.P. (In. W. G.)				.013	.023	.034	.049	.065	.086	.107	.129	.186	.255						
		Throw (ft.)				9	11	14	17	20	23	26	28	34	40						
		Vel. (fpm)				366	488	610	732	854	976	1098	1220	1464	1709						
24″ x 4″ 16″ x 6″ 12″ x 8″	61 64 64	T.P. (In. W. G.)					.020	.031	.044	.059	.078	.098	.116	.170							
		Throw (ft.)					11	14	17	19	22	25	28	33							
		Vel. (fpm)					464	581	697	813	929	1045	1161	1393							
20″ x 5″	66	T.P. (In. W. G.)					.019	.028	.040	.054	.071	.090	.108	.155	.206						
		Throw (ft.)					11	14	16	19	22	24	27	33	38						
		Vel. (fpm)					443	554	665	775	886	997	1108	1329	1551						
18″ x 6″ 22″ x 5″	72 73	T.P. (In. W. G.)					.015	.023	.033	.045	.059	.074	.090	.126	.170	.220					
		Throw (ft.)					10	13	15	18	21	23	26	31	36	41					
		Vel. (fpm)					400	500	600	700	800	900	1000	1200	1400	1600					
14″ x 8″	75	T.P. (In. W. G.)					.014	.021	.031	.041	.054	.069	.084	.118	.160	.203					
		Throw (ft.)					10	13	15	18	20	23	25	30	35	40					
		Vel. (fpm)					384	480	576	672	768	864	960	1152	1344	1536					
24″ x 5″ 20″ x 6″	80 81	T.P. (In. W. G.)						.018	.027	.036	.047	.060	.072	.103	.139	.179	.225				
		Throw (ft.)						12	15	17	20	22	24	29	34	39	44				
		Vel. (fpm)						450	540	630	720	810	900	1080	1260	1440	1620				
16″ x 8″	87	T.P. (In. W. G.)						.016	.023	.031	.040	.051	.062	.090	.120	.150	.195				
		Throw (ft.)						12	14	16	19	21	23	28	33	37	42				
		Vel. (fpm)						414	497	579	662	745	828	993	1159	1324	1490				
22″ x 6″	89	T.P. (In. W. G.)						.015	.021	.029	.038	.049	.059	.086	.113	.146	.185				
		Throw (ft.)						12	14	16	19	21	23	28	32	37	42				
		Vel. (fpm)						404	485	566	647	728	809	971	1133	1294	1456				
24″ x 6″ 18″ x 8″	98 99	T.P. (In. W. G.)							.018	.024	.032	.040	.049	.070	.094	.120	.150	.190			
		Throw (ft.)							13	15	18	20	22	27	31	35	40	44			
		Vel. (fpm)							441	514	588	661	735	882	1029	1175	1322	1469			
20″ x 8″	110	T.P. (In. W. G.)							.014	.019	.025	.031	.038	.055	.074	.095	.118	.145	.215		
		Throw (ft.)							13	15	17	19	21	25	29	33	37	42	50		
		Vel. (fpm)							393	458	524	589	655	785	916	1047	1178	1309	1571		
22″ x 8″	122	T.P. (In. W. G.)								.015	.020	.025	.030	.044	.059	.076	.097	.116	.172	.235	
		Throw (ft.)								14	16	18	20	24	28	32	36	40	47	55	
		Vel. (fpm)								413	472	531	590	708	826	944	1062	1180	1416	1653	
24″ x 8″	134	T.P. (In. W. G.)								.013	.017	.021	.026	.037	.050	.064	.082	.100	.145	.195	
		Throw (ft.)								13	15	17	19	23	26	30	34	38	45	53	
		Vel. (fpm)								376	430	484	537	645	752	860	967	1075	1289	1504	

FINISH: White   Note: For overall outside dimensions add 1½″ to width and height listed.

Figure 5.17 Sidewall Registers. (Courtesy of Lima Register Company)

series 100V

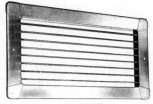

series 100H

STEEL 100 Series

Representative Sizes	Area in Sq. Feet	Air Capacities in CFM										
		250 FPM	300 FPM	400 FPM	500 FPM	600 FPM	700 FPM	750 FPM	800 FPM	900 FPM	1000 FPM	1250 FPM
8 x 4	.163	41	49	65	82	98	114	122	130	147	163	204
10 x 4	.206	52	62	82	103	124	144	155	165	185	206	258
10 x 6	.317	79	95	127	158	190	222	238	254	285	317	396
12 x 4	.249	62	75	100	125	149	174	187	199	224	249	311
12 x 5	.320	80	96	128	160	192	224	240	256	288	320	400
12 x 6	.383	96	115	153	192	230	268	287	306	345	383	479
14 x 4	.292	73	88	117	146	175	204	219	234	263	292	365
14 x 5	.375	94	113	150	188	225	263	281	300	338	375	469
14 x 6	.449	112	135	179	225	269	314	337	359	404	449	561
16 x 5	.431	108	129	172	216	259	302	323	345	388	431	539
16 x 6	.515	129	155	206	258	309	361	386	412	464	515	644
20 x 5	.541	135	162	216	271	325	379	406	433	487	541	676
20 x 6	.647	162	194	259	324	388	453	485	518	582	647	809
20 x 8	.874	219	262	350	437	524	612	656	699	787	874	1093
24 x 5	.652	162	195	261	326	391	456	489	522	587	652	815
24 x 6	.779	195	234	312	390	467	545	584	623	701	779	974
24 x 8	1.053	263	316	421	527	632	737	790	842	948	1053	1316
24 x 10	1.326	332	398	530	663	796	928	995	1061	1193	1326	1658
24 x 12	1.595	399	479	638	798	951	1117	1196	1276	1436	1595	1993
30 x 6	.978	245	293	391	489	587	685	734	782	880	978	1223
30 x 8	1.321	330	396	528	661	793	925	991	1057	1189	1371	1651
30 x 10	1.664	416	499	666	832	998	1165	1248	1331	1498	1664	2080
30 x 12	2.007	502	602	803	1004	1204	1405	1505	1606	1806	2007	2509
36 x 8	1.589	397	477	636	795	953	1112	1192	1271	1430	1589	1986
36 x 10	2.005	501	602	802	1003	1203	1404	1504	1604	1805	2005	2506
36 x 12	2.414	604	724	966	1207	1448	1690	1811	1931	2173	2414	3018

Capacities of sizes not shown above furnished upon request. Steel 100 Series (except 100M) available in any size in 2″ increments.

Figure 5.18 Grills. (Courtesy of Lima Register Company)

5.6. Design of a System

Example 5.3 Using an air system, design an installation for heating and cooling for the Basic Plan, first introduced in Figure 2.16, page 38. The heat loss and heat gain quantities and the cubic feet per minute required for heating and cooling are tabulated in Table 5.4. Refer also to Figure 5.9a and b.

Solution: There can be as many designs as there are designers. In no case could one expect two identical solutions. Differing answers could all be correct and appropriate. Some designs will be better than others. In solving this problem the following approach is *suggested* based on the data in Tables 5.3 and 5.4.

(a) Select a system *type*.
(b) Locate and size supply registers and return grills.
(c) Select and locate furnace, cooling units and thermostat.
(d) Make a layout of ducts and size them.

(a) *Select a system type.* Partial basement and crawl space suggest a basement duct system with floor registers below glass or near exterior doors. Supply ducts and branches in the unexcavated areas will be insulated. The return duct will have a duct liner to reduce blower sounds and also for thermal insulation.

(b) *Locate and size supply registers and return grills.* Table 5.4 indicates the usual situation. The flow rate of air for cooling is greater than that for heating. Registers, grills and ducts will be designed for this rate. Small floor diffusers, which are suitable for this house, deliver roughly 100± cfm per register. The rooms will require one, two or three registers depending on their required airflow rate. Refer to Figure 5.16 and Table 5.4. As an example, bedroom no. 2 requires 246 cfm. Two registers at 125 cfm each will satisfy this demand. Figure 5.16 shows that a 4 × 12 Series 40 will deliver 125 cfm at a register face velocity of 460 fpm. This is within the 500 fpm± that is set as a design standard in Table 5.6. These registers will not be noisy.

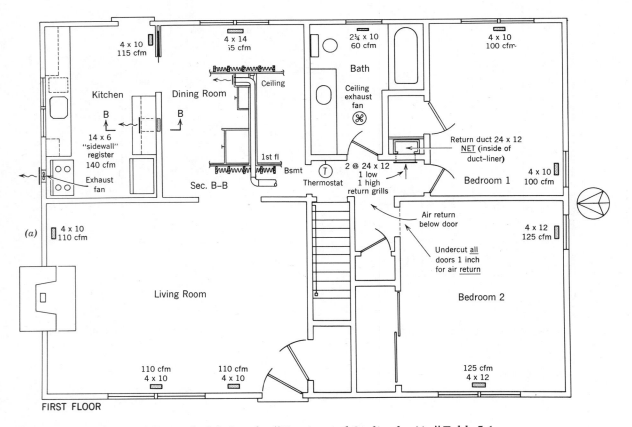

Figure 5.19 Solution to Example 5.3. See also "Heating and Cooling by Air," Table 5.6. *(a)* First floor plan *(b)* Basement air distribution system.

Higher velocities might cause undesirable sound. Similar thinking should be applied to return grills. These are located (one high and one low) in the corridor. The 1270 cfm of air (Figure 5.6) is drawn back through these two 24 × 12 grills. The face velocity is (feet per minute = cubic feet per minute/square feet of grill):

$$\text{feet per minute} = 1270 \div \frac{24 \text{ in.} \times 12 \text{ in.} \times 2}{144}$$

$$= \frac{1270 \times 144}{24 \times 12 \times 2} = 370 \text{ fpm}$$

well below the recommended 500 fpm limit of Table 5.6.

Extreme precision in assigning the total rate of airflow to the house is not necessary. For instance, as a 'rule of thumb' for *cooling*, about 450 cfm/ton of cooling is a common approximation. For the 30,000 Btuh (2½ tons)* (Table 5.6) actually installed, the cubic feet per minute could be

*one ton of cooling is 12,000 Btuh.

$450 \times 2\frac{1}{2} = 1125$ cfm. The engineer's calculations (Table 5.3) shows a total of

$$
\begin{array}{r}
330 \\
256 \\
167 \\
201 \\
\underline{246} \\
1200 \text{ cfm}
\end{array}
$$

If we allow some flow to the bathroom and basement, the total becomes 1330 cfm (Figure 5.19b). An engineer's check on the actual flow based on friction is 1270 cfm (Table 5.6).

It *is* usual, however, for the *cooling* air rate to exceed that for heating. Hence, there are automatic blower controls that regulate the airflow to about 800 cfm for winter and to about 1200 for summer. In *both* seasons, there can be "*continuous* circulation" (at the proper seasonal rate). In winter, the thermostat starts the burner to heat this moving airstream. In summer, it starts the condenser-evaporator to cool the moving airstream.

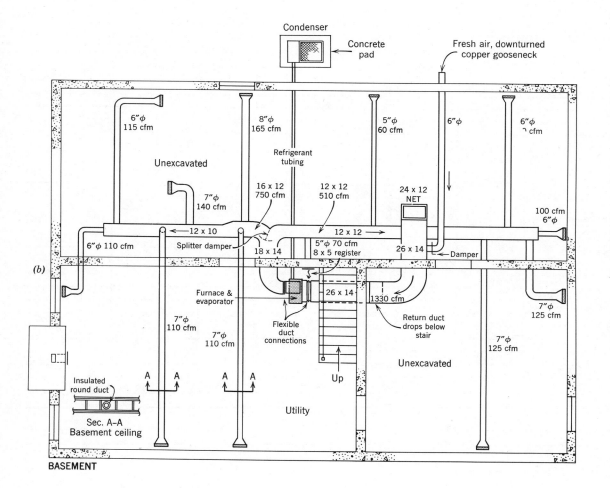

BASEMENT

Table 5.3 Typical Heat Gain Calculations[a]

Form No. CCD-681-L10
Litho U.S.A.

WORKSHEET
LOAD CALCULATIONS FOR RESIDENTIAL AIR CONDITIONING

COOLING CONTRACT _____ ADDRESS _____

NAME OF JOB ___*Basic Plan*___ ADDRESS _____

JOB NO. ___—___ DATE _9/12/1975_ BY _R.P._ DAILY RANGE _Medium_ PEAK LOAD _2:00_ (A.M.-P.M.) LATITUDE _40°N._

FLOOR AREA _1885 S.F._ OUTSIDE DESIGN D.B. _95°_ W.B. _76°_ INSIDE DESIGN D.B. _75°_ W.B. _62°_ DESIGN TEMP. DIFF. _20°_

		TABLE NUMBER	FACTOR	1. *Liv. Rm.* Area or Number	Btuh Gain	2. *Kit.* Area or Number	Btuh Gain	3. *Dining* Area or Number	Btuh Gain	4. *Br #1* Area or Number	Btuh Gain	5. *Br #2* Area or Number	Btuh Gain
1. Room Use and Floor Level													
2. Room Length and Width				16 x 21		9 x 15		12 x 15		11 x 15		13 x 16	
SOLAR Gain Through Glass	North	Incl.											
	East												
	South	Below		Direction House Faces *West*									
	West												
	Horizontal												
NORTH Wall and Glass (Conduction)	Exposed Wall Length			16		15							
	Ceiling Height			8		8							
	Gross Exposed Wall			128		120							
	Sq Ft Glass Area	23		—	—	15	345						
	Net Exposed Wall		3.6	128	460	105	378						
EAST Wall and Glass (Conduction)	Exposed Wall Length					9		12		11			
	Ceiling Height					8		8		8			
	Gross Exposed Wall					72		96		88			
	Sq Ft Glass Area	56				—	—	24	1344	24	1344		
	Net Exposed Wall		3.6			72	259	72	230	64	230		
SOUTH Wall and Glass (Conduction)	Exposed Wall Length									15		16	
	Ceiling Height									8		8	
	Gross Exposed Wall									120		128	
	Sq Ft Glass Area	31								12	372	12	372
	Net Exposed Wall									108	388	116	417
WEST Wall and Glass (Conduction)	Exposed Wall Length			21								15	
	Ceiling Height			8								8	
	Gross Exposed Wall			168								104	
	Sq Ft Glass Area	56		32	1792							28	1568
	Net Exposed Wall		3.6	136	489							76	273
9. Wood Doors			14	21	294	21	294						
10. Warm Partition			2.7										
11. Warm Ceiling			2.9	336	974	135	391	180	522	165	478	208	606
12. Warm Floor			4.6	120	552	—	—	—	—	—	—	—	—
13. Lights (Watts)		Incl.											
14. Appliances (Sensible)		2400					2400						
15. People (Sensible)		300		2	600			2	600	1	300	2	600
16. Outside Air or Infiltration (Sensible)		1.5	276	444	192	288	96	144	208	312	232	348
17. Other Gain (Evap. Motor) (Sensible)													
18. Sub-Total-Sensible					5605		4355		2840		3424		4184
19. Duct Allowance 20%			.20		1121		870		568		684		836
20. Total Sensible					6726		5225		3408		4108		5020
21. Appliances (Latent)		Incl.											
22. People (Latent)		"											
23. Outside Air or Infiltration (Latent)		"											
24. Total Latent (30 %)			.30		2017		1567		1022		1232		1506
25. Total Sensible & Latent					8743		6792		4430		5340		6526
26. Theoretical cfm—CU. FT. AIR					330		256		167		201		246
27. Delivered cfm —DUCT SIZE													

Bldg. Sen. Gain _24,487_ LENNOX Equip. _____

Bldg. Lat. Gain _7,344_

Total Bldg. Gain _31,831_ Btuh Cap. _____

B. Capacity Multiplier, from Table 7	.491

@ _____ °F. S/T Ratio _____

Cfm _____ S.P _____

Rpm _____ Air Changes _____

[a] Calculations by Roger Pinto, Quiet Air, Inc.

Table 5.4 (Example 5.3) Design Data from Engineer's Calculations[a]

Space	Winter Heat Loss, Btuh	Summer Heat Gain, Btuh (Line 25, Table 5.3)	Cubic Feet per minute for Cooling (Line 26, Table 5.3)
Living room	9,300	8,743	330
Dining room	4,300	4,430	167
Bedroom no. 1	5,200	5,340	200
Bedroom no. 2	7,500	6,526	246
Bath	2,900	(Negligible)	—
Kitchen	4,800	6,792	256
Basement	4,900	(Negligible)	—
	38,900	31,831	1199
		(2.65 tons)	

cubic feet per minute for *cooling:* say, 1200 cfm[a]

cubic feet per minute for *heating* $= \dfrac{38,900}{0.075 \times 0.24 \times 60 \times 45}$

$\qquad\qquad\qquad\qquad\qquad = 800$ cfm

[a]Use the greater for duct sizing.

Table 5.5 Recommended and Maximum Duct Velocities for Conventional Systems

Designation	Recommended Velocities, Fpm		
	Residences	Schools, Theaters, Public Buildings	Industrial Buildings
Outdoor Air Intakes[a]	500	500	500
Filters[a]	250	300	350
Heating Coils[a,b]	450	500	600
Cooling Coils[a]	450	500	600
Air Washers[a]	500	500	500
Fan Outlets	1000–1600	1300–2000	1600–2400
Main Ducts[b]	700–900	1000–1300	1200–1800
Branch Ducts[b]	600	600–900	800–1000
Branch Risers[b]	500	600–700	800
	Maximum Velocities, Fpm		
Outdoor Air Intakes[a]	800	900	1200
Filters[a]	300	350	350
Heating Coils[a,b]	500	600	700
Cooling Coils[a]	450	500	600
Air Washers[a]	500	500	500
Fan Outlets	1700	1500–2200	1700–2800
Main Ducts[b]	800–1200	1100–1600	1300–2200
Branch Ducts[b]	700–1000	800–1300	1000–1800
Branch Risers[b]	650–800	800–1200	1000–1600

[a] These velocities are for total face area, not the net free area; other velocities in table are for net free area.
[b] For low velocity systems only.
Reprinted by permission from *Systems and Equipment,* of the American Society of Heating, Refrigerating and Air Conditioning Engineers.

(c) *Select and locate furnace, cooling units and thermostats.* For the warm air upflow furnace, to make up the hourly heat loss of 38,900 Btuh, a Lennox E10Q2-561 unit is selected. Assuming a 208v electric service, it will deliver 42,300 Btuh, about 9% more than required. See Figure 5.6*b*. For cooling we choose

Lennox C4-41FF Evaporator (Figure 5.7)
Lennox HS8-311FF Condenser (Figure 5.8)

This condenser-evaporator combination delivers 30,000 Btuh of cooling. It is about 10% less than the hourly heat gain. This kind of undersizing is recommended for cooling. It assures more continuous cooling and less "build up" of room humidity.

These three elements are located as shown in Figure 5.19*b*. The furnace is electric. For this reason, it does not have to be near the chimney. A position in the corner of the utility room and near the center of the house is most appropriate. The evaporator is in the airstream and on top of the furnace. The condenser unit can be placed adjacent to the house as shown. The sound of its compressor is not loud enough to be annoying. However, if space permits, it could well be placed 30 or 40 ft away from the house. Locations close to bedroom windows should be avoided.

An electronic air cleaner and a humidifier could both be used in this system. For simplicity, they are not shown.

(d) *Duct layout and design.* Low air velocity is good because it results in a quieter system. The lower the velocity, the larger the ducts must be. Systems can be noisy when ducts are made small to lower installation costs. A second consideration is friction of airflow. The lower the velocity, the less is the friction. When ducts are planned for low velocity, the friction in *inches of water* (in. wg.) is small and within the capacity of the blower. See blower data, Figure 5.6*c*. The static head* of our system is 0.07 in. wg at 1270 cfm. See Table 5.6. (Using the duct velocity design standards of Table 5.6, let us check the designer's choice of duct sizes as shown in Figure 5.19*b*.

SOUTH DUCT

12 in. × 12 in.   510 cfm
Duct size 1 sq ft

---

*For explanation of static head, see Figure 5.20.

Air velocity $\dfrac{510}{1 \text{ sq ft}}$ = 510 fpm   OK<800

for this main duct.

NORTH DUCT

16 in. × 12 in.   750 cfm

Duct size $\dfrac{16 \times 12}{144}$ = 1.33 sq ft

Air velocity $\dfrac{750}{1.33}$ = 563 fpm   OK<800

ONE SELECTED BRANCH DUCT

8 in. round to Dining room 165 cfm

Area $\dfrac{\pi 4^2}{144}$ = 0.35 sq ft

Air velocity $\dfrac{165}{0.35}$ = 471 fpm   OK<500

One duct on the suction (return side) of the air system pulls in fresh air. A damper in a convenient place (near an access opening) can partially or entirely close this duct. It can be fully open for freshness or fully closed for the greatest fuel economy. Balance between north and south ends of the house may be adjusted by the splitter damper where the main duct divides. Adjustment of flow from registers can be made by the opposed-blade dampers behind the register. When the flow adjustment is satisfactory, the damper setting can be fixed by a setscrew.

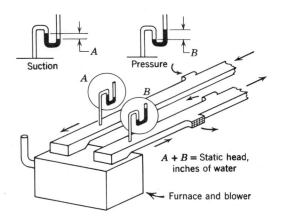

Figure 5.20 Static head is the pressure in inches of water [inches, water gauge (in. wg)] to overcome friction of airflow in an entire duct system.

Table 5.6 (Example 5.3) Heating and Cooling by Air, Basic Plan House. Standards, Equipment and Results

DESIGN STANDARDS		
Air velocities, feet per minute (fpm)		
Main ducts	800 max	Table 5.5
Branch ducts	500 max	Table 5.5
Face of register	500±	Figure 5.16

EQUIPMENT		
Electric service, assumed,	120/208 v	
Ducts, steel		Figures 5.11 and 5.12, Table 5.1
Covering on rectangular ducts	Ultralite	Figure 5.11
Duct liner (in return duct)		
Registers, Lima	Series 40	Figure 5.16
Grills, Lima	Series 100V/H	Figure 5.18
Central equipment, Lennox		
Furnace	E10Q2-561	Figure 5.6*a*
208 v, 12.4 kw	42,300 Btuh	Figure 5.6*b*
Evaporator	C4-41FF	Figure 5.7
Condenser	HS8-311FF	Figure 5.8
Btuh (cooling) 30,000	3800 w	Figure 5.8*b*

RESULTS		
cubic feet per minute delivered by the blower		
Heating	800	
Cooling (engineer's calculation)	1270 at 0.07 in. wg.	

For layout see Figure 5.19. For design analysis see Chapter text.

Problems

5.1. A 48 by 12 in. main air duct of steel in a large house carries 6000 cfm. Is this a satisfactory design? Give reasons for your answer.

5.2. What gage would you select for an 84-in.-wide steel duct?

5.3. A 4 × 10 in. Series 40 Lima floor register delivers 200 cfm. Would it be satisfactorily quiet? Give reasons for your answer.

5.4. A house has a heat loss of 82,000 Btuh. At an air temperature rise of 60 F°, how many cubic feet per minute would have to be circulated?

5.5. Select an E10Q5 Lennox warm air furnace to serve the house in Problem 5.4. State its output in Btuh. The electric service is 208 v.

5.6. A gas-fired stowaway furnace in an attic provides full winter and summer air conditioning. It is served by an outdoor compressor-condenser. Name five connections to the attic unit other than electricity.

5.7. In an air furnace, is the filter located between return duct and the blower or between the furnace and the air-delivery duct?

5.8. In a duct system what does a splitter damper do?

5.9. In addition to electricity, what two connections are there to an outdoor condensing unit?

5.10. Air can be moved by either cylindrical- or propeller-type fans. Which type is usually found in (a) an indoor air furnace and (b) an outdoor condensing unit?

5.11. (a) Is the fresh (outdoor) air supply duct connected to the return duct or to the supply air duct? (b) Give the reason for your answer.

Additional Reading

Sheet Metal Shop Drawing, Howard Betz, Industrial Press. Written by a chief draftsman, this book presents the best in technical drafting, especially for large buildings.

Architectural Drafting, William J. Hornung, Prentice-Hall. The student should refer to the section on duct drafting and the detailing of sheet metal parts.

Fibrous Glass Duct Construction Standards, June 1975, Sheet Metal and Air Conditioning Contractors National Association (SMACNA).

6. Residential Heating-Cooling

In the introduction to Chapter 1, a step-by-step approach is suggested for the development of the technical systems. This four-stage plan consists of:

1. Knowledge of basic principles.
2. Familiarity with equipment.
3. The design.
4. Drawings and specifications.

In Chapter 1, we discuss all of the basic principles of heating, ventilating and cooling. In Chapter 2, we apply the four-stage approach to a heating problem. Heat losses are given for the Basic Plan. These calculations lead in turn to the choice of a system, equipment, the making of a design and, finally, to a summary of the entire scheme.

We use the same procedure for heating in:

A residence, Chapter 3

A building for light industry, Chapter 4

In Chapter 5, a new item is added—cooling. Again, we use the Basic Plan as an exercise.

Now, in Chapter 6, our task is to plan, design and finalize a heating-cooling system for an actual residence. Here energy conservation is most important. In this chapter the heat pump will be used to reduce the energy needed in an electrical heating–cooling system.

Study of this chapter, which introduces the heat pump and *zoned air distribution*, will enable you to:

1. Provide enough air registers to handle high rates of air delivery without causing drafts.
2. Understand the principle of the *heat pump* and the reason for its efficiency.
3. Lay out a two-zone central heating–cooling system.
4. Use (in a basement-less house) an attic as the location for the central heating–cooling plant.
5. Specify minimum clearances of an outdoor compressor–condenser unit from adjacent structure.
6. Select electric resistance heating elements to

supplement the performance of the heat pump at low outdoor temperatures.

7. Read and understand engineering graphs of heating performance.
8. Design and draw plans for an unusual type of heating–cooling job.

9. Locate zone thermostats and be able to discuss their use to control heating and cooling.
10. Specify seven or more typical locations from which air should be exhausted and plan for sufficient "make up" outdoor air to be drawn into the circulation system.

6.1. Adapting to the Structure

The Merker residence is typical of many modern, custom-designed houses. Wasteful basement space is completely eliminated. Attic volume is at a minimum. At the ridge, the height of the small, partial attic is 5 ft. Space for central heating or cooling equipment is limited.

Another consideration is planning the house for the least possible heat loss and heat gain. Good insulation and double-pane (insulating) glass are the best barriers against heat loss. These two items also reduce heat *gain*. More important, however, in the reduction of heat gain is shading much of the exterior glass from the sun. Observe evidence of effective shading in Figure 3.1, page 64. This *summer* photograph taken at the hottest time of the day shows most of the west glass in full shade. These shaded glass areas include the living room, family room, master bedroom and study.

Concerning the lower level rooms, the *only* glass receiving the full impact of the intense western sun is a row of small windows in the bedrooms.

It is sometimes a good decision to cool only *part* of a house. In the house we are now studying, the actual choice was to cool only the upper level rooms. The lower level has only a very small heat gain. The east wall of that story is below grade against the cool earth. The north and south walls have no glass. The west glass (in the family room) is in shade. Finally, the windows and sliding glass doors can be opened up to the cooling breezes from Long Island Sound.

Adding strength to the decision for cooling only the upper level rooms, was the planned occupancy and use of the house. Most of the family living is in the upper level. The lower level is largely for guest use.

A good method of conditioning the upper level is to have central heating–cooling units in the attic.

This makes possible short air-duct runs to ceiling registers. Return air can be picked up at several central locations. This arrangement is closest to the actual scheme used. For this reason, it was possible to photograph a part of the system during construction to show the method of air distribution that we will use. See Figures 6.1 and 6.2. This will be an all-electric job. For heating the lower level, we will retain the electric heating design of Example 3.2 shown in Figure 3.14, page 78. The upper level will be heated and cooled by a *heat pump*. This is an energy-saving unit. It is described in the next section.

6.2. The Heat Pump

This device has been known and used for many years. When, in the early 1970s, the energy crisis began, the heat pump suddenly became very popular. Many manufacturers experienced a large increase in demand. Production rates jumped. See Figure 6.3.

It is essentially a cooling unit that also heats. It uses the compressive refrigeration cycle. We learned about this cycle in Chapter 1. Now review Figures 1.10 to 1.12. You will recall that these illustrations are

Figure 1.10 The principles of cooling, page 16.
Figure 1.11 Through-the-wall cooling unit, page 18.
Figure 1.12 A "split" system with a remote outdoor compressor and an indoor cooling coil, page 19.

In Chapter 5, we used the cooling method of Figure 1.12 (Example 5.3, Figure 5.19, page 130). An electric air-furnace was used for heating. In summer, a coil in the airstream provided cooling. The heat pump has some of the elements of the above system but is quite different in design and operation.

Figure 6.1 Construction photographs, Merker residence. *(a)* View of living room looking South. *(b)* Close-up of air distribution system. Three "two-way throw" curved-blade registers (90 CB-OB 2) will deliver a warm (or cool) air blanket in the region of the glass doors. Two double-deflection wall registers (90 HV-OB) will deliver air horizontally to effect a "throw" and good circulation in the room. Flexible insulated ducts will be trimmed flush with wall before installation of the two wall registers. See Figure 6.4*b*, Section B.

Figure 6.2 Ceiling air supply. Warm or cool air enters all rooms from the ceiling. Except in the living room, "one-way throw" curved blade (CB) registers deliver a flat layer of air into the room, inducing a secondary flow of room air up across the glass (see Section C of Figure 6.4b). In the above photograph the flexible insulated air duct turns down to a metal adapter. Into this square opening will be fitted a register, Series 90 CB-OB 1. An opposed-blade damper (OB) above the curved blades can be adjusted to regulate the flow of air into the room.

Please follow the circuits of Figure 6.3 as we discuss this unique system. It has a compressor and an indoor airstream. There the similarity to Figure 1.12 ends. It has two coils, one in the airstream and one outdoors near the compressor. For *cooling*, the heat pump operates exactly as shown in Figure 1.12. The outdoor coil is the condenser coil and the airstream coil is the evaporator (cooling) coil. The outdoor air is warmed and the indoor air is cooled. Heat is "pumped" *out*. In winter the direction of the refrigerant flow is reversed. The indoor coil becomes the condenser coil and the outdoor coil becomes the evaporator. That coil takes heat from the outdoor air. The heat is "pumped" to the indoor coil where it warms the airstream as the refrigerant condenses and loses its latent heat. Energy in the form of heat is taken from the outdoor air and is delivered to the indoor circulated air. Although ground water is not discussed here, it can also be an energy source.

As the outdoor temperature drops, the efficiency of the heating cycle goes down. At certain low, outdoor temperatures, electrical resistance heating coils turn on. These coils supplement the low-cost heat furnished by the condenser coil. Obviously the heat pump is most efficient in areas where winter temperatures are moderate. However, in the New York area the overall operation of the heat pump, including the use of resistance coils, is about twice as efficient as straight resistance heating. Energy use (and cost) would, therefore, be about one half as much.

6.3. Planning and Designing the System

Example 6.1 Based on the engineer's cubic feet per minute requirements and the heat losses and gains (Table 6.1), design a heating–cooling system for the upper level of the Merker residence (Figures 3.1 to 3.5, page 64 to 68), using heat pumps.

Solution: A suggested approach is to:

(a) Locate the heat pumps and blower coil units.
(b) Select register types and locations.
(c) Make a duct layout.
(d) Determine heat pump specification and controls.
(e) Size ducts and registers.
(f) Determine exhaust ventilation.

(a) *Locate the heat pumps and blower coil units.* A decision is made for two zones. See Figure 6.4b. This has several advantages. Duct runs are much shorter. Control is a little more flexible. Energy savings are possible. Thermostat settings can be reduced in a zone of the house that is not occupied during periods of the day or night. This last fact, together with the natural efficiency of the heat pump, results in economical and energy-saving operation.

In section 6.2 we found that the heat pump has two elements, one outdoors and one indoors. With two zones, four pieces of equipment are required. The two outdoor units (the heat pumps) are placed (Figure 6.9) behind some planting that screens them from view. The indoor equipment, two blower-coil units, are placed in the attic.

Section A in Figure 6.4b shows that there is space to hang the two units from the rafters.

Cooling—

The outdoor unit compressor feeds liquid refrigerant through its coil to the indoor coil. Warm indoor air is forced over the coil by the blower. The liquid refrigerant changes to vapor and absorbs heat, lowering the temperature of the indoor air blowing over the coil.

Refrigerant vapor goes back to the outdoor unit. It is compressed and flows through the outdoor coil where its stored heat is released to the air. This is a continuous process as long as there is a need for cooling.

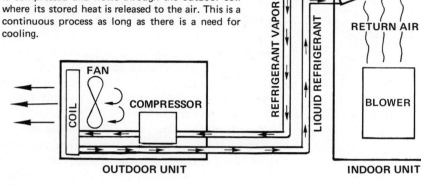

Heating—

Changeover from cooling to heating is controlled by the thermostat. It works through the heat pump controls to reverse the refrigerant flow.

Refrigerant in the outdoor coil absorbs heat from the air (even when the temperature is quite low) and the compressor pumps it, in hot vapor form, to the indoor coil. Heat is picked up from the warm coil by circulating indoor air. Liquid refrigerant returns to the outdoor coil to continue the cycle as long as there is a need for heat.

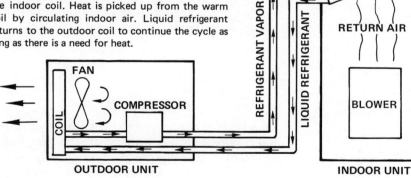

Figure 6.3 Operating principles of the heat pump. (Courtesy of Lennox Industries, Inc.)

Table 6.1 (Example 6.1) Engineering Requirements. Heat Losses and Gains and the Required Cubic Feet per Minute Distribution for Each Season, for Sizing of Ducts, Registers and Equipment[a]

	Winter Heating		Summer Cooling	
Zone 1	Heat Loss (Btuh)	Cubic Feet per Minute for Heating	Heat Gain (Btuh)	Cubic Feet per Minute for Cooling
Living room	28,600	565[b]	10,800	395
Hall/foyer	9,400	180	4,900	180
Dining room	8,000	155	9,000	325[c]
	46,000	900	24,700	900
Zone 2				
Master bedroom	15,000	513[b]	9,400	325
Master bath	2,800	60	1,600	60
Study	3,900	115	3,000	115
Powder room	1,500	30	300	30
Dressing rooms	—	20	600	20
Kitchen	5,100	162	9,600	350[c]
	28,300	900	24,500	900

[a]For heat pump installations, it is usual to assume about 450 cfm per ton of refrigeration. Also, the total cubic feet per minute rate for a *heat pump* is the same for summer and winter operation.

[b]Note increase for *winter* operation.

[c]Note increase for *summer* operation. These changes are made by the motorized splitter dampers (SD) shown in Figure 6.4b.

Spring-type hangers minimize the transmission of vibration to the house. The units are placed high to permit access and service. Access openings are on the *bottom* of the units. There must be a path by which the units can be carried into the attic, or removed, if that should ever be necessary. This can be done through an access door in the wall of the skylight shaft in the master bedroom.

(b) *Select register types and locations.* Of the many types of registers and grills in Figure 6.11, we use 90CB1, 90CB2, 90HV and 90RAH. The 90CB (curved blade) supply register is seen in Figure 6.12a. It is shown in place in Section C of Figure 6.4b. The below-ceiling plane of "primary" air induces "secondary" airflow from the glass. The mild, tempered mixture is such that direct hot or cold primary air does not blow directly on people. Sidewall registers Nos. 4 and 5 are of the 90HV type. All three return grills are 90RAH. All the supply registers and two of the return grills have opposed-blade dampers (OB), which permit adjustment of the airflow rate.

Summarizing, the items are
> 90CB1OB
> 90CB2OB
> 90HVOB
> 90RAHOB
> 90RAH

See Table 6.2.

(c) *Make a duct layout.* The ducts must, of course, serve the registers at their selected locations. The number and location of supply and return registers is a matter of trial, check and re-design. Figure 6.4a shows the final choice and Figure 6.4b shows the duct system.

All rectangular and round supply ducts must have insulative covering. Return ducts are lined with a layer of acoustic, sound absorbing material. This reduces blower noise that reaches the room. The lining also doubles as thermal insulation.

Figure 6.4 is an engineering layout. Before installation, the contractor is required to submit, for approval of the engineer and architect, shop drawings of the duct system. These would resemble the drawing in Figure 5.14, page 123.

(d) *Determine heat pump specification and controls.* Table 6.1 sets the following demands of the heat pump, expressed in Btuh.

	Heating	Cooling
Zone 1	46,000	24,700
Zone 2	28,300	24,500

By advice of the manufacturer's representative, we choose outdoor unit type HP8 and indoor unit CBH8. See Figures 6.5 and 6.6. Figure 6.5b (selector) tells us that an HP8-261FF pump is in a range that provides *about* 23,000 Btuh. The *specific* output (Figure 6.5d) under actual outdoor summer conditions, (95° F air temperature, 67° wet bulb and 900 cfm circulated), is found to be 24,200 Btuh. Air rate and *cooling* capacity both satisfy the requirments of Table 6.1. We specify this heat pump for both zones.

Although these heat pumps, HP8-261FF, together with their indoor units, CBH8-261FF, fully satisfy the cooling needs of both zones, their heating outputs do not. Both require auxiliary electric resistance heating coils at low outdoor temperatures. This is mentioned on Section 6.2. (continued on page 162)

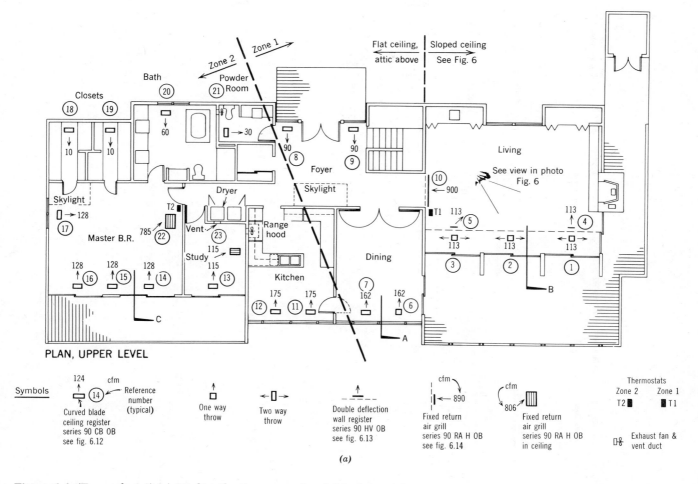

Figure 6.4 (Example 6.1) *(a)* Air distribution, upper level, Merker residence.

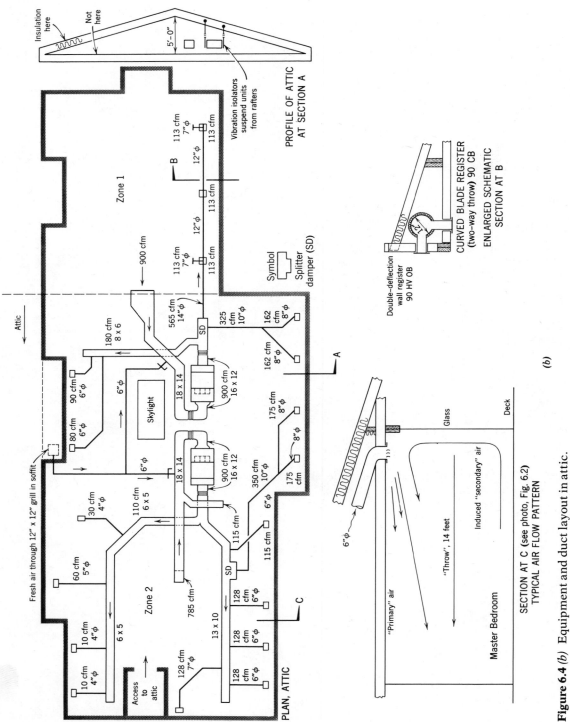

Figure 6.4 (*b*) Equipment and duct layout in attic.

(*b*)

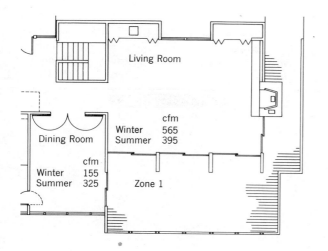

Zone 1	Winter	Summer
Living Room	565	395
Dining Room	155	325
Air *Entering* Damper	720 cfm	720 cfm

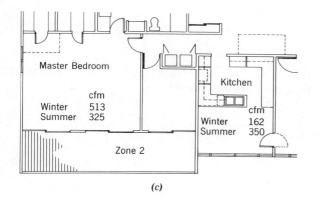

Zone 2	Winter	Summer
Master Bedroom	513	325
Kitchen	162	350
Air *Entering* Damper	675 cfm	675 cfm

(c)

Figure 6.4 *(c)* Performance of splitter dampers. Winter/summer air rates. In Zone 1 under winter conditions, the living room requires about 3½ times as much warm air as the dining room, In summer, however, they require nearly equal rates of *cool* air. In Zone 2, a similar situation exists between the master bedroom and the kitchen. In each case, the rates are adjusted by the motorized splitter damper of the zone. Notice that the rate of air *entering* the damper in either winter or summer is the same. See discussion in chapter text. Please Note: RE: The living room, dining room, master bedroom and kitchen. The ducts and registers are sized for the *larger* airflow rates as shown in Table 6.1 and here in Fig. 6.4(c). The flow rates to *other* spaces are the *same* summer and winter.

Table 6.2 (Example 6.1) Selection of Grills and Registers

Register or Grill Number	Space	Type	Req'd. cfm, Each	Size (In.)	Tabular (CFM)	Throw (Feet)	Face Velocity (Ft/Min)
1, 2, 3	Living	90CBOB[a]	113	8 × 8	100	9	670
4 and 5	Living	90HVOB	113	10 × 6	100	10	421
6 and 7	Dining	90CBOB	162	12 × 10	175	13	630
8 and 9	Foyer	90CBOB	90	8 × 8	100	12	670
10	*Return (L.R.)*	90RAH	900	30 × 18			475±
11 and 12	Kitchen	90CBOB	175	12 × 10	175	13	630
13	Study	90CBOB	115	8 × 8	100	13	630
14, 15, 16, 17	Master bedroom	90CBOB	128	10 × 8	125	14	680
18 and 19	Dressing rooms	90CBOB	10	8 × 4	50	10	670
20	Bath	90CBOB	60	8 × 4	50	10	670
21	Powder room	90CBOB	30	8 × 4	50	10	670
22	*Return (B.R.)*	90RAHOB	785	24 × 18			475±
23	*Return (Study)*	90RAHOB	115	10 × 8			525±

[a]Two-way throw. All other CB Types one-way throw. See Figure 6.12a, b and c. 90 HV OB, vanes set at 45°. 90 RA H OB, vanes set at 40°. CB curved blade, HV horizontal–vertical vanes, OB opposed-blade damper.

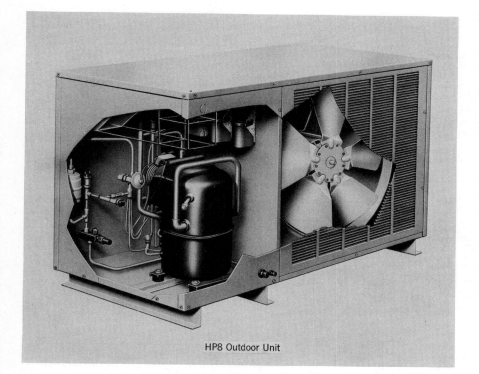

HP8 Outdoor Unit

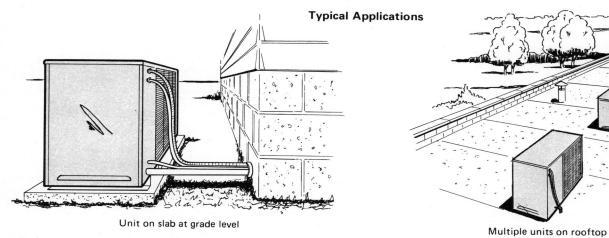

Typical Applications

Unit on slab at grade level

Multiple units on rooftop

NOTE—Specifications, ratings and dimensions subject to change without notice.

(a)

Figure 6.5 (Example 6.1) *(a)* Outdoor unit of the HP8 series heat pump. Note that the compressor is housed in the outdoor unit as in conventional split cooling sets. In the heat pump, however, the function of the outdoor and indoor *coils* reverses seasonally by means of a reversing valve.

	Action of Refrigerant in	
Season	*Outdoor Coil*	*Indoor Coil*
Winter	Evaporation	Condensing
Summer	Condensing	Evaporation

SELECTOR

Lennox Outdoor Unit Model No.	*ARI Standard 240 Certified Ratings						Dehumidifying Capacity	Lennox Indoor Unit Used		
	Btuh Cooling Capacity	Btuh Heating Capacity	Heating Application Capacity (Btuh)	**Total Unit Watts				Up-Flo	Down-Flo	Horizontal
				Cooling	Heating	Htg. Appl.				
HP8-211FF	17,500	19,000	12,000	3000	2600	2200	29%	CBH8-21FF	----	CBH8-21FF
	18,000	19,000	12,000	3100	2600	2200	29%	CB9-21FF	----	----
HP8-261FF	23,000	25,000	15,000	3700	3300	3000	30%	CBH8-26FF	----	CBH8-26FF
(Ex. 6.1)							25%	CB9-26FF	----	----
HP8-311FF	26,000	26,000	15,000	4100	3700	3100	27%	CBH8-31FF	----	CBH8-31FF
	28,000	28,000	16,000	4300	3600	3000	25%	CB9-31FF	----	----
	28,000	28,000	17,000	4300	3500	2900	32%	CB3-41FF CBH3-41FF	CB3-41FF CBH3-41FF	CB3-41FF CBH3-41FF
HP8-411FF †HP8-413FF	34,000	37,000	22,000	5000	4300	3500	32%	CB3-41FF CBH3-41FF	CB3-41FF CBH3-41FF	CB3-41FF CBH3-41FF
†HP8-461FF †HP8-463FF	40,000	41,000	25,000	6000	5100	4500	27%	CB3-51FF CBH3-51FF	CB3-51FF CBH3-51FF	CB3-51FF CBH3-51FF

*Ratings are in accordance with ARI Standard 240:   †Derate 1000 Btuh for 208 volt operation.

  Cooling Ratings—450 cfm indoor coil air volume per ton of cooling capacity, 95F outdoor air temperature and 80db/67wb entering indoor coil air and 25 ft. of connecting refrigerant lines.

  Heating Ratings—450 cfm indoor coil air volume, 45F outdoor air temperature and 70db entering indoor coil air and 25 ft. of connecting refrigerant lines.

  Heating Application Ratings—450 cfm indoor coil air volume, 20F outdoor air temperature and 70db entering indoor coil air and 25 ft. of connecting refrigerant lines.

**Wattage for blower motor included in total unit watts listed.

DIMENSIONS (inches)

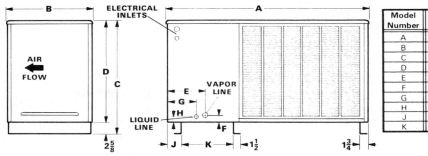

Model Number	HP8-211FF HP8-261FF	HP8-311FF	HP8-411FF HP8-413FF	HP8-461FF HP8-463FF
A	36-3/4	42-7/8	48-1/8	48-1/8
B	17-3/16	18-3/8	21-3/8	21-3/8
C	25-5/8	25-5/8	28-1/8	28-1/8
D	23	23	25-1/8	25-1/8
E	8	8	7-1/4	12-5/8
F	1-1/2	1-1/2	1-5/8	2-1/2
G	6-1/4	6	4-1/2	10-5/8
H	1-3/8	1-3/8	4-1/4	2-1/2
J	3	3	2-1/2	2-1/2
K	10	11	9-3/4	9-3/4

REFRIGERANT LINE KITS

Outdoor Unit Model No.	Line Set Model No.	Vapor Line Length (ft.)	Liquid Line Length (ft.)
HP8-211FF HP8-261FF	L2-26-10FF	10	10
	L2-26-18FF	18	35
	L2-26-25FF	25	
	L2-26-30FF	30	
	L2-26-35FF	35	
	L2-26-40FF	40	50
	L2-26-45FF	45	
	L2-26-50FF	50	
HP8-311FF HP8-411FF HP8-413FF	L6-41-10FF	10	10
	L6-41-20FF	20	50
	L6-41-30FF	30	
	L6-41-40FF	40	
	L6-41-50FF	50	
HP8-461FF HP8-463FF	L6-46-20FF	20	50
	L6-46-30FF	30	
	L6-46-40FF	40	
	L6-46-50FF	50	

Use correct line kit model number when ordering.

INSTALLATION CLEARANCES

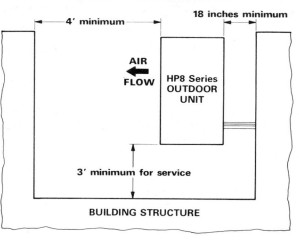

(b)

(b) Selector, dimensions, refrigerant line kits, installation clearances.

ELECTRICAL DATA

Model Number		HP8-211FF	HP8-261FF	HP8-311FF	HP8-411FF		HP8-413FF	HP8-461FF	HP8-463FF
Line voltage data		208-230v 60cy/1ph	208-230v 60cy/1ph	208-230v 60cy/1ph	208v 60cy/1ph	230v 60cy/1ph	208-240v 60cy/3ph	208-230v 60cy/1ph	208-240v 60cy/3ph
Compressor	Full load amps	13.3	16.0	18.1	25.0	22.0	13.7	25.0	14.0
	Power factor	.92	.92	.92	.92	.92	.85	.92	.85
	Locked rotor amps	51.0	71.0	76.0	113.0	103.0	72.0	115.0	87.0
Outdoor Coil Fan Motor	Full load amps	1.4	1.4	2.6	3.2	3.2	3.2	3.2	3.2
	Locked rotor amps	2.9	2.9	5.4	6.6	6.6	6.6	6.6	6.6
*Minimum circuit ampacity		19.0	22.4	25.8	35.2	31.1	21.1	35.2	21.5

*Refer to National Electrical Code manual to determine wire, fuse and disconnect size requirements.
NOTE—Extremes of operating range are plus and minus 10% of line voltage. 208-230v models are plus 10% and minus 5% of line voltage.

FIELD WIRING

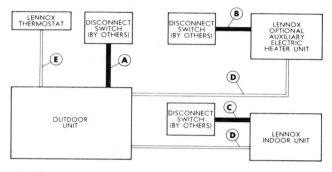

A — Two or three wire power (See Electrical Data table for size)

B — Two wire power (Size to heater capacity)

C — Two wire power (Size to indoor coil blower motor)

D — † Two wire low voltage (18 ga. minimum)

E — † Five wire low voltage (18 ga. minimum)

NOTE — All wiring to conform to NEC and local electrical codes.
†If electrical code permits may be class 2 wiring.

RATINGS

HP8-211FF RFC HEAT PUMP OUTDOOR UNIT COOLING CAPACITY

Indoor Unit Model No.	Entering Wet Bulb Degrees (F)	Total Air Volume (cfm)	Total Cooling Capacity (Btuh) 85	Sensible To Total Ratio (S/T) 85	Comp. Motor Watts Input 85	Total Cooling Capacity (Btuh) 95	Sensible To Total Ratio (S/T) 95	Comp. Motor Watts Input 95	Total Cooling Capacity (Btuh) 105	Sensible To Total Ratio (S/T) 105	Comp. Motor Watts Input 105	Total Cooling Capacity (Btuh) 115	Sensible To Total Ratio (S/T) 115	Comp. Motor Watts Input 115
CBH8-21FF	63	600	17,300	.82	2170	16,200	.84	2260	15,400	.86	2330	14,400	.88	2410
	63	675	17,800	.86	2200	16,700	.88	2310	15,900	.90	2400	14,900	.93	2490
	63	750	18,400	.89	2240	17,400	.91	2370	16,600	.93	2490	15,600	.96	2590
	67	600	19,200	.67	2300	18,000	.69	2420	17,000	.70	2540	15,800	.72	2620
	67	675	19,700	.69	2330	18,600	.71	2480	17,600	.72	2620	16,400	.74	2710
	67	750	20,300	.72	2370	19,200	.73	2530	18,300	.75	2700	17,100	.77	2810
	71	600	20,600	.51	2390	19,600	.51	2580	18,800	.52	2770	17,700	.53	2900
	71	675	21,100	.53	2430	20,100	.54	2620	19,200	.55	2830	18,100	.56	2970
	71	750	21,600	.55	2460	20,700	.56	2670	19,800	.57	2900	18,700	.58	3050
CB9-21FF	63	600	18,200	.84	2230	17,300	.87	2360	16,400	.89	2460	15,400	.92	2550
	63	675	18,600	.85	2250	17,600	.88	2380	16,700	.90	2500	15,600	.94	2600
	63	750	18,900	.87	2280	17,900	.89	2410	17,000	.91	2540	15,900	.95	2640
	67	600	19,400	.68	2310	18,600	.70	2480	17,900	.71	2660	17,000	.73	2800
	67	675	19,800	.69	2340	18,900	.71	2510	18,100	.73	2690	17,200	.74	2820
	67	750	20,200	.70	2370	19,300	.72	2540	18,400	.74	2720	17,400	.76	2860
	71	600	20,500	.54	2390	19,800	.54	2590	19,100	.55	2810	18,300	.55	3000
	71	675	20,900	.56	2410	20,100	.56	2620	19,400	.57	2850	18,500	.57	3030
	71	750	21,300	.58	2440	20,400	.58	2640	19,600	.58	2880	18,700	.59	3060

(Note: The four capacity groups above correspond to Air Temperature Entering Outdoor Coil (F) = 85, 95, 105, and 115, with Indoor Unit 80F Dry Bulb.)

(c)

(c) Electrical data, field wiring, ratings.

RATINGS

HP8-261FF RFC HEAT PUMP OUTDOOR UNIT COOLING CAPACITY

Indoor Unit Model No.	Indoor Unit 80F Dry Bulb Entering Wet Bulb (F)	Total Air Volume (cfm)	85 Total Cooling Capacity (Btuh)	85 Sensible To Total Ratio (S/T)	85 Comp. Motor Watts Input	95 Total Cooling Capacity (Btuh)	95 Sensible To Total Ratio (S/T)	95 Comp. Motor Watts Input	105 Total Cooling Capacity (Btuh)	105 Sensible To Total Ratio (S/T)	105 Comp. Motor Watts Input	115 Total Cooling Capacity (Btuh)	115 Sensible To Total Ratio (S/T)	115 Comp. Motor Watts Input
CB9-26FF	63	800	22,400	.89	2850	21,300	.92	2940	20,100	.95	3110	18,600	.98	3270
	63	900	22,800	.90	2870	21,700	.93	2960	20,500	.97	3140	19,000	.99	3300
	63	1000	23,300	.92	2900	22,200	.95	2990	21,000	.98	3170	19,600	1.00	3340
	67	800	24,600	.71	2970	23,400	.73	3070	22,100	.76	3230	20,500	.79	3410
	67	900	25,100	.73	3000	24,000	.75	3100	22,700	.77	3270	21,100	.80	3460
	67	1000	25,600	.75	3020	24,400	.77	3130	23,100	.79	3290	21,600	.82	3490
	71	800	25,900	.51	3050	25,200	.52	3170	24,200	.53	3360	23,000	.55	3590
	71	900	26,300	.57	3070	25,500	.57	3190	24,600	.58	3380	23,300	.59	3620
	71	1000	26,600	.58	3080	25,900	.59	3210	24,900	.60	3400	23,800	.61	3650
CBH8-26F (Ex. 6.1)	63	800	22,700	.86	2870	21,600	.88	2950	20,300	.90	3120	18,900	.93	3290
	63	900	23,200	.88	2890	22,000	.90	2980	20,700	.92	3150	19,300	.95	3320
	63	1000	23,600	.90	2910	22,400	.92	3010	21,200	.94	3180	19,700	.97	3350
	67	800	24,800	.68	2980	23,800	.69	3090	22,600	.71	3260	21,000	.74	3450
	67	900	25,100	.71	3000	24,200	.73	3110	23,000	.75	3290	21,400	.77	3480
	67	1000	25,600	.74	3030	24,600	.76	3140	23,500	.78	3320	22,000	.80	3520
	71	800	26,900	.50	3100	25,900	.51	3220	24,800	.52	3390	23,300	.53	3620
	71	900	27,200	.54	3120	26,300	.55	3240	25,100	.56	3420	23,700	.57	3650
	71	1000	27,500	.58	3140	26,600	.59	3260	25,500	.60	3440	24,100	.61	3680

HP8-211FF RFC HEAT PUMP OUTDOOR UNIT HEATING CAPACITY

Indoor Unit Model No.	Indoor Coil Air Volume (cfm) 70F db	65 Total Heating Capacity (Btuh)	65 Comp. Motor Watts Input	45 Total Heating Capacity (Btuh)	45 Comp. Motor Watts Input	25 Total Heating Capacity (Btuh)	25 Comp. Motor Watts Input	5 Total Heating Capacity (Btuh)	5 Comp. Motor Watts Input
CB9-21FF CBH8-21FF	600	23,800	2270	18,850	2010	12,800	1670	8350	1500
	675	23,900	2250	19,000	2000	12,900	1650	8500	1370
	750	24,000	2230	19,150	1980	13,000	1630	8650	1200

NOTE—Heating capacities include the effect of defrost cycles in the temperature range where they occur.

HP8-261FF RFC HEAT PUMP OUTDOOR UNIT HEATING CAPACITY

Indoor Unit Model No.	Indoor Coil Air Volume (cfm) 70F db	65 Total Heating Capacity (Btuh)	65 Comp. Motor Watts Input	45 Total Heating Capacity (Btuh)	45 Comp. Motor Watts Input	25 Total Heating Capacity (Btuh)	25 Comp. Motor Watts Input	5 Total Heating Capacity (Btuh)	5 Comp. Motor Watts Input
(Ex. 6.1) CB9-26FF CBH8-26FF	800	29,000	2890	24,500	2710	16,900	2450	11,200	2210
	900	30,000	2880	25,000	2700	17,000	2440	11,300	2200
	1000	30,500	2860	25,500	2680	17,100	2420	11,400	2180

NOTE—Heating capacities include the effect of defrost cycles in the temperature range where they occur.

HP8-211FF HEATING PERFORMANCE
at 675 Cfm Coil Air Volume (CB9-21FF & CBH8-21FF)

*Outdoor Temperature (Degree F)	Comp. Motor Watts Input	Total Output (Btuh)
65	2250	23,900
60	2190	22,800
55	2130	21,800
50	2070	20,800
45	2000	19,000
40	1830	16,000
35	1770	14,800
30	1710	13,900
25	1650	12,900
20	1600	12,000
15	1520	10,800
10	1440	9,700
5	1370	8,500
0	1280	7,300

*Outdoor temperature at 85% relative humidity.
Indoor temperature at 70°.

HP8-261FF HEATING PERFORMANCE
at 900 Cfm Coil Air Volume (CB9-26FF & CBH8-26FF)

*Outdoor Temperature (Degree F)	Comp. Motor Watts Input	Total Output (Btuh)
65	2880	30,000
60	2830	29,000
55	2790	27,600
50	2740	26,300
45	2700	25,000
40	2650	21,100
35	2560	19,800
30	2500	18,400
25	2440	17,000
20	2380	15,500
15	2320	13,700
10	2260	12,500
5	2200	11,300
0	2140	10,200

*Outdoor temperature at 85% relative humidity.
Indoor temperature at 70°.

(d)

(d) Ratings and heating performance, CBH8-26FF (Example 6.1). (Courtesy of Lennox Industries, Inc.) (See Figures 6.5*(b)-(d)* on the following pages.)

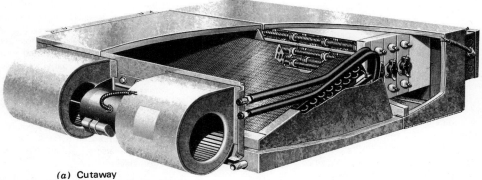

(a) Cutaway
With Factory Installed Electric Heat

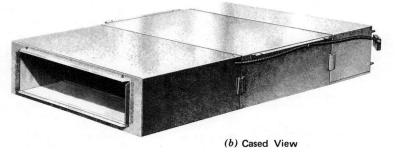

(b) Cased View

(c) Typical Applications

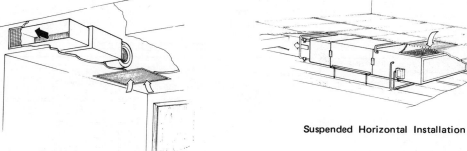

Furred-in Horizontal Installation

Suspended Horizontal Installation

Figure 6.6 (Example 6.1) *(a)* Indoor heating/cooling (blowe̷ ̷ ̷il) unit CBH8-261FF. *(b)* Cased view. Filter section at the left. The right section encloses hot water coils, sometimes used instead of electricity *(not* in Example 6.1). *(c)* Typical applications. *(d)*Features, specifications, electric heat data. *(e)* Blower data, hot water data. *(f)* Dimensions. Note that access for service or replacement is from the *bottom*. Our attic unit must be placed high enough to allow for this. (Courtesy of Lennox Industries, Inc.) (See Figures 6.6*(d)-(f)* on the following pages.)

FEATURES

Lennox Evaporator Coil—Lennox designed and built coil has ripple-edge aluminum fins machine fitted to seamless copper tubes for maximum strength and heat transfer. Pressure leak tested at 450 to 500 psi. Each joint is silver soldered resulting in leak proof joints.

Powerful Dual Blowers—Units are equipped with Lennox designed and built twin direct drive blowers. Each blower assembly is statically and dynamically balanced before it is installed in the unit. This special attention to design details and assembly balancing provides an extremely quiet and efficient operating blower assembly. Double shaft motor is resiliently mounted. The entire blower assembly can be removed from cabinet for ease of service. A choice of three blower speeds is available. See blower performance charts. Change of blower speeds is easily accomplished by a simple change in wiring.

Durable Cabinet—Constructed of unpainted heavy gauge galvanized steel and completely insulated with thick fiberglass insulation. Suspending angles on cabinet provide means for suspending unit in horizontal applications or securing to a wall in up-flo applications. Quick connect clamps with drive cleats provide quick and positive joining of add-on electric or hot water heat cabinets and filter cabinet to blower-coil cabinet. Locating lugs in blower-coil cabinet provide sure alignment and support of optional add-on cabinets in suspended horizontal applications.

Wiring Junction Box—Furnished as standard equipment for DX cooling applications. Length of blower motor wiring harness permits field mounting of the box on the blower-coil unit or in a convenient location near the unit. Wiring harness is factory connected from blower motor to junction box.

Blower Cooling Relay—Furnished as standard equipment and field installs on wiring junction box in DX cooling applications.

Drain Pan—Rugged construction and corrosion resistant. Equipped with two 3/4 mpt drain connections for added protection from overflow. Interchangeable overflow plug is furnished in left side drain nipple for auxiliary drain connection in horizontal applications.

Refrigerant Line Connections—Suction and liquid lines are equipped with flare fittings and are accessible for easy connections. See dimension drawings for refrigerant line and drain connection location.

Electric Heat (Optional)—Adaptable for heating-cooling applications or where reheat is required for humidity control. Factory installed add-on ECBH8 electric heat models are available in matching cabinets and several KW sizes. See Electric Heat table for heating capacities. Nichrome heating elements are exposed directly in the air stream resulting in instant heat transfer. Cabinet is constructed of unpainted heavy gauge galvanized steel and completely insulated with thick fiberglass insulation. Hinged access panel provides complete service access to controls and elements. Access panel is located at bottom of cabinet in horizontal applications and adequate access for service must be provided. Supply air opening is flanged for ease of duct connection. Factory installed and wired controls consist of: heating delay relay(s), transformer, terminal block and limit controls. A factory installed electrical make-up box is also furnished. Units are U.L. Listed for zero installation clearances.

Hot Water Heat (Optional)—Field installed add-on CHW8-70 hot water coil unit is available for heating-cooling applications or where reheat is required for humidity control. Furnished in a separate matching cabinet that matches the blower-coil cabinet. Cabinet is constructed of unpainted heavy gauge galvanized steel and completely insulated with thick fiberglass insulation. Removable panel provides service access. Panel is located at bottom of unit in horizontal applications and adequate access for service must be provided. Coil is constructed of ripple-edge aluminum fins machine fitted to seamless copper tubes. Water lines are stubbed outside of cabinet for easy connection. See specification table for coil data. A factory installed freeze protection control is furnished as standard equipment. Control terminates compressor operation if water coil temperature approaches 32°F. A three-way water valve is available as optional equipment and must be ordered extra. Order number P-8-10121.

Filter Section (Optional)—Field installed add-on FH8-31 filter section is available in a separate cabinet that matches the blower-coil cabinet. Filter cabinet is constructed of unpainted heavy gauge galvanized steel and completely insulated with thick fiberglass insulation. Interchangeable cabinet panels allow a choice of return air entry. One inch thick frame type filter with oil impregnated fiberglass media is furnished. Knockouts are provided in the cabinet for a choice of piping and electrical line entry. See dimension drawings for location.

SPECIFICATIONS

Model No.		CBH8-21FF	CBH8-26FF	CBH8-31FF
Nominal cooling capacity (tons)		1-1/2	2	2-1/2
Evaporator Coil	Net face area (sq. ft.)	1.70	2.72	2.72
	Tube diam. (in.)	3/8	3/8	3/8
	No. of rows	3	3	3
	Fins per inch	13	13	16
	Suction line o.d. (in.)	5/8 FF	5/8 FF	3/4 FF
	Liquid line o.d. (in.)	3/8 FF	3/8 FF	1/2 FF
Refrigerant		R-22	R-22	R-22
Condensate drain (mpt) in.		3/4	3/4	3/4
Blower wheel nom. diam. x width (in.)		(2) 6 x 6	(2) 6 x 7	(2) 6 x 7
Blower motor hp		1/8	1/5	1/5
Net Weight (lbs.)—1 Pkg.		67	74	76
Electrical Characteristics		230 volt—60 hertz—1 phase		
Optional Hot Water Section Model No.		CHW8-70		
Heating Capacity Range (Btuh)		45,000—100,000		
Hot Water Coil	Net face area (sq. ft.)	1.67		
	Tube diam. (in.)	1/2		
	No. of rows	3		
	Fins per inch	13		
	Inlet connection o.d. (in.)	5/8 Sweat		
	Outlet connection o.d. (in.)	5/8 Sweat		
	Net weight (lbs.)—1 Pkg.	32		
Optional Filter Section Model No.		FH8-31		
No. & size of filter (in.)		(1) 12 x 24 x 1		
Net weight (lbs.)		25		

ELECTRIC HEAT DATA

Blower-Coil Model No.	Electric Unit Model No. & Net Weight	No. Of Steps	Number of Elements	Volts Input	KW Input	Btuh Output	*Minimum Circuit Ampacity
CBH8-21FF CBH8-26FF CBH8-31FF	ECBH8-161 (48 lbs.)	1	1	208	3.5	11,800	25.6
				220	3.9	13,200	
				230	4.3	14,400	
				240	4.6	15,700	
	ECBH8-261 (54 lbs.)	2	2	208	5.6	19,200	40.7
				220	6.3	21,500	
				230	6.9	23,500	
				240	7.5	25,600	
	ECBH8-311 (54 lbs.)	2	2	208	6.9	23,600	49.5
				220	7.7	26,400	
				230	8.5	28,800	
				240	9.2	31,400	
CBH8-26FF CBH8-31FF	ECBH8-471 (57 lbs.)	3	3	208	10.3	35,400	73.5
				220	11.6	39,600	
				230	12.7	43,300	
				240	13.8	47,100	

*Refer to National Electrical Code manual to determine wire, fuse and disconnect size requirements. Use wires suitable for at least 75C

BLOWER DATA

CBH8-21FF BLOWER PERFORMANCE

External Static Pressure (in. wg)	Air Volume (cfm) @ Various Speeds		
	High	Medium	Low
0	960	765	650
.05	935	745	630
.10	910	725	610
.15	885	705	585
.20	855	680	560
.25	825	655	535
.30	800	630	505
.40	735	570	445
.50	660	495	----
.60	565	----	----

CBH8-26FF BLOWER PERFORMANCE

External Static Pressure (in. wg)	Air Volume (cfm) @ Various Speeds		
	High	Medium	Low
0	1160	990	850
.05	1150	980	845
.10	1135	965	835
.15	1120	950	825
.20	1100	935	810
.25	1080	920	795
.30	1055	900	775
.40	1000	850	725
.50	935	790	670
.60	850	720	----

CBH8-31FF BLOWER PERFORMANCE

External Static Pressure (in. wg)	Air Volume (cfm) @ Various Speeds		
	High	Medium	Low
0	1160	975	845
.05	1140	965	835
.10	1120	950	825
.15	1100	935	810
.20	1080	915	795
.25	1055	895	780
.30	1030	875	755
.40	970	820	710
.50	895	760	----
.60	805	----	----

OPTIONAL ELECTRIC HEAT, HOT WATER HEAT, AND FILTER SECTION AIR RESISTANCE

Air Volume (cfm)	Total Resistance (inches water gauge)		
	Electric Heat Section	Hot Water Heat Section	Filter Section
400	.02	.06	.08
500	.03	.08	.11
600	.04	.11	.14
700	.05	.15	.19
800	.07	.19	.26
900	.08	.25	.35
1000	.10	.30	.42
1100	.13	.37	----

HOT WATER DATA
CHW8-70 HEATING CAPACITY CHART

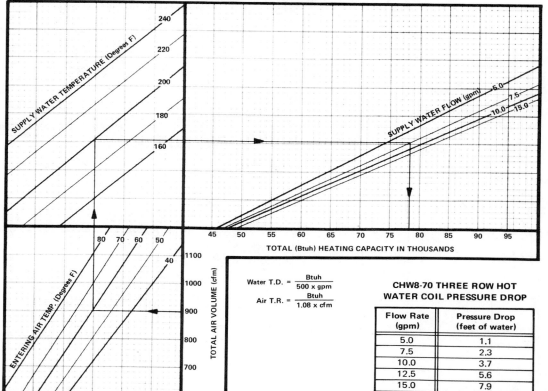

$$\text{Water T.D.} = \frac{\text{Btuh}}{500 \times \text{gpm}}$$

$$\text{Air T.R.} = \frac{\text{Btuh}}{1.08 \times \text{cfm}}$$

CHW8-70 THREE ROW HOT WATER COIL PRESSURE DROP

Flow Rate (gpm)	Pressure Drop (feet of water)
5.0	1.1
7.5	2.3
10.0	3.7
12.5	5.6
15.0	7.9

(e)

DIMENSIONS (in.)
BLOWER-COIL UNIT WITH OPTIONAL ELECTRIC HEAT SECTION

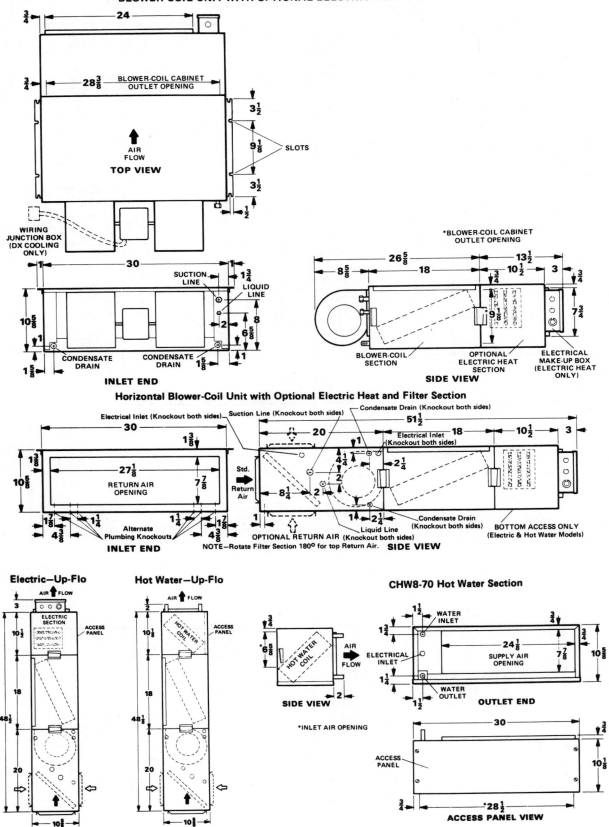

(f)

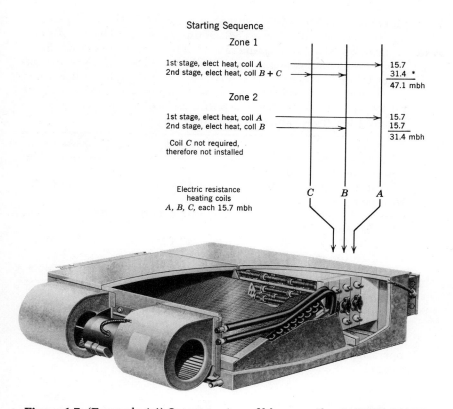

Starting Sequence

Zone 1

1st stage, elect heat, coil A	15.7
2nd stage, elect heat, coil $B + C$	31.4 *
	47.1 mbh

Zone 2

1st stage, elect heat, coil A	15.7
2nd stage, elect heat, coil B	15.7
	31.4 mbh

Coil C not required,
therefore not installed

Electric resistance
heating coils
A, B, C, each 15.7 mbh

C B A

Figure 6.7 (Example 6.1) Cutaway view of blower coil unit CBH8-261FF, two of which are used for the heating and cooling of the upper level rooms. This indoor air-handling unit has, in the position shown, an airstream from left to right. The air leaving the return ducts passes first through a filter section, not shown. It is then drawn in by the twin centrifugal blowers. It passes over the refrigerant coils. In these coils the refrigerant evaporates in summer and condenses in winter. At lower outdoor temperatures additional heat is needed. This is added by electric resistance coils A, B and C* in the amounts shown above the illustration. The air then passes out to a supply duct, which will be connected at the right. (*Shown in Figure 6.6d as ECBH8-471.) (Courtesy of Lennox Industries, Inc.)

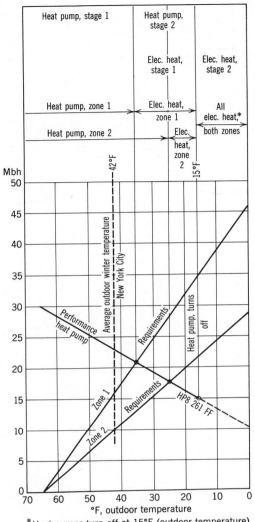

*Heat pumps turn off at 15°F (outdoor temperature)

Figure 6.8 (Example 6.1) Heating performance, HP8-261FF heat pump. (See Table on Figure 6.5d). Output is indicated by the designation HP8-26. The numeral, 1, following this shows that the compressor motor is single phase. The letters, FF, call for flare fittings. When the heat pump alone cannot meet the heating demand, electric resistance coils turn on to assist the heat pump. At 15° F the heat pump turns off. Then *additional resistance heating comes one. For* Example 6.1, these values are as follows:

	Zone 1	Zone 2
Second stage Heat pump (first stage resistance)	Add 15.7 Mbh	Add 15.7 Mbh
Heat pump off (second stage resistance)	Add *31.4* more	Add *15.7* more
Electric resistance below 15° F	47.1 Mbh	31.4 Mbh
House requirements at 0° F (See Table 6.1)	46.0 Mbh	28.0 Mbh

The temperature in New York City (Example 6.1) drops below 15° F only about 5% of the time. The *average* winter temperature in that area is 42° F. These two facts show how seldom electric resistance needs to be used. Under the conditions shown above, the heat pump operating alone is about twice as efficient as electric resistance heating.

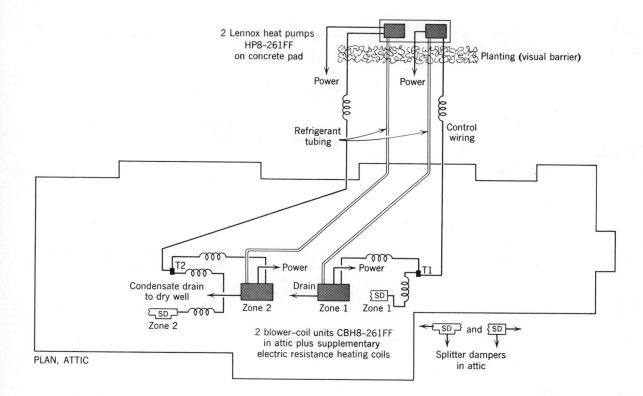

Figure 6.9 (Example 6.1) Electrical power and control requirements for heating/cooling of upper level rooms of Merker residence. Motorized splitter dampers (SD) adjust the cubic feet per minute flow in four rooms for summer and winter rates. See Figure 5.10*b*, page 120 for sketch of splitter damper.

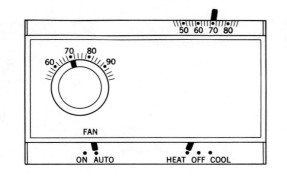

Thermostat for the control of heat pump and blower-coil unit.

One for each zone system.

A temperature anticipator senses even slight changes and starts the heat pump to compensate.

Heat or *cool* setting, in addition to starting the heat pump, also adjusts the motorized splitter damper for seasonal airflow.

Room temperature at the thermostat   Room temperature desired

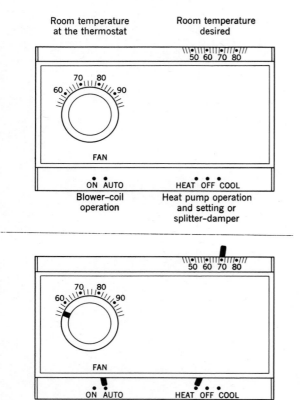

Blower-coil operation   Heat pump operation and setting or splitter–damper

Blank diagram identifying the several controls. Typical settings and operations as shown below in (*a*) through (*e*).

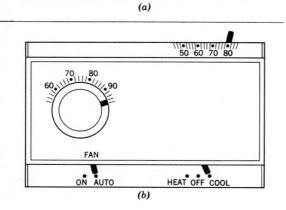

(*a*)

Turn on for Heating. First cold day of winter. Room temperature 55° F. Desired temperature set at 70° F. Heat pump set for *heat.* Blower set for *automatic.* When the heat pump starts, blower turns on automatically. When room temperature reaches 70° F, both turn off.

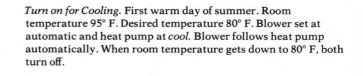

(*b*)

Turn on for Cooling. First warm day of summer. Room temperature 95° F. Desired temperature 80° F. Blower set at automatic and heat pump at *cool.* Blower follows heat pump automatically. When room temperature gets down to 80° F, both turn off.

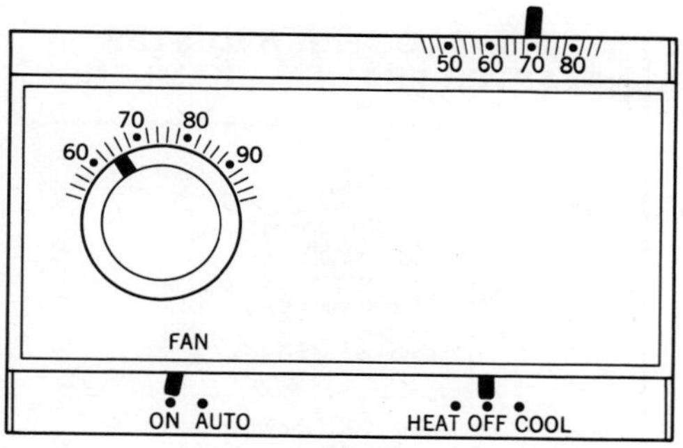

(c)

Ventilation. In mild weather—spring or fall when neither heating nor cooling is necessary—the blower-coil unit will circulate air if set at *on* and, leaving the heat pump at *off*, *desired temperature* is inactive and room temperature is not affected.

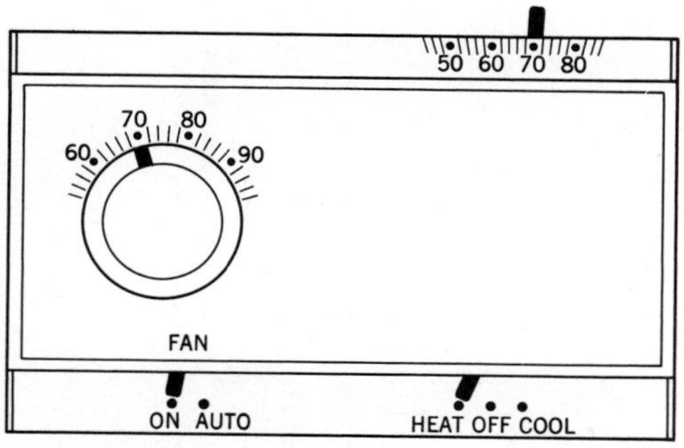

(d)

Continuous Circulation—Heating. Blower is set at *on*. It operates 24 hr a day. Heat pump is set at *heat*. Desired temperature is set at 70° F. Whenever the room temperature drops below 70° F the heat pump turns on to restore heat in the room. When the room temperature reaches 70° F, the heat pump turns off but the blower continues to run.

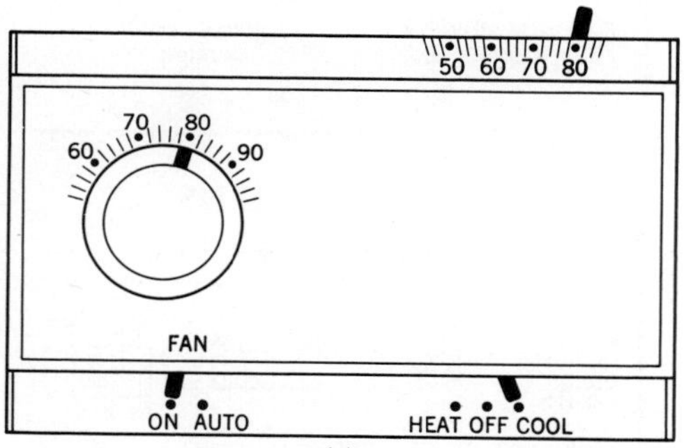

(e)

Continuous Circulation—Cooling. As above, the blower runs continuously. The heat pump operates intermittently to maintain the desired room temperature setting (80° F).

Figure 6.10 (Example 6.1) Thermostat and control system. Table 6.1 tells us that two rooms in each zone should have different airflow rates in summer and winter. This is done by a motorized splitter damper in the duct. Control of this is an addition to the basic operation of the thermostat. Contractor should be required to furnish a shop drawing showing the wiring diagram. (Courtesy of Lennox Industries, Inc.)

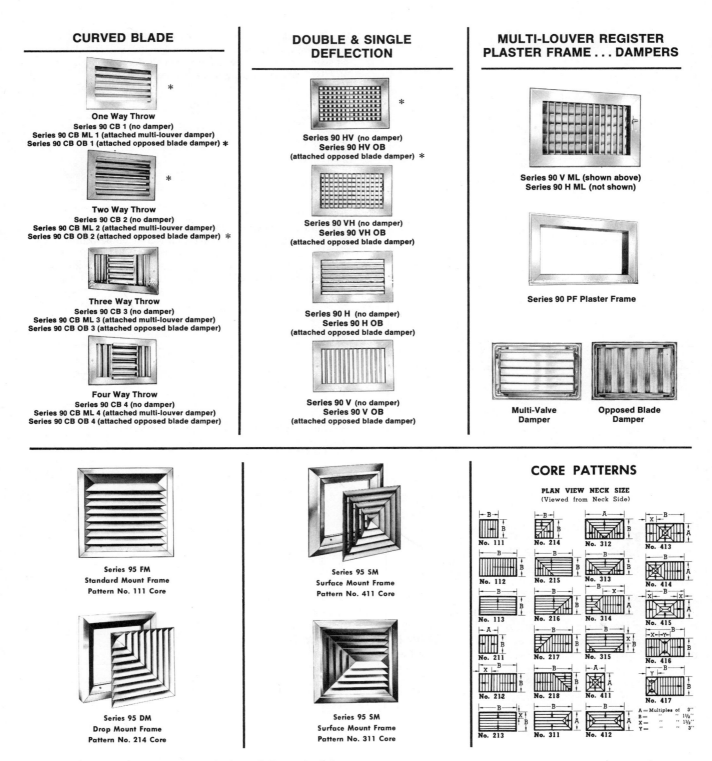

CURVED BLADE

One Way Throw
Series 90 CB 1 (no damper)
Series 90 CB ML 1 (attached multi-louver damper)
Series 90 CB OB 1 (attached opposed blade damper) *

Two Way Throw
Series 90 CB 2 (no damper)
Series 90 CB ML 2 (attached multi-louver damper)
Series 90 CB OB 2 (attached opposed blade damper) *

Three Way Throw
Series 90 CB 3 (no damper)
Series 90 CB ML 3 (attached multi-louver damper)
Series 90 CB OB 3 (attached opposed blade damper)

Four Way Throw
Series 90 CB 4 (no damper)
Series 90 CB ML 4 (attached multi-louver damper)
Series 90 CB OB 4 (attached opposed blade damper)

DOUBLE & SINGLE DEFLECTION

Series 90 HV (no damper)
Series 90 HV OB
(attached opposed blade damper) *

Series 90 VH (no damper)
Series 90 VH OB
(attached opposed blade damper)

Series 90 H (no damper)
Series 90 H OB
(attached opposed blade damper)

Series 90 V (no damper)
Series 90 V OB
(attached opposed blade damper)

MULTI-LOUVER REGISTER PLASTER FRAME . . . DAMPERS

Series 90 V ML (shown above)
Series 90 H ML (not shown)

Series 90 PF Plaster Frame

Multi-Valve Damper

Opposed Blade Damper

Series 95 FM
Standard Mount Frame
Pattern No. 111 Core

Series 95 DM
Drop Mount Frame
Pattern No. 214 Core

Series 95 SM
Surface Mount Frame
Pattern No. 411 Core

Series 95 SM
Surface Mount Frame
Pattern No. 311 Core

CORE PATTERNS

PLAN VIEW NECK SIZE
(Viewed from Neck Side)

No. 111, No. 214, No. 312, No. 413
No. 112, No. 215, No. 313, No. 414
No. 113, No. 216, No. 314, No. 415
No. 211, No. 217, No. 315, No. 416
No. 212, No. 218, No. 411, No. 417
No. 213, No. 311, No. 412

A = Multiples of 3″
B = " " 1½″
X = " " 1½″
Y = " " 3″

*Indicates items used in the solution of Example 6.1

Figure 6.11 Extruded-anodized aluminum registers, grills, and diffusers. *Indicates items used in the solution of Example 6.1. (Courtesy of Lima Register Company)

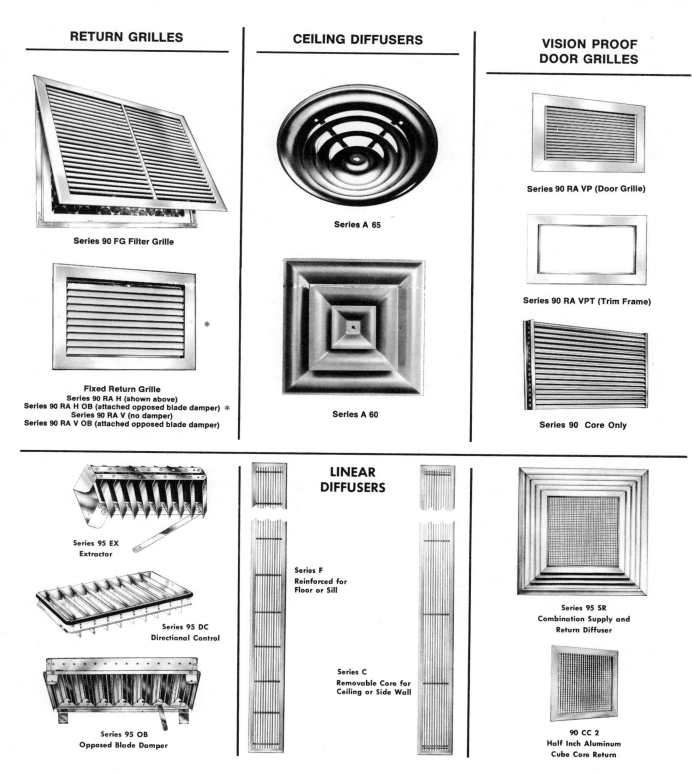

RETURN GRILLES

Series 90 FG Filter Grille

Fixed Return Grille
Series 90 RA H (shown above)
Series 90 RA H OB (attached opposed blade damper) *
Series 90 RA V (no damper)
Series 90 RA V OB (attached opposed blade damper)

CEILING DIFFUSERS

Series A 65

Series A 60

VISION PROOF DOOR GRILLES

Series 90 RA VP (Door Grille)

Series 90 RA VPT (Trim Frame)

Series 90 Core Only

Series 95 EX Extractor

Series 95 DC Directional Control

Series 95 OB Opposed Blade Damper

LINEAR DIFFUSERS

Series F Reinforced for Floor or Sill

Series C Removable Core for Ceiling or Side Wall

Series 95 SR Combination Supply and Return Diffuser

90 CC 2 Half Inch Aluminum Cube Core Return

Notice, in Figure 6.5d that the heat pump performance drops from 30,000 Btuh at 65° F (outdoors) to 10,200 Btuh at 0° F. At that temperature (0° F), we need 46,000 Btuh for Zone 1 and 28,300 Btuh for Zone 2. We plot this in Figure 6.8 and explain the auxiliary resistances that must come on (in two stages) as the heating power of the heat pump drops. In Zone 1, for instance, at 35° F outdoors, the heat pump just carries the Zone 1 heating requirements. The resistance coils must come on at this temperature or slightly *before* it drops to 35°. A careful study of Figure 6.7 and 6.8 will make this clear.

The control of both heating and cooling is as shown in Figures 6.9 and 6.10. Thermostats T1 and T2 control their respective zones. In Figures 6.10d and e continuous circulation is mentioned. In this method, air is circulated through the house at all times. The heating or cooling, each in its own season, comes on as required by the thermostat *termperature desired* setting. This is considered to be the best system for comfort. One of the reasons is that in between the periods of operation of the heat pump, the continuous distribution of air makes for uniformity. In summer, it prevents the buildup of increased temperature and humidity in the house at areas of high heat gain. In winter, chilly spots do not develop. Although the power to run the blower is not great, the movement for the conservation of energy now suggests that a little comfort may be sacrificed for a saving in energy.

(e) *Size ducts and registers.* If you add up the cubic feet per minute written next to each duct branch of Zone 1, you will find that the total is 1070 cfm. There is a reason why this number exceeds the 900 cfm that is circulated by the blower coil unit. The cubic feet per minute noted at each duct supplying the living room and the dining room is the *maximum* cubic feet per minute supplied. It is higher in winter for the living room and higher in summer for the dining room. See Figure 6.4c. The total supplying the two rooms is always the same, but the *division* changes seasonally. However, the *duct* size must be based on the maximum air speed allowed. During the other season the air speed will be less. The maximum speeds in this design are

> Main ducts 800 ft/min
> Branch ducts 600 ft/min

These standards vary to some extent. It is important, though, to maintain low velocities to minimize noise and to keep the frictional resistance within the capacity of the blower.

Registers must be large enough to deliver the air assigned to them but also must be sized to "throw" the air a reasonable distance into the room. Face velocities are kept less than the manufacturer's suggested maximum of 750 fpm. Selection tables are Figures 6.12c and 6.13b and c.

(f) *Determine exhaust ventilation.* Sources of odor include the kitchen range, laundry dryer, laundry room, garage and bathrooms. By exhausting air from these spaces to the outdoors these odors are reduced and some humidity is eliminated. See Figures 6.15 to 6.18. This air which is drawn out of the house during seasons of heating or cooling, must be replaced by outdoor air drawn *in* and conditioned by the central equipment. Figure 6.4b shows how this is done. In both zones, fresh air is admitted to the suction side of the blower coil unit. Its rate of flow may be adjusted by volume dampers in the fresh air duct near each unit.

Custom designed houses like this often have unusual features. One such feature is the exhausing of air from the upper-level rooms *down* through ducts instead of up through the roof. This arrangement reduces the number of items that protrude through the roof.

Figure 6.12 *(a)* (Facing page) Curved-blade ceiling registers and grills. In Example 6.1, all supply registers and return grills will be equipped with opposed-blade dampers for control. Three- and four-way throw units are not shown here nor are they used in Example 6.1. (Courtesy of Lima Register Company)

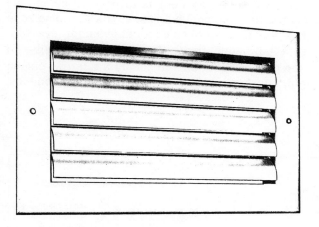

one way throw

SERIES 90 CB 1 (no damper)

SERIES 90 CB ML 1 (attached multi-louver damper)

SERIES 90 CB OB 1 (attached opposed blade damper)

Lima Aluminum Curved Blade Registers and Grilles are alike in design . . . the only difference is the positioning of the streamlined blades for one, two, three or four way throw. When you select a Lima Curved Blade Grille, you may specify grille only without damper, grille with an attached multi-louver damper or with an attached opposed blade damper. The face and blades on each curved blade grille are constructed of the finest Lima Aluminum extrusions and each has an unusually attractive anodized finish. Horizontal blades shown are standard. Vertical blades on one and two way throw models are available on special order.

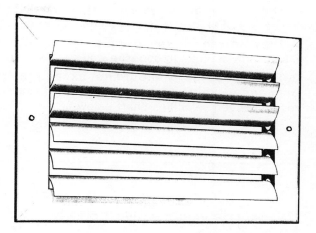

two way throw

SERIES 90 CB 2 (no damper)

SERIES 90 CB ML 2 (attached multi-louver damper)

SERIES 90 CB OB 2 (attached opposed blade damper)

The streamlined curved blades on this (and all 4 styles) curved blade registers are individually mounted and secured with press-fit nylon bushings. This allows complete freedom of adjustment of each blade without loosening or spinning of individual vanes. There are no wires to break and rattle. All four Lima Aluminum curved blade styles are recommended for both heating and cooling systems . . . you won't find a more attractive or more functional aluminum line.

Any style face available with multi-louver or attached opposed blade damper.

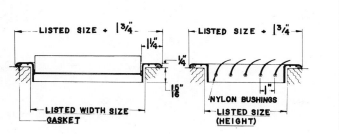

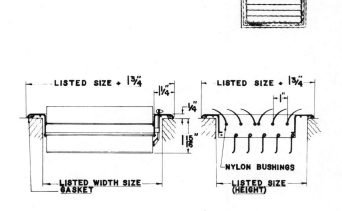

(a)

SELECTION PROCEDURE:

1. The grille or grilles selected should deliver the necessary CFM for the area to be conditioned.

2. The throw should reach approximately ¾ of the distance from outlet to opposite wall.

3. The face velocity should not exceed the recommended velocity for the application. (See recommended velocities chart below.)

4. Use correct ceiling heights to prevent air stream from dropping into occupied zone. The occupied zone is generally thought of as 6 feet above floor level.

5. After determining CFM requirements, consult chart for the proper outlet size to satisfy the throw, ceiling height, total pressure and face velocity requirements.

RECOMMENDED DELIVERY VELOCITIES FOR VARIOUS APPLICATIONS

The sound caused by an air outlet in operation varies in direct proportion to the velocity of the air passing through it. The air velocity can be controlled by selecting outlets of proper sizes. The following recommended outlet velocities are within safe sound limits for most applications.

Application	Recommended Face Velocities
Broadcasting studios	500 FPM
Residences	500 to 750 FPM
Apartments	500 to 750 FPM
Churches	500 to 750 FPM
Hotel bedrooms	500 to 750 FPM
Legitimate theatres	500 to 1000 FPM
Private offices, acoustically treated	500 to 1000 FPM
Motion Picture Theatres	1000 to 1250 FPM
Private offices, not treated	1000 to 1250 FPM
General offices	1250 to 1500 FPM
Stores, upper floors	1500 FPM
Stores, main floors	1500 FPM
Industrial buildings	1500 to 2000 FPM

(b)

Figure 6.12 *(b)* Series 90 CB. Aluminum curved-blade ceiling registers and grills. Engineering data. Maximum face velocity, residences 750 fpm.

FIN SETTINGS:

Set 60% of fins at ⅜″ setting.
Set 15% of fins at ⁵⁄₁₆″ setting.
Set 25% of fins at ½″ setting.

Performance data compiled with fins set in pattern shown below.

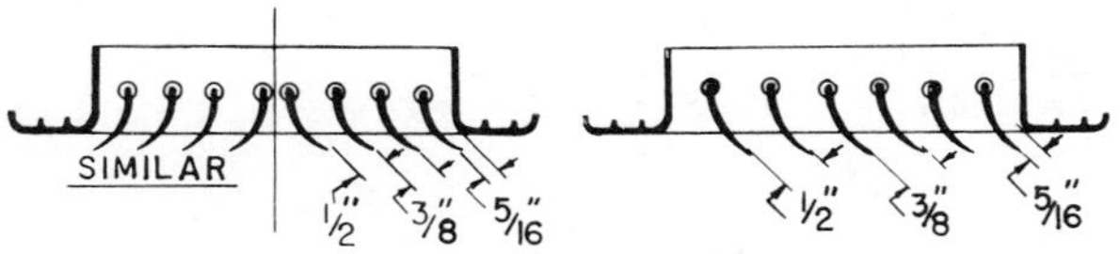

C.F.M.	DIRECTIONAL THROW	8×4 — 1	2	3	4	8×5 / 10×4 / 6×6 — 1	2	3	4	8×6 / 12×4 / 10×5 / 14×4 — 1	2	3	4	10×6 / 12×5 — 1	2	3	4	8×8 / 12×6 / 14×5 — 1	2	3	4	10×8 / 14×6 — 1	2	3	4	10×10 / 12×8 / 16×6 — 1	2	3	4	18×6 / 14×8 — 1	2	3	4	12×10 / 20×6 — 1	2	3	4
50	TOTAL PRESS.	.028				.018				.013																												
	FACE VELOCITY	670				535				445																												
	THROW	10	8	7	7	9	7	6	6	8	6	5	5																									
	MIN. CEILING HGT.	8	8	8	8	8	8	8	8	8	8	8	8																									
75	TOTAL PRESS.	.058				.046				.037				.018				.012																				
	FACE VELOCITY	1000				880				770				545				500																				
	THROW	14	12	10	10	12	10	8	7	10	8	6	6	9	7	5	5	9	6	4	4																	
	MIN. CEILING HGT.	8	8	8	8	8	8	8	8	8	8	8	8	8	8	8	8	8	8	8	8																	
100	TOTAL PRESS.	.106				.084				.048				.030				.021				.019				.013												
	FACE VELOCITY	1350				1140				900				725				670				545				450												
	THROW	17	14	11	10	16	13	10	9	15	12	10	9	12	10	8	7	12	9	6	5	11	9	6	6	9	7	6	6									
	MIN. CEILING HGT.	8	8	8	8	8	8	8	8	8	8	8	8	8	8	8	8	8	8	8	8	8	8	8	8	8	8	8	8									
125	TOTAL PRESS.					.101				.072				.048				.034				.027				.019				.016								
	FACE VELOCITY					1350				1100				900				840				680				570				510								
	THROW					18	14	12	11	16	13	10	9	15	12	9	8	14	11	8	7	14	11	8	8	12	9	7	7	10	8	7	7					
	MIN. CEILING HGT.					8	8	8	8	8	8	8	8	8	8	8	8	8	8	8	8	8	8	8	8	8	8	8	8	8	8	8	8					
150	TOTAL PRESS.					.19				.155				.070				.049				.038				.027				.022				.019			
	FACE VELOCITY					1600				1300				1080				1000				800				670				620				545			
	THROW					25	20	18	15	20	17	13	12	16	14	11	9	16	14	11	9	15	12	10	10	13	11	8	8	12	10	8	8	12	10	8	8
	MIN. CEILING HGT.					9	8	8	8	9	8	8	8	9	8	8	8	9	8	8	8	9	8	8	8	9	8	8	8	9	8	8	8	9	8	8	8
175	TOTAL PRESS.													.094				.066				.053				.038				.030				.023			
	FACE VELOCITY													1260				1175				950				790				700				630			
	THROW													18	16	12	10	18	15	12	10	18	14	12	10	15	12	10	9	14	12	10	10	13	11	9	9
	MIN. CEILING HGT.													9	8	8	8	9	8	8	8	9	8	8	8	9	8	8	8	9	8	8	8	9	8	8	8
200	TOTAL PRESS.													.123				.085				.070				.049				.038				.031			
	FACE VELOCITY													1440				1340				1080				900				790				720			
	THROW													23	18	15	13	23	17	12	10	20	16	14	13	17	14	12	11	16	14	12	11	15	13	10	10
	MIN. CEILING HGT.													9	8	8	8	9	8	8	8	9	8	8	8	9	8	8	8	9	8	8	8	9	8	8	8
225	TOTAL PRESS.													.156				.108				.086				.061				.049				.039			
	FACE VELOCITY													1615				1520				1200				1020				900				850			
	THROW													27	22	16	14	25	19	14	13	23	17	15	14	20	17	14	13	18	15	13	12	17	14	12	11
	MIN. CEILING HGT.													9	8	8	8	9	8	8	8	9	8	8	8	9	8	8	8	9	8	8	8	9	8	8	8
250	TOTAL PRESS.																	.133				.105				.075				.060				.048			
	FACE VELOCITY																	1690				1350				1130				1000				900			
	THROW																	29	23	17	15	25	20	16	15	22	18	15	14	20	17	15	14	19	16	13	12
	MIN. CEILING HGT.																	10	8	8	8	10	8	8	8	10	8	8	8	10	8	8	8	10	8	8	8
275	TOTAL PRESS.																	.161								.090				.069				.061			
	FACE VELOCITY																	1840								1240				1100				990			
	THROW																	32	27	23	20					24	20	17	16	23	19	16	15	21	18	15	14
	MIN. CEILING HGT.																	10	9	8	8					10	9	8	8	10	9	8	8	10	9	8	8
300	TOTAL PRESS.																									.106				.087				.070			
	FACE VELOCITY																									1350				1200				1080			
	THROW																									26	22	19	18	25	20	17	16	23	19	16	15
	MIN. CEILING HGT.																									10	9	8	8	10	9	8	8	10	9	8	8
325	TOTAL PRESS.																									.125				.100				.085			
	FACE VELOCITY																									1460				1300				1160			
	THROW																									29	24	21	20	27	22	19	18	25	21	18	17
	MIN. CEILING HGT.																									10	9	8	8	10	9	8	8	10	9	8	8
350	TOTAL PRESS.																									.147				.120				.094			
	FACE VELOCITY																									1575				1440				1260			
	THROW																									30	25	22	21	28	23	20	19	27	22	19	18
	MIN. CEILING HGT.																									11	9	8	8	11	9	8	8	11	9	8	8
375	TOTAL PRESS.																													.130				.110			
	FACE VELOCITY																													1500				1350			
	THROW																													30	24	21	20	29	23	20	19
	MIN. CEILING HGT.																													11	9	8	8	11	9	8	8
400	TOTAL PRESS.																																	.122			
	FACE VELOCITY																																	1450			
	THROW																																	31	24	21	20
	MIN. CEILING HGT.																																	11	9	8	8

SIMILAR   ½″   ⅜″   ⁵⁄₁₆″   ½″   ⅜″   ⁵⁄₁₆″

(c)

Figure 6.12 (c) Series 90 CB. Aluminum curved-blade ceiling registers and grills. Engineering data and specifications. Partial excerpt from manufacturer's catalog. (Courtesy of Lima Register Company)

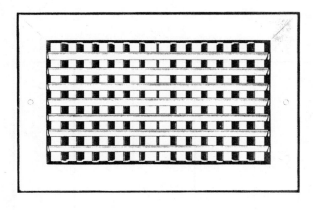

SERIES 90 HV (no damper)

SERIES 90 HV OB (attached opposed blade damper)

Lima Aluminum double deflection registers and grilles provide complete four way directional control of air in both horizontal and vertical planes.

Individual adjustments of horizontal and vertical louvers can be made as desired. No matter how frequently this occurs the individual vanes will not loosen or spin . . . this is prevented by press-fit nylon bushings on Lima Aluminum registers and grilles.

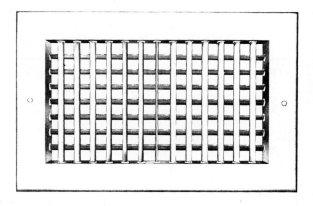

SERIES 90 VH (no damper)

SERIES 90 VH OB (attached opposed blade damper)

Series 90 VH has the same quality design and construction features as Series 90 HV. The difference between the two is that Series 90 VH has vertical front vanes and a second row of horizontal vanes (both adjustable).

Volume control on the cam-operated opposed blade valve can be quickly and easily regulated with a screwdriver.

Any style face available with attached opposed blade damper.

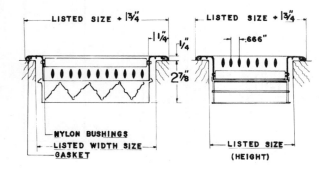

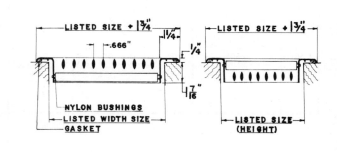

(a)

Figure 6.13 (a) Double deflection registers and grills known as "airfoil vane" types. For engineering data (size selection and specifications), see (c).

SIZE	CALCULATED AREA IN SQ. FT. 0°	40°	250 F.P.M. 0°	40°	300 F.P.M. 0°	40°	400 F.P.M. 0°	40°	500 F.P.M. 0°	40°	600 F.P.M. 0°	40°	700 F.P.M. 0°	40°	800 F.P.M. 0°	40°	900 F.P.M. 0°	40°	1000 F.P.M. 0°	40°	1250 F.P.M. 0°	40°	1500 F.P.M. 0°	40°
10 x 6	.281	.156	70	39	84	47	112	62	140	78	169	94	197	109	225	125	253	140	281	156	337	195	421	234
12 x 6	.343	.217	86	54	103	65	137	87	171	108	206	130	240	159	274	174	309	195	343	217	429	271	514	325
10 x 8	.385	.243	96	61	115	73	154	97	192	121	231	145	269	170	308	194	346	219	385	243	481	304	577	364
12 x 8	.47	.297	117	74	141	89	188	118	235	148	282	178	329	208	376	238	423	267	470	297	587	371	705	445
18 x 6	.525	.33	131	82	157	99	210	132	262	165	315	198	367	231	420	264	472	297	525	330	656	412	787	495
12 x 12	.75	.46	187	115	225	138	300	184	375	230	450	276	525	322	600	368	675	414	750	460	937	575	1125	690
18 x 12	1.11	.704	277	176	333	211	444	281	555	352	666	422	777	493	888	563	999	634	1110	704	1387	880	1665	1056
24 x 12	1.53	.95	382	237	459	285	612	380	765	475	918	570	1071	665	1224	760	1377	855	1530	950	1912	1187	2295	1425
18 x 18	1.69	1.08	422	270	507	324	676	432	845	540	1014	648	1183	756	1352	864	1521	972	1690	1080	2112	1350	2535	1620
30 x 12	1.88	1.23	470	307	564	362	752	490	940	615	1125	738	1316	861	1540	984	1692	1107	1880	1230	2350	1537	2820	1845
24 x 18	2.28	1.46	570	365	684	438	912	584	1140	730	1368	876	1596	1022	1824	1168	2052	1314	2280	1460	2850	1825	3420	2190
30 x 18	2.87	1.87	717	467	861	561	1148	748	1435	935	1722	1122	2009	1309	2296	1496	2583	1683	2870	1870	3587	2337	4305	2805
24 x 24	3.15	2.00	787	500	945	600	1260	800	1575	1000	1890	1200	2205	1400	2520	1600	2835	1800	3150	2000	3937	2500	4725	3000
36 x 18	3.47	2.20	867	550	1041	660	1388	880	1735	1100	2082	1320	2429	1540	2776	1760	3123	1980	3470	2200	4337	2750	5205	3300
30 x 24	3.96	2.48	990	620	1188	744	1584	992	1930	1240	2376	1488	2772	1736	3168	1984	3564	2232	3960	2480	4950	3100	5940	3720
36 x 24	4.78	3.00	1195	750	1434	900	1912	1200	2330	1500	2868	1800	3446	2100	3824	2400	4302	2700	4780	3000	5975	3750	7170	4500
36 x 30	5.87	3.62	1467	905	1761	1086	2348	1448	2935	1810	3522	2172	4109	2534	4696	2896	5283	3258	5870	3620	7337	4525	8805	5430
48 x 24	6.42	4.00	1605	1000	1926	1200	2568	1600	3210	2000	3852	2400	4494	2800	5136	3200	5778	3600	6420	4000	8025	5000	9630	6000
48 x 30	7.86	5.00	1965	1250	2358	1500	3144	2000	3930	2500	4716	3000	5520	3500	6288	4000	7074	4500	7860	5000	9825	6250	11790	7500
48 x 36	9.40	6.00	2350	1500	2820	1800	3760	2400	4700	3000	5640	3600	6580	4200	7520	4800	8460	5400	9400	6000	11750	7500	14100	9000

RECOMMENDATIONS FOR SELECTING OUTLETS

1. The grille or grilles selected should deliver the necessary cfm for the area to be conditioned.

2. The throw should reach approximately ¾ of the distance from outlet to opposite wall.

3. The face velocity should not exceed the recommended velocity for the application. (See recommended velocities chart on page 24).

4. The drop should be such that the air stream will not drop into the occupied zone. The occupied zone is generally thought of as 6 feet above floor level.

HOW LIMA SERIES NUMBERS are Keyed by Letters for Ordering

V—Indicates vertical bars (air foil)
H—Indicates horizontal bars (air foil)
RA—Indicates fixed deflection (air foil)
CB—Indicates curved blade
VP—Indicates vision proof (V-type vanes)
OB—Opposed blade damper
ML—Multi louver damper

In the Series number of any double deflection Grille, the first letter following "Series 90" refers to the front bank of louvers. The second letter refers to the second or back bank of vanes. For example, the Series 90VH Grille has vertical louvers in front with horizontal louvers behind.

(b)

Figure 6.13 *(b)* Aluminum airfoil vane registers and grills. Size selection chart. For double deflection registers 90 HV and also for fixed air return grill, 90 RA H.

CFM	SIZE / DEFLECTION	8 x 4 — 0	22	45	8 x 5 / 10 x 4 — 0	22	45	8 x 6 / 10 x 5 / 12 x 4 — 0	22	45	14 x 4 — 0	22	45	10 x 6 / 12 x 5 / 16 x 4 — 0	22	45	8 x 8 / 14 x 5 / 18 x 4 — 0	22	45	SIZE / DEFLECTION	CFM
50	THROW	10	9	7	9	8	6													THROW	50
	DROP	3.7	3.4	1.8	3.7	3.4	1.8													DROP	
	VELOCITY	354	401	451	276	314	340													VELOCITY	
	TOTAL P.	.011	.013	.016	.008	.009	.011													TOTAL P.	
75	THROW	14	12	10	13	11	9	11	10	8	11	9	8							THROW	75
	DROP	4.2	3.9	2.0	4.2	3.9	2.0	4.4	4.0	2.1	4.7	4.4	2.3							DROP	
	VELOCITY	532	602	679	414	471	529	338	379	428	286	323	388							VELOCITY	
	TOTAL P.	.021	.026	.031	.014	.017	.021	.010	.012	.014	.009	.010	.012							TOTAL P.	
100	THROW	19	16	14	17	14	11	15	13	11	14	12	10	13	11	10	12	11	7	THROW	100
	DROP	4.8	4.5	2.0	4.9	4.6	2.4	5.0	4.6	2.4	5.0	4.7	2.4	4.6	4.5	2.3	4.8	4.5	2.2	DROP	
	VELOCITY	709	811	911	552	632	708	450	499	560	382	440	491	334	379	421	294	340	397	VELOCITY	
	TOTAL P.	.036	.044	.055	.023	.028	.035	.015	.018	.021	.013	.015	.018	.010	.012	.014	.008	.010	.012	TOTAL P.	
125	THROW	23	20	17	20	18	15	18	16	13	17	15	13	16	14	12	15	13	9	THROW	125
	DROP	4.9	4.6	2.3	5.1	4.8	2.5	5.2	4.9	2.4	5.4	5.0	2.6	5.2	4.8	2.6	5.1	4.7	2.5	DROP	
	VELOCITY	887	1014	1142	691	794	891	562	630	704	478	550	618	417	458	538	368	428	498	VELOCITY	
	TOTAL P.	.054	.067	.085	.034	.042	.053	.022	.027	.033	.018	.022	.027	.014	.017	.021	.012	.014	.018	TOTAL P.	
150	THROW	27	23	20	24	21	18	22	18	16	21	18	15	19	17	14	18	16	11	THROW	150
	DROP	5.0	4.7	2.4	5.4	5.0	2.6	5.6	5.2	2.7	5.8	5.4	2.8	5.6	5.2	2.8	5.6	5.2	2.6	DROP	
	VELOCITY	1064	1201	1365	830	945	1076	675	759	850	573	658	732	500	578	660	442	510	545	VELOCITY	
	TOTAL P.	.077	.095	.121	.048	.059	.075	.031	.037	.047	.025	.030	.038	.019	.024	.030	.016	.020	.025	TOTAL P.	
175	THROW	32	27	23	28	24	20	25	21	18	24	20	17	22	19	16	20	18	13	THROW	175
	DROP	5.3	4.9	2.6	5.6	5.2	2.7	5.8	5.4	2.8	6.0	5.6	2.9	5.9	5.5	2.9	5.9	5.5	2.3	DROP	
	VELOCITY	1240	1400	1592	968	1098	1238	788	882	991	668	767	864	583	683	765	514	600	675	VELOCITY	
	TOTAL P.	.103	.128	.163	.064	.079	.111	.040	.049	.063	.033	.040	.050	.026	.031	.040	.027	.026	.033	TOTAL P.	
200	THROW	36	31	26	32	27	23	28	24	21	27	23	20	23	21	15	22	20	15	THROW	200
	DROP	5.6	5.2	2.7	5.8	5.4	2.8	6.0	5.6	2.9	6.2	5.8	3.0	5.9	5.5	2.9	6.1	5.8	3.0	DROP	
	VELOCITY	1423	1625	1832	1102	1262	1502	902	999	1197	764	878	984	668	774	881	589	690	791	VELOCITY	
	TOTAL P.	.133	.166	.211	.083	.103	.131	.052	.064	.087	.041	.051	.065	.033	.041	.052	.027	.034	.042	TOTAL P.	
225	THROW	41	35	29	36	31	26	32	27	23	30	26	22	27	24	15	25	22	15	THROW	225
	DROP	5.9	5.3	2.8	6.0	5.6	3.0	6.2	5.9	3.0	6.5	6.1	3.1	6.5	6.1	3.1	6.5	6.1	3.0	DROP	
	VELOCITY	1589	1798	1998	1242	1401	1598	1070	1124	1252	860	973	1100	750	853	997	662	780	890	VELOCITY	
	TOTAL P.	.168	.211	.268	.104	.129	.165	.065	.080	.102	.052	.064	.081	.041	.051	.065	.034	.042	.053	TOTAL P.	
250	THROW	45	38	32	40	34	29	35	30	25	34	29	24	29	26	16	27	24	17	THROW	250
	DROP	5.9	5.5	2.9	6.2	5.8	3.1	6.5	6.0	3.1	6.7	6.2	3.3	6.6	6.1	3.1	6.6	6.2	3.2	DROP	
	VELOCITY	1767	1980	2190	1381	1590	1765	1123	1265	1397	955	1089	1215	864	977	1100	735	849	972	VELOCITY	
	TOTAL P.	.207	.258	.330	.128	.159	.203	.079	.097	.124	.064	.078	.100	.050	.063	.080	.041	.051	.065	TOTAL P.	
275	THROW				44	37	31	39	33	28	37	31	26	32	29	18	30	27	18	THROW	275
	DROP				6.4	6.0	3.1	6.6	6.1	3.2	6.8	6.3	3.3	6.8	6.3	3.2	7.0	6.5	3.4	DROP	
	VELOCITY				1520	1745	1968	1240	1392	1570	1050	1200	1330	975	1098	1238	800	857	972	VELOCITY	
	TOTAL P.				.154	.201	.245	.095	.118	.150	.076	.094	.120	.064	.079	.111	.045	.057	.078	TOTAL P.	
300	THROW				48	41	34	42	36	30	40	34	29	35	31	22	31	27	19	THROW	300
	DROP				6.6	6.2	3.2	6.6	6.3	3.3	7.0	6.5	3.4	7.1	6.6	3.4	7.1	6.6	3.5	DROP	
	VELOCITY				1659	1890	2101	1352	1497	1685	1142	1315	1476	1000	1145	1305	883	1028	1192	VELOCITY	
	TOTAL P.				.183	.228	.291	.113	.140	.178	.089	.111	.141	.072	.090	.114	.059	.073	.093	TOTAL P.	
350	THROW							49	42	35	47	40	33	40	35	25	37	33	23	THROW	350
	DROP							7.0	6.5	3.6	7.2	6.7	3.5	7.4	6.9	3.6	7.6	7.1	3.7	DROP	
	VELOCITY							1576	1767	1972	1336	1520	1734	1168	1354	1545	1028	1200	1382	VELOCITY	
	TOTAL P.							.152	.189	.242	.121	.150	.191	.097	.121	.154	.079	.099	.125	TOTAL P.	
400	THROW							56	48	40	53	45	38	45	40	27	42	37	26	THROW	400
	DROP							7.3	6.7	3.8	7.4	6.9	3.6	7.8	7.3	3.8	8.0	7.5	3.9	DROP	
	VELOCITY							1800	2015	2260	1530	1760	1958	1334	1523	1732	1176	1348	1526	VELOCITY	
	TOTAL P.							.198	.246	.315	.157	.195	.249	.126	.157	.20	.103	.128	.164	TOTAL P.	
450	THROW										60	51	43	50	44	30	46	41	28	THROW	450
	DROP										7.5	7.0	3.7	8.3	7.9	4.1	8.3	7.7	4.1	DROP	
	VELOCITY										1720	2000	2190	1500	1793	1935	1324	1617	1817	VELOCITY	
	TOTAL P.										.198	.247	.315	.160	.199	.254	.130	.162	.206	TOTAL P.	
500	THROW													55	44	30	51	45	31	THROW	500
	DROP													8.6	7.9	4.1	8.5	7.9	4.2	DROP	
	VELOCITY													1670	1905	2201	1470	1701	1962	VELOCITY	
	TOTAL P.													.196	.244	.312	.158	.227	.254	TOTAL P.	

(c)

Figure 6.13 (c) Aluminum airfoil registers and grills, Specifications. (Courtesy of Lima Register Company)

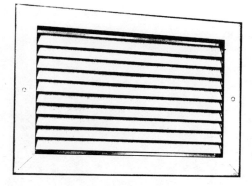

SERIES 90 RA H (shown above)

SERIES 90 RA H OB (attached opposed blade damper)

SERIES 90 RA V (no damper)

SERIES 90 RA V OB (attached opposed blade damper)

Deflection Fixed at either 40° or 0°

The streamlined air foil louvers on this Lima Aluminum unit reduce resistance thereby increasing the efficiency of the grille. It is available without damper or with an attached opposed blade damper. It is also available with either horizontal or vertical vanes . . . when ordering, specify RA H or RA V and desired angle of deflection.

Any style face available with the attached opposed blade damper.

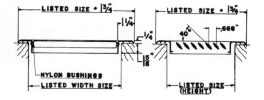

Figure 6.14 Fixed return air grill. For size selection see Figure 6.13b.

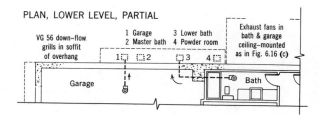

PLAN, LOWER LEVEL, PARTIAL

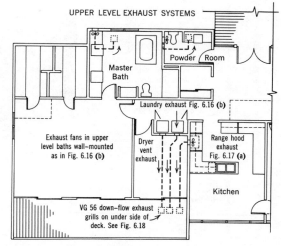

PLAN, UPPER LEVEL, PARTIAL

Product Specification

Exhaust, three bathrooms and garage: 8310 (Figure 6.16a)
Range hood: V-80 AG 30 (Figure 6.17a)
Hood blower: VP-400 (Figure 6.17b)
Switch, hood blower, integral with hood. NuTone catalog numbers.

Figure 6.15 (Example 6.1) Exhaust ventilation. This design avoids units or ducts that would protrude through exterior walls or roofs. Interior ducts 10 in. × 3¼ in. between studs or joists carry all exhausted air to inconspicuous downflow soffit grills. Dryer vent is self-powered by the fan which is part of the dryer unit.

8'' MODEL 8310 Extension–type hanger bars are included for 16'' or 24'' o.c. joists or studs. Housing with duct collar 13 1/16'' × 11 5/8'' × 3 3/16''. Use Model 834 Permanent Washable Filter, extra.

(a)

Model number	Type	Grille finish and size	Duct size	Blade size	Certified test data: Air delivery	Certified test data: Sound level
RF-68	Roof-attic cooling fan	—	—	14''	1250 CFM	
RF-58	Roof-attic cooling fan	—	—	12''	1020 CFM	
WF-57	Wall-attic cooling fan	—	—	16''	2090 CFM	Exterior Fans. All motor sound is outside the home!
RF-1N	Roof-blower	—	8''	—	607 CFM	
WF-1N	Wall-blower	—	8''	—	625 CFM	
RF-40	Roof-blower	—	10''	—	1000 CFM	
RF-35	Roof-blower	—	7''	—	300 CFM	
RF-17	Roof-impeller	—	7''	6⅜''	185 CFM	
WF-35	Wall-blower	—	7''	—	300 CFM	
WF-17	Wall-impeller	—	7''	6⅜''	185 CFM	
QT-110	Ceiling-blower	White pebble-grained polystyrene 10⅞'' x 13''	4''	—	110 CFM	2.5 sones
QT-80	Ceiling-blower	White pebble-grained polystyrene 10⅞'' x 13''	4''	—	80 CFM	1.5 sones
8812	Ceiling-blower	White pebble-grained polystyrene 10⅞'' x 13''	4''	—	100 CFM	4.0 sones
8662	Ceiling-blower	Silver anodized aluminum 13½'' dia.	4''	—	90 CFM	3.5 sones
8672	Ceiling-blower	Silver anodized aluminum 13½'' sq.	4''	—	90 CFM	3.5 sones
8833	Ceiling/wall blower	Silver anodized aluminum 8 9/16'' x 10 7/16''	3''	—	80 CFM	3.5 sones
8832	Ceiling/wall blower	Silver anodized aluminum 8 9/16'' x 10 7/16''	3''	—	80 CFM	3.5 sones
8510	Ceiling/wall impeller	Silver anodized aluminum 13½'' dia.	3¼'' x 10''	10''	300 CFM	7.5 sones
8310 *	Ceiling/wall impeller	Silver anodized aluminum 11'' dia.	3¼'' x 10''	8''	180 CFM	5.5 sones
8220	Ceiling-impeller	Silver anodized aluminum 11'' dia.	8''	8''	170 CFM	4.0 sones

(d)

Figure 6.16 (Example 6.1) *(a)* Exhaust fan type chosen for ventilation of baths, laundry and garage. *(b)*, *(c)*, Alternate uses of fans from NuTone's "Idea Book." *(d)* Exhaust fan selector chart (excerpt from a more complete list of products). (*Selected for use in Example 6.1.) (Courtesy of NuTone, Division of Scovill)

V-80 SELECT-A-MATIC RANGE HOOD SELECTOR CHART—Each Hood includes built-in power unit housing with filter, light assembly and switches.

Finish	30″ wide model	36″ wide model	42″ wide model
Harvest gold enamel	V–80HG30	V–80HG36	V–80HG42
Avocado green enamel *	V–80AG30 *	V–80AG36	V–80AG42
White enamel	V–80WE30	V–80WE36	V–80WE42
Colonial copper enamel	V–80CC30	V–80CC36	V–80CC42
Stainless steel	V–80SS30	V–80SS36	V–80SS42

*Poppy red enamel: Available on special order at extra charge. Allow four weeks from date order is received for delivery.

POWER UNIT SELECTOR CHART—V-80 SELECT-A-MATIC

			Certified Test Data			
			Air delivery:		Sound level:	
Model No.	Description	Fan Type	Horizontal	Vertical	Horizontal	Vertical
VP-200	Interior-mounted power unit	Single blower	210 CFM	210 CFM	5.5 sones	6.0 sones
VP-300	Interior-mounted power unit	Twin blower	330 CFM	320 CFM	5.0 sones	5.5 sones
VP-400 *	Interior-mounted power unit	Twin blower	410 CFM	420 CFM	6.5 sones	7.0 sones
RF-17	Exterior-mounted Roof Fan	Mixed-flow impeller	185 CFM			
WF-17	Exterior-mounted Wall Fan	Mixed-flow impeller	185 CFM		All motor sound is outside the home!	
RF-35	Exterior-mounted Roof Fan	Centrifugal blower	300 CFM			
WF-35	Exterior-mounted Wall Fan	Centrifugal blower	300 CFM			

(c)

Figure 6.17 (Example 6.1) *(a)* Range hood and *(b)* blower suitable for kitchen exhaust. *(c)* Hood and blower selector charts. (*Selected for Example 6.1.) (Courtesy of NuTone, Division of Scovill)

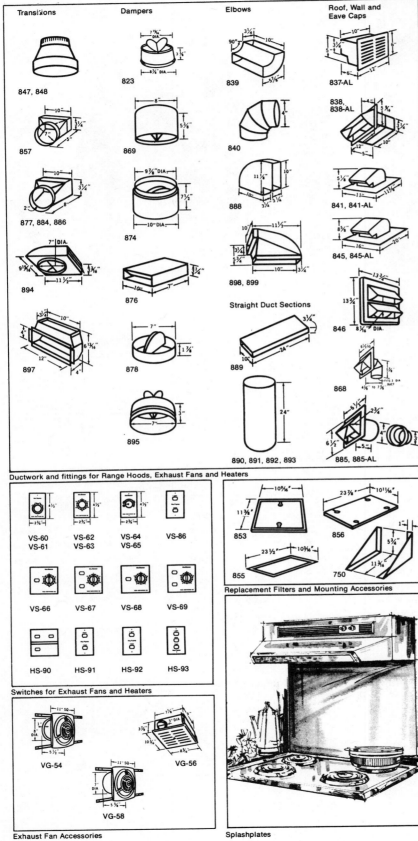

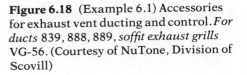

Figure 6.18 (Example 6.1) Accessories for exhaust vent ducting and control. *For ducts 839, 888, 889, soffit exhaust grills VG-56.* (Courtesy of NuTone, Division of Scovill)

Problems

6.1. According to the selector chart for heat pumps, what is the Btuh cooling output of an HP8-41FF heat pump?

6.2. If an HP8 series heat pump is located close to a structure, what clearances must be maintained on:
(a) The side through which the air is drawn?
(b) The airflow discharge side?
(c) The end, for service?

6.3. Heat pump HP8-261FF is operating at an outdoor temperature of 25° F. What is the *minimum* value in Btuh of auxiliary electric resistance heating that needs to be supplied to carry the requirements of Zone 1 (Example 6.1)?

6.4. (a) An aluminum curved blade one-way throw, 90CB register 8 in. × 8 in. delivers 175 cfm. What is the
(1) Face velocity?
(2) Throw?
(3) Minimum ceiling height?
(b) Give three reasons why this performance would not be satisfactory in a small residence.

6.5. A series 90 RAH return grill 30 in. × 12 in. is used. If the selected feet per minute through the grill is 500, how many cubic feet per minute can be handled if:
(a) Vanes are set at 0° to the airstream?
(b) Vanes are set at 40° to the airstream?

6.6. In a heat pump, the compressor operates whenever heating *or* cooling is needed.
(a) Where does *evaporation* of the refrigerant take place in summer: Outdoors or indoors?
(b) Where does *condensing* of the refrigerant take place in winter: Outdoors or indoors?

6.7. Name four locations in a residence from which it is desirable to have exhaust ventilation.

6.8. In a house with an *air* heating–cooling system (Problem 6.7), how is the air that is drawn *out* of the house replaced?

6.9. When a heat pump is in use for *heating* and the outdoor temperature drops from 50° F to 30° F, does the heat pump become *less* efficient or *more* efficient?

6.10. Refer to Figure 6.5. An HP8-463FF outdoor heat pump has been specified. It will be placed in a U-shaped alcove of the building. The architect requires that all clearances be 50% larger than minimum. Make a plan drawing to scale of the unit and this outdoor alcove. Show all dimensions, including those of the unit, the clearances and the alcove.

6.11. Detail the reinforced concrete pad for the two HP8-261FF outdoor heat pump units (Figures 6.9 and 6.5*b*.) Show plan, two elevations and all dimensions. Pad is 10 in. thick, projects 2 in. above the ground and is reinforced with rods 2 in. above its bottom. Edge of pad is to form a 6-in. margin around the units. Maintain minimum clearance for service between units. The *edge* of the planting is to be kept at a distance twice the allowed minimum for airflow.

Additional Reading

Applied Heat Pump Systems, ASHRAE Handbook and Product Directory, 1973, Chapter 11. An advanced engineering treatment of heat pump applications.

The Idea Book, Nutone, Division of Scovill, Cincinnati, OH. A very well illustrated book of equipment for the home, including residential exhaust fans.

7. Nonresidential Heating-Cooling

The terms *centralized*, *decentralized* and *zoned* can be applied to heating systems and to heating–cooling systems.

The possibility of adjusting the conditions in any part of a building or turning off any section of it is a great advantage. The comfort of occupants is usually best when local control is possible. Turning off any part of a system always saves energy.

Now that we have designed a number of systems, we can classify them. Example 2.1, page 40, illustrates a central system with one thermostat. Control of the heat in each room is possible only by manually adjusting the damper at each hot water baseboard convector element. In no room can the heat be shut off entirely. Example 2.2, page 54, affords some improvement. There is a thermostat and switch in each room. With the switch on, the temperature of that room can be controlled by the thermostat. Throwing off the switch cuts off all heat supply to the room. We consider this system decentralized. Example 3.1, page 69, has other advantages. There are two zones and each is controlled by its own thermostat and circulator. At each baseboard, convector, there is a valve *and* a damper. Adjustment or full shutoff are both possible. This is a centralized system. The solution to Example 3.2, page 75, is a *de*centralized system resembling that of Example 2.2.

Example 4.1, page 90, is a three-zone centralized system. Two of these zones are two-pipe circuits. In these, a shutoff valve at each convector allows both adjustment *and* shutoff.

Example 5.3, page 130, and example 6.1, page 140, are central air systems. Example 6.1 is a two-zone arrangement. Each zone has its own central unit which is an energy-saving heat pump. Both of these systems (Examples 5.1 and 6.1) can be balanced but individual room adjustment is not possible.

In this chapter, *full* decentralization is shown. The *incremental* units that are used include heating, cooling, ventilation and adjustable control at each unit.

Study of the incremental units described in this chapter will enable you to:

1. Understand the principle of all-electric (incremental) self-sufficient heating–cooling units.
2. Select, from five or more products, a unit suitable in type and output for a given purpose.
3. Read and understand "exploded" and "cutaway" views of a complex unit.
4. Obtain manufacturer's dimensional data by which the structure of the building can be made to adapt.
5. Accept and understand full control *at the unit* instead of at a remote thermostat.
6. Understand the operation of the heat–cool–low–high–off control at each incremental unit.
7. Design a system and select units for a specific use in a building for light manufacturing.
8. Make a plan and a section drawing by which the system can be installed.
9. Specify the incremental system for energy-saving operation in situations where individual units or groups of them can be turned down or off as the demand decreases.
10. Specify the unit you have selected by style, output designation, current model, voltage rating, and heater code rating.

7.1. Incremental Units

These conditioning units are of the through-the-wall type. See Figure 7.1. There are two kinds. One style uses hot water (or steam) from a central boiler for heating, together with electric cooling at the unit. The other type is all-electric, using electricity for heating and also for cooling, both *at the unit*. In this chapter, we deal only with the second type. It is all-electric and completely decentralized. This kind of conditioner offers a complete heating–cooling and ventilating service. The unit is brought to the site, built into the exterior wall, plugged into a suitable electric circuit and is ready for action.

The compressor-condenser (high pressure) side of the cooling cycle is part of the unit as shown in Figure 1.11, page 18, instead of being isolated or "split" as shown in Figures 1.12, page 19, and 5.19*b*, page 131. Because the compressor and condenser must be cooled by outdoor air, this part of the incremental unit constitutes a "breathing wall." Some of the characteristics that distinguish incre-

mentals from most conventional through-the-wall "room" air conditioners are:

(a) Greater cooling capacities.
(b) Inclusion of the heating feature.
(c) Styles adaptable to industrial and commercial buildings.

In distinction to *centralized* cooling and heating systems which use circulated air, incrementals are *not ducted*. Instead, each operates locally. One, several or many units are employed, depending on the size of the room or space to be conditioned.

7.2. Selection of a Type

Items that affect the choice of the type of unit include:

Heating–cooling outputs
Adaptability to the architecture
Cost

Sometimes more than one type could be acceptable. In Example 7.1 to follow, our choice is for the EA type. See Figure 7.2 and Section 7.5.

7.3. Installation and Operation of the EA Model

Figures 7.2 to 7.5. Basic to the installation is the wall sleeve or cabinet (3) (Figure 7.2), that is built into the wall and accepts the heating–cooling chassis (2). This chassis can be conveniently removed for service when the electrical power plug is disconnected after removing the front panel (1).

Operation of the assembled unit is clear from a study of Figure 7.3. Weather barrier I separates the outdoor* side of the unit from the indoor elements. Outdoor air is drawn in through the right side of the louver, L, by condenser fan, M. The air passes over the compressor, J, cooling it. A small amount of this *fresh* outdoor air is blown into the indoor circulation airstream through fresh air inlet, D. The bulk of the outdoor air, however, is blown across the condenser coil, O, and out through the left side of the louver. In passing through the condenser coil, the air draws

---

*Exposed to outdoor conditions through the louver.

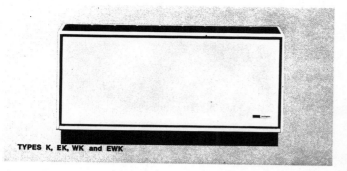

TYPES K, EK, WK and EWK

INCREMENTAL is adaptable . . . There's a type for every building. Those with prefix "E" heat electrically; all others with hot water or steam.

TYPES K, EK, WK, and EWK—Types K and EK are for through-wall installation where the outdoor air opening is at or near the floor. Cooling: 6500-18,500 Btuh. Heating: To 24,000 Btuh. Types WK and EWK are water-cooled, making them suitable for core use. Ventilation from a central source may be introduced through the EWK. Cooling: 9000-17,200 Btuh. Heating same as types K and EK. Room cabinets are available with top air discharge (as shown) or angle discharge.

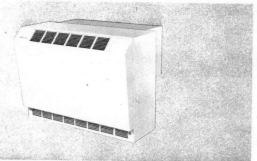

TYPES J AND EJ

TYPES J AND EJ—Have their outdoor air connections at the top, as often required at building offsets, with inverted spandrel beams or where louvers and windows are combined. Cooling: 6500 and 11,000 Btuh. Heating: Up to 18,700 Btuh.

TYPE EA

TYPE EA—For use where budgets are "tight." Cooling: 7100 to 17,300 Btuh in five sizes. Heating: Up to 18,750 Btuh. Heating and cooling chassis are combined. Can be mounted on floor, sub base or in wall. Positive pressure ventilation with concealed manual or motorized ventilation damper operator.

TYPES KG & EKG

TYPES KG & EKG—Used in larger offices and in classrooms which have minimum air requirements. Cooling: 15,900 to 18,500 Btuh. Heating: Up to 24,000 Btuh.

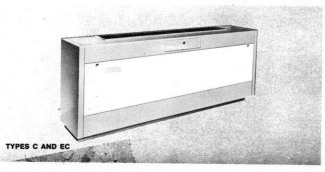

TYPES C AND EC

TYPES C AND EC—Equipment of greater capacity specifically designed for classrooms, laboratories, libraries, conference rooms and other densely occupied areas. Ventilation is adjustable to meet state codes. Cooling: 21,000 and 27,000 Btuh. Heating: Up to 32,700 Btuh.

Figure 7.1 Incremental heating-cooling equipment. See literature of The Singer Company, Climate Control Division, for detailed specifications that aid in the selection of a proper style.

1. *Front panel design* permits four-way room air return around its periphery.

2. *Heating-cooling chassis* is easily removed for service, without the use of tools. Slide-in, plug-in design means replacement in less than a minute.

3. *Room cabinet dual design* permits the cabinet also to function

as a wall sleeve—finished to resist weather on the outside, and to complement decor on the inside.

4. *Outdoor anodized aluminum louver* is flush-stamped to blend with building exteriors. (Optional extruded louver shown.)

5. *Base* (optional) (Not used in Example 7.1).

Figure 7.2 *(a)* Type EA incremental conditioner. Exploded diagram. *(b)* View, room side, front removed. (Courtesy of The Singer Company, Climate Control Division)

A Cabinet-Wall Sleeve
B Discharge Grille
C Front Panel
D Fresh Air Inlet
E Sub-Base (Optional)
F Evaporator Coils
G Evaporator Fan
H Electrical Receptacle (Supplied by others)
I Weather Barrier
J Compressor
K Control Box
L Outdoor Louver
M Condenser Fan
N No Frost Full Capacity Control Valve
O Condenser Coil
P Heating Element

Figure 7.3 Type EA incremental conditioner. Cutaway view of assembled unit. See discussion of function in Chapter text. (Courtesy of The Singer Company, Climate Control Division)

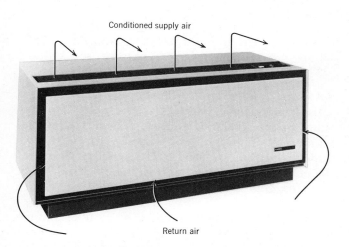

Conditioned supply air

Return air

Figure 7.4 Airflow pattern. Type EA incremental conditioner. (Courtesy of The Singer Company, Climate Control Division)

Front Panel

Four-way room air return around its periphery eliminates unsightly dust-catching grilles, and permits flat on the floor installation. Self-latching, the panel tilts forward for easy access to the permanent, cleanable aluminum mesh filter mounted on its reverse side. The panel can be readily removed from its hinging device for full access to the heating-cooling chassis.

Four-position Steel Discharge Grilles

Located at the top of the cabinet, the grilles can be easily arranged to provide the proper air distribution for the room. Four-position decorative extruded aluminum or continuous aluminum discharge grilles are optionally available.

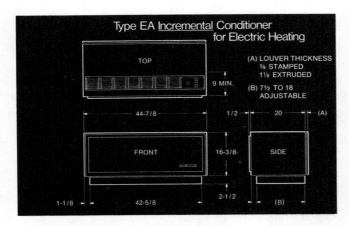

Figure 7.5 Dimensions, the EA incremental conditioner. (Courtesy of The Singer Company, Climate Control Division)

heat out of the hot, compressed refrigerant gas, causing it to condense to a liquid under pressure.

The *indoor* airstream is as follows. Return air from the room is drawn in around the edges of front panel, C, by evaporator fan, G. In passing the outdoor air inlet, D, it picks up some fresh air for ventilation. The return air plus a small amount of ventilation air is blown through both the evaporator (cooling) coils, F, *and* the heating element, P. Only one (*either* F or P) operates, depending on the season. The cooled (or warmed) air then passes back into the room through grills, B. This grill assembly consists of seven square, one-directional grills. Each can be raised, rotated and dropped back into a selected position. According to its position, each can deliver air directly toward the room *or* to the left or right *along* the wall *or against* the wall or the glass above. The airflow pattern is thus very controllable. In winter, the electric resistance heating coils turn on. In summer, the compressor-condenser combination operates to deliver high pressure liquid refrigerant to the evaporator coils. There it evaporates to a gas, absorbing heat from the circulated room air.

Two items not shown in Figure 7.3 are nevertheless part of the equipment. They are

1. Filtration
2. Fresh air control

Both the return air from the room and the fresh air from duct, D, are filtered before entering the evaporator fan at G. The front panel, C, is actually a

sandwich. The rear layer of the panel is a filter. The fresh air drawn in can be varied or shut off completely. For this purpose a small, manually operated lever at the right side of the front panel controls a damper in the fresh air duct.

7.4. Unit Control

The control box in each EA conditioner unit is of the off–heat–cool–high–low type. The low button permits reduced output for either cooling or heating. For cooling (with the *cool* button depressed) the *low* button causes the evaporator fan to run at lower speed. For heating (with the *heat* button depressed), the *low* button not only reduces the fan speed but also disconnects half the electrical resistance coils. Pressing the off button at any time terminates all action.

7.5. Design of a System

Example 7.1 A heating–cooling system is required for the work area of the building for light industry. Refer to Figure 4.3. The requirements are:

Cooling   66,200 Btuh
Heating   52,600 Btuh
Ventilation—outdoor air required

All-electric incremental units are to be used. Voltage is 230 v.

Solution: Singer units type EA are selected as a satisfactory choice. Outputs are suitable. Also they adapt to the structure and are reasonable in cost. Subbase, E, (Figure 7.3) is not used. Figure 7.5 shows dimensions that are used in Figure 7.8 to illustrate the position of a typical unit in the exterior wall of the work area. For summary and layout, see Figure 7.7.

The next step is to determine the number of units to be used, their location and the output to be used per unit for both heating and cooling.

As a trial, assume the use of six EA units (output to be determined). They can be located in the same six positions chosen for the hot water convectors that are used in this room for Example 4.1 and that are shown in Figure 4.12, page 94.

If six are used, we can find the required output for heating and cooling by dividing the total room requirements by 6.

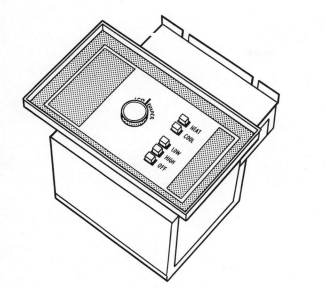

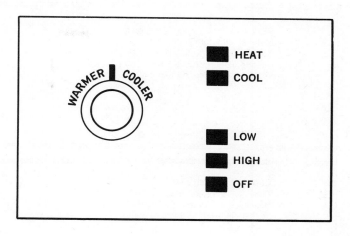

Figure 7.6 Control for the EA unit. See (K) in Figure 7.3. (Courtesy of The Singer Company, Climate Control Division)

Cooling   66,200 Btuh ÷ 6 = 11,000 Btuh per unit
Heating   52,600 Btuh ÷ 6 = 8700 Btuh per unit
Ventilation All EA units provide fresh air ventilation. Its rate must be checked against the needs of the workers and the requirements of codes and ordinances.

From Table 7.1, Performance Data, we see that an EA 12 will have a cooling output of 11,900 Btuh. OK > than 11,000. Table 7.1, Electrical Data, lists the EA 12, using "heater Code" B to produce 10,000 Btuh. OK > than 8700.

OVERDESIGN

Cooling   11,900 ÷ 11,000 = 1.08   (8% margin)
Heating   10,000 ÷ 8700 = 1.15   (15% margin)

These values, although slightly more than the requirements, are satisfactory and an EA 12 unit is specified. When both heating and cooling are required, it is difficult (and unnecessary) to meet *exactly* the engineer's requirements.

The EA unit is called out as noted in Figure 7.7 as EA 12 C 3 B:
EA Style (see Figure 7.1)
12 Output designation (Table 7.1)
C   Latest up-dated style (A and B were earlier and different)
3   230 v unit
B   Heater Code rating

7.6. System Performance

Economy of operation is an important feature of the incremental method chosen for the work area. At peak load for heat gain, all units can operate at cool-high. This could occur under conditions of:
Maximum outdoor temperature and humidity.
Full staff working and all lights on.
Power equipment in use.
Ventilation at a maximum.
A reduction in any or all of the above thermal loads would permit several or all of the units to operate at cool-low.

Similarly, for heating in winter, all units in mild weather could operate at heat-low.

Obviously, for *extremely* light demand for cooling or heating, a number of the six units could be turned off completely.

7.7. Rule of Thumb for Cooling

In the early stages of any technical design, some shortcuts are useful in estimating cooling "ton-

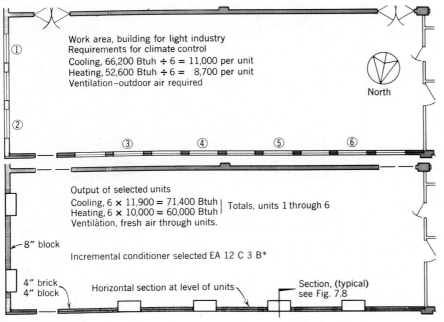

Work area, building for light industry
Requirements for climate control
Cooling, 66,200 Btuh ÷ 6 = 11,000 per unit
Heating, 52,600 Btuh ÷ 6 = 8,700 per unit
Ventilation–outdoor air required

North

Output of selected units
Cooling, 6 × 11,900 = 71,400 Btuh ⎱ Totals, units 1 through 6
Heating, 6 × 10,000 = 60,000 Btuh ⎰
Ventilation, fresh air through units.

Incremental conditioner selected EA 12 C 3 B*

8″ block

4″ brick
4″ block   Horizontal section at level of units   Section, (typical) see Fig. 7.8

PLAN, PARTIAL

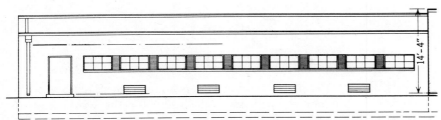

14′-4″

NORTH ELEVATION, PARTIAL

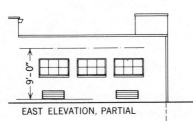

9′-0″

EAST ELEVATION, PARTIAL

Figure 7.7 (Example 7.1) Layout of the solution to Example 7.1 Heating and cooling of work space in the Building for Light Industry.
(*Manufacturer's equipment designation: EA is for the model identification, 12 identifies its output ratings, C means that is the third style change, 3 states that it operates at 230V, B is its "heater code"; see Table 7.1).

Architects/Engineers Specifications

Outside Air Louver

Shall be stamped anodized aluminum (or extruded) in natural finish. Louver shall be easily installed from inside of building after Room Cabinet-Wall Sleeve is set in place.

(Louvers furnished by others must be approved as to free area and design by Air Conditioner manufacturer.)

Room Cabinet-Wall Sleeve

Shall be entirely constructed of zinc-coated, phosphatized steel. Top and sides shall be 18-gauge with baked epoxy corrosion-resistant finish. Base pan shall be 16-gauge dipped in EHBP (environmental hydrous baking polymer) thermo-setting plastic and baked to form a continuous film of corrosion protection.

Installed height of Room Cabinet-Wall Sleeve shall not exceed 16-3/8". (Installed height of Room Cabinet-Wall Sleeve with sub-base shall not exceed 18-7/8".)

Discharge grilles shall be four-position stamped (extruded aluminum) to adjust to conditioner air delivery pattern without use of tools.

Front panel shall be capable of being opened and/or removed without the use of tools. Filter service shall not require removal of front panel.

Heating-Cooling Chassis

Shall be a slide-in, plug-in chassis with self-contained refrigerant cycle consisting of compressor, condenser fan and coil, evaporator fan and coil, refrigerant tubing and controls, electrical and operating controls, pressure ventilation system and condensate removal system.

Chassis shall be ready installable in and removable from the wall sleeve without the use of tools.

Compressor shall be welded hermetic, internally and externally vibration isolated, with permanent split capacitor motor and overload protection.

Refrigerant metering device shall consist of capillary restrictor, supplemented by a constant pressure automatic expansion valve for full cooling capacity at ambient temperatures down to 35°F. without evaporator coil freeze-up, compressor short-cycling, or slugging.

Chassis shall be constructed of zinc-coated, phosphatized steel parts dipped in thermo-setting plastic for corrosion protection.

All electrical components and controls shall be located in the conditioned air stream, except the hermetically sealed compressor.

Evaporator and condenser fans shall be forward-curved, aluminum centrifugal, statically and dynamically balanced. Fan assembly shall be driven by a three-speed, permanent split capacitor, permanently lubricated fan motor located in conditioned air stream. Motor shall be provided with oilers for life extension relubrication.

Condensate shall be removed by re-evaporation on the condenser coil surface without drip, splash, or spray. Condensate shall not come in contact with fan or fan motor. Slinger rings and propeller fans are not acceptable.

Forced, filtered ventilation air shall be available year-round. Conditioner shall be equipped with concealed manual (motorized) ventilation damper operator.

Electric heating elements shall be the quick response, low mass type with a high limit cut-out. (Hot water heating element shall be one row serpentine coil. Coil shall be controlled by motorized, normally open valve.) (Steam heating element shall be one row serpentine coil. Coil shall be controlled by motorized, normally closed valve.)

Operating Controls

Shall be provided in a separable, plug-in module as part of the heating-cooling chassis.

Control module shall consist of self-contained adjustable thermostat, with OFF-HEAT-COOL-HIGH-LOW selector switches.

A three-speed fan motor is supplied; two higher fan speeds are used on cooling mode; two lower on heating mode.

The electric heater capacity is controlled by the high-low pushbuttons in conjunction with the fan speed. Low heat capacity is approximately 50% of high heat.

Performance Data

FUNCTION	EA7	EA10	EA12	EA15
Cooling, Btuh (1)	6,500	9,300	11,900	14,200
Hot Water Heating, Btuh (2)	11,900	11,900	14,500	14,500
Steam Heating, Btuh (3)	16,000	16,000	19,000	19,000
Total Air/Vent Air, CFM Cooling—Hi-Fan	205/60	205/60	290/70	290/70
Cooling—Lo-Fan	180/50	180/50	255/60	255/60
Heating—Hi-Fan	190/50	190/50	280/60	280/60
Heating—Lo-Fan	170/45	170/45	250/55	250/55
Net Shipping Weight — lbs.	198	202	206	210

(1) **Based on ASHRAE and ARI test conditions of 95° DB/75° WB outside; 80° DB/67° WB inside.**

(2) **Based on 200° entering water, 180° leaving water, 65° entering air.**

(3) **Based on 2 psig steam, 65° entering air.**

Electrical Data

Rated In Accordance With ARI Standard 310-70

Cooling Load, Total	EA-7			EA-10			EA-12			EA-15		
	208V	230V	277V	208V	230V	277V	208V	230V	277V	208V	230V	277V
Watts	1,330	1,320	1,320	1,680	1,660	1,660	2,100	2,080	2,080	2,560	2,540	2,540
Full Load Amps	6.5	5.9	5.0	9.6	8.2	7.1	10.7	9.9	8.5	13.5	12.2	9.8
Locked Rotor Amps	28.3	23.5	24.5	41.0	36.0	35.0	44.0	43.0	44.0	59.5	56.0	49.0
Power Factor	98%	97%	95%	84%	88%	84%	94%	91%	88%	91%	91%	94%
Electric heat* Btuh												
A	7,800	7,400	7,400	7,800	7,400	7,400	8,000	7,500	7,500	8,000	7,500	7,500
B	11,400	9,900	11,400	11,400	9,900	11,400	11,600	10,000	11,500	11,600	10,000	11,500
C	—	—	—	—	—	—	15,000	14,000	14,000	15,000	14,000	14,000
Electric Heat* Amps												
A	11.2	9.5	7.9	11.2	9.5	7.9	11.3	9.6	8.0	11.3	9.6	8.0
B	16.1	12.7	12.1	16.1	12.7	12.1	16.4	12.8	12.2	16.4	12.8	12.2
C	—	—	—	—	—	—	21.1	17.8	14.8	21.1	17.8	14.8
Electric Heater Watts												
A	2,050	1,900	1,900	2,050	1,900	1,900	2,050	1,900	1,900	2,050	1,900	1,900
B	3,100	2,600	3,100	3,100	2,600	3,100	3,100	2,600	3,100	3,100	2,600	3,100
C	—	—	—	—	—	—	4,100	3,800	3,800	4,100	3,800	3,800

*Includes Fan Motor

Specifications subject to change without notice.

NOTE: Circles and arrows relate to Ex. 7.1

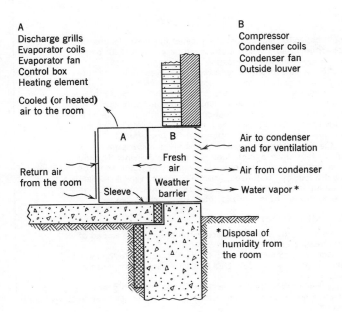

A
Discharge grills
Evaporator coils
Evaporator fan
Control box
Heating element

Cooled (or heated)
air to the room

B
Compressor
Condenser coils
Condenser fan
Outside louver

Air to condenser
and for ventilation

Return air
from the room

Fresh
air

Air from condenser

Weather
barrier

Water vapor *

Sleeve

*Disposal of
humidity from
the room

Figure 7.8 (Example 7.1) Incremental unit in place. Schematic cross section of one of the six self-sufficient heating-cooling-ventilating units powered by electricity at 230 v. Condensed room moisture from the surface of the evaporator coil is vaporized and blown outdoors together with the air from the condenser coil. See also Figures 1.11, page 18, 7.2, 7.3, and 7.4.

nages." Experience can help in setting standards for such estimates. Often they are expressed as "square feet of occupied area that can be cooled by one ton (12,000 Btuh) of refrigeration."

Since we have already designed three cooling systems, we can calculate values that might be useful on other jobs. The values of square feet divided by tonnage of *installed* cooling for our three jobs are:

$$\text{Basic Plan } \frac{1355}{2.5} = 542 \text{ sq ft per ton}$$

$$\text{Merker (upper level) } \frac{2060}{3.83} = 538 \text{ sq ft per ton}$$

$$\text{Industry (work area) } \frac{1656}{5.95} = 278 \text{ sq ft per ton}$$

For the first two (residential) cases, above, we can check against a well-informed source. A spokesman for an electric utility company in Long Island, New York states that about 550 sq ft per ton of refrigeration is a commonly used standard for residences.

Our figure of 278 sq ft per ton for the industrial building fits in well with the range of figures for other busy public areas in the following list. It is quoted by permission of the author, Mr. Herbert L. Laube, and his publisher, Business News Publishing Company, Birmingham, Michigan, from the book *How To Have Air Conditioning and Still Be Comfortable.*

Type of Store	Floor Area per Ton
Drugstore with lunch counter	120 sq ft
Drugstore or jewelry shop	160 sq ft
Camera shop or candy store	180 sq ft
Clothing store	200 sq ft
Furrier's shop	250 sq ft
Ma and Pa market	300 sq ft

Figure 7.9 Other uses of incrementals. In addition to their use in industrial and office buildings these units find wide application in motels, residences (as shown here), and also in hospitals, schools and other institutional buildings.

Problems

7.1. A large residence with 6500 sq ft of occupied area is to be cooled. As a preliminary approximation, how many tons of refrigeration will be needed?

7.2. What are the principal reasons that one ton of cooling serves only about 200 sq ft in stores but more than 500 sq ft in residences?

7.3. A 10,000 sq ft clothing store is being planned. About how many tons of refrigeration will be required to cool it?

7.4. A classroom requires 28,000 Btuh for cooling and 20,000 Btuh for heating. Two EA incrementals will be used. Select a size and the proper (A, B, or C) "heater code" for the electric coils.

Voltage is 230 v. What is the actual total output of the two units for heating? Cooling?

7.5. Describe the flow of air and water vapor through the louver on the external side of the weather barrier of an incremental unit.

7.6. Name three items on the external side of the weather barrier and three on the room side.

7.7. Describe the settings you would select on the control of an EA unit on an extremely hot day.

7.8. What is the principal difference in the action and performance of the all-electric incremental unit and the all-electric heat pump.

Additional Reading

How To Have Air Conditioning and Still Be Comfortable, Herbert L. Laube, Business News Publishing Company, Birmingham, Mich. This excellent book is written for laymen *and* technical planners by an early associate of Dr. Willis H. Carrier, the "father" of air conditioning.

Incremental Comfort Conditioners, Singer Climate Control Division, Auburn, N.Y. Guide book for architects and engineers for the selection and specification of incremental units.

Systems, ASHRAE Handbook and Product Directory, 1973, Chapter 6, "Multiple Unit or Unitary Systems." A broad development of the method described in our Chapter 7.

8. Principles of Plumbing

In our approach to this subject, it is most important that we postpone actual design considerations and put emphasis on:

What plumbing *does*.
The names of materials and equipment.
Methods of making connections.
The appearance of the systems.
How they adapt to the structure.

Thus, before we discuss the code requirements, tables and calculations in Chapters 9 and 10, we give attention to the following.

Water services
Domestic hot water heating
Sanitary drainage
Drain, waste and vent materials
Plumbing fixtures
Private sewage treatment
Storm drainage

The study of this chapter will enable you to:

1. Sketch a hot and cold water supply piping system to serve all the required water functions of a building.
2. Understand the differences in joining pipe-to-fittings, using steel, cast iron, copper or plastic.
3. Specify enough water valves and valves of the correct type.
4. Provide for the support of all vertical and horizontal pipes.
5. Distinguish between tank-type and tankless hot water coils, and between internal tankless and high output external tankless heaters.
6. Distinguish between plumbing fixtures with built-in (integral) traps and fixtures requiring external traps.
7. Draw a plumbing section.
8. Specify prefabricated assemblies of drainage, waste and vent piping when conditions permit.
9. Obtain and specify "roughing dimensions" for the guidance of your plumber.

10. Know that you may call for septic tank plus seepage pit sewage disposal *only* when there is no nearby sanitary sewer.

8.1. Water Services

The water system in Figure 8.1 serves two bathrooms, a lavatory, kitchen and laundry. Water closet tanks and garden hose bibbs are supplied with cold water; all other fixtures are supplied with both hot and cold water. The entire system is under pressure from the street main. When faucets at the fixtures are opened, the street pressure causes flow. When a *hot* water faucet is opened, the pressure acts *through* the hot water tank to deliver hot water at the faucet.

The water service entry pipe is valved at three locations, at the water main, below the curb box and inside the house. The pressure at the street main is usually about 50 psi, which is enough to raise the water to upper levels, overcome friction in the tubing and to assure proper flow at the fixtures. In buildings of three or four stories, there is seldom need for pumps or roof tanks. Taller buildings often require both.

In addition to faucets for the normal operation of the fixtures, there is usually a shutoff valve on each water pipe below the fixture. It can be closed against flow to permit replacement of the washer in the faucet. In cold climates, an unoccupied house should be drained to prevent freezing and bursting of tubing. This can be done by shutting off the valve below the curb box and opening the drain at the meter and at all other low points in the system.

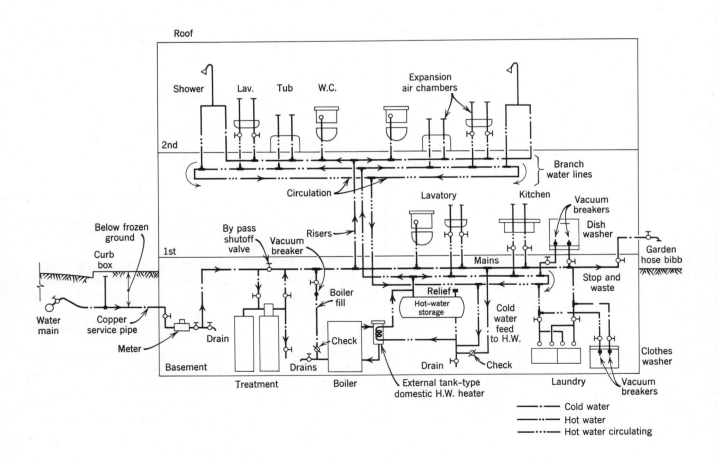

Figure 8.1 Upfeed water distribution by pressure in street mains. Schematic section, the water services of a residence.

In normal flow after the meter, the first step is passing the water through a treatment unit. This unit is optional. When it is used, its purpose is to neutralize acidity, which would corrode pipes and equipment, or to eliminate hardness, which would form a deposit in the system, clogging it.

Next the boiler must be supplied with water, usually automatically when boiler pressure drops. The balance of the cold water tubing supplies the garden hose bibbs, all fixtures and the domestic hot water heater–storage combination. This feed connection to the hot water heating is below the heater–storage. At the top of the hot water storage tank, the mains of the hot water system begin. Hot and cold water mains and branch lines supply the fixtures through fixture "runout" tubing.

Expansion of hot water is absorbed by the expansion air chambers, above each hot water faucet. These chambers (hot *and* cold) also minimize rattling of tubing when faucets are shut off abruptly.

The *direction* of water flow must be controlled at a number of places. Hot boiler water must not flow back into the cold water main. This is prevented by the check valve and a vacuum breaker. There is another check valve in the cold water feed line to the domestic hot water system. This prevents domestic hot water from flowing back into the cold water main. Polluted water in the dishwasher and clothes washer must not get into the "potable" water system. Vacuum breakers on hot and cold supplies of these machines prevent a suction that might allow this to happen.

At the supply to an exterior garden hose bibb, there is a "stop and waste" valve. In cold climates, this is closed in winter to prevent freezing of water in the hose bibb. When the stop and waste is shut off, water between it and the hose bibb automatically drains out at the indoor location of the stop and waste.

An important feature is the operation of the domestic hot water system. It is always circulating whether or not a hot water faucet is opened. There are three distinct circulation patterns.

(a) Hot boiler water flows into the jacket of the external tank–type domestic hot water heater. There it cools off as it delivers its heat to the (potable) water in the coil. This cool jacket water then descends to the (cool) bottom of the heating boiler. This cycle is continuous, day and night.

(b) The second continuous flow is between the coil and the hot water storage tank. The hot water from the coil rises to the top of the tank. The cooler water at the bottom of the tank drops back to the bottom of the coil for reheating.

(c) The third continuous circulation is between the tank and the hot water mains. The warmer water rises from the tank to the mains. As it gradually cools off it drops back to the bottom of the tank through the "circulation" lines.

During all of this motion, there need be no hot water demand. It goes on anyway. When, after a period of nondemand, a hot water faucet is opened, hot water at full temperature has only a few feet to flow from the main to the fixture—"Instant hot water." The circulation pipe for (c) above is an advantage but is not always installed. In that case, it may take half a minute or more for the water at full design temperature to arrive.

a. Pipe, Tubing and Fittings

The terms "pipe" and tubing can be used interchangeably. However, copper is usually referred to as tubing, and other materials are referred to as pipe.

For water distribution, possible choices are galvanized steel, brass (not frequently used), copper and plastics. Because of a tendency to corrosion, galvanized steel is used much less than copper and plastics. Notice that the plural form, plastics, is used rather than the singular, plastic. The reason is that a number of differing plastic materials are used. They are listed below.

Abbreviations used in this book to identify plastics pipe materials are those commonly accepted in the plastics industry. They are:

ABS   Acrylonitrile-Budadiene-Styrene
PB   Polybutylene
PE   Polyethylene
PP   Polypropylene
PVC   Polyvinyl Chloride
CPVC   Chlorinated Polyvinyl Chloride

For cold water piping, PVC is a good choice. For hot water piping, CPVC is chosen. It was especially developed to handle water at 180° F. If copper tubing is used for water distribution, type M is preferred of the three (K, L and M) that are available. One should refer to the standards and information of the Plastics Pipe Institute and the Copper Development Association before selecting or using any of these materials.

Pipes or tubes that run parallel carrying cold and hot water should have space-separation to prevent thermal interchange. All should be properly supported to minimize sagging.

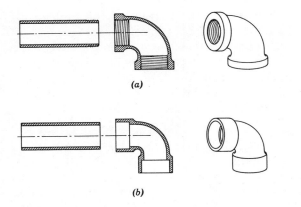

(a)

(b)

Figure 8.2 Methods of connecting pipes and fittings, and tubes and fittings. *(a)* Threaded: for ferrous pipe and fittings and for "iron pipe size" (IPS) brass. *(b)* Soldered: for copper tubing and fittings. A sliding fit similar to that of *(b)* is used for the solvent weld of plastic connections.

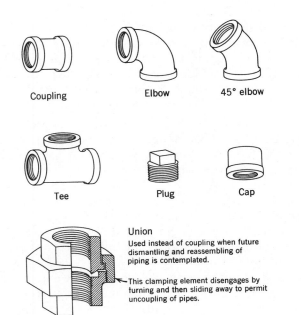

Coupling   Elbow   45° elbow

Tee   Plug   Cap

Union
Used instead of coupling when future dismantling and reassembling of piping is contemplated.

—This clamping element disengages by turning and then sliding away to permit uncoupling of pipes.

Figure 8.3 Examples of *threaded* pipe fittings for ferrous or brass pipe. A few of the many fittings used in water piping. These and all fittings are also available for solder-joint (copper) or solvent weld (plastic) connections and usually for transition from one system of material to another.

b. Fittings and Valves

Study Figures 8.2 to 8.4. Steel fittings and pipe are connected by threads (Figure 8.2*a*). Copper tubing and its fittings are connected by a sliding fit as shown in Figure 8.2*b*. The contacting surfaces in copper are soldered with a special solder. In plastic joints (also a sliding fit), a solvent weld or heat fusion is used.

Valves are shown in Figure 8.4. Type (a), the gate valve, is used where the valve is usually *open*. For valves that are commonly *closed*, (b) and (d) are used. Check valve (c) limits flow to one direction.

c. Expansion and Shock

In a closed system, an increase in water temperature could burst a pipe or open the hot water relief valve. See Figure 8.5. Methods are shown that solve this problem. Capped air chambers are used. Small bubbles of air are trapped in *all* water. They collect and remain in the chambers. This air provides a cushion against expansion and also against dynamic shock.

d. Pipe Support and Expansion

Study Figure 8.6. This illustration shows pipe sup-

port methods in heavy construction. In wood frame construction perforated metal straps are suitable as shown in Figure 8.17*b*. Allowance must be made for thermal expansion of pipe. Relative values for several materials are shown in Figure 8.23.

8.2. Domestic Hot Water Heating

There are many methods of heating, storing and distributing domestic hot water. One system already described was shown in Figure 8.1. An external tank-type heater utilized boiler warmth to transfer heat to the domestic water. By continuous circulation, a reservoir of hot water was created in the tank and drawn on as required.

Another method, a tankless coil that eliminates the tank, is illustrated in Figure 8.7. Depending on the size and rating of the coil, it can deliver between

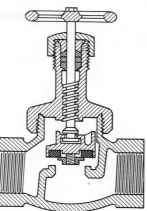
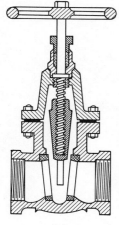

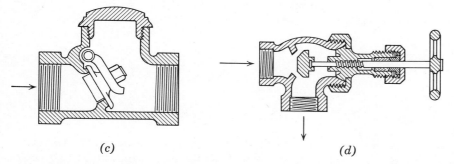

Figure 8.4 Typical valves for water systems. *(a)* Gate valve. *(b)* Globe valve. *(c)* Check valve. *(d)* Angle valve.

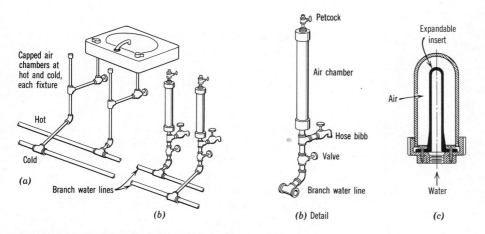

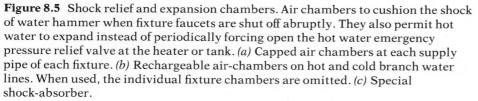

Figure 8.5 Shock relief and expansion chambers. Air chambers to cushion the shock of water hammer when fixture faucets are shut off abruptly. They also permit hot water to expand instead of periodically forcing open the hot water emergency pressure relief valve at the heater or tank. *(a)* Capped air chambers at each supply pipe of each fixture. *(b)* Rechargeable air-chambers on hot and cold branch water lines. When used, the individual fixture chambers are omitted. *(c)* Special shock-absorber.

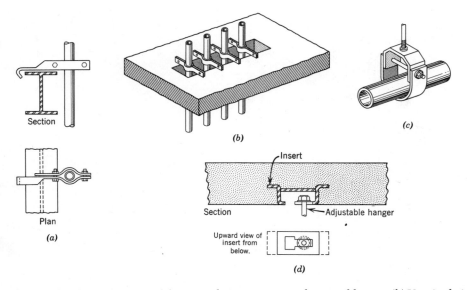

Figure 8.6 Pipe supports. *(a)* Vertical riser supported at steel beam. *(b)* Vertical riser groups supported at slot in concrete slab. *(c)* Horizontal pipe hung from slab above by adjustable-length clevis hanger. *(d)* Typical metal insert in soffit of concrete to receive hanger-rod.

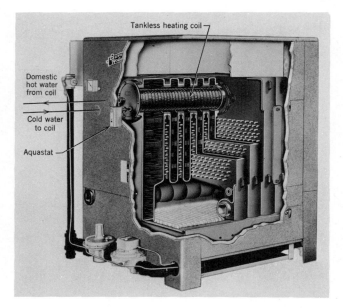

Figure 8.7 Internal tankless heating coil for domestic hot water, immersed in the jacket water of a gas-fired hot water heating boiler. Approximate capacity range 3 to 15 gpm at 100 F° rise. (Courtesy of the Burnham Corporation)

3 to 15 gpm at 100 F° rise in temperature. An aquastat in the boiler water at the location of the coil turns on the gas fire if the water temperature drops. Aquastat setting could be 180° F. The domestic water would then be delivered at about 160° F, a 100 F° rise from the probable 60° F cold water entering the coil.

By using an *external* tankless heater, greater heating rates are possible. Figure 8.8 indicates a yield of 4 to 25 gpm at 100 F° rise. In the cutaway view, a finned copper tube carrying the domestic water is seen. Hot boiler water enters at the left and returns to the boiler from the bottom connection in this insulated unit. Experience shows that an internal or external tankless of about 6 gpm capacity will usually take care of hot water needs of a house with two baths, a kitchen and a laundry.

A third method is illustrated in Figure 8.9. This is a separately fired unit. The ROF-50 type *stores* 50 gal of hot water which could serve a sudden heavy demand period. In the time between such demands the oil burner would heat a make-up supply for the tank at a rate of 120 gal/hr.* The tank effectiveness could thus be restored in 50/120 × 60 min. = 25 min.

*At a *firing* rate of 1.0 gallons of oil per hour.

TANKLESS · EXTERNAL

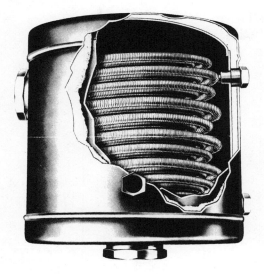

ALL-COPPER

Integral finned tubing in unique whirl-pool coil design assures maximum heat transfer at low cost. Fiberglas insulation reduces fuel consumption. Rugged octagon wrench grips permit tight connections. Compact design allows easy one-man installation. Unit easily drained. Attractive exterior finish. Use of General mixing valve is recommended with this heater.

No vent required.

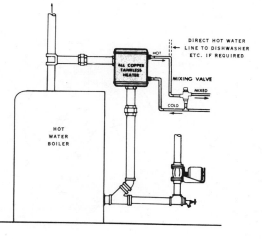

MODEL NUMBER	GPM 40°—140°	WATER CONNECTIONS INCHES		DIMENSIONS, INCHES		SHIP. WGT., LBS.
		SERVICE	BOILER	HEIGHT	DIAMETER	
400	4	½	2	14⅛	11⅝	31
500	5	½	2	14⅛	11⅝	33
501	6	½	2	14⅛	11⅝	35
502	8	1	2	16¼	11⅝	40
503	10	1	2	18	11⅝	48
504	12	1¼	3	21⁷⁄₁₆	13	70
505	16	1¼	3	22¹⁵⁄₁₆	13	80
506	20	1½	3	24⁷⁄₁₆	13	90
507	25	1½	3	24⁷⁄₁₆	13	100

Capacity ratings based on boiler water at 200°F.
Maximum working pressure: shell — for water, 30 lbs; for steam, 15 lbs; coils — 150 lbs.

Figure 8.8 High output external tankless heater. This unit for heating domestic hot water is similar to the internal tankless coil shown in Figure 8.7, but it operates on boiler water pipes to the *external* coil. Output ratings 4 to 25 GPM. (Courtesy of General Fittings Company)

Duraclad OIL WATER HEATER

STANDARD FEATURES

1. **FINEST GLASS-LINED TANK** from the inventors of the glass-lined water heater. High carbon steel coated with A. O. SMITH's exclusive corrosion-resistant glass-lining formula.

2. **PATENTED HYDRASTEEL** construction-glass fused permanently to steel.

3. **FACTORY-INSTALLED OIL BURNER** is clean and quiet. Advanced design flame retention principle ideal for water heating purposes. Combines high efficiency with easy-service features. The fuel unit includes pump, strainer, pressure regulating valve and ignition transformer. UL listed for use with No. 1 and 2 commercial grade fuel oil.

4. **PRIMARY BURNER CONTROL.** Equipped with cadmium cell. relay to automatically shut off power if ignition or flame fails.

5. **EFFICIENT COMBUSTION CHAMBER.** Pre-cast, high temperature combustion chamber made of alumina silica ceramic fiber. Engineered for maximum heat transfer. Unique design assures more complete • combustion by stabilizing flame pattern.

6. **FLAME OBSERVATION PORT** permits easy inspection of flame.

7. **100% SAFETY SHUT-OFF** with built-in temperature limiting device.

8. **FULLY AUTOMATIC THERMOSTAT** provides even and accurate control to a maximum of 160°F on ROF-30, and 180°F on ROF-50.

9. **FLUE BAFFLE** scientifically designed for maximum thermal efficiency.

10. **HIGH-DENSITY ANODE** of high-purity extruded magnesium for longer tank life.

11. **DIP TUBE** — stainless steel on ROF-50; non-metallic on ROF-30.

12. **NON-CORRODING NIPPLES.**

13. **FULL INSULATION** Glassfiber blanket prevents costly heat loss; non-sagging and vermin-proof.

14. **HEAVY-DUTY EXTERIOR FINISH.** All surfaces bonderized and finished with baked enamel.

15. **PEDESTAL BASE** for easy cleaning.

16. **RELIEF VALVE OPENING** for ease of installation. Working pressure 150 psi . . . tested at 300 psi.

¾" **DRAIN VALVE** at tank bottom for fast, convenient draining. Brass on ROF-50. Master-Flo on ROF-30.

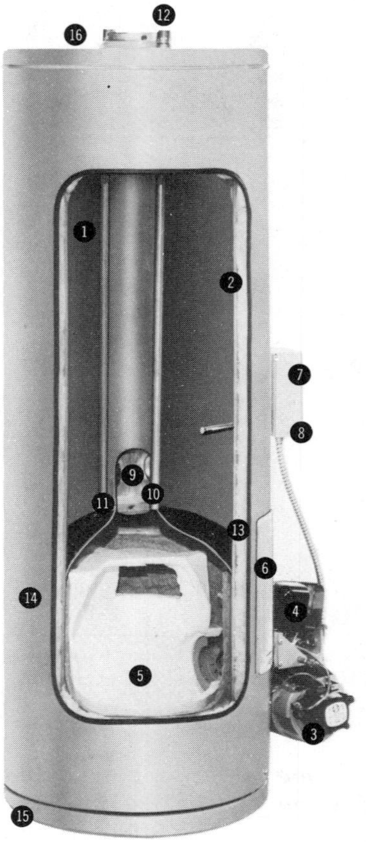

OPTIONAL FEATURES

- Two-stage pump — for use with underground oil storage tanks.
- Burner-mounted cadmium cell relay for 30-second flame detection; (45-second safety timing switch standard).
- Hand hole cleanout — on ROF-50 only.
- Solenoid valve for fast shutdown.

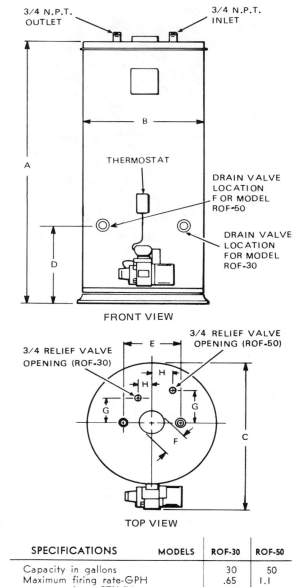

FRONT VIEW

TOP VIEW

SPECIFICATIONS MODELS	ROF-30	ROF-50
Capacity in gallons	30	50
Maximum firing rate-GPH	.65	1.1
Input-BTU/Hr.	91,000	154,000
Recovery-GPH/100° rise	78	131
Recommended firing rate-GPH	.60	1.0
Input-BTU/Hr.	84,000	140,000
Recovery-GPH/100° rise	72	120
Motor HP	1/7	1/7
Approx. Ship. Wt., lbs.	184	335
DIMENSIONS: A	54⅝	59½
B	18¾	25⅜
C	26	32⅝
D	19⅞	21½
E	8	16
F	4	5
G	3½	2
H	5¾	5¾

Figure 8.9 Oil-fired domestic hot water heater. Excerpt from the manufacturer's brochure describing features and specifications. The ROF-50 would be suitable for Example 3.1, page 69, as an optional separate hot water heater in the boiler room if the tankless heater in the Smith Pac–12 boiler were not used. See Figure 3.12a, page 74. (Courtesy of A.O. Smith)

8.3. Sanitary Drainage

It has been said that today's water supply is tomorrow's sewage problem. Although this statement is trite, it does have some value for us. It reminds us that there are two distinct systems and that "never the twain shall meet." The water system distributes clean potable water. At each fixture, we *destroy* this cleanliness. The "sanitary" drainage system carries away the foul fluids and solids created at the fixture. The drainage system is not really very sanitary. It *does*, if effective, assure sanitation for the occupants.

An important precaution, already mentioned, is that sewage or polluted "effluent" must not be drawn into the water system. To prevent this from occurring, the water is usually delivered *above* the water level of the fixture. The lavatory (washbasin) is an example of this. Because of the high position of the water faucet, an accidental suction in the faucet *cannot* draw polluted water from the basin into the clean water pipes. If the water *must* be delivered *below* the water level of a fixture (flush-valve-type water closet, clothes washer and dishwasher), a

vacuum breaker is installed (Figure 8.1). This device, placed in the water supply pipes, breaks the suction that would cause backflow of sewage into the water piping.

a. Traps and Vents

Another important item comes to mind. The house should not smell. More important, odor-laden air and *germ*-laden air should not rise out of the *un*-sanitary drainage system pipes. There must be enough water pressure at the fixture and proper flow of water to wash away all of the foul material. Second, the water must flow freely enough to leave a measure of *clean* water to seal the fixture trap.

Now Figure 8.10 becomes important. If there were no trap (a), there would be no seal against an unsanitary condition. Sketch (b) adds the trap. This is still unsatisfactory. The last of the water *siphons* out. It just keeps going, leaving the trap dry. But this can be prevented. If an air-filled vent is attached at the horizontal runout (c), we can break the siphon and allow clean water to drop back into the trap.

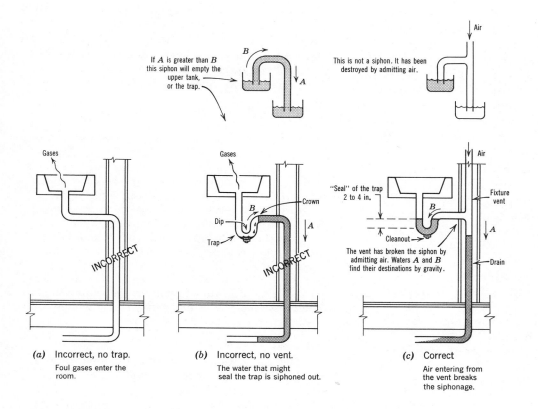

Figure 8.10 The function of a trap and one of the several functions of a vent (preventing siphonage).

Another function of the venting system (not detailed here) is to prevent foul gases from bubbling through the trap and passing into the room. This could be caused by pressure in the soil and waste pipes that results from hydraulic flow. The pressure must be relieved through a network of vents, *not* through the traps. *Every* fixture must be trapped. The traps are not always apparent. Some are part of the fixture as in a water closet. A lavatory trap is usually below the fixture (Figure 8.24). Occasionally two fixtures may use one trap [stack (b) in Figure 8.11].

b. Sanitary Drainage Systems

Study Figures 8.11 to 8.13. Figure 8.11 is an aid in learning the names generally used for parts of the system. It is also a clear example of several classes of drainage combined into one system. They are: *storm*, *waste* and *soil*. Respectively, they comprise:

Storm   Rainwater
Waste   Minor pollutants: lavatories, sink, laundry trays, showers
Soil   Major pollutants; body wastes—water closets and urinals

The sewer is therefore a "combined" sewer. It combines sewage and storm drainage. When a separate municipal storm sewer exists (which is seldom) rain runoff piping should connect to it and *not* to a sanitary sewer. This makes it easier for city sewage treatment plants to treat the heavy pollutants separately. When combined storm and sanitary systems are unavoidable, the storm leader is trapped so that house drain odors are stopped at its base. Cleanouts are located at the base of stacks, at the start of the house drain and at the start of the house sewer.

Although this diagram shows the traditional plumbing system, which is still very much in use, many changes are being adopted. They include:

Elimination of the house trap and fresh air inlet.

"Loop venting" which reduces the extent of vent piping.

Return of storm water to adjacent ground instead of through storm sewers to nearby waterways.

Treatment of the products of soil lines separate from that of waste lines.

Total elimination of vents by employing the ventless "Sovent" system (See Additional Reading at the end of this chapter).

In Figure 8.12, we find two of the above listed developments; there is no house trap or fresh air

inlet and no storm drainage combined with sewage. The floor drain, an item seldom used, is trapped before it connects to the house drain. The trap can be seen just behind the house drain. Private sewage treatment is shown. The septic tank provides the major purification while *its* effluent is returned to the ground through one or more seepage pits.

Figure 8.13 gives us a close-up view of the details of drainage and vent piping for a single bathroom and for two bathrooms back to back.

8.4. Materials for Drainage, Waste and Vent (DWV)

a. Cast Iron

Cast iron, a very durable material, has long been a standard for sanitary and storm drainage and for waste piping and vents. Typical fittings are shown in Figure 8.14. They are of the conventional type to be connected by a lead and oakum joint. The lead and oakum joint and two other joints (see Figure 8.15), using rubber gaskets instead of lead and oakum, are described in the following paragraphs.

The Cast Iron Soil Pipe Joint and Its Characteristics

"The cast-iron soil pipe joints [Figure 8.15] are semi-rigid, water and gas tight connections of two or more pieces of pipe or fittings in a sanitary system. These joints are designed to give rigidity under normal conditions, and still permit sufficient flexibility under adverse conditions, such as ground shift, footing settlement, wall creepage, building sway, etc. to allow pipe movement without breakage or joint leakage. Properly installed, the joints have equal longevity with the cast iron soil pipe, and can be installed in walls, under ground, and in other inaccessible places and forgotten. The joints seal by compression of the sealing material and will give a good tight joint even under zero pressure.

The Lead and Oakum Joint

"The conventional cast-iron soil pipe joint is made with oakum fiber and molten lead [Figure 8.15a]. This provides a waterproof joint that is strong, flexible and root-proof. The waterproofing characteristics of oakum fiber have long been recog-

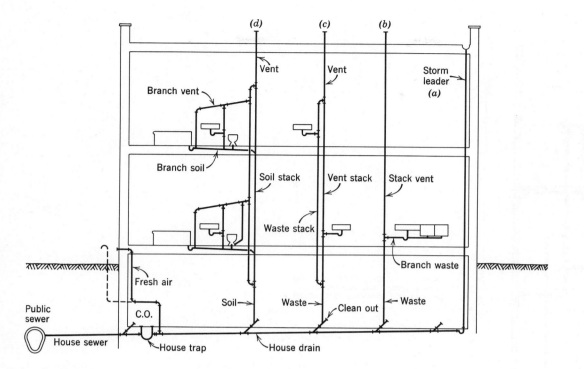

Figure 8.11 Typical plumbing layout. This diagram, known as a plumbing "section," is for a combined system (sanitary plus storm drainage). When so combined, the storm leader (a) must be trapped at its base. Stacks (b) and (c) are known as wastes. Stack (d), serving water closets, is designated as a soil stack.

nized by the plumbing trades, and when molten lead is poured over the oakum in a cast iron soil pipe joint, it completely seals and locks the joint. This is due to the fact that the hot metal fills a groove in the bell end of the pipe firmly anchoring the lead in place after cooling. When the lead has cooled slightly, it is rammed into the joint with a caulking tool to form a solid metal insert. The result is a locktight soil pipe joint with excellent flexural characteristics.

The Compression Joint

"The compression joint is the result of research and development pursued by a number of foundries to provide an efficient, lower-cost method for joining cast iron soil pipe and fittings. The joint is relatively new only in application to cast iron soil pipe, since similar compression-type gaskets have been successfully used with watermain for more than thirty years. The compression joint [Figure 8.15b] uses hub and spigot pipe as does the lead and oakum joint. The major differences are the one-piece rubber gasket and the spigot end of the pipe which is always plain or without a bead. When the spigot end of the pipe or fitting is pushed or drawn into the gasketed hub, the joint is sealed by displacement and compression of the rubber gasket. The resultant joint is leak-proof, root-proof and pressure-proof, absorbs vibration and can be deflected up to 5 degrees without leakage or failure.

The No-Hub Joint

"The no-hub joint for cast iron soil pipe and fittings is a new plumbing concept which supplements the lead and oakum and compression-type hub and spigot joints by providing another and more compact arrangement without sacrificing the quality and permanence of cast iron. As can be seen [in Figure 8.15c], the system uses a one-piece neoprene gasket and a stainless steel shield and retaining clamps. The great advantage of the system is that it permits joints to be made against a ceiling or in any limited-access area. In its 2-inch and 3-inch sizes it will fit into a standard 2-by-4 inch partition without furring. Installation is fast and efficient.

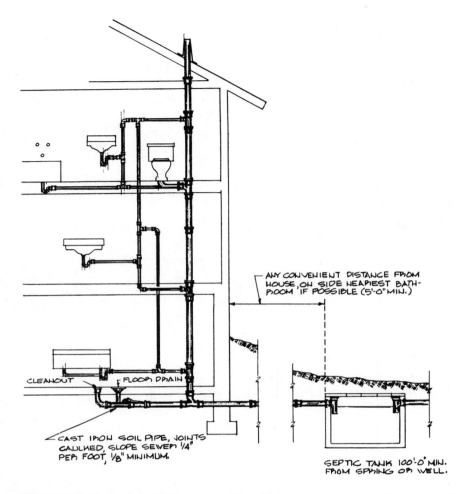

ANY CONVENIENT DISTANCE FROM HOUSE, ON SIDE NEAREST BATH-ROOM IF POSSIBLE (5'-0" MIN.)

CLEANOUT

FLOOR DRAIN

CAST IRON SOIL PIPE, JOINTS CAULKED, SLOPE SEWER 1/4" PER FOOT, 1/8" MINIMUM.

SEPTIC TANK 100'-0" MIN. FROM SPRING OR WELL.

Figure 8.12 Typical piping layout and details for septic tank use. (Courtesy of Cast Iron Soil Pipe Institute)

"The 300 series stainless steel shield which is used with no-hub joint was selected as a result of soil corrosion testing conducted by the National Bureau of Standards and on test data supplied by the International Nickel Company, Armco Steel Corporation and Crucible Steel Company of America. It is noncorrosive and resistant to oxidation, warping and deformation. It offers rigidity under tension with a minimum tensile strength of 165,000 psi, and yet provides sufficient flexibility. The shield is corrugated in order to grip on the gasket sleeve and give maximum compression distribution. The stainless steel worm gear clamps compress the neoprene gasket to make a permanent watertight, gastight joint. The gasket absorbs shock vibration and completely eliminates galvanic action between the cast iron pipe and the stainless steel shield.

Soundproofing Qualities of Cast Iron with Rubber Gasket Joints

"One of the most significant features of both the compression and no-hub joints is that they assure a quieter plumbing drainage system. The problem of noise is particularly acute in multiple dwelling units and although soundproofing has become a major concern in construction design, certain plumbing products have been introduced which not only carry noise but in some cases actually amplify it. The use of rubber gaskets and cast iron soil pipe reduces noise and vibration to an absolute

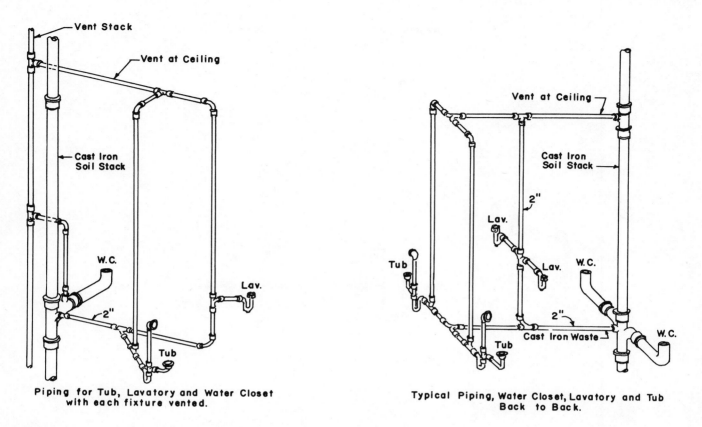

Figure 8.13 Two typical piping arrangements for a water closet, lavatory and tub. (Courtesy of Cast Iron Soil Pipe Institute)

minimum. Because of the weight and wall thickness of the pipe, sound is muffled rather than transmitted or amplified, and the rubber gaskets separate the lengths of pipe and the units of fittings so that they cushion any contact-related sound. The result is a home which is more livable and of greater value."*

*These paragraphs are reprinted verbatim from "Cast-Iron Soil Pipe and Fittings Handbook" by permission of the Cast-Iron Soil Pipe Institute.

b. Plastics for DWV

The technical development of plastics for use in drainage waste and vent applications has brought these materials to complete suitability in their prescribed uses. They have widespread acceptance in the plumbing field. Codes have given approval.

A list of six plastic materials was given in Section 8.1. It refers to the use of plastics for water piping in plumbing systems. Two of them, ABS and PVC, are particularly suited for DWV use.

Figure 8.16 lists and describes the identification to be expected on plastic DWV pipe and fittings. Figure 8.17 shows pipe supports and one step in joining

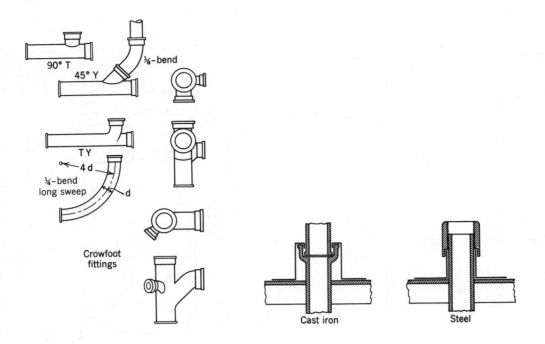

Figure 8.14 Cast-iron fittings. Principal types and method of flashing at roofs.

pipe and fittings. Figure 8.18 illustrates the diverse use of plastics in a plumbing system. It also provides a good picture of a complete installation and its services. Items to be noted are:

Position of traps. (Water closet trap is *in* the fixture.)

Combined view of fixtures, drainage, venting and water services.

Gas-fired domestic hot water heater. Note relief valve above it.

Underground building sewer to municipal street sewer.

Water and gas service from street mains to building.

In *any* DWV assembly, vent connections can be made at 90-deg, but 45-deg slopes are preferred for fluid flow.

Figures 8.19 and 8.20 indicate the lightness of plastic piping and the way it fits into the wood framing of residential construction. Soil piping (regardless of material) is usually 4 in. in diameter when one or several water closets are served. Smaller sizes are adequate for lavatories and other "non-soil" uses (see vertical pipes at right side of Figure 8.20).

c. Copper for DWV

Some advantages are claimed for copper in plumbing systems. Copper is considered to be:

Faster to assemble.
Easier to cut.
When preassembled, to hold together better.

An example of efficient preassembly is shown in Figure 8.21. It is referred to as a "plumbing tree" and is used in the manufactured kitchen-bath unit shown in the background. The tree can be factory assembled in one-half hour. Assembly in place in the field would take a half day.

In the photograph you will find that, on either face of the DWV assembly, there are water-distributing assemblies strapped on for shipping. Their positions are upside down in the picture. One can see the expansion chambers, which are of somewhat larger diameter than the water supply tubing.

A typical example of an all-copper installation in a split-level residence is illustrated in Figure 8.22. The legend describing the system is self-explanatory. Note, however, that the drains are capped and remain so prior to the connection of the plumbing fixtures. This closes the tubing system to permit

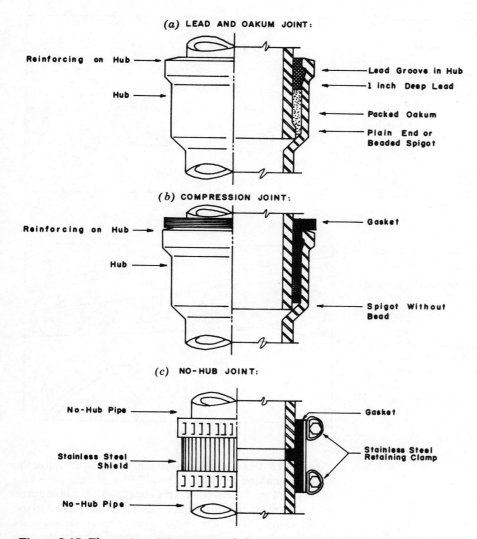

Figure 8.15 The various joints presently being used to connect cast-iron soil pipe and fittings. (Courtesy of Cast Iron Soil Pipe Institute)

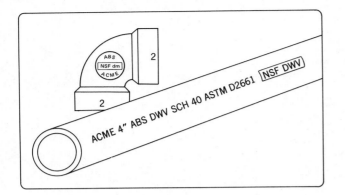

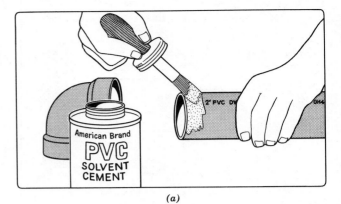

(a)

ACME	The name of the manufacturer.
4 in.	Diameter of the pipe.
ABS	Acrylonitrile-Butadiene-Styrene, the material.
DWV	Suitable for drainage waste and vent.
SCH 40	Schedule 40. This identifies the wall thickness of the pipe.
ASTM D2661	"Standards Number" assigned by the American Society for Testing Materials.
NSF DWV	Tested by the National Sanitation Foundation Testing Laboratory. The pipe meets or exceeds the current standards for sanitary service.

Figure 8.16 Typical identification symbols on plastic pipe. (Courtesy of Plastics Pipe Institute)

(b)

Figure 8.17 Details in the use of plastic pipe. (a) One of the steps in making a "solvent weld" of a plastic pipe to a plastic fitting. (b) In wood frame construction, plastic pipe assemblies can be supported by metal straps nailed to the wood joists. Flexibility of the plastic material suggests that the supports be more closely spaced than in the case of metal piping. (Courtesy of Plastics Pipe Institute)

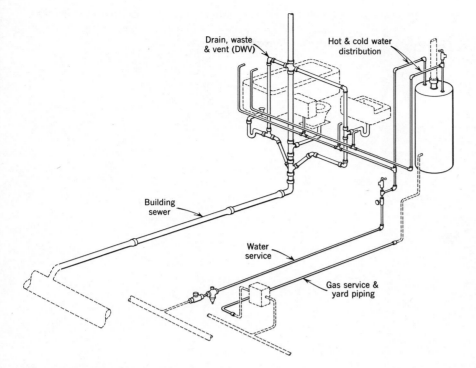

Figure 8.18 Plastic piping (solid lines) for water service, gas service, hot and cold water lines and for drainage, waste and vent. Gas service *below grade* can be PE, PB or PVC. (Courtesy of Plastics Pipe Institute)

Figure 8.19 This photograph indicates the lightness of plastic DWV materials (Courtesy of Plastics Pipe Institute. Photo by Richards Studio)

Figure 8.20 Plastics lend themselves to pre-assembly of sections of DWV piping. Materials used for DWV are: polyvinyl chloride (PVC), either tan or white; and acrylonitrile-butadiene-styrene (ABS), black.

The architect or engineer should check structure that may be cut to accommodate piping. Notice in this picture that the notched joists are deeper and more closely spaced than other floor beams. (Courtesy of Plastics Pipe Institute. Photo by Richards Studio.)

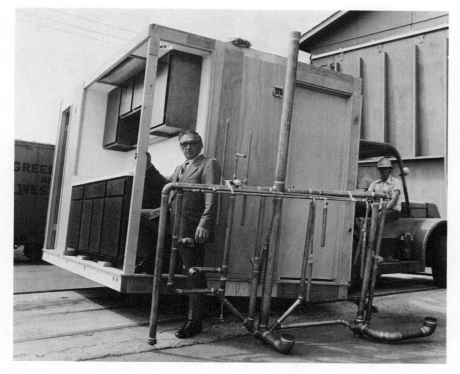

Figure 8.21 Pre-assembly. Raleigh Fisher, Chief Mechanical Designer for Wausau Homes, Inc., is shown with a Wausau copper plumbing "tree," which includes use of Copper Drainage Tube DWV. Wausau is one of the nation's largest builders of manufactured housing. (Photograph courtesy of Copper Development Association, Inc.)

Figure 8.22 Drainage and vent piping in a frame residence. This installation in copper "drainage waste and vent tubing" (DWV) serves two bathrooms at the upper level behind the 6-in. stud partition and a kitchen sink and laundry tray at the lower level, which are on this side of the partition. In the bathrooms the roughing serves, from left to right, a lavatory, water closet and bathtub and a lavatory, shower and water closet. Bathtub and shower traps can usually be accommodated within the joist depth. The bend below the water closets, however, often leads to a horizontal branch exposed or furred-in below the joists. Some codes permit this branch from a water closet to be 6 to 10 ft long before joining a vent. The water piping is not yet entirely in place. (Courtesy of Copper Development Association)

testing and correction of possible leaks. After this step the wall surfacing may be applied to the studs. The method of joining DWV copper tubing and fittings is similar to methods that are suggested in Figure 2.8, page 32.

d. Thermal Expansion of DWV Pipe

For detailed information on the changes in the length of DWV pipe, you should refer to Technical Report No. 21, "Thermal Expansion and Contraction of Plastic Pipe," by the Plastics Pipe Institute. Plastic materials vary in their coefficients of expansion. One diagram from literature of the Institute is shown in Figure 8.23. It indicates greater unit expansion of PVC than occurs in metals from which DWV piping is manufactured. For long straight runs of drainage pipe that may carry away hot water from fixtures, allowance must be made for changes in pipe length.

e. Choice of a Material (cast iron, plastic or copper)

It should be understood that the selection of a kind of piping material for a specific DWV installation cannot easily be made from the very brief introduction to qualities and installation methods described above.

Such a choice would have many criteria. They could include the requirements of the project, comparative prices at the location of the job and availability of workmen who are familiar with the technique of installation.

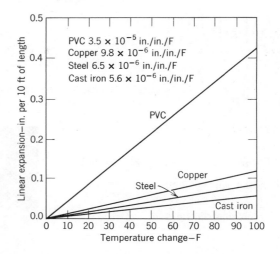

Figure 8.23 Thermal expansion of DWV pipe. (Courtesy of Plastics Pipe Institute)

Now the pipe caps are removed and the fixtures are installed *if they fit* the location of the pipes. To be sure that the fixtures *will* fit, the plumber had previously located the runouts and drains from diagrams that were furnished him by the architect. These diagrams prepared by the fixture manufacturer included "roughing" dimensions. For instance, in Figure 8.24 the water pipes are 6 in. apart and 22¼ in. above the finished floor. The waste line pipe is similarly located as is the position of the hanger that is built into the wall to support this Ledgewood lavatory.

8.5. Plumbing Fixtures

In the preceding sections we discuss water piping systems and DWV systems. Both systems are tested and corrected before being closed in by wall board, plaster or other surfacing. Following the closing-in stage, one sees water pipes and drainage pipes sticking out through a finished wall. The pipes are all capped to prevent dirt and unwanted obstructions from entering the systems.

Later in the construction timing-schedule, the plumbing fixtures arrive. They were chosen before the plumbing piping was installed. They were selected by the client who was guided by the architect.

8.6. Private Sewage Treatment

Many suburban areas do not have sewers or sewage treatment plants. In these locations it is necessary to provide local digestion (partial purification) of the sewage and disposal of the partially purified effluent to the ground. When there is adequate property area around a house or building and the soil is absorbent, like sand, the problem is easy. On small lots with soil of clay the problem is not easy. In densely populated areas with houses on 40 × 100 ft lots the situation is very difficult. County and village health authorities usually consider a private treat-

AMERICAN STANDARD

Ledgewood lavatory

Ledgewood lavatory with factory mounted Heritage faucet—pop-up drain—wall hanger—enameled cast iron

catalog number

19 x 17"

☐ **4300.042** wall hanger and Heritage faucet

22 x 19"

☐ **4300.117** wall hanger and Heritage faucet

(P 4300 previous plate number)

nominal dimensions

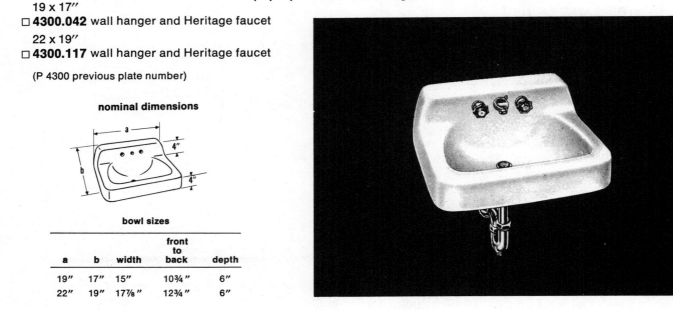

bowl sizes

a	b	width	front to back	depth
19"	17"	15"	10¾"	6"
22"	19"	17⅞"	12¾"	6"

Figure 8.24 Fixture information furnished by manufacturer. Photograph, fixture dimensions, roughing dimensions for plumbers location of water supply and waste pipes.

LEDGEWOOD LAVATORY

ENAMELED CAST IRON

2303.154 SUPPLIES — 4401.014 "P" TRAP

4300.042/.075
4300.117/.141

POP-UP DRAIN

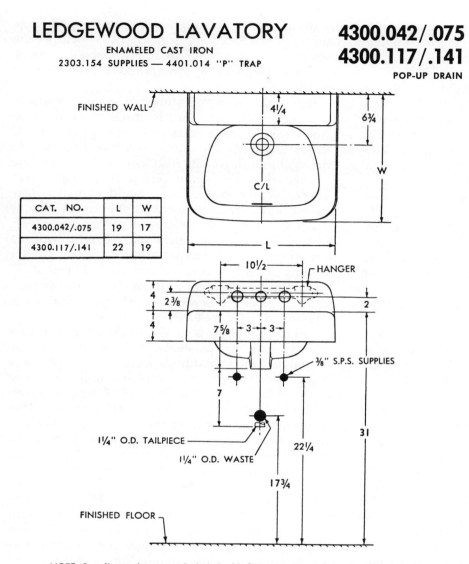

CAT. NO.	L	W
4300.042/.075	19	17
4300.117/.141	22	19

NOTE: Supplies and trap not included with fixture and must be ordered separately.

PLUMBER NOTE — Provide suitable reinforcement for all wall supports.

IMPORTANT: Dimensions of fixtures are nominal and may vary within the range of tolerances established by Commercial Standards CS77.

These measurements are subject to change or cancellation. No responsibility is assumed for use of superseded or voided leaflets.

AMERICAN STANDARD

PLUMBING / HEATING

L190

JUNE, 1971

ment plant as a temporary measure. They sometimes hold the privilege of requiring a complete replacement in time unless a municipal sewer replaces the private disposal system.

The general scheme of a private treatment system is relatively simple. The sewage is retained in a submerged, tightly enclosed septic tank of concrete or steel. Solids are diverted to the bottom of the tank. Bacterial action breaks up the solids and aids in purifying the fluids. A very small amount of sludge slowly builds up at the bottom of the tank and a scum forms at the top surface of the contents. The outflow pipe is located and protected in a way that prevents its being clogged by these materials. The septic tank needs to be pumped out at intervals, sometimes not oftener than every 5 or 10 years.

The fluid discharge goes out to one of two systems, see Figure 8.25a and b. Neither arrangement may perforate the groundwater level, since the outflow might pollute it. This requirement often makes system (b) preferable because it is flat and shallow, a quality that is appropriate above a high water table. However, (b) requires a great deal of area. The discharge of raw sewage into a pit or cesspool (c) is fast becoming illegal.

An efficient septic tank is shown in Figure 8.26. Cast-iron pipe fittings at both ends serve important functions. At the inlet, solids are directed to the bottom. The vertical pipes prevent surface scum from fouling either horizontal pipe. Flow is assured by placing the inlet pipe 3 in. above the outlet. Tight-fitting covers provide well-located access for inspection and servicing. The tank projects above ground. This is a convenience, but a tank may be placed slightly below grade if access is maintained. Several variations in piping arrangements are possible as shown in Figure 8.27.

A long time ago tanks and pits were built in place, but the prefabricated concrete and steel items of Figure 8.28 have largely superseded this technique.

8.7. Storm Drainage

Roofs and paved areas collect rainwater. If roof water is allowed to run off the edge of the structure, several disadvantages occur. Leakage to the interior is possible if walls, windows, doors and foundation walls are not completely watertight. Earth and topsoil may be eroded and washed away. Before rainwater can cause these troubles, it should be collected and led away by gutters and leaders to a point of disposal.

Before the advent of municipal sewage treatment plants, it was customary to discharge this storm water into storm sewers or combined storm and sanitary sewers. This flow then found its way to nearby natural streams and rivers. Public or private sewage treatment is now considered most essential. For this reason, storm water should *not* join sewage to overburden the treatment plant in times of heavy rain. Another new development has appeared. Natural water levels below the ground have been dropping, to the disadvantage of water-supply wells. Currently it is much preferred to "recharge" the ground with storm water *at the site* rather than connect to storm sewers. Suggested methods are illustrated in Figure 8.29

In large buildings a *number* of vertical leaders are usually necessary as shown in Figure 8.30. In city locations, where buildings meet at their property lines, these leaders are best placed indoors. Figure 8.31 and 8.32 show this. Figure 8.32 suggests a possible (if legal) connection to a combined sewer. This is usually undesireable; but it is sometimes unavoidable. In cities, disposal of storm water to the ground locally is not practical because of congested construction and very little open ground space. It also happens to be a fact that very few cities have separate sanitary and storm sewers. Hence, in many instances, it is impossible to *avoid* sending storm water into the combined system and overburdening the treatment plants. In our current condition of dense population, the separation of sanitary and storm sewers is most urgent. It is a difficult problem and one that is high on the municipal agenda of most cities. In a recent summary by states, this separation of existing combined sewers throughout the United States was found to represent a multi-*billion* dollar expense.

When leaders are indoors (Figure 8.33), they carry, in winter, cold rainwater through warm humid indoor space. This makes the outside surface of the metal leader cold. Humidity condenses on the cold surface and the pipe *drips*. These leaders, especially the horizontal runs, should be insulated to prevent this.

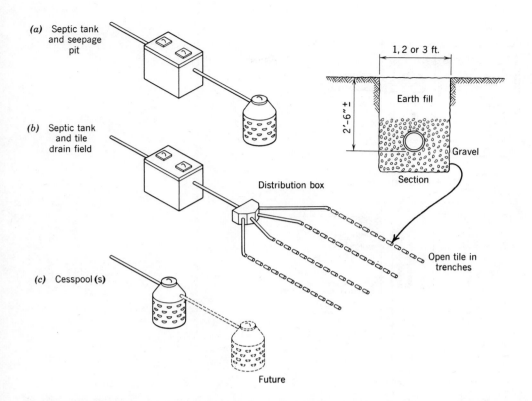

(a) Septic tank and seepage pit

(b) Septic tank and tile drain field

(c) Cesspool(s)

Distribution box

1, 2 or 3 ft.

Earth fill

2'-6"±

Gravel

Section

Open tile in trenches

Future

Figure 8.25 Private sewage treatment. *(a)* This method is most suitable in porous soil and where the groundwater level is low. *(b)* This method finds its best use in less pervious soils and where the groundwater level is high *(c)* This cesspool disposal is now discouraged and in some locations illegal.

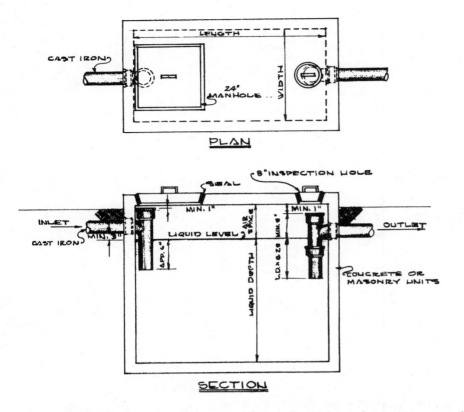

Figure 8.26 Typical piping layouts and details for septic tank use. (Courtesy of Cast Iron Soil Pipe Institute)

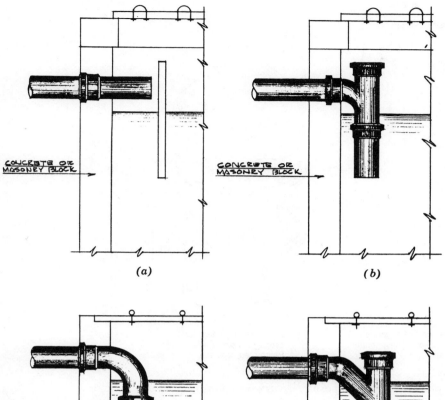

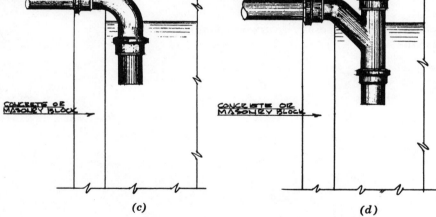

Figure 8.27 Typical piping layouts and details for septic tank use. (Courtesy of Cast Iron Soil Pipe Institute)

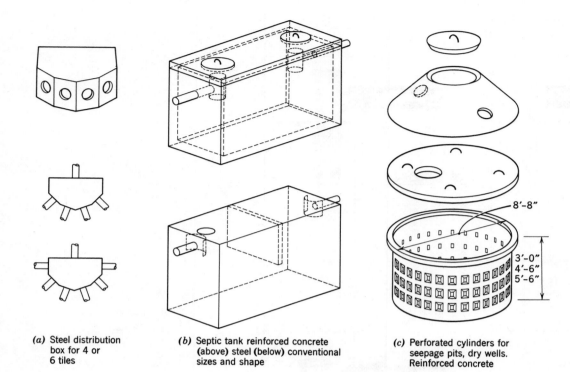

(a) Steel distribution box for 4 or 6 tiles

(b) Septic tank reinforced concrete (above) steel (below) conventional sizes and shape

(c) Perforated cylinders for seepage pits, dry wells. Reinforced concrete

Figure 8.28 Prefabricated elements for private sewage disposal. As a dry well (c) is also used for dispersal of storm drainage to the earth.

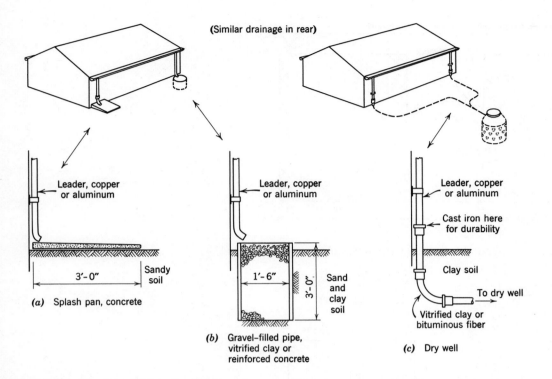

(Similar drainage in rear)

Leader, copper
or aluminum

Leader, copper
or aluminum

Leader, copper
or aluminum

Cast iron here
for durability

Clay soil

To dry well

Vitrified clay or
bituminous fiber

3'-0"

Sandy
soil

1'-6"

3'-0"

Sand
and
clay
soil

(a) Splash pan, concrete

(b) Gravel–filled pipe,
vitrified clay or
reinforced concrete

(c) Dry well

Figure 8.29 Roof drainage for houses. Gutters and leaders are sized with the aid of Tables 9.16 and 9.17, page 246. Method *(a)* is suitable for low rates of flow introduced into very pervious soil. When denser soil is encountered, *(b)* is used to get the water into the ground and thus to avoid surface erosion. For heavy flow or to lead the water further from the structure, *(c)* may be used with one or several dry wells.

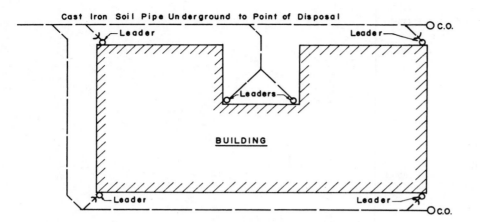

Figure 8.30 Roof leaders and drain outside building. (Courtesy of Cast Iron Soil Pipe Institute)

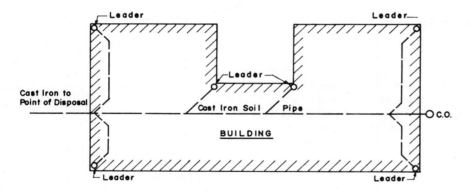

Figure 8.31 Roof leaders and drain inside building. (Courtesy of Cast Iron Soil Pipe Institute)

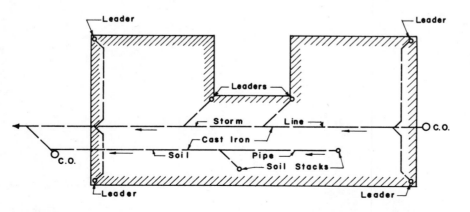

Figure 8.32 Combination sewer, sanitary and storm (where permitted by Code). (Courtesy of Cast Iron Soil Pipe Institute)

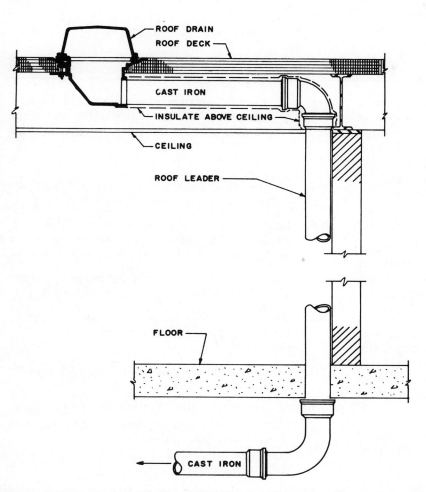

Figure 8.33 Typical roof drain and roof leader. (Courtesy of Cast Iron Soil Pipe Institute)

Problems

8.1. A large house requires domestic hot water at the rate of 16 gpm at a temperature rise of 100 F°. Select a General Fittings copper tankless heater. Give the following information:

Model Number   Height
Gallons per minute   Diameter
Water connections, inches: Weight
   Service
   Boiler

8.2. An A. O. Smith ROF-30 domestic hot water heater has been out of service. When it is turned on, how long will it be before a full tank of hot water is ready for use? The firing rate is 0.60 gallons of oil per hour.

8.3. A 150-ft run of DWV PVC plastic drainage line is suddenly subjected to a flow of hot water from an industrial process. The effective temperature of the pipe rises to 120° F from the factory temperature of 60° F. By how many inches will it increase in length.

8.4. A cast-iron pipe under the same conditions as in Problem 8.3 will lengthen by how many inches?

8.5. Make a water supply plan diagram for each of:
(a) A specific residence.
(b) A public toilet.
All diagrams must be based on an actual installation that you have inspected and sketched. Omit pipe sizes.

8.6. (a) In a domestic hot water system using a circulation pipe, does the circulation return pipe lead the water to the *top* or the *bottom* of the domestic hot water storage tank?
(b) Why?

8.7. Name two reasons for a valve below the fixture in a water supply line to a lavatory.

8.8. Describe two methods, other than the relief valve, of taking care of the expansion of water in the piping of a domestic hot water system.

8.9. Name three low points in the water system of a residence where you would require a drain.

8.10. Describe the method of connecting:
(a) Iron pipe to fittings.
(b) Copper tube to fittings.

8.11. Describe briefly three methods of joining lengths of cast-iron pipe.

8.12. In disposing of the effluent from a septic tank, what conditions of ground and groundwater would cause you to specify:
(a) Open tile drains in trenches?
(b) Seepage pits?

8.13. Make a plumbing section for:
(a) A specific residence.
(b) A public toilet.
All diagrams must be based on an actual installation that you have inspected and sketched. Omit pipe sizes.

Additional Reading

Modern Piping With Plastics
Plastics Piping for Plumbing
These publications and numerous "technical reports" are useful information published by the Plastics Pipe Institute.

Cast Iron Soil Pipe and Fittings Handbook. A complete manual of information and standards about cast-iron pipe by the Cast Iron Soil Pipe Institute.

Copper Tube Handbook for Plumbing, Heating, Air Conditioning and Refrigeration. Sovent Single Stack Plumbing System Design Handbook. The Copper Development Association, Inc.

9. Residential Plumbing

In a residence the owner determines the nature and extent of plumbing facilities. The architect incorporates them into the general plan. Local codes and health authorities require certain drawings to assure a legal and sanitary installation. The least of their requirements is that a plumbing "section" be submitted. This drawing must include the piping arrangements with all sizes shown. In large projects, a consulting engineer may be retained to complete the more intricate design work. He will prepare drawings that are included in the general contract documents. The actual positioning of the pipes cannot always be entirely foreseen. The plumbing contractor may be required to submit shop drawings of parts of the system.

Plumbers are licensed and, therefore, are under obligation to follow local statutes. This makes them associates of the architect-engineer team.

In applying the principles of Chapter 8, the study of Chapter 9 will enable you to:

1. Use the National Standard Plumbing Code and some engineering principles in sizing pipes for water service and distribution, sanitary drainage and storm drainage.
2. Identify all proposed fixtures or items that require:
   (a) Cold water only
   (b) Hot and cold water
   (c) Sanitary, waste or storm piping
3. Select an electric domestic hot water heater of proper storage capacity and recovery rate.
4. Call for furring and ceilings below horizontal piping when required, leaving access for valves and cleanouts.
5. Provide headers (mains) for hot and cold water distribution piping.
6. Draw a plumbing section and establish all pipe sizes.
7. Select plumbing fixtures suitable for a residence.
8. Design a private sewage treatment system.

9. Design the storm drainage systems for a roof and for a paved driveway area.
10. Make a mechanical site drawing.

9.1. Codes

The regulations relating to plumbing have, in the past, often been quite local in nature. This made it necessary for architects and others to vary their designs to suit the officials of the various regions in which they practiced. There is a strong trend now for greater uniformity of code requirements. One code that has been widely adopted is the National Standard Plumbing Code. When adopted, it could apply to both residential and nonresidential work.

Permission has been given to us to quote from the National Standard Plumbing Code. For this permission we are indebted to the Cosponsoring organizations. They are the National Association of Plumbing-Heating-Cooling Contractors and the American Society of Plumbing Engineers. Tables and quotations from the code that are reprinted in this chapter and in Chapter 10 are identified in each instance. We begin with two quotations from the code. They are "Introductory Note" and "Contents."

Introductory Note

"The material presented in this Code does not have legal standing unless it is adopted by reference, or by inclusion, in an act of state, county, or municipal government. Therefore, administration of the provisions of this Code must be preceded by suitable legislation at the level of government where it is desired to use the Code.

"In some places in the Code, reference is made to 'Administrative Authority.' The identity of an Administrative Authority will be established by the act which gives legal standing to the Code provisions.

"Meetings for purposes of review and revision are scheduled for three times a year with proper public notices.

"Suggestions and requests for revisions can be made by any interested party and should be submitted on the special forms provided by the Committee.

"Personal appearance before the Committee for a hearing on any Code matter can be had by interested parties after a request in writing.

"Throughout the body of the Code an asterisk (*) is shown at those sections where local changes may be desired."

We learn from the foregoing "Introductory Note" that the National Standard Plumbing Code is a "Recommended Code" and must have the authorization of a local or regional Administrative Authority. The "Contents," which is reprinted below, is to acquaint the reader with the scope of the code. The notations, V, BP-1, 1-1, 2-1, etc., refer to chapters *in the code manual,* only a few brief excerpts of which are reprinted in this book.

Contents

Introductory Note		V
Basic Principles		BP-1
Chapter I	—Definitions	1-1
Chapter II	—General Regulations	2-1
Chapter III	—Materials	3-1
Chapter IV	—Joints and Connections	4-1
Chapter V	—Traps and Cleanouts	5-1
Chapter VI	—Interceptors, Separators, and Backwater Valves	6-1
Chapter VII	—Plumbing Fixtures	7-1
Chapter VIII	—Hangers and Supports	8-1
Chapter IX	—Indirect Waste Piping and Special Waste	9-1
Chapter X	—Water Supply and Distribution	10-1
Chapter XI	—Sanitary Drainage Systems	11-1
Chapter XII	—Vents and Venting	12-1
Chapter XIII	—Storm Drains	13-1
Chapter XIV	—Medical Care Facility Plumbing	14-1
Chapter XV	—Tests and Maintenance	15-1
Chapter XVI	—Regulations Governing Individual Sewage Disposal Systems for Homes and Other Establishments where Public Sewage Systems are not Available	16-1
Chapter XVI	—Potable Water Supply System Pumps	17-1
Chapter XVIII	—Mobile Home and Travel Trailer Park Plumbing Standards	18-1
Appendix A	—Maximum Rates of Rainfall for Various Cities	A-1
Appendix B	—Sizing the Building Water Supply System	B-1
Appendix C	—	Deleted
Appendix D	—Alphabetical Index	D-1

The examples of Chapters 9 and 10 are solved in *general* conformity to the code. At this point a word of caution is necessary. When in actual practice under an Administrative Authority that has adopted the code, the reader should conform *strictly* to its requirements.

9.2. Water Service

When water is supplied from a street main, it is the responsibility of the owner to connect to the main. The *size* of the water service pipe is important. It must be large enough to assure proper flow at the fixtures. If it is too small, the pressure at the street main can be largely wasted in overcoming the friction between the water and the pipe through which it is flowing. Therefore, we must take into account the probable flow in gallons per minute under heavy demand, the maximum pressure required at a fixture, the height to which the water must be raised, and the pressure lost through friction of flow in the pipe that is selected.

Example 9.1 Select a copper tube water service pipe to carry water from the town water main in the street to the Merker residence. See Figures 9.1 and 9.2. The following data are given:

Pressure at the street main	50 psi
Total *equivalent* length of tube from main to the house	175 ft
Height through which the water must be lifted	12 ft

Water closets use flush tanks

Solution: We need to know:

(a) The "demand" (gallons per minute of flow).
(b) Maximum pressure needed at any faucet.
(c) Height that water must be raised above the main (12 ft, as noted above).

When (a), (b) and (c) are known, we can go ahead to the selection of a tube size.

The first step in finding the load is to find the total of the water supply "fixture units" for the fixtures found in Figure 9.2. Fixture units are listed in Table 9.1. For this house, we have added them in Table 9.3 to find that there are 34 fixtures units in this system. Since *fixture* pipe sizes are listed in Table 9.2, we have also shown *them* on our Table 9.3.

Using 34 fixture units, which is a measure of the probable flow, we look at Table 9.4 to find that the flow for this "load" will be about 22.5 gpm maximum under conditions of typical water use in the house.

Let us pause to study Table 9.4. It is based on experience in water use. If there are twice as many fixtures units, there will *not* be twice the flow. The reason for this can be stated as follows: the more fixtures there are in a system, experience shows the less chance there is that many of them will be used at the same time. An example of this (for systems with predominantly flush tanks) follows.

Fixture Units	Gallons per Minute
100	43.5
200	65, not 2 × 43.5 = 87
300	85, not 3 × 43.5 = 130.5

The gallons per minute quantity is *not* directly proportional to the number of fixture units.

Let us get back to our design. Item (b) is the maximum pressure required at any faucet. Table 9.5 tells us that 30 psi is necessary for a garden hose bibb. If we design for *this* pressure, other fixtures, *most* of which require less pressure, will function properly. If pressure at those other fixtures is too *high*, it can be reduced by the valve below the fixture.

Although the street main is supposed to provide 50 psi, it is safe to assume a lesser pressure, 45 psi.

Before we work in the numbers, we set up an equation: A, main pressure = B, fixture pressure + C, pressure required to raise the water 12 ft + D, pressure lost in friction in the tubing. We need to know D to select a tubing size that will be large enough to assure 30 psi at the hose bibb. Thus we can transpose and solve for D. Transposing, therefore, we have

D   Pressure loss in the tubing = A, main pressure minus:
B   Pressure to raise the water 12 ft minus:
C   Pressure at the fixture.

Substituting numbers, we have

		psi
A	Main pressure,	45
B	12 ft × 0.433 psi/ft. of height,*	− 5
		40
C	Fixture pressure (hose bibb),	−30
D	Pressure loss in the tube,	10

The tube must be selected for a friction loss per 100 ft of 10/175 × 100 = 5.71 psi *or less* at a flow of 22.5 gpm to assure 30 psi at the hose bibb.

Figure 9.3, adapted from Figure 2.17, page 41, shows the relationship of pressure drop per hundred feet of tube, water flow rate in gallons per minute and tube size. We select 1¼ in. tubing. At 22.5 gpm, the expected flow rate, the pressure drop per 100 ft of tube is found to be 3.5 psi (less than 5.71). Using the 1¼-in. tubing at 3.5 psi, a revised pressure study is shown on Figure 9.3. It indicates that the actual pressure at the hose bibb will be 34 psi, which is satisfactory. Type M copper tube of 1¼-in. diameter is chosen to run from the street main to the house. An actual photograph if this water service tube entering the house at the southeast corner of the garage is seen in Figure 9.4. In plan it is shown on the completed site drawing of Figure 9.17.

---

*0.433 psi is needed to raise water 1 ft in height.

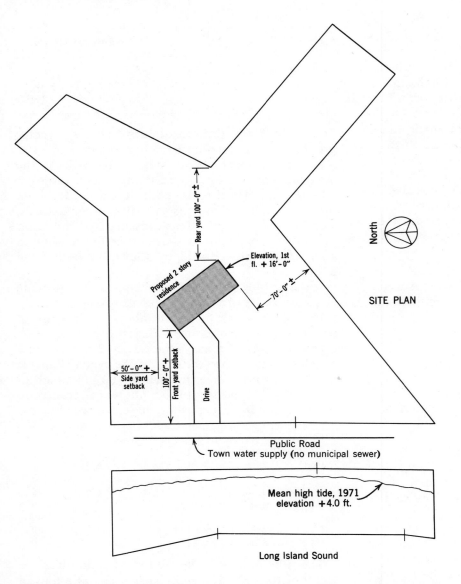

Figure 9.1 Site plan, Merker residence. This illustration is adapted from the surveyor's findings and the architect's layout. Because of water service from the street main, private sewage treatment and storm water disposal, the site plan is important in the plumbing design.

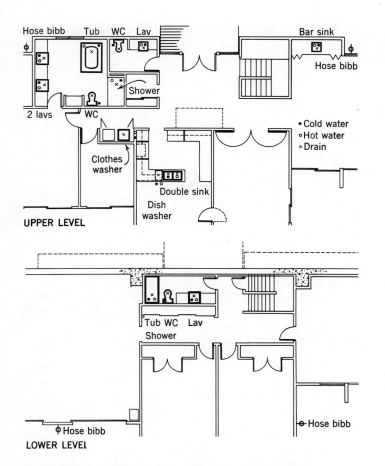

Figure 9.2 Plumbing requirements, water supply and sanitary drainage.

Table 9.1 Sizing the Water Supply System

Fixture	Occupancy	Type of Supply Control	Load in Fixture Units
Bathroom group	Private	Flush valve for closet	8
Bathroom group	Private	Flush tank for closet	6
Bathtub	Private	Faucet	2
Bathtub	Public	Faucet	4
Clothes washer	Private	Faucet	2
Clothes washer	Public	Faucet	4
Combination fixture	Private	Faucet	3
Kitchen Sink	Private	Faucet	2
Kitchen sink	Hotel, restaurant	Faucet	4
Laundry trays (1 to 3)	Private	Faucet	3
Lavatory	Private	Faucet	1
Lavatory	Public	Faucet	2
Separate shower	Private	Mixing valve	2
Service sink	Office, etc.	Faucet	3
Shower head	Private	Mixing valve	2
Shower head	Public	Mixing valve	4
Urinal-pedestal	Public	Flush valve	10
Urinal-stall or wall	Public	Flush valve	5
Urinal-stall or wall	Public	Flush tank	3
Water closet	Private	Flush valve	6
Water closet	Private	Flush tank	3
Water closet	Public	Flush valve	10
Water closet	Public	Flush tank	5

Water supply outlets for items not listed above shall be computed at their maximum demand, but in no case less than:

	Number of Fixture Units	
Fixture	Private Use	Public Use
⅜ in.	1	2
½ in.	2	4
¾ in.	3	6
1 in.	6	10

Source. National Standard Plumbing Code.

1. For supply outlets likely to impose continuous demands, estimate continuous supply separately and add to total demand for fixtures.

2. The given weights are for total demand. For fixtures with both hot and cold water supplies, the weights for maximum separate demands may be taken as three-fourths the listed demand for the supply.

3. A bathroom group for the purposes of this table consists of not more than one water closet, one lavatory, one bathtub, one shower stall or not more than one water closet, two lavatories, one bathtub or one separate shower stall.

Table 9.2 Minimum Sizes of Fixture Water Supply Pipes

Type of Fixture or device	Nominal Pipe Size (Inches)
Bath tubs	½
Combination sink and tray	½
Drinking fountain	⅜
Dishwasher (domestic)	½
Electric drinking water cooler	⅜
Kitchen sink, residential	½
Kitchen sink, commercial	¾
Lavatory	⅜
Laundry tray 1, 2 or 3 compartments	½
Shower (single head)	½
Sinks (service)	½
Sinks (flushing rim)	¾
Urinal (flush tank)	½
Urinal (direct flush valve)	¾
Water closet (tank type)	⅜
Water closet (flush valve type)	1
Hose bibb	½
Wall hydrant	½

Source. National Standard Plumbing Code.

Table 9.3 (Example 9.1) Water Supply Fixture Units. Fixture Pipe Sizes

	Fixture Units (Table 9.1)	Pipe Size, Inches (Table 9.2)
Bar sink	2	½
Dishwasher	2	½
Kitchen sink[a]	2	½
Guest lavatory	1	⅜
Guest water closet	3	⅜
Clothes washer	2	½
Master bath	6	L ⅜ T ½
2 lavatories, tub, water closet		WC ⅜
Shower stall in master bath (extra)	2	½
Downstairs bath	6	T ½ SH ½
tub, shower, water closet, lavatory		WC ⅜ L ½
Hose bibbs[b] (4) 4 × 2	8	½
	34	

[a]Swing faucet serves either sink.

[b]See note near bottom of Table 9.1. A hose bibb has a size of ½ in. (Table 9.2); thus, two fixture units.

Table 9.4 Estimating Demand

Supply Systems Predominantly for Flush Tanks		Supply Systems Predominantly for Flush Valves	
Load (Water Supply Fixture Units)	Demand Gallons per Minute	Load (Water Supply Fixture Units)	Demand Gallons per Minute
6	5		
8	6.5		
10	8	10	27
12	9.2	12	28.6
14	10.4	14	30.2
16	11.6	16	31.8
18	12.8	18	33.4
20	14	20	35
25	17	25	38
30	20	30	41
35	22.5	35	43.8
40	24.8	40	46.5
45	27	45	49
50	29	50	51.5
60	32	60	55
70	35	70	58.5
80	38	80	62
90	41	90	64.8
100	43.5	100	67.5
120	48	120	72.5
140	52.5	140	77.5
160	57	160	82.5
180	61	180	87
200	65	200	91.5
225	70	225	97
250	75	250	101
275	80	275	105.5
300	85	300	110
400	105	400	126
500	125	500	142
750	170	750	178
1,000	208	1,000	208
1,250	240	1,250	240
1,500	267	1,500	267
1,750	294	1,750	294
2,000	321	2,000	321
2,250	348	2,250	348
2,500	375	2,500	375
2,750	402	2,750	402
3,000	432	3,000	432
4,000	525	4,000	525
5,000	593	5,000	593
6,000	643	6,000	643
7,000	685	7,000	685
8,000	718	8,000	718
9,000	745	9,000	745
10,000	769	10,000	769

Source. National Standard Plumbing Code.

Table 9.5 Proper Flow and Pressure Required During Flow for Different Fixtures

Fixture	Flow Pressure[a]
Ordinary basin faucet	8
Self-closing basin faucet	12
Sink faucet—⅜ in.	10
Sink faucet—½ in.	5
Bathtub faucet	5
Laundry tub cock—¼ in.	5
Shower	12
Ball cock for closet	15
Flush valve for closet[b]	10–20
Flush valve for urinal	15
Garden hose, 50 ft, and sill cock	30

[a]Flow pressure is the pressure psig in the pipe at the entrance to the particular fixture considered.

[b]Wide range due to variation in design and type of flush-valve closets.

Source. Copyright by the American Society of Heating, Refrigerating and Air-Conditioning Engineers, Inc. Reprinted by permission from *ASHRAE Guide and Data Book, Systems Equipment, 1967.*

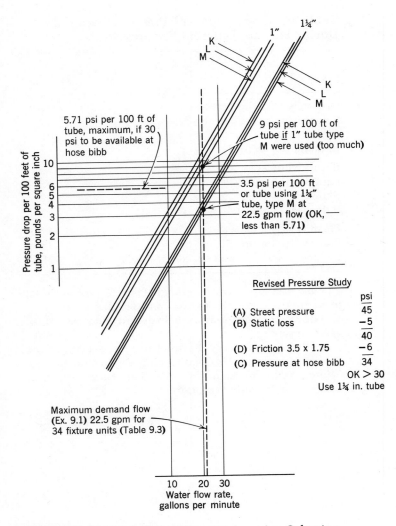

Figure 9.3 (Example 9.1) Sizing water service. Selection of a 1¼-in. type M tube as suitable to assure 30+ psi at hose bibb when street pressure is 45 psi.

Note. Basic diagram is a partial transcript of Figure 2.17, page 41.

9.3. Domestic Hot Water Heating

Very often it is an advantage to use the same energy source for domestic hot water heating as for general heating in the house.

For this reason we select an A. O. Smith 2-element Pen 52 *electric* hot water heater, (Figure 9.5). It would be part of either of the electric heating installations previously designed in Examples 3.2 or 6.1, pages 75 and 140.

In Chapter 8, as part of the general discussion of domestic hot water heaters, a separate oil-fired domestic hot water heater is suggested *if* an oil-fired hot water system were used in the Merker house. That oil-fired unit or the Pen 52 electric unit differ in both storage capacity and recovery rate. Yet either would be suitable for the probable occupancy of the house.

9.4. Water Distribution

The sizes of tubing for the individual hot and cold water branches of fixtures are listed in Table 9.3. The master water service, calculated in Section 9.2 is 1¼ in. in diameter. See Figures 9.6 and 9.7.

Example 9.2 Develop a water distribution system for the Merker residence.

Solution: Unlike systems in high rise buildings or other very large structures, the distribution tubing in houses is usually sized by experience and judgment rather than by flow and pressure calculations. The *fixture* supply sizes listed in Table 9.2, although named as minimal, are usually followed. It will be noted that they compare with the sizes called out in manufacturer's roughing dimensions for plumbing fixtures.

Figure 9.6 is a picture of the runouts to the several locations of water use. Their connections to the hot and cold water service can be as shown in Figure 9.7. Valves control the flow. These valves can make partial shutdown for repairs easy or they can be used to regulate pressure. Expansion chambers are located on the mains. You can see in Figures 9.9 and 9.11 that they are sometimes not used (or needed) at fixtures if central chambers on mains are installed.

Routing of tubing through the structure is often at the discretion of the plumber. He must see that it does not conflict with the architecture. For houses, it

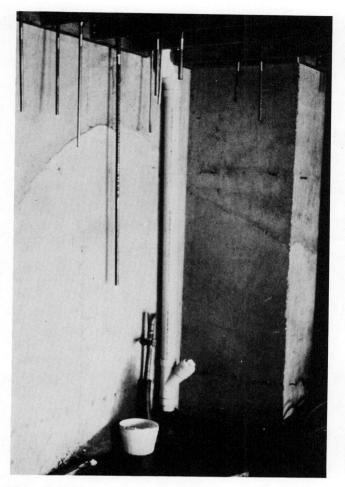

Figure 9.4 Construction stage, Merker residence, water and drainage. View at rear south corner of garage. The 1¼-in. copper water service enters the house. The 4-in. plastic soil stack carries sewage (but not roof drainage) away to septic tank. Heater for domestic hot water will be placed where the pail stands. Branch water lines, above, will be connected to water service and water heater. Main cleanout is seen at base of soil stack. Drainage from lower level joins the sewage flow at a point below the garage floor slab.

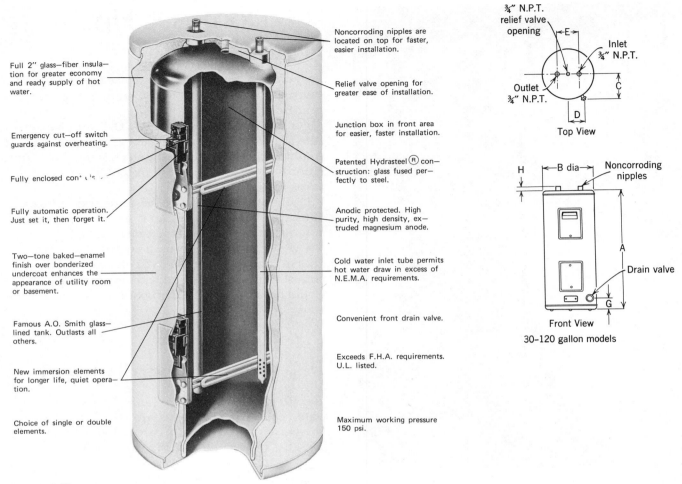

Full 2" glass–fiber insula—tion for greater economy and ready supply of hot water.

Emergency cut—off switch guards against overheating.

Fully enclosed con†r⌐ls ⌐

Fully automatic operation. Just set it, then forget it.

Two—tone baked—enamel finish over bonderized undercoat enhances the appearance of utility room or basement.

Famous A.O. Smith glass—lined tank. Outlasts all others.

New immersion elements for longer life, quiet opera—tion.

Choice of single or double elements.

Noncorroding nipples are located on top for faster, easier installation.

Relief valve opening for greater ease of installation.

Junction box in front area for easier, faster installation.

Patented Hydrasteel ® con—struction: glass fused per—fectly to steel.

Anodic protected. High purity, high density, ex—truded magnesium anode.

Cold water inlet tube permits hot water draw in excess of N.E.M.A. requirements.

Convenient front drain valve.

Exceeds F.H.A. requirements. U.L. listed.

Maximum working pressure 150 psi.

¾" N.P.T. relief valve opening

Inlet ¾" N.P.T.

Outlet ¾" N.P.T.

Top View

Noncorroding nipples

Drain valve

Front View

30–120 gallon models

A. O. Smith

PEN MODELS

Models	Pen 30		Pen 40		Pen 52		Pen 66		Pen 80		Pen 120	
Number of Elements	1 or 2		1 or 2		1 or 2		1 or 2		1 or 2		1 or 2	
Max. kw @ 120 v	3	6	3	6	3	6	3	6	3	6	3	6
Max. kw @ 208v	5	8	6	8	6	8	6	8	6	8	6	8
Max. kw @ 240 v	5	9	6	9	6	9	6	9	6	9	6	9
Max. kw @ 277 v	5	9	6	9	6	9	6	9	6	9	6	9
Max. kw @ 480 v	5	9	6	9	6	9	6	9	6	9	6	9
Size in U.S. gallons	30		40		52		66		80		120	
Est. shipping weight (lb)	104		126		148		188		226		355	
Dimensions A	48		46¾		58¼		62¾		61¾		62⅛	
B	18		20½		20½		21¾		24		29⅜	
C	7¼		9⅜		9⅜		10½		11¾		14¾	
D	8⅞		8¾		8¾		8⅝		8½		8¼	
E	8		8		8		8		8		16	
F	—		—		—		—		—		—	
G	4¼		4¼		4¼		4¼		4¼		4¼	
H	1⅛		1⅛		1⅛		1⅛		1⅛		1⅛	
J	—		—		—		—		—		—	

Figure 9.5 (Facing page) Electric hot water heater, Merker residence. 240 v, Pen 52 Model, two elements, 9 kw. Capacity 52 gal, recovery rate 4.1 gal/hr/kw at 100 F° temperature rise = 9 × 4.1 = 36.9 gal/hr. (Courtesy of A.O. Smith)

Figure 9.6 Horizontal branches of copper water system and plastic drainage system are located in the furred space between the garage ceiling and the floor joists of the upper level. Construction stage photograph.

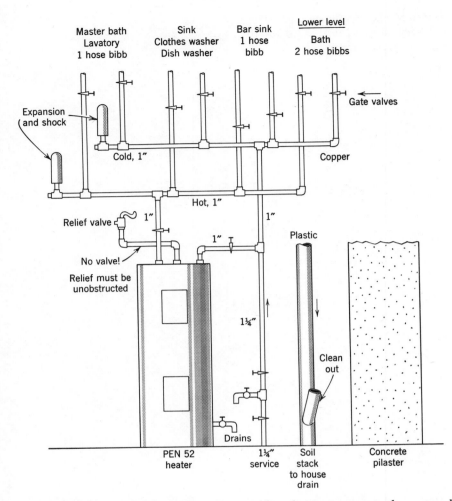

Figure 9.7 (Example 9.2) Suggested water distribution center on the east wall of the garage. Runouts from 1-in. mains could be ¾-in. All sizes are a bit on the generous size but would assure adequate pressure if several fixtures were in use at the same time. The cold location in the garage would suggest covering on all tubing.

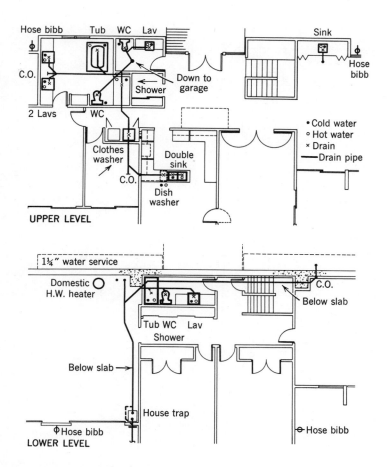

Figure 9.8 (Example 9.3) Sanitary drainage plan.

Figure 9.9 (Example 9.3) Water supply and drainage roughing for the two lavatories in the master bathroom, Merker residence. View looking North. Copper water tubing. Plastic (ABS and PVC) waste and vents. Vertical stack is a vent above and a waste below. The two branch waste pipes lead in to the stack through ⅛ bends and "Y" fittings.

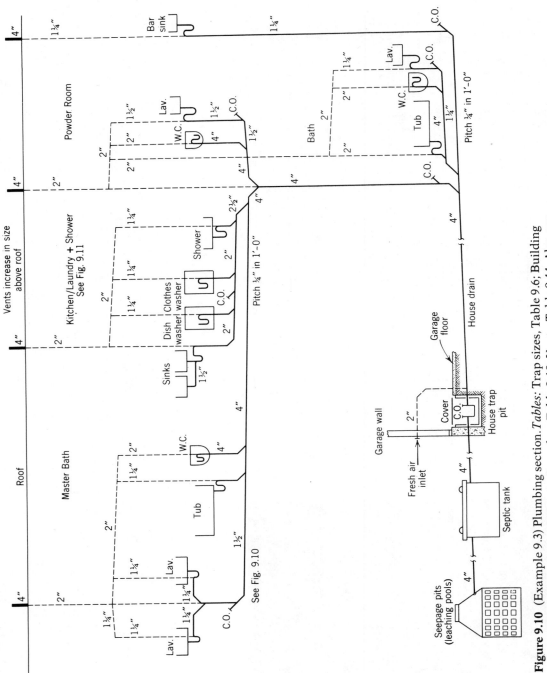

Figure 9.10 (Example 9.3) Plumbing section. *Tables*: Trap sizes, Table 9.6; Building Drain, Table 9.9; Horizontal Fixture Branches, Table 9.10; Vents, Table 9.11. Also, see discussion in chapter text.

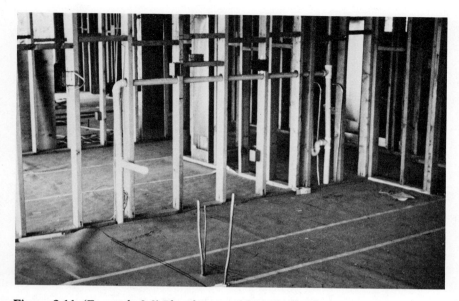

Figure 9.11 (Example 9.3) Plumbing roughing for kitchen, Merker residence. View looking North. Waste branch at left takes the runoff of sinks and dishwasher. Trap at the right receives the discharge of the clothes washer. Vents are joined and run out through the roof. Hot and cold water tubes are seen in the foreground together with a special high-temperature water tube for the preparation of items such as instant coffee.

Note. Concerning the distribution of air for heating and cooling, a vertical return air duct with a grill opening at the bottom is seen in the wall of the study. The back of a similar duct is to the left in the master bedroom.

This arrangement offers some advantage over our return air system in Chapter 6. With return grills at both top *and* bottom, the lower grill can draw in cool air that could, in winter, collect near the floor. It is a plus for comfort.

is usually not necessary to make a drawing such as Figure 8.1, page 188. That is a schematic diagram for instructional purposes.

You will notice that no hot water circulation lines are called for. The reason is that hot water runs are *short*. Circulation becomes more important with greater distances.

9.5. Sanitary Drainage

Refer to Figures 9.2, 9.4 and 9.8 to 9.11 and to Tables 9.6 to 9.11.

Example 9.3 Design a sanitary drainage system for the Merker residence based on the fixture requirements as shown in Figure 9.2.

Solution: Lay out a drainage piping plan (Figure 9.8) and a plumbing section drawing (Figure 9.10). On the latter drawing, we will put sizes when they are determined. Plan and section should conform as far as possible. The plumber who installs the system may vary the pipe routing. Since he is licensed, it will be his responsibility to make minor changes in sizes as required.

The next step is to list the fixtures (Table 9.8) and to assign drainage fixture unit values from Table 9.7. Setting the house drain slope at ¼ in./ft of length, we can select a size for the house drain and house sewer (also called "building drain and building sewer"). Although Table 9.9 would, under certain conditions, allow 3-in. piping (48 fixture units), we make a more conservative choice.

A 4-in. size is chosen for these pipes. This is a common choice as a minimum size and also as the size of soil branches from any water closet. Note that this choice permits 216 fixture units; a number generously greater than our 28 units.

As we consult Table 9.10 for the sizes of horizontal fixture branches, we begin to suspect that this table (and indeed Table 9.11 for vents) is scaled for very large buildings. However, our horizontal sizes are selected from the left-hand column of Table 9.10 and are entered in Figure 9.10. In no case should any pipe be smaller than the trap size (see Table 9.6) of a fixture that it drains. Vents come next. We find that there are three criteria in selecting vent sizes. They are: fixture units connected, size of soil or waste stack vented and length of vent. Our choices are for vents of 2 in. and less. Length of vent is not significant. Our longest vent does not exceed 20 ft. A

Table 9.6 Size of Nonintegral Traps for Different Type Plumbing Fixtures

Plumbing Fixture	Trap Size in Inches
Bathtub (with or without overhead shower)	1½
Bidet	1¼
Combination sink and wash (laundry) tray	1½
Combination sink and wash (laundry) tray with food waste grinder unit	1½[a]
Combination kitchen sink, domestic, dishwasher, and food waste grinder	2
Dental unit or cuspidor	1¼
Dental lavatory	1¼
Drinking fountain	1¼
Dishwasher, commercial	2
Dishwasher, domestic (nonintegral trap)	1½
Floor drain	2
Food waste grinder—Commercial use	2
Food waste grinder—Domestic Use	1½
Kitchen sink, domestic, with food waste grinder unit	1½
Kitchen sink, domestic	1½
Kitchen sink, domestic, with dishwasher	1½
Lavatory, common	1¼
Lavatory (barber shop, beauty parlor or surgeon's)	1½
Lavatory, multiple type (wash fountain or wash sink)	1½
Laundry tray (1 or 2 compartments)	1½
Shower stall or drain	2
Sink (surgeon's)	1½
Sink (flushing rim type, flush valve supplied)	3
Sink (service type with floor outlet trap standard)	3
Sink (service trap with P trap)	2
Sink, commercial (pot, scullery, or similar type)	2
Sink, commercial (with food grinder unit)	2

Source. National Standard Plumbing Code.
[a]Separate trap required for wash tray and separate trap required for sink compartment with food waste grinder unit.

2-in. vent 20 ft long will serve 500 fixture units or a 4-in. soil stack. Water closets will have 2-in. vents as will the house trap. Vent stacks and major horizontal vent branches will be 2 in. Fixture vents are seen to be 1¼ and 1½ in. By custom, vent stacks increase to 4 in. above the roof line.

Table 9.7 Drainage Fixture Unit Values for Various Plumbing Fixtures

Type of Fixture or Group of Fixtures	Drainage Fixture Unit Value (d.f.u.)
Automatic clothes washer (2 in. standpipe)	3
Bathroom group consisting of a water closet, lavatory and bathtub or shower stall:	
Flushometer valve closet	8
Tank type closet	6
Bathtub[a] (with or without overhead shower)	2
Bidet	1
Clinic Sink	6
Combination sink-and-tray with food waste grinder	4
Combination sink-and-tray with one 1½ in. trap	2
Combination sink-and-tray with separate 1½ in. traps	3
Dental unit or cuspidor	1
Dental lavatory	1
Drinking fountain	½
Dishwasher, domestic	2
Floor drains with 2 in. waste	3
Kitchen sink, domestic, with one 1½ in trap	2
Kitchen sink, domestic, with food waste grinder	2
Kitchen sink, domestic, with food waste grinder and dishwasher 2 in. trap	3
Kitchen sink, domestic, with dishwasher 1½ in. trap	3
Lavatory with 1¼ in waste	1
Laundry tray (1 or 2 compartments)	2
Shower stall, domestic	2
Showers (group) per head	2
Sinks:	
Surgeon's	3
Flushing rim (with valve)	6
Service (trap standard)	3
Service (P trap)	2
Pot, scullery, etc	4
Urinal, pedestal, syphon jet blowout	6
Urinal, wall lip	4
Urinal, stall, washout	4
Urinal trough (each 6-ft section)	2
Wash sink (circular or multiple) each set of faucets	2
Water closet, tank-operated	4
Water closet, valve-operated	6
Fixtures not listed above:	
Trap size 1¼ in. or less	1
Trap size 1½ in.	2
Trap size 2 in.	3
Trap size 2½ in.	4
Trap size 3 in.	5
Trap size 4 in.	6

Source. National Standard Plumbing Code.
[a]A shower head over a bathtub does not increase the fixture unit value.

Table 9.8 (Example 9.3) Drainage Fixture Units

	Units
Bar sink	2
Kitchen sink and dishwasher	3
Lavatory	1
Water closet	4
Clothes washer	3
Master bath, lavatory, water closet, tub	6
Extra lavatory	1
Shower	2
Lower bath, lavatory, water closet, tub and shower	6
	28

Values are from Table 9.7
Hose bibb drainage to ground
Roof drainage to dry wells
Blower-coil units condensate to dry wells

Table 9.9 Building Drains and Sewers[a]

Diameter of Pipe	Maximum Number of Fixture Units That May Be Connected to Any Portion of the Building Drain or the Building Sewer Including Branches of the Building Drain			
	Fall per Foot			
	1/16 in.	1/8 in.	1/4 in.	1/2 in.
Inches				
2			21	26
2½			24	31
3		36[b]	42[b]	50[b]
4		180	216	250
5		390	480	575
6		700	840	1,000
8	1,400	1,600	1,920	2,300
10	2,500	2,900	3,500	4,200
12	2,900	4,600	5,600	6,700
15	7,000	8,300	10,000	12,000

Source. National Standard Plumbing Code.
[a]On site sewers that serve more than one building may be sized according to the current standards and specifications of the Administrative Authority for public sewers.
[b]Not over two water closets or two bathroom groups.

Table 9.10 Horizontal Fixture Branches and Stacks

Maximum Number of Fixture Units that may be Connected to:

Diameter of Pipe	Any Horizontal Fixture Branch[a]	Stack Sizing For 3 Stories in Height or 3 Intervals	Stack Sizing For More Than 3 Stories in Height	
			Total for Stack	Total at One Story or Branch Interval
Inches				
1½	3	4	8	2
2	6	10	24	6
2½	12	20	42	9
3	20[b]	48[b]	72[b]	20[b]
4	160	240	500	90
5	360	540	1,100	200
6	620	960	1,900	350
8	1,400	2,200	3,600	600
10	2,500	3,800	5,600	1,000
12	3,900	6,000	8,400	1,500
15	7,000			

Source. National Standard Plumbing Code.
[a]Does not include branches of the building drain.
[b]Not more tha 2 water closets or bathroom groups within each branch interval nor more than 6 water closets or bathroom groups on the stack.

Stacks shall be sized according to the total accumulated connected load at each story or branch interval and may be reduced in size as this load decreases to a minimum diameter of ½ of the largest size required.

Table 9.11 Size and Length of Vents

Size of Soil or Waste Stack	Fixture Units Connected	Diameter of Vent Required (Inches)								
		1¼	1½	2	2½	3	4	5	6	8
		Maximum Length of Vent (Ft)								
Inches										
1½	8	50	150							
1½	10	30	100							
2	12	30	75	200						
2	20	26	50	150						
2½	42		30	100	300					
3	10		30	100	100	600				
3	30			60	200	500				
3	60			50	80	400				
4	100			35	100	260	1000			
4	200			30	90	250	900			
4	500			20	70	180	700			
5	200				35	80	350	1000		
5	500				30	70	300	900		
5	1100				20	50	200	700		
6	350				25	50	200	400	1300	
6	620				15	30	125	300	1100	
6	960					24	100	250	1000	
6	1900					20	70	200	700	
8	600						50	150	500	1300
8	1400						40	100	400	1200
8	2200						30	80	350	1100
8	3600						25	60	250	800
10	1000							75	125	1000
10	2500							50	100	500
10	3800							30	80	350
10	5600							25	60	250

Source. National Standard Plumbing Code.

9.6. Plumbing Fixtures

For installation in both bathrooms, the powder room, kitchen and the bar, plumbing fixtures are needed. Roughing dimensions accompany the catalog illustrations. These are used by the plumbing contractor. Installation of fixtures is at a late stage in the construction schedule. Until then, they remain crated and stored away to prevent damage.

Fixtures needed are:

Lavatory	Kitchen sink
Bathtub	Bar sink
Toilet	Shower pan

a. Lavatory

The "Merrilyn" style is chosen. See Figure 9.12. In both bathrooms and in the powder room they are built-in to custom-type vanity counters. Separate faucet handles permit blending of hot and cold water to achieve the desired temperature at the single spout.

b. Bathtub

The "Ultra Bath" bathing pool shown in Figure 9.13 is sometimes used as a sunken pool, its 16-in. depth submerged below the finished tile floor. In Figure 9.2, it is seen to be an island, some distance away from both adjacent walls. Its actual use in the master bath was in a position with its top edge 12 in. above the floor. These 12-in.-high spaces around its entire perimeter were enclosed with ceramic tile. Since there was a separate shower stall in the room, the shower head and control at the bathing pool were, obviously, omitted. In its "island" position, special detailing of the faucets and supply piping were required. Not shown here is the conventional bathtub selected for the lower level bathroom.

c. Toilet

All three water closets are of the Luxor "Vent-away" type with integral low-level flush tank. See Figure 9.14.

d. Kitchen Sink

For this, we select the "Explorer" double-bowl stainless steel sink. See Figure 9.15. In custom built-in installations like this, the fixture dimensions are of interest to the kitchen cabinet subcontractor as well as to the plumber. The single swing spout serves either sink, and flow is controlled by a combination hot/cold, on/off lever.

e. Bar Sink and Shower Pan

These two items are not illustrated here. For the bar, a single stainless sink would be suitable. The shower pan could be a prefabricated terrazzo unit. This is an item not usually made by a fixture manufacturer. It is purchased separately.

9.7. Private Sewage Treatment

Example 9.4 Design and draw a sewage treatment plant for the Merker residence.

System:	Septic tank and seepage pits as shown in Figure 8.25a, page 211.
House category:	luxury residence
Bedrooms:	three
Occupancy:	Six people
Soil absorption category test result:	Two minutes for 1-in. drop.

Solution:

(a) *Septic tank.* Based upon Table 9.12, we determine the proper size for a septic tank. The house has three bedrooms, but it would be better to select from Table 9.12 a tank that is one size larger than the minimum 1000 gal. A 1200-gal tank is preferred. Its general arrangement will conform to Figures 8.26 and 8.27, pages 212 and 213. After several trial sets of dimensions, the tank in Figure 9.16 is approved. Its fluid capacity is

$$\frac{4 \text{ ft} \times 8 \text{ ft} \times 5 \text{ ft} \times 1728}{231} = 1196 \text{ gal}$$

1728 is the number of cubic inches in one cubic foot.

231 is the number of cubic inches in one gallon.

(b) *Seepage pits.* The effective absorption area of a seepage pit is the product of the perimeter and the height. The area at the bottom is not included. The effective area that we must provide will depend on the gallons per day of sewage flow and the porosity of the soil. Before calculating these values, we must describe how the soil porosity is tested. The method is called the *per-*

MERRILYN LAVATORY

VITREOUS CHINA — FOR COUNTERTOP INSTALLATION
2248.045 FITTING — 2303.154 SUPPLIES
4401.014 "P" TRAP

0140.178
(POP-UP DRAIN)

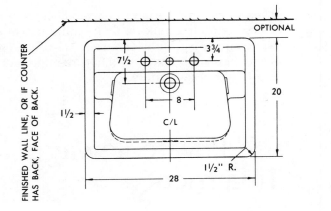

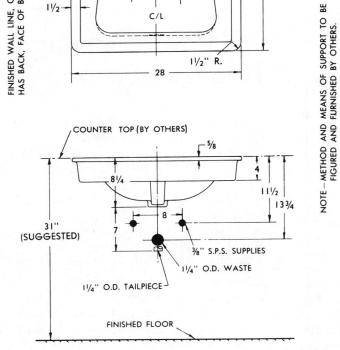

NOTE: FITTINGS NOT INCLUDED WITH FIXTURE AND MUST BE ORDERED SEPARATELY.

IMPORTANT: Dimensions of fixtures are nominal and may vary within the range of tolerances established by Commercial Standards CS20.

These measurements are subject to change or cancellation. No responsibility is assumed for use of superseded or voided leaflets.

Figure 9.12 Merrilyn lavatory. Fixture dimensions and roughing dimensions. (Courtesy of American Standard)

ULTRA BATH-BATHING POOL

ENAMELED CAST IRON
1390.186 SUPPLY FITTING
560. SER. OR 1561. SER. C.D.&O.

2640.019

CONVENTIONAL TILING-BEAD AT OUTLET END OF TUB ONLY.
<u>CAUTION</u>: DISTANCE BETWEEN STUDDING & TILING BEAD MUST NOT EXCEED ⅛"

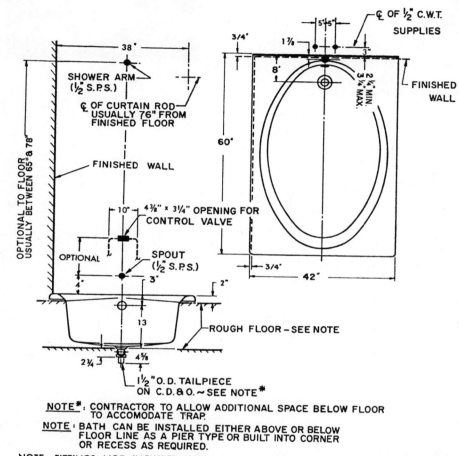

<u>NOTE</u>**: CONTRACTOR TO ALLOW ADDITIONAL SPACE BELOW FLOOR TO ACCOMODATE TRAP.

<u>NOTE</u>: BATH CAN BE INSTALLED EITHER ABOVE OR BELOW FLOOR LINE AS A PIER TYPE OR BUILT INTO CORNER OR RECESS AS REQUIRED.

NOTE: FITTINGS NOT INCLUDED WITH FIXTURE AND MUST BE ORDERED SEPARATELY.

PLUMBER NOTE — Provide suitable reinforcement for all wall supports.

IMPORTANT: Dimensions of fixtures are nominal and may vary within the range of tolerances established by ANSI Standards — A112.19.1.
These measurements are subject Jo change or cancellation. No responsibility is assumed for use of superseded or voided leaflets.

Figure 9.13 Ultra Bath bathing pool. Fixture dimensions and roughing dimensions. (Courtesy of American Standard)

LUXOR TOILET

VITREOUS CHINA — ONE PIECE
WITH ½" S.P.S. SUPPLY FURNISHED

2003.010
2003.036

PLUMBER NOTE: THIS COMBINATION IS DESIGNED TO ROUGH-IN AT MINIMUM DIMENSION OF 12" FROM FINISHED WALL TO C/L OF OUTLET.

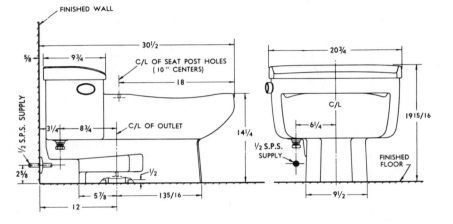

NOTE: 2003.010 – FURNISHED WITH VENT-AWAY TOILET VENTILATOR.
2003.036 – LESS VENT-AWAY.

NOTE – **30 P.S.I. MIN. WORKING PRESSURE REQUIRED** AT WATER CONTROL

NOTE: ½" supply pipe included with toilet.

IMPORTANT: Dimensions of fixtures are nominal and may vary within the **range of tolerances** established by Commercial Standards CS20.
These measurements are subject to change or cancellation. No responsibility is assumed for use of superseded or voided leaflets.

Figure 9.14 Luxor toilet. Fixture dimensions and roughing dimensions. (Courtesy of American Standard)

EXPLORER SINKS

STAINLESS STEEL — SELF-RIMMING
STRAINERS

3258.233
3258.241

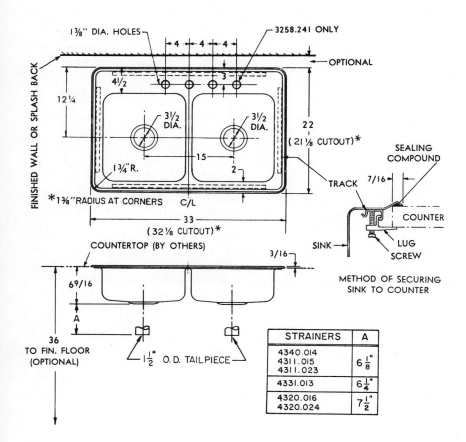

STRAINERS	A
4340.014 4311.015 4311.023	$6\frac{1}{8}$"
4331.013	$6\frac{1}{4}$"
4320.016 4320.024	$7\frac{1}{2}$"

NOTE: STRAINERS NOT INCLUDED WITH FIXTURE AND MUST BE ORDERED SEPARATELY

These measurements are subject to change or cancellation. No responsibility is assumed for use of superseded or voided leaflets.

Figure 9.15 Explorer sinks. Fixture dimensions and roughing dimensions. (Courtesy of American Standard)

colation test. We now quote from the National Standard Plumbing Code:

Percolation Test

"Percolation test to determine the absorption capacity of soil for septic tank effluent shall be conducted in the following manner:

"When subsurface irrigation is contemplated, test pit shall be prepared 2 feet square but not less than 1 foot deep. At the time of conducting the percolation test, a hole 1 foot square and 1 foot deep shall be prepared in the test pit.

"The hole shall be filled with water to a depth of 7 inches. For pre-wetting purposes, the water level shall be allowed to drop to 6 inches before time of recording is started.

"The time required for the water level to drop 1 inch from 6 inches to 5 inches in depth is noted and the length of tile in the subsurface irrigation system can be obtained from Table 16.5.4* In no case, however, shall less than 100 feet of tile be installed when 1 foot trenches are used.

"When seepage pits are contemplated, test pits approximately 5 feet in diameter to permit a man entering the pit by means of a ladder and to such depth as to reach a porous soil shall be prepared. In the bottom of this pit a 1 foot square by 1 foot deep hole is made at the time of testing and the percolation test conducted as indicated under paragraphs mentioned above.

"The absorption area of a seepage pit required can be obtained from Table 16.5.7.* In no case, however, shall the absorption area in the porous soil be less than 125 square feet. The bottom of the pit shall not be considered part of the absorption area."

We return now to the gallons per day sewage flow and the absorption area required. Table 9.13 tells us that, for a luxury residence, the sewage flow is 150 gal per day per person. At times there are six people in residence. The flow will therefore be 900 gpd.

The problem states that our percolation test results were: 2 min for a 1-in. drop. This indicates a very porous soil. We skip over Table 9.14, which is for a tile drain field (Figure 8.25*b*, page 211). We go ahead to Table 9.15, which is for a seepage pit installation (Figure 8.25*a*). The absorption area for a 2-min drop is 40 sq ft for each 100 gpd of flow. We have nine such units (900 gpd). The area that we need is $9 \times 40 = 360$ sq ft.

Author's note. Tables 16.5.4 and 16.5.7 referred to above are Tables 9.14 and 9.15 respectively, in this book.

Table 9.12 Capacity of Septic Tanks[a]

Single Family Dwellings– Number of Bedrooms	Multiple Dwelling Units or Apartments– One Bedroom Each	Other uses; Maximum Fixture Units Served	Minimum Septic Tank Capacity in Gallons
1 to 3		20	1000
4	2 units	25	1200
5 or 6	3	33	1500
7 or 8	4	45	2000
	5	55	2250
	6	60	2500
	7	70	2750
	8	80	3000
	9	90	3250
	10	100	3500

Extra bedroom, 150 gal each.
Extra dwelling units over 10, 250 gal each.
Extra fixture units over 100, 25 gal per fixture unit.

Source. National Standard Plumbing Code.
[a]*NOTE:* Septic tank sizes in this table include sludge storage capacity and the connection of domestic food waste disposal units without further volume increase.

Table 9.13 Sewage Flows According to Type of Establishment

Type of Establishment	Gallons per Day per Person
Schools (toilet and lavatories only)	15 gal per day per person
Schools (with above plus cafeteria)	25 gal per day per person
Schools (with above plus cafeteria and showers)	35 gal per day per person
Day workers at schools and offices	15 gal per day per person
Day camps	25 gal per day per person
Trailer parks or tourist camps (with built-in bath)	50 gal per day per person
Trailer parks or tourist camps (with central bathhouse)	35 gal per day per person
Work or construction camps	50 gal per day per person
Public picnic parks (toilet wastes only)	5 gal per day per person
Public picnic parks (bathhouse, showers and flush toilets)	10 gal per day per person
Swimming pools and beaches	10 gal per day per person
Country clubs	25 gal per locker
Luxury residences and estates	150 gal per day per person
Rooming houses	40 gal per day per person
Boarding houses	50 gal per day per person
Hotels (with connecting baths)	50 gal per day per person
Hotels (with private baths—two persons per room)	100 gal per day per person
Boarding Schools	100 gal per day per person
Factories (gallons per person per shift—exclusive of industrial wastes)	25 gal per day per person
Nursing homes	75 gal per day per person
General Hospitals	150 gal per day per person
Public Institutions (other than hospitals)	100 gal per day per person
Restaurants (toilet and kitchen wastes per unit of serving capacity)	25 gal per day per person
Kitchen wastes from hotels, camps, boarding houses, etc. Serving three meals per day	10 gal per day per person
Motels	50 gal per bed space
Motels with bath, toilet, and kitchen wastes	60 gal per bed space
Drive-in theaters	5 gal per car space
Stores	400 gal per toilet room
Service stations	10 gal per vehicle served
Airports	3-5 gal per passenger
Assembly halls	2 gal per seat
Bowling alleys	75 gal per lane
Churches (small)	3-5 gal per sanctuary seat
Churches (large with kitchens)	5-7 gal per sanctuary seat
Dance halls	2 gal per day per person
Laundries (coin operated)	400 gal per machine
Service stations	1000 gal (first bay) 500 gal (each add. bay)
Subdivisions or individual homes	75 gal per day per person
Marinas—Flush toilets	36 gal per fixture per hour
Urinals	10 gal per fixture per hour
Wash basins	15 gal per fixture per hour
Showers	150 gal per fixture per hour

Source. National Standard Plumbing Code.

Table 9.14 The Tile Length for Each 100 Gal of Sewage per Day[a]

Time in Minutes for 1-In. Drop	Tile Length for Trench Widths of:		
	1 ft	2 ft	3 ft
1	25	13	9
2	30	15	10
3	35	18	12
5	42	21	14
10	59	30	20
15	74	37	25
20	91	46	31
25	105	53	35
30	125	63	42

Source. National Standard Plumbing Code.
[a]Table 16.5.4 in the National Standard Plumbing Code.

Table 9.15 Effective Absorption Area in Seepage Pits for Each 100 Gal of Sewage per Day[a]

Time in Minutes for 1-In. Drop	Effective Absorption Area (Square Feet)
1	32
2	40
3	45
5	56
10	75
15	96
20	108
25	139
30	167

Source. National Standard Plumbing Code.
[a]Table 16.5.7 in the National Standard Plumbing Code.

Since we may 'not allow the pit to penetrate deep enough to pollute the groundwater we must use a shallow unit. This is shown in the sketch in Figure 9.16. A 4 ft 6 in ring will fit the site conditions (Figure 9.16) with its lower edge well above the water level. Calculations for a 4 ft 6 in. by 8 ft 8 in. ring are found in Figure 9.16. The absorption area of one such ring is 122.5 sq ft. Three rings of that size fulfill the 360 sq ft requirement. Details are summarized in Figures 9.16 and 9.17.

If you have occasion to design a tile drain field you should consult the National Standard Plumbing Code. There you will find requirements about trench length and spacing and items relating to method of construction.

9.8. Storm Drainage and Disposal

Example 9.5 Design a storm drainage and disposal system for the Merker Residence. See Tables 9.16 to 9.19.

Solution

(a) *Gutters, leaders and drains.* The projected area of the roof is 2060 sq ft. Since rainwater in each gutter flows to the leader at the corner of the house, each gutter must have a size that properly drains one fourth of the projected (flat projection) roof area, or 2060/4 = 515 sq ft. Simi-

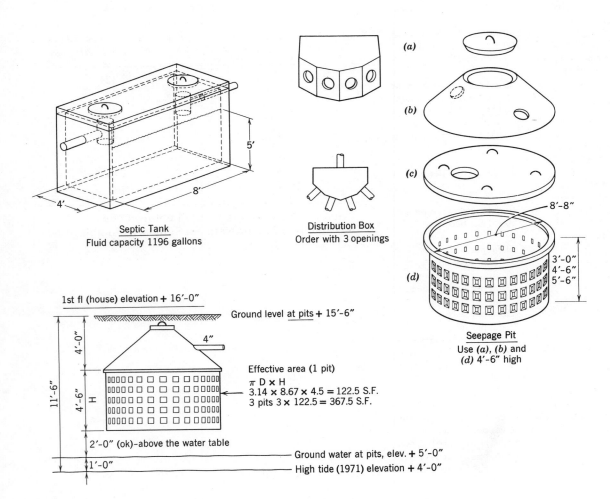

Septic Tank
Fluid capacity 1196 gallons

Distribution Box
Order with 3 openings

Seepage Pit
Use (a), (b) and
(d) 4'-6" high

Effective area (1 pit)
$\pi\, D \times H$
$3.14 \times 8.67 \times 4.5 = 122.5$ S.F.
3 pits $3 \times 122.5 = 367.5$ S.F.

1st fl (house) elevation + 16'-0"
Ground level at pits + 15'-6"
2'-0" (ok)-above the water table
Ground water at pits, elev. + 5'-0"
High tide (1971) elevation + 4'-0"

Figure 9.16 (Example 9.4) Field components, sewage treatment system.

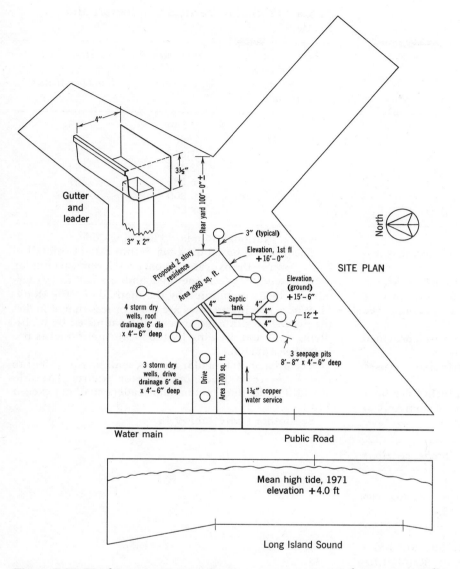

Figure 9.17 Site layout, water service, sewage treatment and storm water disposal.

Table 9.16 Size of Roof Gutters[a]

Diameter of Gutter[b]	Maximum Projected Roof Area for Gutters	
	1/16 in. Slope[c]	
Inches	Square Feet	Gallons per Minute
3	170	7
4	360	15
5	625	26
6	960	40
7	1380	57
8	1990	83
10	3600	150

Source. National Standard Plumbing Code.

[a]This table is based on a maximum rate of rainfall of 4 in./hr for a 5-min. duration and 10-yr return period. Where maximum rates are more or less than 4 in./hr, the figures for drainage area shall be adjusted by multiplying by 4 and dividing by the local rate in inches per hour. See Appendix A.[d]

[b]Gutters other than semicircular may be used provided they have an equivalent cross-sectional area.

[c]Capacities given for slope of 1/16 in./ft shall be used when designing for greater slopes.

[d]*Author's note.* The wide variation in inches per hour is indicated by these three examples selected from Appendix A of the Code (not reprinted in full here).

	Inches per Hour
Juneau, Alaska	1.7
New York, N.Y.	6.6 (See Table 9.19)
Mobile, Ala.	8.4

Rates given are intensities for a 5-min. duration and a 10-yr period, from Technical Paper No. 25, Rainfall Intensity-Duration-Frequency Curves, U.S. Department of Commerce, Weather Bureau.

Table 9.17 Size of Vertical Conductors and Leaders[a]

Size of Leader or Conductor[b]	Maximum Projected Roof Area	
	Square Feet	Gallons per Minute
Inches		
2	544	23
2½	987	41
3	1,610	67
4	3,460	144
5	6,280	261
6	10,200	424
8	22,000	913

Source. National Standard Plumbing Code.

[a]This table is based on a maximum rate of rainfall of 4 in./hr and on the hydraulic capacities of vertical circular pipes flowing between one third and one half full at terminal velocity, computed by the method of NBS Mono. 31. Where maximum rates are more or less than 4 in./hr, the figures for drainage area shall be adjusted by multiplying by 4 and dividing by the local rate in inches per hour. See Appendix A.[c]

[b]The area of rectangular leaders shall be equivalent to that of the circular leader or conductor required. The ratio of width to depth of rectangular leaders shall not exceed 3 to 1.

[c]See author's note, table 9.16.

Table 9.18 Size of Horizontal Storm Drains[a]

Diameter of Drain, Inches	Maximum Projected Area for Drains of Various Slopes					
	1/8-in. Slope		1/4-in. Slope		1/2-in. Slope	
	Square Feet	Gallons per Minute	Square Feet	Gallons per Minute	Square Feet	Gallons per Minute
3	822	34	1160	48	1644	68
4	1880	78	2650	110	3760	156
5	3340	139	4720	196	6680	278
6	5350	222	7550	314	10700	445
8	11500	478	16300	677	23000	956
10	20700	860	29200	1214	41400	1721
12	33300	1384	47000	1953	66600	2768
15	59500	2473	84000	3491	119000	4946

Source. National Standard Plumbing Code.

[a]This table is based upon a maximum rate of rainfall of 4 in./hr. Where maximum rates are more or less than 4 in./hr, the figures for drainage area shall be adjusted by multiplying by 4 and dividing by the local rate in inches per hour. See Appendix A.[b]

[b]See author's note, Table 9.16.

Table 9.19 (Example 9.5) Author's Adaptation of Tables 9.16 to 9.18, which are based on 4 in. of rainfall per hour. Recalculated here for New York City rainfall, 6.6 in./hr

Roof gutters, Diameter, inches (1/16-in. slope)	Maximum Projected Roof Area in Square Feet	
	Table Values	New York Region (Table Value × 4/6.6)
3	170	103
4	360	218
5	625	378
6	960	581
7	1380	836
8	1990	1205
10	3600	2181
Vertical conductors and leaders, diameter in inches		
2	544	329
2½	987	598
3	1610	975
4	3460	2096
5	6280	3805
6	10200	6181
8	22000	13332
Horizontal storm drains, 1/4-in. slope, diameter in inches		
3	1160	703
4	2650	1606
5	4720	2860
6	7550	4575
8	16300	9878
10	29200	17695
12	47000	28482
15	84000	50902

larly, each leader and each horizontal subsurface drain serves the same (515) square feet of area.

The first footnote of Tables 9.16 to 9.18 states that the roof area given for each gutter, leader or drain is based on an assumed maximum rainfall of 4 in./hr. Since the Merker residence is in the New York City area where the maximum rainfall is 6.6 in./hr, we cannot use these tables directly. For this reason, Table 9.19 has been developed for the New York area. We will use the third column of Table 9.19. Please examine Tables 9.16 to 9.18, including their footnotes, before using Table 9.19.

We find that the sizes required by Table 9.19 to handle not less than 515 sq ft are:

    Gutter (semicircular), diameter of 6 in.
    Leader (circular), diameter of 2½ in.
    Drain, diameter of 3 in.

A semicircular gutter of 6-in. diameter has an effective area of $\frac{1}{2} \times \pi r^2$, or $\frac{1}{2} \times 3.14 \times (3)^2 = 14$ sq in. Footnote (b) of Table 9.16 permits a gutter of rectangular section of equivalent area. A gutter 4 in. wide and 3½ in. deep would give us 14 sq in. Rectangular gutters *and* leaders are common for houses. The area required for our *circular* leader is

$$\pi r^2 \text{ or } 3.14 \times (1.25)^2 = 4.9 \text{ sq in.}$$

A common size for a rectangular house leader is 2×3 in. $= 6$ sq in. This would satisfy the 4.9-sq-in. requirement.

Aluminum gutters and leaders could be used. The horizontal drain from house to each dry well could be 3-in. plastic (ABS or PVC).

(b) *Dry wells for storm drainage.* Dry wells are designed differently from gutters, leaders and horizontal drains. The pipes that lead the rainwater to the dry wells are designed for maximum flow. The dry wells, however, are constantly discharging much of this water to the earth.

The "storage volume" of the dry well is only a portion of the actual maximum hourly rainfall. By judgment and experience in the Long Island, New York area, a storm dry well should store a volume of water equal to 2 in. of rainfall in 1 hr. Two inches falling on 515 sq ft of roof area would be

$$\frac{2 \text{ in.}}{12 \text{ in.}} \times 515 \text{ sq ft} = 86 \text{ cu ft}$$

The storage volume for each of the three wells that serve the driveway should be

$$\frac{2 \text{ in.}}{12 \text{ in.}} \times \frac{1700 \text{ sq ft}}{3} = 94 \text{ cu ft}$$

In a manufacturer's catalog, we find that a 6-ft. diameter precast drainage ring that is 4 ft-6 in. in height has a volume of 100.8 cu ft. This is based on an *interior* diameter. We order seven of these rings, four for the roof drainage and three for the driveway drainage.

For site layout of services, see Figure 9.17.

Note that the effective absorption area of a seepage pit taking septic tank effluent is an *exterior* surface but storm dry wells are rated on *interior volume*. Always consult catalogs for dimensions, rated areas and rated volumes.

Problems

9.1. How many water supply fixture units are assigned to the following fixtures in (a) a private house and (b) a public building?
>Bathtub
>Clothes washer
>Kitchen sink
>Lavatory
>Shower head
>Water closet, flush valve

9.2. What size water supply pipes would you specify for the following?
(a) A lavatory.
(b) A kitchen sink, commercial.
(c) A water closet, flush-valve type.

9.3. You are planning a public toilet. It will have a total of 120 water supply fixture units. What would be the demand in gallons per minute if you specified:
(a) Flush tank water closets?
(b) Flush valve water closets?

9.4. A type M copper water service pipe will carry 10 gallons per minute. It must be selected for a pressure drop, due to friction, of 2.4 psi/100 ft of total equivalent length. What size would you select?

9.5. What trap size should be used for the following fixtures?
(a) Lavatory, common.
(b) Food waste grinder, commercial use.
(c) Sink, service type with floor outlet trap standard.

9.6. What would be the total drainage fixture units of a toilet room equipped with the following?
(a) 10 water closets, flush valve type.
(b) 5 urinals, pedestal, syphon jet blowout.
(c) 5 lavatories, 1¼ in. waste.
(d) 1 service sink (trap standard).

9.7. The toilet room in Problem 9.6 is drained by a branch of the building drain. This branch has a fall of ⅛ in./ft.
(a) Select a size. (b) Select a material.

9.8. An apartment house in the country will require a septic tank. What tank capacity would you select for this building of eight apartments each having one bedroom?

9.9. A building has a daily sewage flow rate of 1200 gal. Soil test records 5 min for a 1-in. drop. How many seepage pit rings, 10 ft outside diameter and 5 ft deep would serve the septic tank?

9.10. A 10,000 sq ft flat roof has four conductors (vertical pipes) to carry away the storm drainage. The location has a maximum hourly rainfall of 3 in. Select a pipe size.

9.11. Make detailed drawings for the construction of the 1196 gal septic tank called for in the Merker residence. Show all required views, dimensions, pipe and access details as suggested in Figures 8.26 and 8.27, pages 212 and 213.

Additional Reading

National Standard Plumbing Code. This manual cosponsored by National Association of Plumbing-Heating-Cooling Contractors and the American Society of Plumbing Engineers is the basis for codes adopted by many municipal "Administrative Authorities."

Time Saver Standards, J. H. Callender, Fifth Edition, McGraw-Hill. This book contains an excellent section on plumbing.

10. Nonresidential Plumbing

The study of Nonresidential plumbing will help you to anticipate the differences in plumbing systems that are subject to "public use." Larger buildings and greater usage of the facilities characterize the problems in industrial and commercial buildings. On completing your study of the chapter, you will be able to:

1. Decide when to specify the "flush valve."
2. Allow for the greater water supply fixture units that result from the operation of the flush valve.
3. Understand the effect of geographic location and the slope of the pipe in sizing horizontal storm drains.
4. Avoid spacing fixtures too closely.
5. Make a mechanical "site drawing."
6. Call for roof slopes to permit the most efficient storm drainage.
7. Plan a method of disposing of the water when hot and cold water systems are drained.
8. Size a horizontal fixture branch for a large public toilet room.
9. Plan the "entry services" that connect to municipal street mains.
10. Select fixtures that hang from the wall, leaving the toilet-room floor clear for better cleaning.

10.1. Planning and Layout

In industrial planning the number of fixtures such as water closets, urinals and lavatories must bear a relationship to the number of workers, male and female. Minimum standards are found in the National Standard Plumbing Code. State labor laws and other ordinances written for the welfare of workers must also be observed. Since we are not defining the staff of this building, the layout in Figure 10.1 is shown. It is the *actual* arrangement of the facilities in the building for light industry. See

Figures 4.1 to 4.4, pages 84 to 87. It is evident that there are more female than male workers in this textile factory.

Minimum clearances shown in Figure 10.2 guide the spacing of fixtures and the dimensions of the toilet enclosures. Layout of the hot and cold water piping is also shown in Figure 10.1. The method of disposing of water from these supply piping systems (when drained) is shown in Figure 10.3. It is O.K. to put this water directly into the ground through a dry well, since the water contains no soil or waste material. Figure 10.4 summarizes the sanitary drainage system. Cleanouts below the slab can be accessible through metal floor boxes with hinged covers.

10.2. Water Service

Example 10.1 Establish a size for the water service pipe and the sizes of distribution tubing in the building for light industry.

Solution: It is assumed that this building, in a city area and directly on a street, is served by public services for gas, water, sanitary sewer and separate storm sewer. The water piping from main to fixtures is a much shorter run than it was for the Merker residence. Also, unlike in the Merker residence, the water is delivered on a level and not up a hill and then to a second story. It is still important to calcu-

late the water supply fixture units. They are summarized in Table 10.1 from values in Table 9.1, page 224. Note that the value of units per fixture is greater than for the Merker residence because this is a *public* building. Fixture use is greater. Using the total units (54), we find in Table 9.4, page 225, that the demand is between 51.5 and 55 gpm. These values are found in the last column of that table and are for systems using predominantly *flush valves*. The exact value of the flow in gallons per minute is found by interpolation as follows:

$$55 \quad - 51.5 \ = \ 3.5 \ \text{gpm}$$
$$60 \quad - 50 \quad = \ 10 \ \text{fixture units}$$
$$54 \quad - 50 \quad = \quad 4 \ \text{fixture units}$$

then

$$\frac{x}{3.5} \ = \ \frac{4}{10} \quad \text{and} \quad x \ = \ 1.4 \ \text{gpm}$$

and

$$51.5 + \ 1.4 \ = \ 52.9, \quad \text{say, 53 gpm}$$

In the residence problem, we use values that are for flush *tanks*. Now it is necessary to know the pressure in the street main, the maximum pressure needed at a fixture, the height of the system, and the total equivalent length of tubing. The hose bibb requires 30 psi. We can assume that the street main carries a minimum of 45 psi. The TEL is 125 ft and the height to which the water is raised above the main is 8 ft.

To find the pressure to be lost in the tubing, we deduct the fixture pressure and the static pressure from the street main pressure:

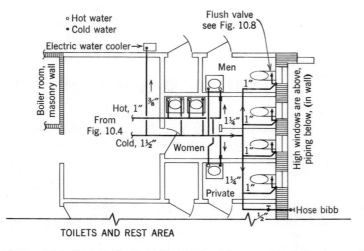

Figure 10.1 (Example 10.1) Hot and cold water. Distribution to fixtures. All pipes are below the slab.

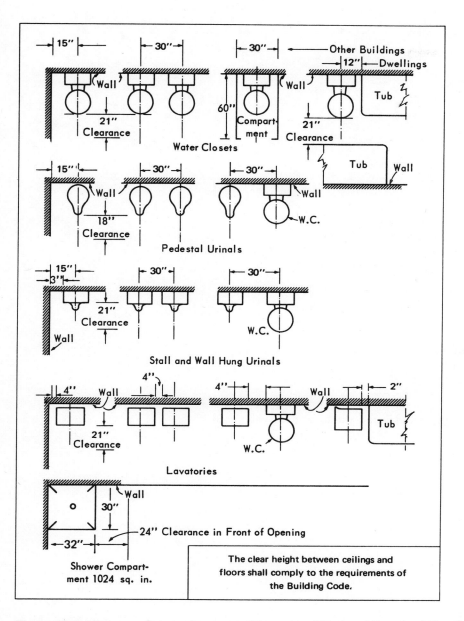

Figure 10.2 Minimum fixture clearances. (Courtesy of National Standard Plumbing Code)

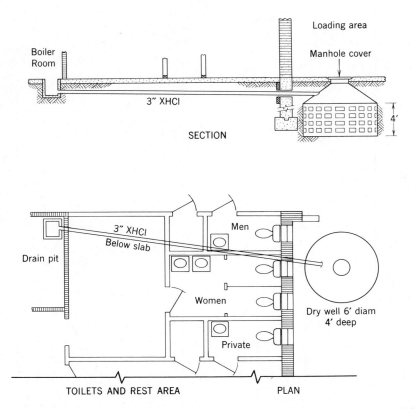

Figure 10.3 (Example 10.1) Drainage for hot and cold water systems and for water used in heating system.

Table 10.1 (Example 10.1) Water Supply Fixture Units

	Fixture Units	
	Per Fixture[a]	Total
4 lavatories, public	2	8
4 water closets, public, flush valve type	10	40
1 drinking fountain, public (⅜ in supply)	2	2
1 hose bibb, public (½-in. supply)	4	4
		54

[a]From Table 9.1, page 224.

Demand (Table 9.4, page 225), supply system predominantly for flush valves) is 53 gpm for 54 fixture units. See text discussion of interpolation.

Main pressure	45	psi
Fixture pressure	−30	
	15	
Static pressure 8 ft × 0.433	− 3.5	
Pressure loss, friction in tubing	11.5	psi

The pressure loss per 100 ft of TEL is found as follows:

$$\frac{x}{100\,\text{ft}} = \frac{11.5\,\text{psi}}{125\,\text{ft TEL}} \quad \text{and} \quad x = 9.2\,\text{psi}$$

Now refer to Table 2.17, page 41. A horizontal line at 9.2 psi crosses a vertical line at 53 gpm on the diagonal line for 1½-in. tubing. The fluid velocity is nearly 10 ft/sec (sloping line). The footnote below the diagram cautions against velocities in excess of 8 ft/sec. Therefore, we choose a 2-in. service.

Within the distribution system (Figure 10.2), 1 in. (hot) and 1½ in. (cold) are satisfactory. The cold water header is larger because it serves four flush valves.

10.3. Domestic Hot Water Heater

If "process" hot water were necessary (see glossary), a high capacity heater would be used together with a storage tank. In this building for light industry there are no "processes" that create this kind of demand. For the lesser needs of staff and employees the gas-fired heater shown in Figure 10.5a is suitable. Its storage capacity and its recovery rate are shown in Figure 10.5b.

10.4. Sanitary Drainage

Example 10.2 Lay out and size the sanitary drainage system.

Solution: The number of drainage fixture units must be known. See Figures 10.4 and 10.6. We consult Table 9.7, page 235. The units are listed in Table 10.2. The total is 28½ units. A look at Table 9.9, page 235, shows that a 4-in. building drain and a 4-in. building sewer at ¼ in./ft slope are satisfactory.

From the drainage *plan* of Figure 10.4, a plumbing section (Figure 10.6) is developed. Notice the omission of a house trap and fresh air inlet. Some codes do not require these two items. Table 9.10, page 236, is a guide to fixture branch sizes. A 2-in. branch will carry 6 fixture units and a 4-in. branch is all right for 160. Both sizes are quite large enough for our needs. Usual practice is to use 4-in. piping from any water closet. The vents from water closets are usually not less than 2 in. This size can serve as a vent for our entire system (Figure 9.11, page 236). Fixtures such as lavatories can be vented by 1¼-in. pipes.

10.5. Plumbing Fixtures

In industrial buildings it is well to keep the floors of toilet rooms clear of fixture pedestals or legs. See Figures 10.7 to 10.9. This makes it easy for the mainenance staff to keep floors and base clean. Fixtures are hung from the wall. This is seen in the illustrations of the Roxalyn lavatory and the Afwall toilet. Figure 10.9 shows the type of *carrier* that is built into the masonry wall to support the toilet. A built-in support is also used for the lavatory, which is carried by arms projecting from the support.

In Figure 10.8 you will see a flush valve. This method of flushing the fixture is most suitable for public toilets. It is more rugged and effective than a flush *tank* and requires less maintenance.

10.6. Storm Drainage

Example 10.3 Layout and size the components of a storm drainage system for the roof and paved parking area of the building for light industry.

Solution: For the purpose of this example, we are choosing an interior system of vertical and horizontal storm drains within the building. It is therefore assumed that the steel open web joists and the roof structure above them can be sloped to the location of the roof drains. A cutaway view of a typical roof drain fixture is shown in Figure 10.11. Also in that figure the vertical leaders are shown close to columns or walls. These locations minimize their possible obstruction to manufacturing activity. The leaders carry the roof water from the drain fixtures to the horizontal drain pipes below the slab.

The four roof drains (and leaders) that drain the main (high) roof area take care of

$$35 \times 110 \times 2 = 7700 \text{ sq ft}$$

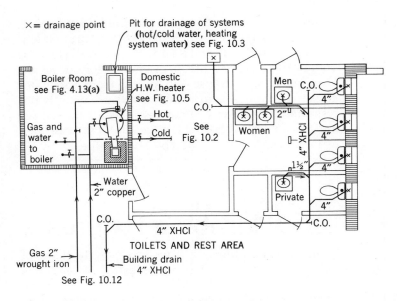

Figure 10.4 (Examples 10.1 and 10.2) Gas and water. Services to boiler room. Sanitary drainage. All pipes are below the slab.

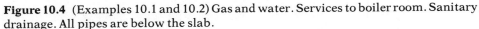

RECOVERY CAPACITIES

MODEL	Approx. Gal. Cap.	Type of Gas*	Input Rating BTU/H	TEMPERATURE RISE — DEGREES F — GALLONS PER HOUR											
				30	40	50	60	70	80	90	100	110	120	130	140
B-180	9	1, 2, 3	50,000	140	105	84	70	60	52	46	42	38	35	32	30
		4	45,000	126	94	75	63	54	47	42	38	34	31	29	27
BT-65	50	1, 3	65,000	183	137	109	91	78	68	61	55	50	46	42	39
		2	53,000	148	111	89	74	64	56	49	45	40	37	34	32
		4	50,000	140	105	84	70	60	53	47	42	38	35	32	30
BT-70	9	1, 2, 3	70,000	196	158	118	98	84	74	65	59	54	49	45	42
		4	60,000	168	126	101	84	72	63	56	50	48	42	39	36
BT-85	75	1, 3	85,000	238	178	143	119	102	89	79	71	65	59	55	51
		2	76,000	212	160	128	106	91	80	71	64	59	53	49	46
		4	70,000	196	158	118	98	84	74	65	59	54	49	45	42
BT-100	100	1, 3	100,000	280	210	168	140	120	105	93	84	77	70	65	60
		2	76,000	212	160	128	106	91	80	71	64	59	53	49	46
		4	80,000	224	168	135	112	96	84	75	67	61	56	52	48
BT-155	89	1, 2, 3, 4	155,000	434	326	260	217	186	163	145	130	118	108	100	93
BT-197	100	1, 2, 3, 4	197,000	552	414	331	276	237	207	184	166	151	138	128	118
BT-199	86	1, 2, 3, 4	199,000	560	420	335	280	240	210	186	167	152	140	129	120
BT-251	84	1, 2, 3, 4	251,000	704	529	422	352	302	264	234	211	192	176	163	151
BT-270	100	1, 2, 3, 4	270,000	757	567	454	378	324	284	252	227	206	189	175	162
BT-365	75	1, 2, 3, 4	365,000	1020	768	613	510	439	384	340	307	279	225	236	219
BT-500	69	1, 5	500,000	1400	1050	839	700	602	526	467	420	381	350	324	301

*1) Natural   2) Manufactured   3) Mixed   4) LP   5) Propane

(b)

Figure 10.5 (Example 10.1) (a) Typical BT Series gas-fired domestic hot water heater. BT 65 is chosen for this example. Storage capacity approximately 50 gal. Recovery in gallons per hour for 100 F° temperature rise is 55, 45 or 42 depending on the type of gas used. (b) Ratings, BT Series. (Courtesy of A.O. Smith)

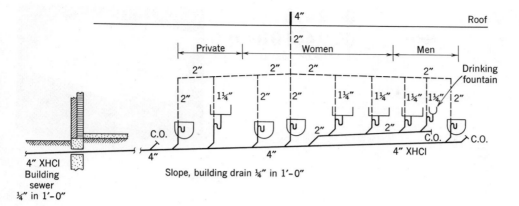

Figure 10.6 (Example 10.2) Plumbing section.

Table 10.2 (Example 10.3) Drainage Fixture Units

	Fixture Units	
	Per Fixture (Table 9.7, page 000)	Total
4 lavatories	1	4
4 water closets, flush valve operated	6	24
1 drinking fountain	½	½
		28½

ROXALYN LAVATORY

VITREOUS CHINA — FOR EXPOSED ARM SUPPORT
2248.045 FITTING — 2303. SER. SUPPLIES
4401. SER. "P" TRAP

0194.019
0194.100
(POP-UP DRAIN)

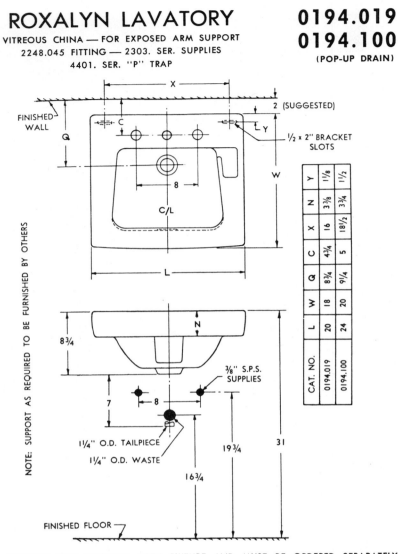

NOTE: SUPPORT AS REQUIRED TO BE FURNISHED BY OTHERS

CAT. NO.	L	W	Q	C	X	N	Y
0194.019	20	18	8¾	4¾	16	3⅜	1⅛
0194.100	24	20	9¼	5	18½	3¾	1½

NOTE: FITTINGS NOT INCLUDED WITH FIXTURE AND MUST BE ORDERED SEPARATELY.

PLUMBER NOTE — Provide suitable reinforcement for all wall supports.

IMPORTANT: Dimensions of fixtures are nominal and may vary within the range of tolerances established by Commercial Standards CS20.

These measurements are subject to change or cancellation. No responsibility is assumed for use of superseded or voided leaflets.

Figure 10.7 Wall-hung lavatory fixture. Fixture dimensions and roughing dimensions. (Courtesy of American Standard)

AFWALL TOILET

2477.016

VITREOUS CHINA

EXPOSED FLUSH VALVE {
SLOAN 110 YV
DELANY 402 VBQ
}

NOTE: TO COMPLY WITH AREA CODE GOVERNING THE HEIGHT OF VACUUM BREAKER ON FLUSH VALVE, THE PLUMBER MUST VERIFY DIMENSIONS SHOWN FOR SUPPLY ROUGHING.

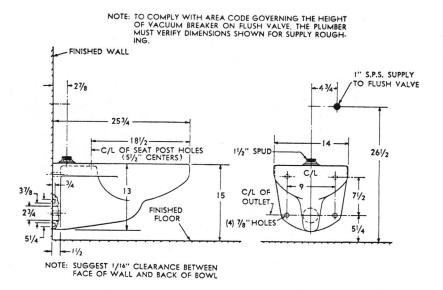

NOTE: SUGGEST 1/16" CLEARANCE BETWEEN FACE OF WALL AND BACK OF BOWL

NOTE: CARRIER FITTING AS REQUIRED TO BE FURNISHED BY OTHERS.

NOTE: Flush valve not included with toilet and must be ordered separately.

PLUMBER NOTE — Provide suitable reinforcement for all wall supports.

IMPORTANT: Dimensions of fixtures are nominal and may vary within the range of tolerances established by Commercial Standards CS20.

These measurements are subject to change or cancellation. No responsibility is assumed for use of superseded or voided leaflets.

Figure 10.8 Wall-hung toilet fixture, fixture dimensions and roughing dimensions. (Courtesy of American Standard)

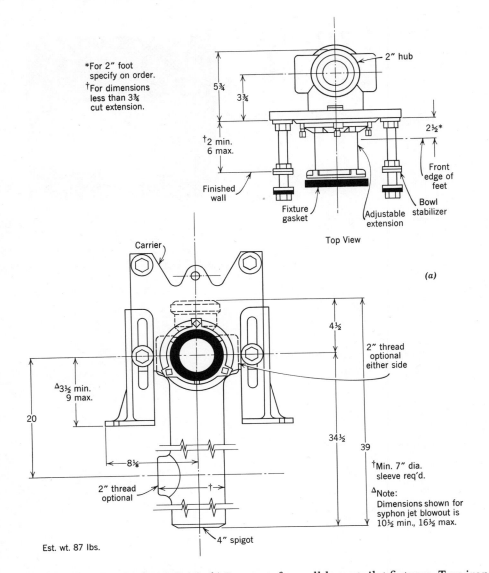

*For 2" foot specify on order.

†For dimensions less than 3¾ cut extension.

5¾

3¾

2" hub

2½*

†2 min. 6 max.

Finished wall

Fixture gasket

Adjustable extension

Bowl stabilizer

Front edge of feet

Top View

(a)

Carrier

4½

2" thread optional either side

△3½ min. 9 max.

20

34½

39

8⅛

2" thread optional

†Min. 7" dia. sleeve req'd.

△Note: Dimensions shown for syphon jet blowout is 10½ min., 16½ max.

Est. wt. 87 lbs.

4" spigot

Figure 10.9 (Example 10.1) *(a)*, *(b)* Supports for wall-hung toilet fixtures. Two iron standards are built into the masonry wall. Carrier plate can be slid up or down to adjust 4-in. drain to the height of the water closet outlet. Elbow [see *(b)*] carries soil discharge down to 4-in. spigot. Topside a 2-in. hub receives a 2-in. vent pipe. Bolts either side of the 4-in. gasketed inlet hold the water closet in cantilever position. (Courtesy of Josam Manufacturing Company)

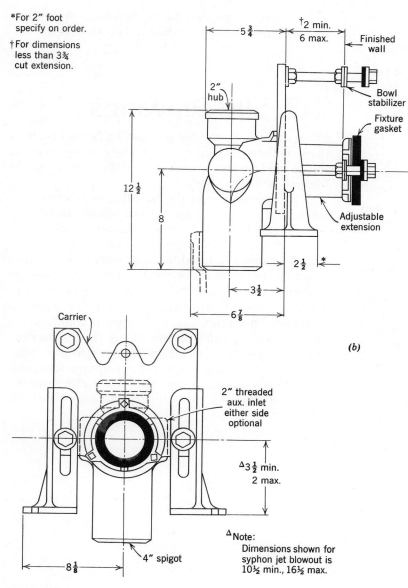

*For 2″ foot
  specify on order.

†For dimensions
  less than 3¾
  cut extension.

2″ hub

12½

8

2½

*

3½

6⅞

5¾

†2 min.
6 max.

Finished wall

Bowl stabilizer

Fixture gasket

Adjustable extension

(b)

Carrier

2″ threaded aux. inlet either side optional

Δ3½ min.
2 max.

ΔNote:
Dimensions shown for syphon jet blowout is 10½ min., 16½ max.

4″ spigot

8⅛

Est. wt. 60 lbs.

The area served by each leader is

$$\frac{7700}{4 \text{ leaders}} = 1925 \text{ sq ft}$$

The two roof drains that drain the *lower* roof take care of

$$25 \times 96 = 2400 \text{ sq ft}$$

The area served by each leader is

$$\frac{2400}{2 \text{ leaders}} = 1200 \text{ sq ft}$$

Table 9.19, page 247, indicates the need for 5-in. leaders each of which would handle 2096 sq ft.

Although a slope of ¼ in./ft is a very common choice for horizontal sanitary drains, we select ½ in./ft for our *storm* drains. As you can see in Table 9.18, page 247, a greater slope of pipe allows any specific size of pipe to serve a greater roof area. As branches of the horizontal system join the main flow, pipe sizes increase. Figure 10.11 shows the increasing roof areas served. Pipe sizes are selected from the right-hand column of Table 10.3 for this building in the New York region. Starting directly below the slab, the total drop for the longest run (about 130 ft) is

$$\frac{130 \text{ ft} \times \text{½ in./ft}}{12} = 5.4 \text{ ft}$$

The ½-in. slope can be used only if *this* storm sewer is about 6 ft lower than the finish floor.

10.7. Site Drawings

The locations shown for the four utility services in the street are schematic. In practice it is necessary to locate their exact position and elevations. Study Figures 10.10 and 10.12.

If we consider that the building is heated by a forced hot water system, like that shown in Figures 4.12, page 94, and 4.13, page 95, there is already a great total length of tubing in or below the slab. In *this* chapter we have added a network of additional pipes for hot water, cold water, sanitary drainage and storm drainage. This amount of piping is impressive. Special drawings known as "composite drawings" are often made to prevent conflicts. On these drawings, the piping of all trades may be shown, including, sometimes, electric and telephone conduits. One of the many important considerations is that cold water pipes be remote from the unwanted heating effect of heating pipes and do-

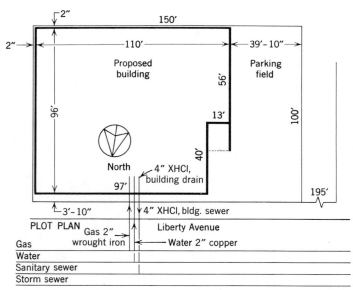

Figure 10.10 Mechanical site drawing. Street services and connections into building. Positions of street services are to be established at the site. For storm drainage see Figures 10.11 and 10.12.

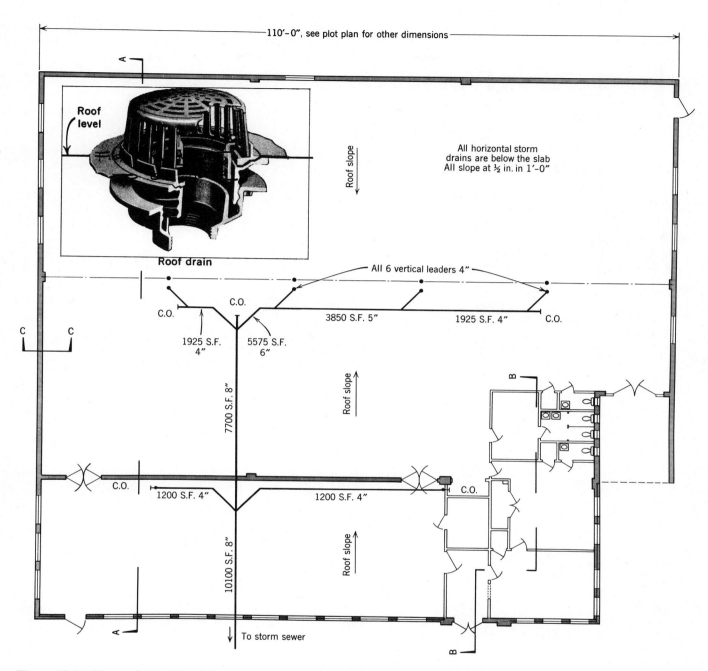

Figure 10.11 (Example 10.3) Roof drainage. Cast-iron drainage pipe sizes are based on areas of roof drained. Table 10.3 is used. That table carries sizes for a (New York) maximum hourly rainfall of 6.6 in. Traps are unnecessary but cleanouts are desirable. (Roof drain, courtesy of Josam Manufacturing Company)

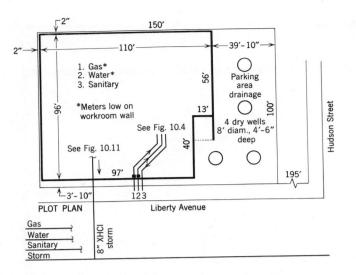

Figure 10.12 (Example 10.3) Mechanical site drawing. Storm sewer connection for roof drainage. Meter locations for water and gas. Dry well storm drainage for parking area.

mestic hot water pipes. Limited space in this book and the complexity of composite drawings prevent us from illustrating these drawings here.

The area of the paved parking lot is

$$40 \times 100 = 4000$$
$$13 \times 40 = \underline{520}$$
$$4520 \text{ sq ft}$$

At the 2 in./hr rainfall* to set the volume of the dry wells, their volume must be

$$4520 \times \frac{2}{12} = 750 \text{ cu ft}$$

If four dry wells are used, the capacity of each must be

$$\frac{750}{4} = 187 \text{ cu ft}$$

A manufacturer's catalog shows that an 8-ft diameter ring 4 ft 6 in. deep provides 190.1 cu ft of volume. These are specified.

*See Chapter 9, page 248.

Table 10.3 Adaptation[a] of part of Table 9.18, page 000, which is based upon 4 in. of rainfall per hour.
Recalculated here for New York City rainfall, 6.6 in./hr

Size of Horizontal Storm Drains	Maximum Projected Roof Area in Square Feet	
½-in. slope (Diameter, in.)	Table value (4 in./hr)	New York Region (Table value × 4/6.6)
3	1644	996
4	3760	2280
5	6680	4050
6	10700	6480
8	23000	13900
10	41400	25000
12	66600	40400
15	119000	72000

[a]By author.

10.8. Space for Roughing

See Figures 10.13*a* and *b*. In large toilet rooms, the piping "behind the wall" takes up a lot of room. Unlike the roughing in a residence which might fit into a 4- or 6-in. stud partition, industrial piping may require several feet. In some jobs, "walk-in access" might be necessary for inspection, repairs or changes. Large-scale detail drawings or shop drawings may prove useful.

Figure 10.13 Installation methods for a large battery of wall-hung water closets in a toilet room of an industrial plant. *(a)* Roughing. Four-inch closet runouts discharge directly into a 5-in. horizontal soil branch. Two-inch vertical vents connect to a horizontal vent branch. The 1¼-in. cold water runouts will connect to flush valves to be installed. *(b)* Two closets in place. Closet carrier is of a different type from that of Figure 10.9. Closet runouts and soil branch are in same horizontal plane. Traps are integral with the fixtures. Notice the wide space behind the fixture line required to accommodate the piping. (Courtesy of Zurn Industries, Inc.)

Problems

10.1. Concerning minimum fixture clearances, what is:
  (a) The size of a toilet compartment?
  (b) The spacing on centers of water closets?
  (c) The clearance in front of a water closet?
  (d) The clearance in front of a lavatory?
  (e) The clearance in front of a shower compartment opening?

10.2. Name materials suitable for the following services below ground.
  (a) Gas piping.
  (b) Sanitary building sewer.
  (c) Water service.

10.3. (a) At 100 F° rise what is the recovery rate of an A. O. Smith BT 500 gas-fired heater for domestic hot water?
  (b) How many kinds of gas can it use?

10.4. You have specified an A. O. Smith B 180 gas-fired domestic hot water heater. What is its recovery rate (at 100 F° rise) for:
  (a) Gas types—natural, manufactured or mixed?
  (b) L.P.?

10.5. You find that there is a 500-ft run to the nearest storm sewer. The horizontal storm drain for a 40,000 sq ft roof at a maximum hourly rainfall of 4 in. is to be selected. You have three choices. Determine the size of drain for:
  (a) A ⅛ in./ft slope.
  (b) A ¼ in./ft slope.
  (c) A ½ in./ft slope.

10.6. In the 500-ft run of Problem 10.5 what will be the drops in feet of elevation for cases (a), (b), and (c)?

10.7. The "building for light industry" is being built in Juneau, Alaska. Its roof area is 10,100 sq ft. Roof drain fixture collects rain that is led away in a single horizontal drain at ½-in. slope. (a) Select the correct pipe size. (b) Would you enclose the drain in thermal insulation?

10.8. A public toilet room is equipped as follows:
  20 water closets, flush valve operated
  10 urinals, pedestal, syphon jet blowout
  10 lavatories, 1¼-in. waste
  (a) Calculate the drainage fixture units.
  (b) Select the size for a horizontal fixture branch.

Additional Reading

Standard Plumbing Engineering Design, Louis S. Nielsen, McGraw Hill. A complete development of the engineering aspects of plumbing systems.

11. Introduction to Electricity

The daily work of the technologist involves the layout, assembly and connection of electrical equipment. This equipment includes common items such as residential lighting and appliances, plus less common items such as commercial heating and refrigeration equipment. It is not enough to know that a motor runs and a lamp lights when connected. The competent electrical technologist, design draftsman or junior designer must have a sound knowledge of how and why electrical circuitry and equipment work as they do. Only in this way can he increase his usefulness in the technical office in which he is employed. This first chapter is devoted to the basics of electricity and circuitry.

The study of the chapter will enable you to:

1. Be familiar with the basic electric quantities of voltage, current and resistance.
2. Do circuit calculations in d-c and a-c circuits, in series and parallel arrangements.
3. Calculate power and energy in electric circuits.
4. Understand voltage levels and their uses.
5. Define ampacity and its application.
6. Know how electric quantities are measured and how the meters work.
7. Understand the differences between d-c and a-c, and apply that knowledge to circuit calculation.
8. Understand the differences in nature and application between single phase and three phase a-c.
9. Acquire a working vocabulary of electric circuit terms for d-c and a-c.
10. Draw basic circuit diagrams.

11.1. Electric Energy

Energy has historically been made available for useful work by burning a natural fuel such as coal or oil. That is, the energy in the fossil fuel has been released in the form of heat. This heat, in turn, is used as we wish, to warm our houses, prepare our food, and to generate steam to drive turbo-generators which produce electricity. This last use is relatively recent, since commercial electric power is barely a century old.

Electricity constitutes a form of energy itself, which occurs naturally only in unusable forms such as lightning and other static discharges, or in the natural galvanic cells, which cause corrosion. The primary problem in the utilization of electric energy is that unlike fuels or even heat, it cannot be stored and, therefore, must be generated and utilized in the same instant. This requires in many respects an entirely different concept of utilization than, say, a heating system with its burner, piping and associated equipment. In the following sections the concepts of electric circuiting and application will be illustrated and explained so that the reader will obtain a thorough basis in the practical application of electric power. Our initial discussion will be on direct current (d-c), which was developed earlier than alternating current (a-c), and which can be used to illustrate the principles of circuitry and power more readily than a-c. Subsequent sections will introduce a-c, the understanding of which is

vital, since the overwhelming proportion of commercial electricity in use today is a-c.

11.2. Batteries

The interested reader can follow in any book on the history of electricity, the development of electric power from its origins in the work of Leyden, Galvani and Volta through Faraday, Darcy and Ampere (note the custom of perpetuating the scientist's name by attaching it to an aspect of his work). He will find it an interesting and often fascinating tale.

To summarize briefly, it was noted that when dissimilar metals were joined by a conducting solution such as salt water or a weak acid, a current was caused to flow between the electrodes. From this rudimentary beginning, the battery was developed to its present state. See Figure 11.1. The limitations of batteries as sources of electric power are fairly obvious. Since the voltage of a cell is determined entirely by its chemical components (1.5 v for a zinc-carbon cell), to obtain a workable voltage many cells must be placed in additive connection. Simi-

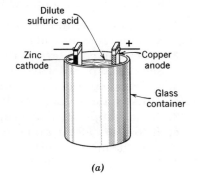

(a)

Figure 11.1 *(a)* Rudimentary battery; a wet cell. (From *Modern Physics*, by H.C. Metcalf, J.E. Williams, and C.E. Dull, Copyright © 1964, Holt, Rinehart and Winston. Used with permission of Holt, Rinehart and Winston.)

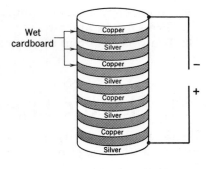

(b)

Figure 11.1 *(b)* Voltaic pile battery, consisting of alternate discs of silver and copper, separated by discs of cardboard saturated in brine. This represents the beginning of the "dry" cell.

larly, to obtain reasonable amounts of power, cells must be stacked. Thus the basic limitations become size and weight. Furthermore, cells discharge and can therefore only supply power for a short time before gradually dropping in voltage and capacity. They are therefore, not suitable for continuous power supply. They are however, irreplaceable as sources of power for standby electric service, telephone equipment power, railway signaling, transportation applications and many types of portable devices.

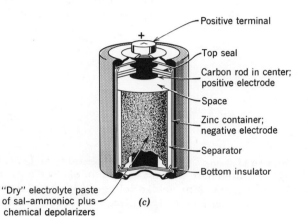

Positive terminal

Top seal

Carbon rod in center; positive electrode

Space

Zinc container; negative electrode

Separator

Bottom insulator

"Dry" electrolyte paste of sal-ammonioc plus chemical depolarizers

(c)

Figure 11.1 *(c)* Construction of common flashlight battery. Courtesy of Power Magazine.

Figure 11.1 *(d)* Typical lead-acid battery construction similar to common automobile battery. Courtesy of Power Magazine.

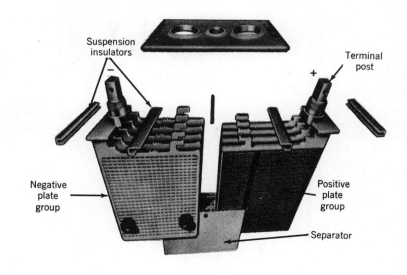

Suspension insulators

Terminal post

Negative plate group

Positive plate group

Separator

Container filled with dilute sulphuric acid

Circuit Basics

11.3. Electric Power Generation

As we stated above, the chemical action of a battery in producing electricity is limited by the chemical capacity of the cells. Non-rechargeable batteries obviously use up their internal capacity and become discharged; rechargeable cells need another source of power to resupply theirs. The need for a means of electric power generation that is independent of chemical action was met with the rotating electric generator. The action basic to all electric generators is illustrated in Figure 11.2. As a result of the work of Oersted and Ampere the electromagnet was developed. Soon afterward, Faraday in 1831 performed the crucial experiments that led to his dis-

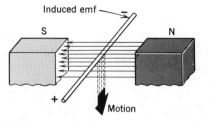

Figure 11.2 *(a)* The action fundamental to all generators is illustrated here. When a conductor of electricity moves through a magnetic field, a voltage is produced in the conductor, with polarity shown.

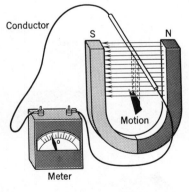

Figure 11.2 *(b)* The existence of voltage can be demonstrated by connecting a meter to the conductor, and noting current flow.

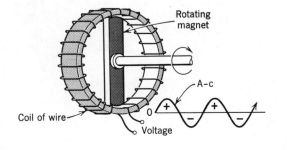

Figure 11.2 *(c)* It does not matter whether the conductor moves and the field is stationary or vice versa, as long as there is relative motion between the two. If the wire or the field (magnet) is *rotated*, an alternating current is produced. The illustration shows the field rotating.

covery of electromagnetic induction. The principal involved is demonstrated in Figure 11.2. A wire moved through a magnetic field has a voltage induced in it. Thus we can see that if the wire is formed into a loop, an alternating polarity voltage will be produced at the loop terminals whenever the loop is rotated in a magnetic field. If many of these loops are wound onto a rotor and the assembly rotated, we have the makings of a generator. The voltage produced will be alternating (see Section 11.16), but can easily be changed, or rectified, into d-c. With the development of the electric generator the search for a continuous source of electric power ended. The basic generating unit described above has changed only in detail and sophistication to the present day.

11.4. Voltage

The electric 'pressure' (potential) produced by a battery is called voltage and is measured in units of *volts*. A simple dry cell produces an electric potential of approximately 1.5 v. This potential, or voltage, is constant in amplitude and direction (polarity) and is therefore designated d-c voltage. (The term d-c originated as an abbreviation for "direct-current," but because of the oddity of an expression such as "direct-current voltage," or the repetition in "direct-current current", the terms universally accepted are "d-c voltage" and "d-c current.") A d-c voltage of 1.5 units and positive polarity is shown in

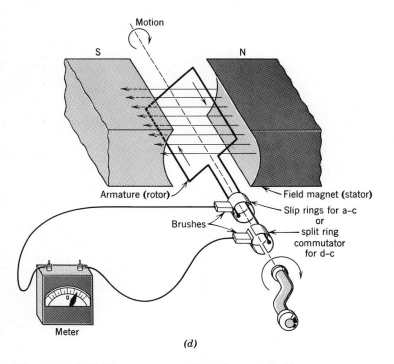

Figure 11.2 *(d)* The same a-c voltage is produced by rotating a coil of wire in a *stationary* magnetic field. By using *commutators* in lieu of slip rings to collect the voltage, d-c is produced. (*(a)*, *(b)* and *(d)* from *Modern Physics*, by H.C. Metcalf, J.E. Williams, and C.E. Dull, Copyright © 1964, Holt, Rinehart and Winston. Used with permission of Holt, Rinehart and Winston.)

Figure 11.3*a*. A voltage of negative polarity is also shown. In physical terms the voltage of a battery can be likened to the water pressure of an hydraulic system. The potential to supply current (water in the analagous hydraulic system) exists, but does not materialize until a circuit is completed, after which current begins to flow. Refer to Figure 11.4. Polarity is similar to flow direction; reversal of polarity causes current (analogy—water) to flow in the opposite direction.

11.5. Current

When a circuit is formed comprising a complete loop, and containing all the required components, a current will flow. The required components are:

(a) A source of voltage.
(b) A closed loop of wiring.
(c) An electric load.
(d) A means of opening and closing the circuit.

The electric circuit shown in Figure 11.4 fulfills all the above requirements and, therefore, an electric current will flow. Just as in the hydraulic circuit the amount of water flowing is proportional to the pressure and inversely proportional to the friction, so in the electric circuit the current is proportional to the voltage and inversely proportional to the circuit resistance, or load. The higher the voltage, the larger the current. The higher the resistance, the lower the current. This relationship is expressed by the equation

$$I = \frac{V}{R}$$

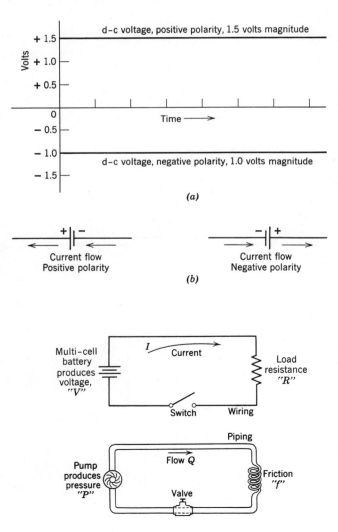

Figure 11.3 *(a)* Graphical representation of d-c voltage with positive and negative polarity. *(b)* Circuit symbol representation of a battery source. The longer bar is positive. By convention, current flows from positive to negative around the circuit, and − to + within the battery.

Figure 11.4 Electric-hydraulic analogy. The circuits show that voltage is analogous to pressure, current to flow, friction to resistance, wire to piping, and switches to valves. As a result of Ben Franklin's wrong guess, current is assumed to flow from + to −. This convention is still used today.

known as Ohm's law, where I is current, V is voltage and R is resistance. We will return to this equation repeatedly, and it is therefore imperative that it be clearly understood. The letters I, V, and R are normally used to represent current, voltage and resistance, and should not be used for other electrical factors, to avoid confusion. The unit of current is the *ampere*, abbreviated amp or simply "*a*."

11.6. Resistance

The flow of fluid in a hydraulic system is impeded by friction; the flow of current in an electric circuit is resisted (impeded) by resistance, which is the electrical term for friction. In a d-c circuit this factor is called resistance. The unit of measurement of resistance is the *ohm*. Materials display different resistance to the flow of electric current. Metals generally have the least resistance and therefore are called conductors because they easily conduct electricity through them. The best conductors are the precious metals such as silver, gold and platinum with copper and aluminum being slightly inferior. Conversely, materials that tend to prevent (resist) the flow of current, displaying high resistance to electricity, are called insulators. Glass, mica, rubber, oil, distilled water, porcelain and certain synthetics such as phenolic compounds have this insulating property. They are therefore used to insulate electric conductors. Common examples are rubber covering on wire, porcelain cable supports, porcelain lamp sockets, glass pole-line insulators and oil-immersed electric switches.

11.7. Ohm's Law

As mentioned above, the relationship in a d-c circuit between current in amperes, voltage in volts and resistance in ohms, is expressed by Ohm's law which is most frequently written

$$V = IR$$

but much more logically is written $I = V/R$, showing the relationship of current to voltage and resistance. We strongly recommend that this second form be remembered, instead of the mathematical relationship that volts = amperes × ohms, which has no logical basis. Physically, we start with a voltage and resistance and produce current. The equation

$$I = \frac{V}{R}$$

demonstrates this. A few examples will show the logic of Ohm's law. All the examples refer to Figure 11.4.

Example 11.1 In the circuit of Figure 11.4, assume that the voltage is supplied by a 12 v automobile battery to a headlight (R) with a resistance of 1.2 ohms. Find the circuit current.

Solution

$$I = \frac{V}{R} = \frac{12 \text{ volts}}{1.2 \text{ ohms}} = 10 \text{ amperes}$$

Example 11.2 A telephone system load of 10 ohms is fed from a 48 v battery. What is the current drawn?

Solution

$$I = \frac{48 \text{ volts}}{10 \text{ ohms}} = 4.8 \text{ amperes}$$

Example 11.3 Instead of the switch shown, a 10-amp fuse is used, and the circuit resistance is 12 ohms. What is the maximum voltage that can be applied without blowing the fuse?

Solution

$$I = \frac{V}{R} \quad \text{or} \quad V = IR$$
$$V = 10 \text{ amperes (max)} \times 12 \text{ ohms}$$
$$V = 120 \text{ volts}$$

Example 11.4 Assume a 120 v d-c source, which can be either a multicell battery or a d-c generator, that feeds a 720 watt electric toaster with a resistance of 20 ohms. Find the current in the circuit.

Solution: For *this* example, the wattage of the toaster is an unnecessary piece of information. To find the current

$$I = \frac{V}{R} = \frac{120 \text{ volts}}{20 \text{ ohms}} = 6 \text{ amperes}$$

We will show later the relation of wattage rating to V, I and R. In fact, the wattage is generally of more concern than the resistance.

Circuit Arrangements

11.8. Series Circuits

Obviously circuit components can be arranged in many ways, as for instance in the complicated network shown in Figure 11.5. Still, even that maze can be reduced to two fundamental types of connections—*series* and *parallel*. In a series connection a single path exists for current flow. That is, the elements are arranged in a series, one after the other, with no branches. In this arrangement, voltage and resistance add. To illustrate, study Figure 11.6. As we explain in Section 11.2, the voltage of a single flashlight battery is about 1.5 v. Since this voltage is too low to use economically, four batteries are arranged in series, with additive polarity, to make a 6 v source. The circuit is then completed with a lamp load and a switch. It is not necessary that a circuit be complete to have a series connection. In Figure 11.7, batteries and resistors are shown in series connection, as separate groupings. The only requirement for a series connection is that only a single current path exist. It is apparent from the

above and from the illustrations shown thus far that component values simply add when connected in series, Hence,

$$V_{\text{TOTAL}} = V_1 + V_2 + V_3$$

and

$$R_{\text{TOTAL}} = R_1 + R_2 + R_3, \text{etc.}$$

Thus, in Figure 11.4, the multicell battery can comprise any number of similar cells in series. Indeed, to supply 120 v emergency power for hospital operating room lights, this is exactly what is done. Some 86 to 95 nickel cadmium cells, each with a voltage of approximately 1.3 v are connected in series to make the 120 v total. Similarly, a large resistance can be composed of several smaller resistances. An example illustrates this clearly.

Example 11.5 In Figure 11.8, the source is a 12 v automobile battery, and the load comprises two auto headlights connected in series, each with a resistance of 1.2 ohms. What is the current flowing in the circuit?

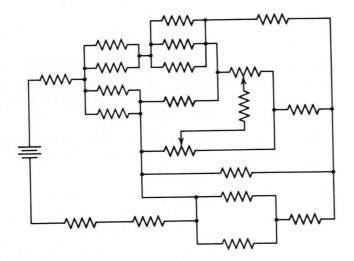

Figure 11.5 A maze of resistances connected in combinations of series and parallel.

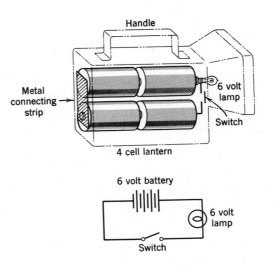

Figure 11.6 Phantom view of a standard four-cell, lantern-type flashlight showing the series (additive) connection of the cells.

Solution

$$V = 12 \text{ volts}$$
$$R_{\text{TOTAL}} = R_1 + R_2 = 1.2 + 1.2 = 2.4 \text{ ohms}$$
$$I = \frac{12 \text{ volts}}{2.4 \text{ ohms}} = 5 \text{ amperes}$$

Note that although such an arrangement—of two 6 v lights in series on a 12 v service—is possible, it is not used often, since the failure of one unit will cause both to go dark. This is the principal disadvantage of a series connection feeding more than a single load. Since there is only a single current path, a failure in any one unit causes a break in the circuit and thereby kills the entire circuit. Strings of Christmas tree lights are occasionally made in this fashion to reduce costs, since only a single wire is required around the circuit. See Figure 11.9. Another disadvantage is that when a single lamp goes out, not only does the entire string go dark, but the location of the fault is unknown. This is a great nuisance and ac-

counts for the disappearance of this type of light string. To avoid this problem a parallel connection is used. This allows multiple loads to be fed without a failure in one disrupting the others.

11.9. Parallel Circuits

In *parallel* or *multiple* connection, the loads being served are all placed across the same voltage and in effect constitute separate circuits. In the hydraulic analogy, the connection is equivalent to a branching piping arrangement. See Figure 11.10. Figure 11.11 illustrates this multiple–circuit idea. Obviously, as in Figure 11.12, if the number of devices exceeds the capacity of the circuit, the fuse will blow (to the surprise of the uninformed housewife). This aspect of circuitry will be discussed in the section on circuit protection. For now, observe in Figure 11.12 that unlike the series circuit, 2 wires are required plus,

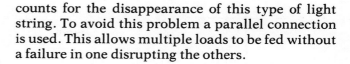

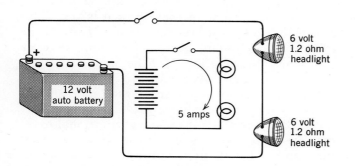

Figure 11.8 Physical and graphic representation of a possible d-c circuit.

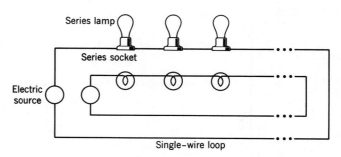

Figure 11.9 Physical and graphic representation of a series lamp circuit. Loss of one lamp disables the entire circuit. Furthermore, the point of fault is not obvious and requires individual testing of lamps.

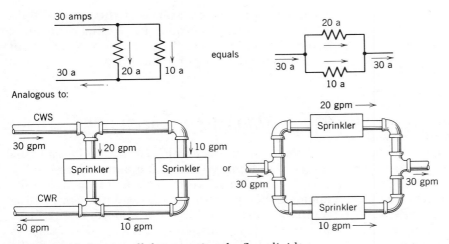

Figure 11.10 In a parallel connection the flow divides between the branches, but the pressure is the same across each branch.

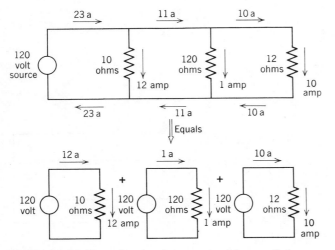

Figure 11.11 Note that loads connected in parallel are equivalent to separate circuits superimposed into a single connection. Each load acts as an independent circuit, unrelated to, and unaffected by, the other circuits.

in proper and safe practice, a separate ground wire. For clarity this ground wire is not shown in the illustration.

The parallel connection is the standard arrangement in all house wiring. A typical lighting and receptacle arrangement for a large room is shown in Figure 11.13. Here the lights constitute one parallel grouping, and the convenience wall outlets constitute a second parallel grouping. The fundamental principal to remember is that loads in parallel are additive for *current*, and that each has the same voltage imposed. A single example will be sufficient.

Refer to Figure 11.11. Notice that the total current flowing in the circuit is the sum of all the branches, but that the current in each branch is the result of a separate Ohm's law calculation. Thus in the 10-ohm load a 12-amp current flows, and so forth. Now study this diagram until the figures and their derivation are perfectly clear.

One additional point is important to appreciate. If we examine Ohm's law again, we note, as previously stated, that current is inversely proportional to resistance. Thus as resistance drops, current rises. Now look at the circuit of Figure 11.13. Under ordinary conditions that circuit will carry 10 amperes and will operate normally. But, if by some chance, a

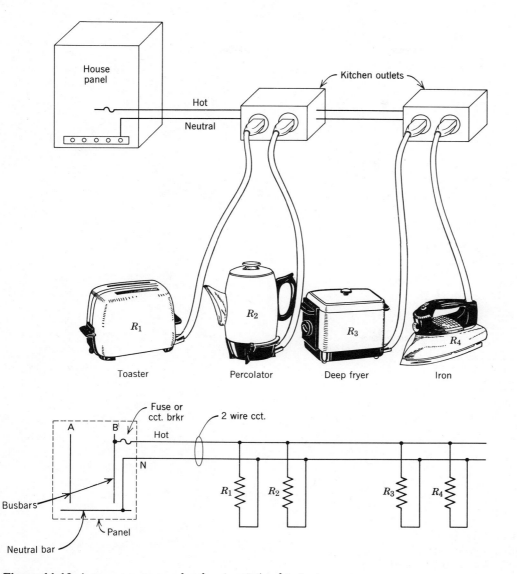

Figure 11.12 A sure way to overload a circuit (and to test the effectiveness of the panel fuse or circuit breaker). (From *Modern Physics*, by H.C. Metcalf, J.E. Williams, and C.E. Dull, Copyright © 1964, Holt, Rinehart and Winston. Used with permission of Holt, Rinehart and Winston.)

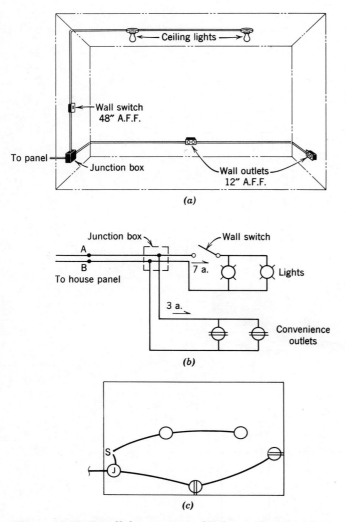

(a)

(b)

(c)

Figure 11.13 Parallel groupings of lights and wall outlets are in turn connected in parallel to each other. Circuit is shown *(a)* pictorially, *(b)* schematically and *(c)* as on an architectural plan.

connection appears between points A and B, the circuited is *shortened* so that there is no resistance in the circuit. The current rises instantly to a very high level, and the condition constitutes a *short circuit*. If the circuit is properly protected, the fuse or circuit breaker will open, and the circuit will be disabled. If not, excessive current will probably start a fire.

Circuit Characteristics

11.10. Power and Energy

We assume that the student is familiar with the concepts of power and energy (work) from his physics studies, including scientific terms such as joules, ergs and calories. The world of technology obviously also uses concepts of power and work, but with different units of measurement and a different approach. In electrical work these units are horsepower, watt, kilowatt and kilowatt-hour. We now carefully define and explain these terms, along with their physical concepts, and show how they differ from these same concepts as used in physics. This is necessary for a thorough understanding of the practical uses of these quantities. If we refer to the hydraulic circuit of Figure 11.14a, we see that a source of motion energy is supplied to the shaft of the pump. This energy is transferred to the water in the pump, and is utilized in forcing the water through the piping and any hydraulic devices in the circuit. As is shown in the diagram, an energy transfer takes place from rotary motion in the pump to heat (friction), mechanical work in the hydraulic devices and motion of the water. Thus the work done (energy utilized) in this arrangement consists of transferring energy from one place to another and from one form to another. The longer the system remains in operation, the more energy is utilized. The physics student would calculate the masses and velocities involved and, then, the work (energy) being expended, by multiplying force and distance. He generally is not interested in the system's power, which is the rate of energy utilization.

The technological approach is quite the opposite.

The technologist would determine the flow required in gpm and the head (friction) of the system. With these two quantities and a chart for a particular type of pump, he would select a pump and determine the horsepower required to drive it. He is not primarily interested in the system's energy (power × time of operation), since he assumes a continuously operating system. He would, however, calculate energy to determine operating costs or to check his energy consumption. To do this, he would convert horsepower to kilowatts (kw), would calculate kilowatt-hours and, with the power company's rate schedule, would calculate the cost of operating the system. Thus we see that the engineer's approach is the opposite of the scientist's. The technologist first calculates power and from that energy; the scientist does the reverse. Their approaches differ because their purposes differ.

Electric systems similar to the hydraulic system of Figure 11.14a are shown in Figure 11.14b and c, the difference being the source of power. The energy transfer that would interest the physics student is shown on the diagrams. What would interest the technologist is quite different. He starts with the power rating of each device (its rate of energy utilization) in watts or kilowatts, and works from there. The basic unit of power in electric terms is the watt, abbreviated "w." Since this unit is small for many power applications, we also use a unit 1000 times as large—the kilowatt, abbreviated "kw." Obviously, 1 kw equals 1000 w. It is also handy to remember that one horsepower equals 746 w, or approximately three fourths of a kilowatt.

$$1 \text{ hp} = 746 \text{ w} \approx \sqrt[3]{4} \text{ kw}$$

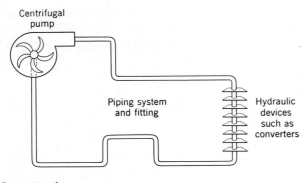

Energy transfer:

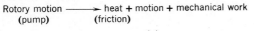

(a)

Figure 11.14 *(a)* Typical hydraulic circuit comprising piping and some mechanical-hydraulic device. Work is done transferring energy from one place to another and one form to another.

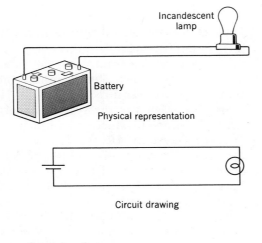

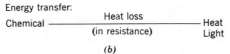

Energy transfer:

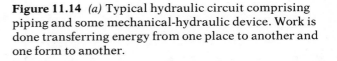

(b)

Figure 11.14 *(b)* In this simple battery circuit, chemical energy is converted to electric in the battery, and then to light and heat in the lamp. The power in the circuit is the rate of energy transfer, that is, the amount of current flowing and, therefore, the rate of heat generated.

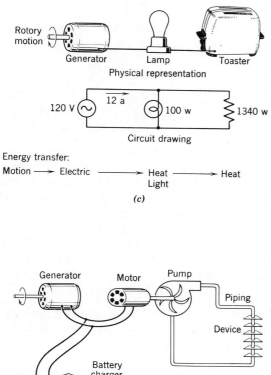

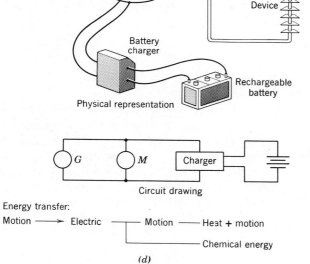

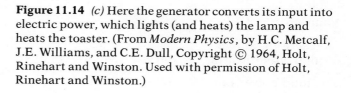

Figure 11.14 (c) Here the generator converts its input into electric power, which lights (and heats) the lamp and heats the toaster. (From *Modern Physics*, by H.C. Metcalf, J.E. Williams, and C.E. Dull, Copyright © 1964, Holt, Rinehart and Winston. Used with permission of Holt, Rinehart and Winston.)

Figure 11.14 (d) Generator supplies power which transfers energy to heat, motion and chemical energy. Only the chemical energy, that is, the recharged battery, is easily recoverable. The amount of current flowing in each circuit depends on its own electric characteristics.

A few examples of the application of these units should help here.

Example 11.6 Referring to the circuit in Figure 11.14c, assume that the lamp is rated 100 w and that the toaster is rated 1340 w. We are interested in the current flowing in the circuit. We would calculate it as follows:

Solution

generator voltage = 120 v
total circuit wattage = 100 w + 1340 w = 1440 w

The total power in a d-c circuit is equal to the product of the current and voltage, that is

$$P = VI$$

therefore, 1440 w = 120 v × I or

$$I = \frac{1440 \text{ w}}{120 \text{ v}} = 12 \text{ amperes}$$

Example 11.7 Refer again to Figure 11.12, and assume the following data:

Toaster 1340 w	Deep fryer 1560 w
Percolator 500 w	Iron 1400 w

Also, assume a conventional house circuit voltage of 120 v. Calculate the circuit current. Assume the circuit is d-c.

Solution

Total circuit power =
    1340 w + 500 w + 1560 w + 1400 w = 4800 w

Knowing that

$$P = VI \quad \text{or} \quad I = \frac{P}{V}$$

We have

$$I = \frac{4800 \text{ w}}{120 \text{ v}} = 40 \text{ amperes!}$$

Since the usual house circuit is sized and fused for 20 amperes, we have a 100% overload. In such a situation, the fuse or circuit breaker must open to prevent overheating and a possible fire.

We stated in Example 11.4 (Section 11.7), that the resistance of the toaster involved did not especially interest us, but that the wattage, or power rating did. Now let us rework that example as it would actually be done.

Example 11.8 Assume a 120 v d-c source feeding a 720 watt toaster. Find the current in the circuit. (This is a practical problem since, as we shall learn, the fuse size is based primarily on the circuit current.)

Solution

$$\text{Power} = \text{voltage} \times \text{current}$$

or

$$\text{current} = \frac{\text{power}}{\text{voltage}}$$

$$I = \frac{720 \text{ w}}{120 \text{ v}} = 6 \text{ amperes}$$

Note that this is the same answer we obtained in Example 11.4 but that this time we obtained it from normally available data. Check the nameplate on a few pieces of kitchen equipment. Notice that the data given are wattage and voltage. Resistance is of little concern, especially since in many instances "hot" resistance is quite different from "cold" resistance, that is, the resistance varies with the temperature of the item.

To review, we have now learned that:

(a) power is the rate at which energy is used, and is expressed in horsepower (hp), watts (w) and kilowatts (kw).

(b) 1 hp = 746 w ≈ $\frac{3}{4}$ kw.

(c) $P = VI$ or $I = \frac{P}{V}$.

As an item of interest, one horsepower is approximately the power capability of a horse for a considerable period of time. This was determined by James Watt, whose name is now used as the basic unit of power. A normal man can exert a horsepower for a short period of time by, for instance, running up a flight of 10 steps in 2 sec. (It has also been estimated that a watt equals one rat-power.)

11.11. Energy Calculation

Since power is the rate of energy use, it follows, as stated above, that energy = power × time. Therfore, the amount of energy used is directly proportional to the power of the system and to the length of time it is in operation. Since power is expressed in watts or kilowatts, and time in hours (seconds and minutes are too small for our use), we have for units of energy, watt-hours (wh) or kilowatt-hours (kwh). Obviously, one watt-hour equals one watt of power in use for one hour, and one kilowatt-hour equals one kilowatt in use for one hour.

Example 11.9

(a) Find the daily energy consumption of the appliances listed in Example 11.7, if they are used as follows daily

Toaster	15 min
Percolator	2 hr
Fryer	½ hr
Iron	½ hr

Solution

Toaster	1340 w = 1.34 kw × ¼ hr = 0.335 kwh
Percolator	500 w = 0.5 kw × 2 hr = 1.00 kwh
Fryer	1560 w = 1.56 kw × ½ hr = 0.78 kwh
Iron	1400 w = 1.4 kw × ½ hr = 0.70 kwh
	Total 2.815 kwh

(b) If the average cost of energy (*not* power) is $0.06 per kilowatt-hour, find the daily operating cost.

Solution: The cost is

$$2.815 \, kwh \times \$0.06/kwh = \$0.1689$$

or approximately 17 cents

Clearly the power being used at any specific time during the day by a residential household varies considerably. If we were to graph the power in use for a typical American household during a normal weekday, the plot might look something like that in Figure 11.15. The average power demand of the household is obviously much lower than the maximum. The ratio between the two is called the overall demand factor. This factor runs about 20% for a typical household.

Example 11.10 It has been estimated that the average power demand of an American household is 1.2 kw. Calculate the monthly electric bill of such a household, assuming a flat rate of $0.035 per kilowatt-hour.

Solution

Monthly energy consumption =

$$1.2 \, kw \times \frac{24 \, hr}{day} \times \frac{30 \, days}{month} = 864 \, kwh$$

Electric power bill =

$$864 \, kwh \times \$0.035/kwh = \$30.24$$

11.12. Circuit Voltage and Voltage Drop

To make certain that the ideas connected with basic electric circuitry are clearly understood, we now analyze several circuits by studying voltage and current division, and voltage drop.

Refer to Figure 11.8. What are the voltages across each component in the circuit? The voltage across each lamp is

$$V = IR = 5 \, amperes \times 1.2 \, ohms$$
$$V = 6 \, volts$$

This is obviously correct, since the drop across the two lamps must be 12 v to equal the supply voltage. This establishes an important principle. *The sum of*

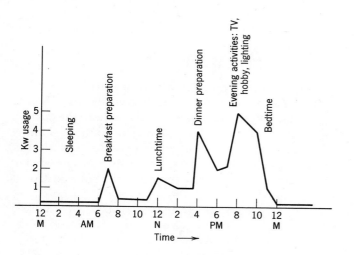

Figure 11.15 Hypothetical graph of power usage for a typical household.

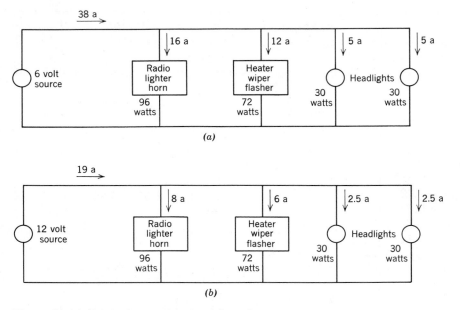

Figure 11.16 Typical auto circuit with same wattage accessories, on different voltages. Note that circuit current is halved when voltage is doubled.

the voltage drops around a circuit is equal to the supply voltage. This principle is most important in series circuits. In parallel circuits, each item has the same voltage across it and constitutes a circuit by itself, as explained above. Refer now to Figure 11.11. Note that each resistance has the same 120 v drop across it, equaling the supply voltage. But also notice that in a parallel arrangement all the currents add. This leads us to the second important fact to remember. *In series circuits, current is the same throughout and voltages differ. In parallel circuits, the voltages are the same and the currents differ.* Look, for example, at Figure 11.12. Note that each appliance has the same voltage imposed, since they are in parallel. Each, however, draws a different current.

Returning to the example of Figure 11.8, we update the information and calculation on the basis of our present knowledge. The lamps would actually be arranged in parallel, as is shown in Figure 11.16*b*. We have added a few of the normal automobile accessories. In Figure 11.16*a* we have drawn the same circuit, but it is fed from a 6 v battery (or generator) as was common a few years ago. Note that the current is double because the voltage is half, since $P = VI$ and the power rating of the devices is the same.

At this point you may well ask what difference it makes what the voltage is, and why most auto manufacturers use the higher voltage. The answer to that question requires an understanding of current-carrying capacity of cables. This characteristic, called *ampacity*, will be discussed more fully in the sections on circuitry, but it is introduced here to demonstrate the effect of circuit voltage, and the reason for the development of a-c.

11.13. Ampacity

Simply defined, ampacity is the ability of a conductor to carry current without overheating. Heat is generated by the resistance of the wire to the current flow. Up to this point we have ignored the resistance of the wiring, assuming perfect conductors, that is, conductors with no resistance. This is obviously not true. When a wire carries current, the voltage drop is

voltage drop in wire =

current carried × resistance of wire

The power loss in the wire can be calculated in the same fashion that we have previously calculated it, that is, as the product of voltage and current:

power loss in wire =
$$\text{circuit current} \times \text{voltage drop}$$
or
$$P = I \times (I \times R) = I^2R$$

We have here derived the law of power loss. Power loss is equal to the component's resistance times the current squared. This wiring power loss is converted to heat which must be dissipated. Since it can be shown that a small diameter wire can safely carry more current in proportion to its weight than a large diameter wire, it follows that less copper is used to carry the same amount of power in a higher voltage circuit. For instance, if it is desired to carry 1440 w (a typical appliance rating), the current flowing is 12 amperes at 120 v, or 6 amperes at 240 v (check these figures using Ohm's law). Since basic wire insulation is good for 300 v, the same amount of power can be carried with less than one half the investment in copper, a very considerable saving! This accounts for the almost universal use of 220 to 240 v for basic circuitry in most of the world except for the United States. Referring back to Figure 11.16, the circuit in (b) is better than the arrangement in (a) because

1) smaller wire can be used, giving economy.
2) power loss is smaller
3) percent voltage drop is smaller

Of course, there are limits to raising the voltage of a circuit, but all other factors being equal, the higher the circuit voltage the more economical the system. The inherent advantages of high voltage for transmission and distribution spurred the search for an easy way to change from one voltage to another. This cannot be done with d-c, but can very easily be accomplished with a-c. This one fact was the major cause of the development of a-c and the almost complete abandonment of d-c for general power purposes. Let us examine a practical situation that illustrates these principles.

Example 11.11 A 5 kw electric swimming pool heater is located 300 ft from the residence from which it is to be fed. The residents have a choice of either 120 v or 240 v feed from the house panel. Which should they choose?

Solution: Let us make parallel solutions for the different voltages.

	120 v	240 v
(a) Current drawn	$\frac{5000\text{ w}}{120\text{ v}} = 41.7\text{ a}$	$\frac{5000\text{ w}}{240\text{ v}} = 20.8\text{ a}$

		120 v	240 v
(b)	Minimum wire size required to carry the current without overheating	No. 8 AWG, copper	No. 12 AWG, copper
(c)	Relative cost of wire	2.2	1.0
(d)	Voltage drop	17.7 v or 14.8%	8.9 v or 3.7%
(e)	Power loss	No. 8 wire $I^2R = 334$ w	No. 12 wire $I^2R = 210$ w

Since a 17.7% voltage drop is obviously unacceptable, the wire size would have to be increased to at least a No. 2 AWG, making the cost relation about *10 to 1*, instead of the 2.2 to 1 shown. This example should make the advantage of higher voltage perfectly obvious.

11.14. Measurement in Electricity

In the preceding sections we have explained the fundamental electric quantities of voltage and current, and have given the units involved as volts and amperes, respectively. As is true of all other physical quantities that are to be used in practical application, the need existed for a simple means of measuring these quantities. This need was met by the development of the galvanometer movement illustrated in Figure 11.17. Everyone at one time or another has felt the repulsion between like poles of two magnets held close together and, conversely, the attraction between opposite poles. This principle is used in the galvanometer. It causes a deflection of the pointer as a result of the repulsion between the field of a permanent magnet and an electromagnet. The electromagnet is formed when current flows in the coil, and its strength is proportional to the amount of current flowing. Thus a strong current causes a larger deflection of the needle and therefore, a higher reading on the dial. To make this very sensitive basic unit usable for large currents (it is intrinsically a microammeter, sensitive to millionths of an ampere), we simply divert, or shunt away, most of the current, allowing only a few microamperes to actually flow in the meter coil. This is illustrated in Figure 11.18.

To use the same unit as a voltmeter, we put a large resistance (*multiplier*) in series with the meter, again limiting the current flowing to a few microamperes. The scale is then calibrated in volts. Note in Figure

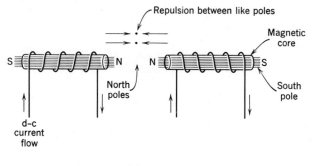

Repulsion between like poles

Magnetic core

S N N S

North poles

South pole

d-c current flow

(a)

Figure 11.17 *(a)* Diagram showing basic electromagnetic principal and interaction between electromagnets. Any iron core becomes an electromagnet when current flows in a coil wound on it, as shown.

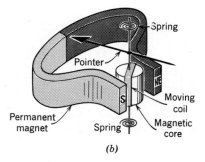

Spring

Pointer

Moving coil

Permanent magnet

Spring

Magnetic core

(b)

Figure 11.17 *(b)* Principal of the electromagnet is used in all d-c meter movements. Current flowing in the coil forms an electromagnet. Its field interacts (see above) with the permanent magnet's field, to cause a deflection proportional to the current flow. (From *Modern Physics*, by H.C. Metcalf, J.E. Williams, and C.E. Dull, Copyright © 1964, Holt, Rinehart and Winston. Used with permission of Holt, Rinehart and Winston.)

11.19 that the meter is precisely the same as the ammeter of Figure 11.18, except that a different method is employed to limit the actual meter current to the few microamperes permissible. All d-c meters are made as shown. Most a-c meters operate on basically the same principle except that instead of a permanent magnet, an electromagnet is used. That way, when the polarity reverses, the deflecting force remains in the same direction. A d-c meter connected to an a-c circuit simply will not read, since inertia prevents the needle from bouncing up and down 60 times a second.

11.15. Power and Energy Measurement

The measurement of current and voltage in practical application is generally not as important as the measurement of power and energy, since the power company regulates voltage very accurately, and current measurements are of more interest to engineers than to technologists. However, power and energy measurements are of great interest in determining loads, costs, consumption and proper system operation. To measure power, we take advantage of the fact learned earlier that power is equal to the product of the voltage and current in the circuit,

$$P = VI$$

Although actual construction is complex, the theory of operation of a wattmeter is simple. We have two coils; a current coil that is similar in connection to an ammeter, and a voltage coil that is similar in connection to a voltmeter. By means of the physical coil arrangement, the meter deflection is proportional to the product of the two, and therefore to the circuit power. The meter can be calibrated as we wish, depending on the size of the shunts and multipliers, from a few milliwatts up to many kilowatts. The schematic arrangement is shown in Figure 11.20. To measure energy, the factor of time must be introduced, since

$$energy = power \times time$$

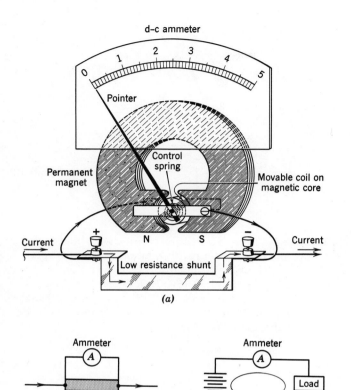

d-c ammeter

Pointer

Permanent magnet

Control spring

Movable coil on magnetic core

N   S

Current

Current

Low resistance shunt

(a)

Figure 11.18 (a) D-C ammeter construction—current enters meter at left. Most of the current is diverted through the shunt. A small portion flows through the meter itself. (From *Modern Physics*, by H.C. Metcalf, J.E. Williams, and C.E. Dull, Copyright © 1964, Holt, Rinehart and Winston. Used with permission of Holt, Rinehart and Winston.)

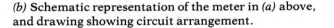

Ammeter

(A)

Shunt

Ammeter

(A)

Load

(b)

(b) Schematic representation of the meter in (a) above, and drawing showing circuit arrangement.

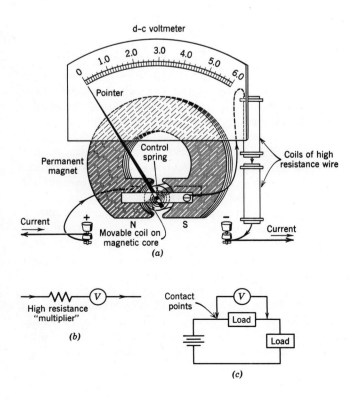

d-c voltmeter

Pointer

Permanent magnet

Control spring

Coils of high resistance wire

Current

Movable coil on magnetic core

N   S

Current

(a)

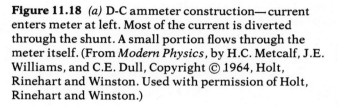

High resistance "multiplier"

(V)

(b)

Contact points

(V)

Load

Load

(c)

Figure 11.19 (a) D-C voltmeter construction. Current is limited by high resistance in series with the meter movement. (From *Modern Physics*, by H.C. Metcalf, J.E. Williams, and C.E. Dull, Copyright © 1964, Holt, Rinehart and Winston. Used with permission of Holt, Rinehart and Winston.) (b) Same as (a) except shown schematically. (c) Note that the voltmeter is not "in" the circuit; it simply touches at the point where measurement is made.

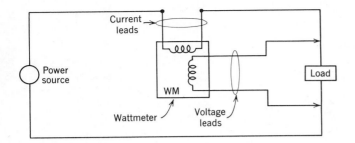

Figure 11.20 Schematic arrangement of wattmeter connections. Note that the circuit coil is in series with the circuit load whereas the voltage leads are in parallel. See also Figure 11.21.

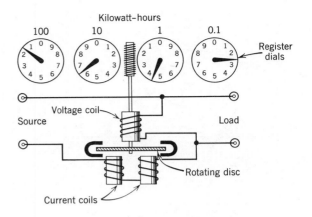

Figure 11.21 Typical induction type a-c watt-hour meter. Dials count and register revolutions of disc, whose speed of revolution is proportional to circuit power. Compare to schematic of Figure 11.20. (From *Modern Physics*, by H.C. Metcalf, J.E. Williams, and C.E. Dull, Copyright © 1964, Holt, Rinehart and Winston. Used with permission of Holt, Rinehart and Winston.)

and energy is measured in watthours or kilowatt-hours. The d-c energy meters are available but are not of general interest because of the rarity of d-c power. The a-c watt-hour meters are basically small motors, whose speed is proportional to the power being used. The number of rotations is counted on the dials, which are calibrated directly in kilowatt-hours. A schematic of an a-c kilowatt-hour meter is shown in Figure 11.21. As can be seen from Figure 11.22, the kilowatt-hour energy consumption can be read directly from the dials. If the numbers involved are too large, or a meter is used with current transformers between it and the line, or for calibration reasons, a multiplying factor is required to arrive at the proper kilowatt-hour consumption. This number is written directly on the meter face or nameplate, and we multiply the meter reading by it to get the actual kilowatt-hours. In the absence of such a number, it can be assumed that the meter reads directly in kilowatt-hours. If we want to know

the energy consumption of a particular circuit and a kilowatt-hour meter is not available, we may also use a wattmeter and a timer to get the same result; we utilize the relation already learned that

$$\text{energy} = \text{power} \times \text{time}$$

However, this is only effective for a constant load such as lighting. Loads that vary in size or turn on and off cannot be measured except with a meter that sums up the instantaneous energy used. This is what a kilowatt-hour meter does. A wattmeter measures instantaneous power, whereas energy, involving time, must be summed. Figure 11.15 illustrates this. Note that the wattmeter measures the amount of power in use at any one time, while the kilowatt-hour meter measures the energy used over a period of time. Thus in Figure 11.15, a wattmeter would read 2 kw at 7 A.M., 0.3 kw at 10 A.M., 4 kw at 4 P.M., and so forth. The *energy* consumed is represented by the area under the curve, and can only be measured by a kilowatt-hour meter.

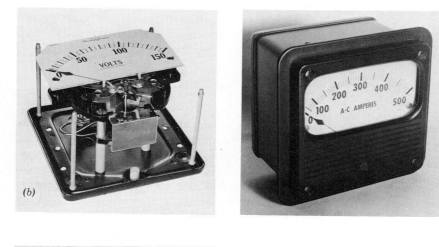

Figure 11.22 Typical electrical meters and instruments. *(a)* 30-amp, 240 v socket type watt-hour meter with kilowatt demand register; *(b)* Some a-c and d-c switchboard instruments. (Courtesy of Westinghouse Electrical Corporation)

Alternating Current

11.16. General

As is amply demonstrated in Section 11.13, circuits operating at elevated voltages have lower power loss, lower voltage drop and are almost always more economical to construct because of savings in copper. Power loss is very important in transmission and distribution, but is much less so in small branch circuits. For these reasons a-c, which allows easy transformation between voltages and much easier generation than d-c, came into favor at the close of the 19th century. A bitter battle ensued between the proponents of d-c electricity such as Edison and the advocates of a-c electricity, including George Westinghouse. Edison opposed a-c on the ground that the high voltages involved in transmission were dangerous. Other opponents derisively labeled a-c as "do-nothing" electricity because the voltage is positive for one-half cycle and negative for one-half cycle yielding, they claimed, a net of zero. That this is nonsense was not appreciated. The a-c alternation can be likened to the strokes of a saw cutting a piece of wood. Just as a saw cuts on the up stroke *and* the down stroke, so a-c does work in both the positive and negative halves of its cycle. See Figure 11.23. References are made to a-c in Section 11.3 and in Figure 11.2. A review of them now would be helpful as an introduction to the detailed discussion of a-c that follows.

11.17. A-C and D-C; Similarities and Differences

The basic characteristics previously discussed for d-c also apply to a-c, with some important differences. In addition, there are certain aspects of a-c that do not apply to d-c, such as frequency. Study Figure 11.23 and note that the a-c current goes through one positive loop and one negative loop to form one complete cycle, which then repeats. The number of times this cycle of a plus and a minus loop occurs per second is called the *frequency* of a-c, and is expressed, logically, in cycles per second. Because of the tendency of people to say simply "cycles,"

which is not the same thing as cycles per second, the electrical profession agreed some years ago to change the expression and at the same time to honor a great physicist who did extensive research in electromagnetism. Therefore, cycles per second are now properly called *hertz* after H. R. Hertz. A correct description of ordinary house current in the United States would be 120 volt, 60 hertz, abbreviated 120 v 60 hz. In Europe, the normal frequency is 50 hz, and in some parts of eastern Europe and Asia, 25 hz. This latter frequency is so low that flicker is easily noticeable in incandescent lamps. The frequency of d-c is obviously zero hertz, since the voltage is constant and never changes polarity.

In a-c the quantity corresponding to resistance in a d-c circuit is called *impedance* and is usually shown by the letter Z. It is a compound of resistance plus an a-c concept called *reactance* but, once given, is treated exactly as resistance is in d-c circuits. That is, Ohm's law for a-c is expressed

$$I = \frac{V}{Z}$$

where Z is impedance, expressed in ohms. For resistive loads, such as incandescent lights and heaters, the impedance is equal to the resistance, and calculation is exactly as for d-c. As we stress above, however, the use of resistance, or impedance, is rare in calculations important to the technologist. Of prime importance is power calculation, and here there is a vital difference between a-c and d-c. You will remember that in d-c, power is the product of voltage and current. That is, for d-c,

watts = volts × amperes

In a-c, the product of volts and amperes gives a quantity called volt-amperes (abbreviated volt-amp or va) which is *not* the same as watts. That is, in a-c;

volts × amperes = volt-amperes.

To convert volt-amperes to watts, or power, we introduce a quantity called *power factor*, abbreviated "pf."

To get power in an a-c circuit:

w = V × I × pf

watts = volts × amperes × power factor

or

w = VI × pf

watts = volts-amperes × power factor

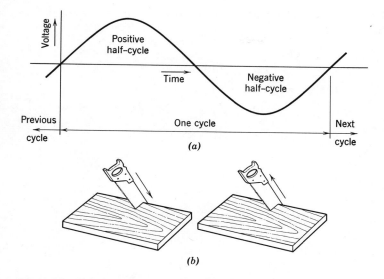

Figure 11.23 Alternating current *(a)* does work in both halves of its cycle, just as the saw in *(b)* cuts in both directions of stroke.

We see, therefore, that power factor relates volt-amperes and watts. With purely resistive loads, the circuit impedance is equal to the resistance, and the circuit power factor is 1.0. With other loads, the power factor varies, and must be given to make power calculation possible. A few examples illustrate this.

Example 11.12 Refer to Figure 11.12 and Example 11.8. Calculate the circuit currents and impedances.

Solution: Since all the appliances listed are basically heating devices for which the impedance equals the resistance, the power factor is 1.0 and the calculations are the same as those given above for d-c.

Example 11.13 Refer to Figure 11.13. Assume the two ceiling lights to be 150 w each, incandescent. Also assume the load connected to one convenience outlet to be a 10-amp hair dryer and blower, with a power factor of 0.80. Calculate the current and power in the two branches of the circuit, and the total circuit current, assuming a 120 v a-c source.

Solution:
(a) In the circuit branch feeding the lights we have

$$P = VI$$
$$300 \text{ w} = 120 \text{ v} \times I$$
$$I = \frac{300}{120} = 2.5 \text{ amperes}$$

If we wished to calculate circuit resistance (which is equal to the impedance, since the load is purely resistive)

$$Z = R = \frac{V}{I} = \frac{120}{2.5} = 48 \text{ ohms}$$

Again we point out that this latter figure is of little practical use to us, and is calculated simply to show technique.

(b) In the second branch we have a 10-amp, 0.8 pf load.

power in watts = volts × amperes × power factor

$$P = 120 \times 10 \times 0.8 = 960 \text{ watts}$$

but the circuit volt-amperes are
$$va = 120 \times 10 = 120 \text{ volt-amperes}$$

This latter figure is significant in sizing equipment, as we shall learn later.

(c) To calculate the *total* current flowing from the panel to *both* branches of the circuit, we must combine a purely resistive current (lamp circuit) with a reactive one (dryer circuit). The exact value of current is the square root of the sum of the squares of the two branch currents. However, in normal practice, the currents are simply added arithmetically. This yields a result that is somewhat higher than actual and is, therefore, on the safe side when we are sizing equipment. Hence,

approximate total current =
$$2.5 + 10 = 12.5 \text{ amps}$$

Actual current is 12.1 amperes; our error in approximating is 3.2%, which is acceptable in branch circuit calculation. The above calculations and techniques will become routine with practice. One further example at this point will demonstrate the importance of power factor in normal situations.

Example 11.14 The nameplate of a motor shows the following data: 3 hp, 240 v a-c, 17 amperes. Assume an efficiency of 90%. Calculate the motor (and, therefore, circuit) power factor.

Solution:
$$1 \text{ hp} = 746 \text{ w}$$
therefore
$$3 \text{ hp} = 3 \times 746 = 2238 \text{ w output}$$
$$\text{efficiency} = \frac{\text{output}}{\text{input}}$$

so
$$\text{power input} = \frac{2238}{0.9} = 2487 \text{ w}$$

but for a-c,
$$\text{power} = \text{volts} \times \text{amperes} \times \text{power factor}$$
so
$$2487 = 240 \times 17 \times \text{power factor}$$
and
$$\text{power factor} = \frac{2487}{240 \times 17} = 0.61$$

Note the large difference between volt-amperes and watts.
$$VI = 240 \times 17 = 4080 \text{ volt-amperes}$$
$$P = \text{as above} = 2487 \text{ watts}$$
The circuit would be sized to carry 17 amperes.

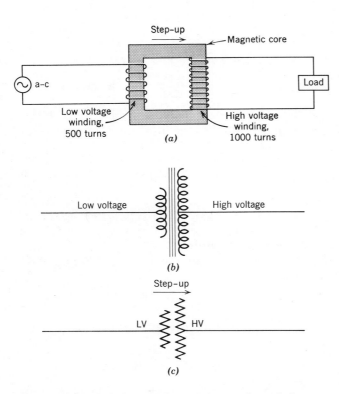

Figure 11.24 Pictorial, diagrammatic and single-line representation of a transformer, in (a), (b) and (c). Representation (c) is most often used in electric construction drawings.

11.18. Voltage Levels and Transformation

We mentioned above that one of the principal reasons for the victory of a-c over d-c was the ease with which voltages could be transformed with a-c. We also pointed out some of the advantages that come from this ability. As part of the same study it is important to understand the application of the different voltage levels in use today. A transformer is a simple static device, consisting of a magnetic core on which are wound primary and secondary windings as shown diagramatically in Figure 11.24. The voltages appearing are in direct proportion to the number of winding turns. Thus in Figure 11.24, if 120 v a-c were connected to the left side, containing 500 turns, 240 v would appear on the right side, containing 1000 turns. The input side is normally termed *primary;* the output is normally termed *secondary.* In this instance, the transformer would be a

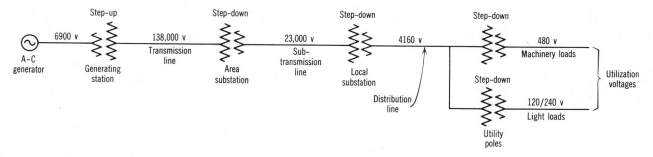

Figure 11.25 One-line diagram of a typical a-c power distribution system, from generation through transmission and distribution, to utilization level.

120/240 v *step-up* transformer with a 120 v primary and a 240 v secondary. The very same transformer can be used to *step-down* by reversing the supply and the load. Thus if a 240 v source were connected to the 1000 turn winding, 120 v would appear on the left, and the unit would be a 240/120 v *step-down* transformer. The 240 v would then be primary and the 120 v would be secondary. Simply stated, transformers are reversible.

Figure 11.25 shows the voltage levels commonly in use and the transformations required to obtain them. Each of the levels was selected after careful engineering and economic studies. The tendency is always to higher levels at all points except utilization. As insulation materials and techniques are improved, voltage levels are raised. Indeed, 345,000 v (345 kv) and 500 kv are becoming common, and voltages up to 750 kv are in use. At the distribution level, 13,200 v is gradually replacing 4160 v, and 46 kv is replacing 23 kv. But at the end of the line, the voltage we find in our house panel will remain 120/240 v for safety reasons and no increase is contemplated.

11.19. Voltage Systems

In all of the preceding material we have assumed a 2 wire circuit. This is generally the case within the building. The usual lighting circuit is 2-wire, 120 v, 60 hz, as is also the normal receptacle circuit that feeds the wall plugs. (In the latter, however, in modern construction, there is a separate ground wire. The entire subject of grounding is discussed in Chapter 13.) However, the service entrance to the house will most probably be a 3-wire circuit, commonly written 3-wire, 120/240 v, 60 hz. This system

is illustrated in Figure 11.26 and has these advantages:

(a) 120 v is available for lighting and receptacle circuits.

(b) 240 v is available for heavier loads such as clothes dryers, air conditioning compressors, and the like.

(c) The service conductors are sized on the basis of 240 v, rather than 120 v, effecting a large saving in conductors, as is explained in Section 11.13.

(d) Voltage drop is low.

Note the custom of identifying the conductors by letter, and by color of wire. The *neutral* conductor is so called because it carries no current when the 120 v loads on both sides of it are balanced (see Figure 11.26d). Also, since it is grounded, it is at neutral, or zero potential, being one-half way in voltage between the hot lines, A and B. The wire color identification system used in construction is:

Neutral, N	white conductor
First hot line, A	black conductor
Second hot line, B	red conductor

The illustrations in Figure 11.26 are to introduce circuit calculation. We shall study these calculations more intensively later in the book.

11.20. Single Phase and 3-Phase

To make life a bit more complicated, the engineers who championed a-c also developed a system of wiring known as 3-phase electricity. The subject is theoretically complex but, in practice, is quite sim-

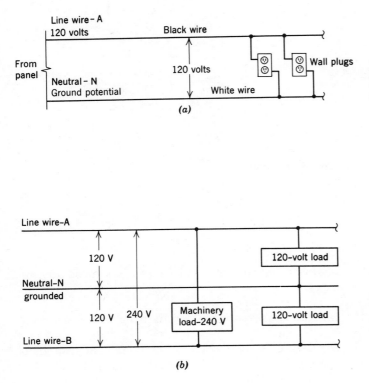

Figure 11.26 *(a)* Typical convenience receptacle (wall plug) circuit. Note designation of lines as A and Neutral, and color coding. Ground connection and separate ground wire are not shown, for clarity.

Figure 11.26 *(b)* In a 3-wire, 120/240 arrangement, which is the common one for residences and other small buildings, the loads are generally arranged as shown. The 120 v loads are lighting and convenience outlets, plus small appliances. An effort is generally made to balance the loads as is explained in the chapter discussion on circuitry. Here, also, the separate ground wire is not shown, for clarity.

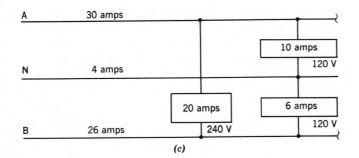

Figure 11.26 *(c)* In a typical 3-wire, 120/240 circuit, the neutral only carries the difference between the 120 v loads on the 2 line wires. This helps to reduce voltage drop and permits savings in wiring. Note that this connection is single phase a-c.

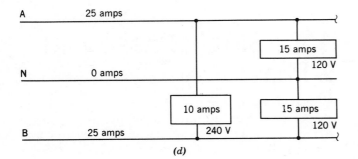

Figure 11.26 *(d)* When the 120 v loads of a 3-wire system are balanced, the neutral carries no current and the line wires A and B carry the entire load. An equivalent 2-wire 120 v system with the same load would carry 50 amperes in each line, or leg, requiring much larger conductors.

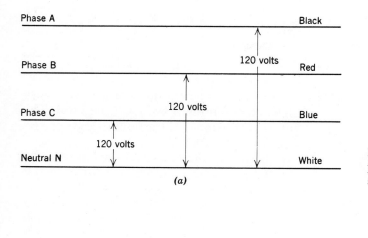

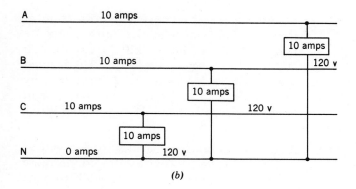

Figure 11.27 (a) Typical 3-phase wiring system, showing phase to neutral voltages. In effect it constitutes three single-phase circuits with one, common, return wire.

Figure 11.27 (b) In a 3-phase system, the neutral carries only unbalanced current; when the phases are balanced the neutral carries no current.

ple. Instead of single phase a-c, which can be either 2 wire or 3 wire as is explained and illustrated above, 3-phase a-c is 4 wire. It consists of three hot legs designated phases A, B and C, plus a neutral N. This is illustrated in Figure 11.27. (*Three*-wire, 3-phase a-c, without a neutral wire was once in common use, but is now used very little—being phased out, one could say.) For our purposes, 3-phase a-c is simply a triple circuit, and can be treated as such. Lighting and outlet loads are connected between any phase leg and neutral, and machinery loads are connected to the phase legs only. This system of wiring is used in all buildings where the building load exceeds approximately 50 kva (kilovolt-amperes), or where it is required for 3-phase machinery. The technologist

need not be concerned with the complexities of phase relationships; instead, he should appreciate the application of 3-phase and single-phase a-c. Ample opportunity to study practical application will be afforded. Notice in Figure 11.27a that the voltage is 120 v between phase and neutral. The line-to-line voltage (between phases) has intentionally been omitted to avoid confusion. It is 208 v, not 240 v as might be expected, but this difference need not disturb us. It is important to note here that the neutral conductor, even though it is common to all three of the phase conductors, does *not* carry triple the phase current. On the contrary, it only carries unbalanced current, and if loads are balanced, it carries *no* current.

Problems

11.1. (a) An automobile with a 12 v electrical system is equipped with the devices listed below. Assuming that all devices were operated with the engine off (power drawn from battery), how long will a 40 amp-hr battery last? (For the purpose of this problem, assume that a 40 amp-hr battery will supply any product of amperes and hours equaling 40; for example 20 amperes for 2 hr, 40 amperes for 1 hr, 5 amperes for 8 hr, etc, even though this is only approximately true.)

Accessories:

Twin headlights	225 w each
Fan	20 w
Radio	10 w

(b) The same automobile has a starting motor that draws 240 amperes when cranking. How long will the battery supply the cranking motor?

11.2. For the circuit shown, calculate the total circuit current and the total circuit power.

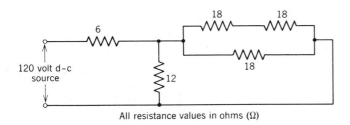

All resistance values in ohms (Ω)

11.3. The following table shows the relationships of current, voltage and resistance for three resistors connected in a battery circuit. Draw a circuit diagram showing the manner in which the resistors are connected, and fill in the missing values in the table.

	R1	R2	R3
Voltage	2 v	10 v	?
Current	?	2 amperes	4 amperes
Resistance	?	5 ohms	?

What is the battery voltage?
(*Hint:* two of the resistors are in parallel.)

11.4. A current of one milliampere (0.001 amp) through the human body can be fatal. The internal resistance of the human body can be considered as 1000 ohms.

(a) If a person contacts 120 v, what must be the minimum contact resistance of the hands to prevent fatal shock?

(b) Dry skin has a contact resistance of about 100,000 ohms; wet hands about 1000 ohms. Calculate the body current with the above voltage, under conditions of wet and dry hands.

11.5. An American moved overseas to a country that uses 240 v household circuit voltage. He had with him a 1440 w, 120 v toaster and a 100 w, 120 v immersion heater. He decided that, since each was a 120 v appliance, by placing both in series he could make them operate on a 240 circuit. Was he right? Why? What happened?

11.6. A rural farm house generates its own power at 120/240 v, a-c. A 2-wire 240 v copper wire line from the generator shed runs 930 ft to the barn, and there supplies a 4.4 kw electric heater. The voltage at the heater is 220 v. Calculate the following:

(a) The line voltage drop.
(b) The line resistance.
(c) The line power loss.
(d) Using Table 12.1 (page 309) determine what size line had been installed.

11.7. The barn of Problem 11.6 needed a second heating unit—this one rated 2200 watts. Since the load of this unit was one half that of the first, the farmer ran another circuit but reduced his wire size to No. 10 AWG. Also, he connected this unit to a 120 v circuit, although the unit was dual rated 120/240 v and could be fed at either voltage. To his surprise the unit did not perform properly. Why? What would you advise?

11.8. The farmer of Problems 11.6 and 11.7 decided to remove all the previous wiring and to run a single circuit for both heaters. He turned to his electric–technologist son with this requirement:
Design a single 240 v circuit of such size that both heaters will produce at least 95% of

their rated nameplate capacity. What size circuit wiring did the son recommend? What is the actual power in each heater?

11.9. A homeowner has recently installed electric baseboard heaters, as listed below, that are individually controlled. The bedroom thermostats have a cycle of 50% on, 50% off. The livingroom and kitchen thermostats cycle their heater 30% on, 70% off.

Livingroom	4.0 kw
Kitchen	1.5 kw
Bedroom No. 1	2.0 kw
Bedroom No. 2	1.5 kw
Bedroom No. 3	1.5 kw

Calculate:

(a) Maximum electric heat power demand.

(b) Monthly electric bill, assuming the thermostat cycle listed above to be continuous day and night, and a flat rate of 2.5 cents per kilowatt-hour.

11.10. A single phase 120 v a-c house circuit supplies the following loads:

Incandescent lighting	300 w
Exhaust fan	2.5 amperes, 0.6 pf
Television set	3 amperes, 0.85 pf

Find:

(a) Circuit volt-amperes

(b) Circuit watts

11.11. Table 430-148 of the 1975 National Electrical Code gives the following motor data for 115 v single phase a-c motors:

Horsepower	Current in Amperes
⅙	4.4
¼	5.8
⅓	7.2
½	9.8
1	16
3	34
5	56
10	100

Assuming an efficiency of 90% for all motors, calculate the power factor of each.

Additional Reading

Electrical Fundamentals for Technicians, R.L. Shrader, McGraw-Hill. This book covers the same material as in the preceding chapter, except in greater depth, and is therefore useful to the student who wishes a deeper level study.

12. Branch Circuits

Now that we have explored the fundamentals of electricity and electric circuits in theory, we are ready to apply this knowledge to practical electric circuitry as it is found in modern construction. All circuits are constructed basically in the same manner but vary in size, application, control and complexity. We begin our study of applied circuitry with the basic building block of electric construction—the branch circuit—and proceed from there to circuits of other types. You will soon learn to recognize the common properties of all circuits, be they lighting, receptacle, power, appliance, special purpose, motor or the like. In this chapter we discuss the characteristics of circuits and demonstrate how they are graphically shown, drawn and wired. We also analyze in detail the materials involved and describe their general and special application.

After studying this chapter you will be able to:

1. Distinguish between the three basic types of electrical drawings.
2. Select the branch circuit wiring method best suited to the type of construction being used.
3. Know the construction, application and properties of branch circuit materials including wire and cable, conduit, surface raceways, connectors, outlet boxes and wiring devices.
4. Understand what an electrical outlet consists of, how to draw and count outlets and how to detail them and specify their materials.
5. Describe wiring devices correctly according to their electrical and mechanical characteristics.
6. Draw basic branch circuits that show all the required electrical information.
7. Prepare a symbol list for equipment in branch circuits, including special wiring devices.
8. Draw details of conduits, fittings, supports, raceways and outlet boxes.
9. Understand commonly used symbols and abbreviations.
10. Gain familiarity with the National Electrical Code.

12.1. National Electrical Code

The National Electrical Code (NEC) is the only nationally accepted code of the electric construction industry. As such, it has become the standard or "bible" of that industry. The NEC, as we will refer to it in this book, is Section 70 of the National Fire Codes published by the National Fire Protection Association. The NEC defines the *minimum* acceptable quality of electrical design and construction practice necessary to produce a safe installation. It is to the NEC that the electrical inspector refers when inspecting a job. It is therefore obvious that the NEC is among the most important reference books in the library of people in the electrical construction industry, and an up-to-date copy should always be at hand. (It is a good idea to save old issues to refer to when you examine plans of buildings built under previous codes.) Some government agencies and some large cities have codes of their own. These generally supplement the NEC but do not replace it. (An exception is the New York City electric code which is, for the most part, more strict than the NEC.) Since the NEC is revised approximately every three years, the current edition must always be consulted for new work. Throughout this book we make frequent reference to NEC provisions. In particular, definitions will be taken word for word from the NEC.

12.2. Drawing Presentation

Before proceeding any further, we must give an explanation of the drawing methods employed in this book, which are the same as those used in construction industry practice. The drawing types are:

(a) Architectural-electrical plan
(b) One-line diagram
(c) Wiring diagram

To illustrate these methods let us consider an area in the conventional small house that you have studied in previous chapters. In Figure 12.1a we have spotted wall receptacles, have circuited them and have shown the circuitry in the manner typically used for this work. The solid lines connecting the receptacles are raceway runs whose purpose is to indicate circuit routing. The number of wires in such a run is generally known from "tic" or hatch marks that are added as needed, although notes may be used instead of tic marking. Such tic marks are used when more than 2 wires are represented. The absence of tics indicates 2 wires, by convention. Symbol lists frequently state that circuit lines indicate "2 wires unless otherwise noted," the "otherwise noted" generally taking the form of tics or notes. In the home-run of Figure 12.1a, both tics and a note are shown, although in practice only one or the other would be used. It is important to note that the type of line used—whether solid, dotted, dashed—is entirely up to the design/drawing staff, and is chosen on the basis of convenience. No rigid convention exists such as in mechanical drafting where solid lines indicate visible edges and dashed lines indicate hidden edges. In circuitry, the solid line is normally used to show the condition most often found, for convenience of drawing, since the solid line is easiest to draw. Since in this house most wiring is concealed, the solid line is chosen for this and a short dash line is used for exposed wiring. The symbols chosen for wiring *must* be shown on the symbol list for each job.

In Figure 12.1b the same electrical work is shown in "one-line" form, so called because multi-wire circuits are indicated by a single line. The actual number of wires is understood by the electrical designer or draftsman. When it is not obvious, the number is shown. The purpose of these diagrams is to show the circuitry at a glance. Here, for instance, one sees at a quick glance that circuits 1 and 2 each serve five duplex receptacles. Nothing is known about location, but that is not the function of a one-line (single-line) diagram. Such a diagram is purely electrical and has no architectural information. Finally, a wiring diagram is shown in Figure 12.1c. This representation shows the number of wires involved and their interconnections. This particular diagram adds nothing to our knowledge in this case and would not be used here. In complex systems, however, a wiring diagram is invaluable. Examples of its usefulness will be shown in later paragraphs. During the course of our discussions we provide the student with symbol lists relating to the material being discussed. These partial symbol groups combine to make a fairly complete and very useful symbol list that will be immediately applicable to practical applications. All materials and methods presented are found commonly in actual practice.

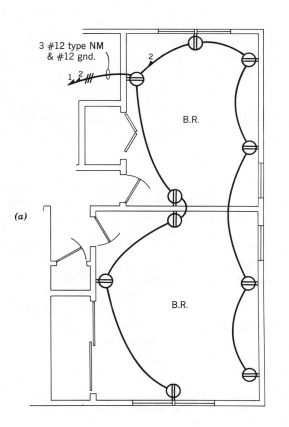

(a)

3 #12 type NM
& #12 gnd.

1 2

2

B.R.

B.R.

Figure 12.1 *(a)* Typical circuitry of basic house bedroom receptacles.

(b)

1

2

Figure 12.1 *(b)* Single-line representation showing circuitry but nothing about arrangement or location.

(c)

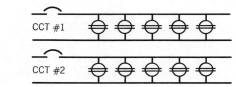

CCT #1

CCT #2

Figure 12.1 *(c)* Wiring diagram. (In this case it adds little information.)

12.3. Branch Circuits

By NEC definition, a branch circuit is "that portion of the wiring system between the final overcurrent device protecting the circuit, and the outlets." You will remember that we stated in the preceding chapter that a circuit consists of a source of voltage, the wiring and the load. After the source of voltage, every practical circuit has some overcurrent protection to protect the circuit components against overloads and short circuits. Overcurrent protection will be considered as a separate subject in a later chapter. The remaining two portions of every circuit are the wiring and the load. This load is called in NEC terminology "the outlets." The branch circuit itself comprises only the wiring, although in everyday trade language a branch circuit is the entire circuit *including* the outlets, and occasionally even the protective device. We will, however, stick to the official NEC definition. See Figure 12.2.

Let us examine the branch circuit, by referring to Figure 12.1. In the drawing, the branch circuit consists of the lines connecting the receptacle outlets in the two bedrooms. In actual practice, the wiring method for the branch circuit can consist of:

(a) Type NM nonmetallic sheathed cables consisting of rubber or plastic insulated wires in a cloth or plastic jacket, commonly called Romex. See Figure 12.3.

(b) Type AC metallic armored cable, usually called BX and comprising two, three or four insulated wires that are covered by an interlocking steel armor jacket. See Figure 12.4.

(c) Individual insulated wires in aluminum, steel or nonmetallic conduit. If the conduit is thin-wall steel, it is called *electric-metallic-tubing* or simply *EMT*. If it is heavy-wall steel, it is called *rigid conduit*. See Figure 12.5. Nonmetallic conduit is called just what it is made of—plastic. Other types of nonmetallic conduit are not used in house wiring.

The wire itself is occasionally aluminum but more frequently is copper, generally No. 12 AWG. Its insulation is rated at least 300 v, and frequently 600 v. The choice of wiring method is left to the designer, but is coordinated with the architect or owner. In general, Romex or BX is used in the stud-wall construction commonly found in private houses, but conduit is used in most other buildings. Let us return now to the two rooms of the small house of Figure 12.1 and see how the branch circuitry is drawn, circuited and installed, depending on the construction and material used.

12.4. Branch Circuit Wiring Methods

As was stated in Section 12.2, the symbol used to represent branch circuit wiring depends on the choice made by the electrical draftsman or designer. A partial symbol list is given in Figure 12.6 showing symbols generally used for branch circuit wiring. We must again emphasize that there is no standard for these symbols and they may, and often do, change from job to job. The thoughtful draftsman will select wiring symbols for clarity of drawing (ease of reading by the contractor) and minimum drawing time. In studying Figure 12.1, let us assume that this basic house plan is typical American frame construction, wired with BX. Since the house is built with an unexcavated crawl space below the bedrooms, and not on a concrete slab, the actual wiring might be run as is shown in Figure 12.7. Horizontal runs above the floor level require drilling of the wood studs, which are normally on 16-in. centers. Runs under the floor require drilling through the sole plate. The contractor will compare the two methods and will make the choice that involves least labor for each run. In this particular case the contractor would most probably choose a combination of underfloor and through-the-studs wiring, feeding the receptacles on the north, south

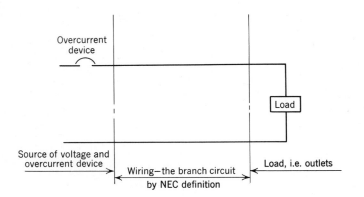

Figure 12.2 Division of electric circuit into its components according to the NEC definitions.

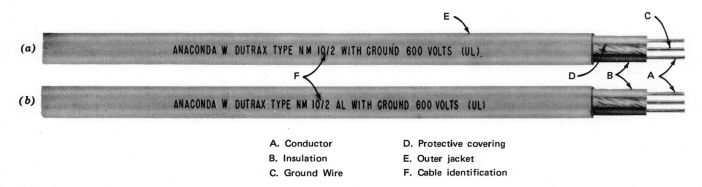

A. Conductor   D. Protective covering
B. Insulation   E. Outer jacket
C. Ground Wire   F. Cable identification

Figure 12.3 Construction of typical NEC type NM Cables. Both cables are 2 conductor No. 10 AWG with ground, insulated for 600 volts. Cable *(b)* uses aluminum wire, and this is clearly shown in the cable identification. Also shown are cable identification *(F)* is the manufacturer, cable trade name, and the letters (UL), which indicate listing of this product by the Underwriters' Laboratories, Inc. Notice that the ground wire *(B)* may be bare or covered, the outer jacket *(E)* may be plastic or cloth braid, and the entire cable may be obtained flat (illustrated), oval, or round. (Courtesy of Anaconda)

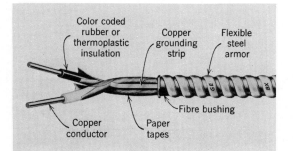

Figure 12.4 Type AC flexible armored cable (BX). Bushing is installed to protect wires from sharp metal edges. (Courtesy of General Electric Co.)

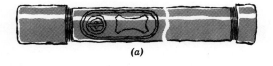

(a)

(b)

(c)

Figure 12.5 Electrical conduits. *(a)* Galvanized, heavy wall, rigid steel conduit. *(b)* Black enameled steel conduit. *(c)* EMT thin-wall steel conduit.

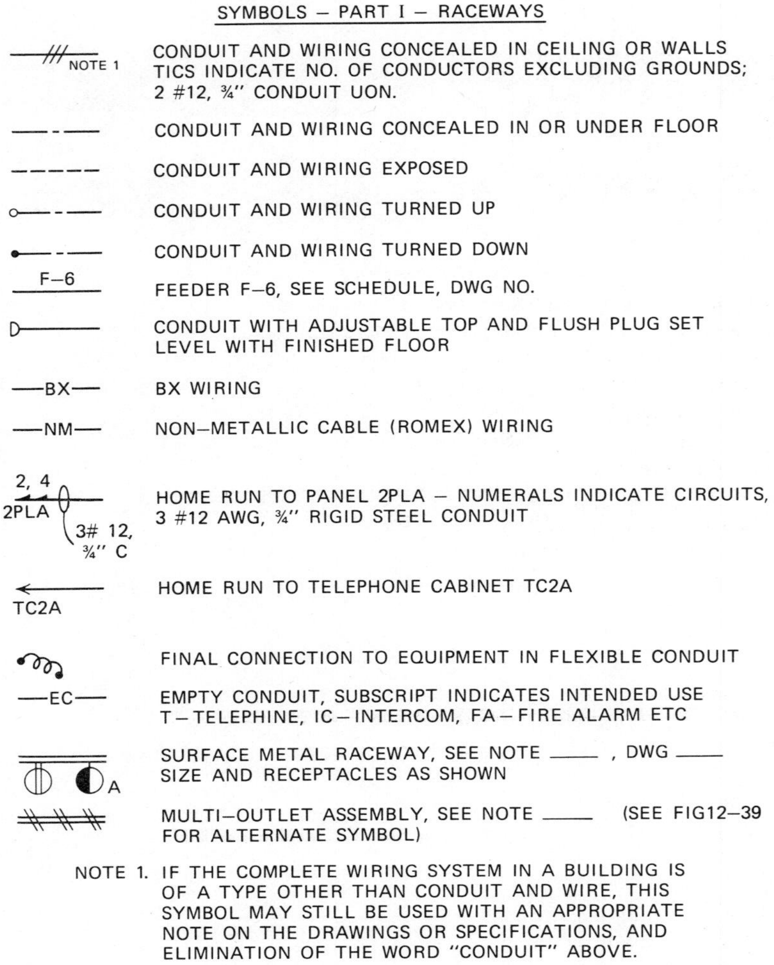

SYMBOLS — PART I — RACEWAYS

─///─ NOTE 1	CONDUIT AND WIRING CONCEALED IN CEILING OR WALLS TICS INDICATE NO. OF CONDUCTORS EXCLUDING GROUNDS; 2 #12, ¾" CONDUIT UON.
─ - ─	CONDUIT AND WIRING CONCEALED IN OR UNDER FLOOR
─ ─ ─ ─	CONDUIT AND WIRING EXPOSED
o─ - ─	CONDUIT AND WIRING TURNED UP
●─ - ─	CONDUIT AND WIRING TURNED DOWN
─F─6─	FEEDER F—6, SEE SCHEDULE, DWG NO.
▷───	CONDUIT WITH ADJUSTABLE TOP AND FLUSH PLUG SET LEVEL WITH FINISHED FLOOR
──BX──	BX WIRING
──NM──	NON—METALLIC CABLE (ROMEX) WIRING
2, 4 / 2PLA / 3# 12, ¾" C	HOME RUN TO PANEL 2PLA — NUMERALS INDICATE CIRCUITS, 3 #12 AWG, ¾" RIGID STEEL CONDUIT
◄─── TC2A	HOME RUN TO TELEPHONE CABINET TC2A
⌒⌒⌒	FINAL CONNECTION TO EQUIPMENT IN FLEXIBLE CONDUIT
──EC──	EMPTY CONDUIT, SUBSCRIPT INDICATES INTENDED USE T — TELEPHINE, IC — INTERCOM, FA — FIRE ALARM ETC
⊕ ◐A	SURFACE METAL RACEWAY, SEE NOTE ____ , DWG ____ SIZE AND RECEPTACLES AS SHOWN
─※※※─	MULTI—OUTLET ASSEMBLY, SEE NOTE ____ (SEE FIG12—39 FOR ALTERNATE SYMBOL)

NOTE 1. IF THE COMPLETE WIRING SYSTEM IN A BUILDING IS OF A TYPE OTHER THAN CONDUIT AND WIRE, THIS SYMBOL MAY STILL BE USED WITH AN APPROPRIATE NOTE ON THE DRAWINGS OR SPECIFICATIONS, AND ELIMINATION OF THE WORD "CONDUIT" ABOVE.

Figure 12.6 Architectural–electrical plan symbol list, Part I, Branch Circuit Wiring.

For other portions of the symbol list, see:
Part II, Outlets Figure 12.37, page 338.
Part III, Wiring Devices, Figure 12.39, page 341.
Part IV, Abbreviations, Figure 12.43, page 345.
Part V, Single Line Diagrams, Figure 13.4, page 354.
Part VI, Equipment, Figure 13.14, page 369.
Part VII, Signaling Devices, Figure 15.34, page 500.
Part VIII, Motors and Motor Control, Figure 16.19, page 527.
Part IX, Control and Wiring Diagrams, Figure 16.22, page 530.

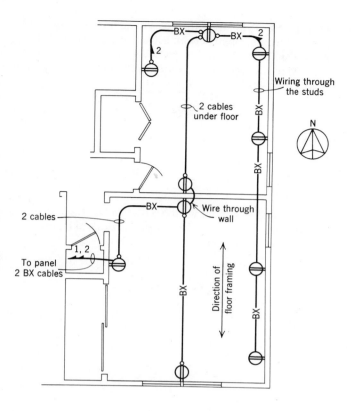

Figure 12.7 One possible wiring solution to layout of Figure 12.1, using BX, and underfloor wiring. A 3-conductor cable can be used instead of two 2-conductor cables.

and west walls with underfloor runs, and the outlets on the east wall with through-the-stud wiring as shown. He would probably not wire in the ceiling space because of the added cost of vertical runs from the receptacle location (12 in. AFF) to the ceiling. The choice of how to wire the last receptacle before the home-run to the panel depends on panel location.

Cable may be fastened to the sides of floor joists and run along them, but under no condition can it be fastened under them for a run across the joist direction. The joists must be drilled and the cable passed through, as shown in Figure 12.8a. The same is true for the through-the-studs wiring that is shown in Figure 12.8b. In this figure we also see a run extending into the ceiling space for wiring of ceiling outlets or for home-runs. As we will discuss later, mixing ceiling and wall outlets on the same circuit is not considered good wiring practice, if avoidable. However, combining separate circuits in the same raceway or multiconductor cable is good practice and is almost always done. In BX cable an interior metal strip acts as a ground. A 3-conductor cable can carry two circuits—two circuit conductors, a common neutral and the built-in ground conductor. Wiring in the ceiling space is similar to basement wiring in ceilings, however, wiring across the ceiling joists

along stiffeners is permitted. See Figure 12.8c. For restrictions on the use of BX and Romex, see the NEC.

If the building were constructed with aluminum studs that come with cutouts for horizontal wiring, the walls would probably be used for all the receptacle wiring. It should be noted, however, that frequently in small residential work the exact wiring paths are left to the electrician, with only the circuitry indicated. This is because the architect assumes that the electrician will select the most economical wiring method. Therefore, as long as the electrician sticks to the outlet layout, the material and the circuitry, little is to be gained in spending the time to do a detailed layout. This is a valid approach for a single house. For a multihouse development, the technologist would be expected to develop detailed *installation* drawings from which the electrician would work. These installation drawings show the exact routing of all wiring. If the same house were built on a concrete slab rather than a crawl space, two practical wiring methods would be available. They are:

(a) To run wiring in the ceiling space and drop down within the walls to wall outlets.

(b) To run wiring in plastic conduit in the floor slab.

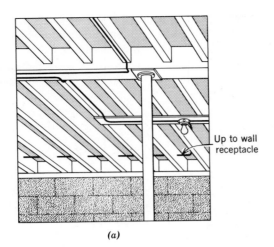

(a)

Figure 12.8 Typical method of wiring nonmetallic cable in wood frame construction. BX wiring is similar. (From *American Electricians' Handbook,* by T. Croft and C. C. Carr, McGraw-Hill Book Company. Used with permission of McGraw-Hill Book Company.)

(a) Typical basement wiring with Romex. BX is similar. Note that wiring is run *along* the directions of beams and joists, but *through* them at right angles.

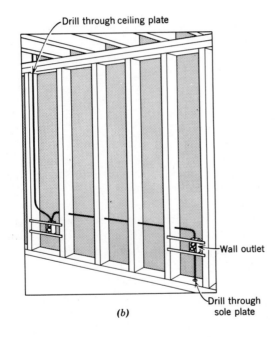

(b)

(b) Wiring at the main level is through-the-studs between devices and through the plates to reach basement and ceiling spaces.

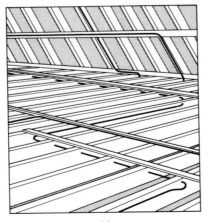

(c)

(c) Wiring in attic spaces is somewhat more flexible because the space is not used and the cable is not subject to impact damage. Horizontal runs along stiffeners are not objectionable.

Other possibilities that are ruled out because of cost are special wiring such as type MI suitable for installation directly in the concrete, or steel conduit. Use of aluminum conduit in concrete is inadvisable and in many localities is prohibited.

Returning to the two practical choices, method (a) would probably be chosen, although preassembled plastic conduit (with wires already inside) might be competitive in price, depending on layout, number of units and labor rates. Plastic conduit is suitable for burial in concrete (in the "mud"), but the additional expense of pulling in wiring would be prohibitive. This is why a preassembled cable-in-conduit might be used if quantities were large as in a large development. The labor cost of installing conduit in the slab is lower than drilling studs and threading BX, and this would offset the higher material cost in a large installation. In addition, any conduit has the great advantage of allowing simple removal of faulty wiring. This is obviously not true with BX or Romex wiring as many a home owner

discovers after the first in-the-wall short circuit. Fortunately such occurences are not common, since most faults occur at outlets.

A layout for the same area as that shown in Figure 12.7, except using plastic conduit in the mud, is shown in Figure 12.9. Note that for a simple case like that of the two bedrooms in the basic house plan, the branch circuiting can use at least three different materials—Type AC (BX), Type NM (Romex), and cable in conduit, and at least as many layouts. You should become familiar with the applications and limitations of all the common branch circuit materials so as to be able to handle layouts with any of them, with ease and assurance.

In large-scale construction, such as high-rise dwelling units, the floor construction is almost always poured concrete and the wiring method is wire and conduit. Some recent multistory dwellings, which are constructed with concrete floor but stud partitions, use combinations of conduit and BX wiring. Two examples of conduit work are shown in

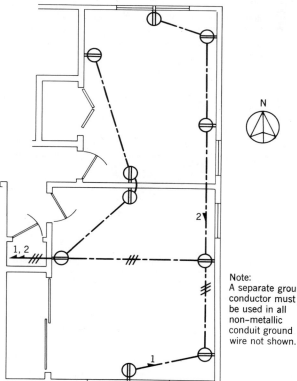

Note:
A separate ground conductor must be used in all non–metallic conduit ground wire not shown.

Figure 12.9 Wiring of spaces in Figure 12.1 assuming a concrete floor slab and plastic conduit in the slab.

Figures 12.10 and 12.11. In Figure 12.10, plastic conduit is used in the floor slab. Notice how the conduit is run directly from outlet box to outlet box. The cylinders in the slab are sleeves through the floor. In Figure 12.11, steel conduit is used. Note that a special telephone cable suitable for direct concrete installation is shown. This method avoids the expense of conduit and has been widely adopted by telephone companies for the prewiring of residential buildings. We will return to special purpose branch circuits after a detailed discussion of outlets and branch circuit materials.

Figure 12.10 Plastic conduit layout for a floor in an apartment building in Ohio. The particular material used is polyvinyl chloride conduit, known on this job by its trade name of "PV-Duit." (Courtesy of Carlon Products)

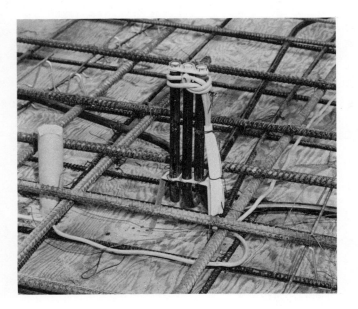

Figure 12.11 An apartment house job using steel conduit and in-the-concrete telephone cable. Note how conduit is bent to avoid sharp angles, for ease of wire pulling. (Courtesy of New York Telephone Co.)

12.5. Wire and Cable

We made brief reference above to some of the common wiring materials used in branch circuitry such as NEC Types NM (Romex), AC (BX) and to some of the wire used in wire and conduit work. It is helpful to examine these types more fully here, to better understand their application.

All building circuit conductors consist of an insulated length of wire, usually copper. Aluminum and copper coated aluminum have made considerable inroads into the electric wire field because of lower weight which leads to lower installation costs. However, to do a proper wiring job with aluminum requires special techniques and men trained in these techniques. Many small contractors prefer to leave well enough alone and to stick with copper, *particularly* for the small sizes encountered in branch circuits where the lower weight advantage of aluminum (cheaper installation labor) is not felt. For this reason we restrict our discussion here to copper wire. In the United States at the present time (prior to any change to metric sizing) the universally used gauge for wire is the American Wire Gauge, called simply AWG. For some good reason that we have yet to discover, the AWG wire numbers proceed about one-half way in reverse order to the conductor size and then switch to an order that proceeds in the logical way. Thus we have the slightly complex situation that, starting with the small size and continuing through the larger sizes, the numbers go from No. 18 AWG to No. 1 AWG, continue from No. 1/0 AWG through No. 4/0 AWG, and then switch to MCM and go from 250 MCM on up. (MCM stands for *thousand circular mil* and is the square of the conductor diameter.) See Table 12.1. This table is by no means complete but is sufficient for our needs. See the NEC for more complete tables. Note that building wire is solid up to No. 8 AWG and is stranded when larger. Also, above No. 8 AWG the wire is often called cable; below that it is called wire. Also, an assembly of two or more conductors in a single jacket is called a cable. For this reason BX and Romex assemblies, being multiconductor, are called cable and not wire, but the individual conductors, being smaller than No. 6 are called wires. All conductors are insulated to prevent contacting each other and short-circuiting. Normal building wire insulation is rated for 600 v, although 300 volt BX and Romex is available and in use. The insulation must be able to withstand the heat generated by the current flow through the wire (I^2R loss). Obviously, the larger the current, the more heat is generated. Therefore, a wire with a heat-resistant insulation such as asbes-

Table 12.1 Physical Properties of Bare Conductors

Size (AWG or MCM)	Area (Circular Mils)	Diameter		d-c Resistance Ohms/1000 ft at 25° C, 77° F (Bare copper)
		Solid	Stranded	
16	2580	0.0508	—	4.10
14	4109	0.0641	—	2.57
12	6530	0.0808	—	1.62
10	10,380	0.1019	—	1.02
8	16,510	0.1285	—	0.64
6	26,240	0.162	0.184	0.41
4	41,740	0.204	0.232	0.26
2	66,360	0.258	0.292	0.16
1	83,690	0.289	0.332	0.13
0 (1/0)	105,600	0.325	0.373	0.10
00 (2/0)	133,100	0.365	0.418	0.081
000 (3/0)	167,800	0.410	0.470	0.064
0000 (4/0)	211,600	0.460	0.528	0.051
250 MCM	250,000	0.500	0.575	0.043
300 MCM	300,000	0.548	0.630	0.036
400 MCM	400,000	0.632	0.728	0.027
500 MCM	500,000	0.707	0.813	0.022

Source. Extracted from the National Electrical Code.

tos can safely carry more current than a thermoplastic (melts when overheated) insulated wire like type TW. And when you look at Table 12.2 that is what you will find. This safe limit, that is, the amount of current a wire will carry safely with a specific insulation, is called its current-carrying-capacity or more simply, its *ampacity*. Table 12.2 is an abbreviated ampacity table. For more complete ones refer to the NEC. The most common types of wire and cable for branch circuitry in small residential work are types BX and Romex cable. Single conductor RHW, TW, THW, THWN and XHHW insulated wires in conduit are used in other types of construction. The choice of which type to use is made by the designer. It is important for the man preparing the drawings to remember that when different types are used on a job, as often happens, they must be clearly marked. It is very common to have type TW branch circuit wiring and type XHHW heavy feeder wiring. The draftsman must remember that these designations are vitally important and must appear either by general note on the drawings, by description in the specification, or along with each wire designation. The most common wire sizes for branch circuit work are Nos. 14, 12 and 10 AWG. This will be dis-

Table 12.2 Allowable Ampacities of Insulated Copper Conductors (Not More Than Three Conductors in Raceway)

Size	Temperature Rating of Conductor (See Table 12.3)[a]			
AWG MCM	60°C (140°F)	75°C (167°F)	90°C (194°F)	110°C (230°F)
	Types T, TW	Types RHW, THW, THWN, XHHW	Types SA, RHH, THHN, XHHW	Type AVA
14[b]	15	15	25[c]	30
12[b]	20	20	30[c]	35
10[b]	30	30	40[c]	45
8	40	45	50	60
6	55	65	70	80
4	70	85	90	105
3	80	100	105	120
2	95	115	120	135
1	110	130	140	160
0	125	150	155	190
00	145	175	185	215
000	165	200	210	245
0000	195	230	235	275
250	215	255	270	315
300	240	285	300	345
350	260	310	325	390
400	280	335	360	420
500	320	380	405	470

Source. Extracted from the National Electrical Code

[a]These ampacities relate only to conductors described in Table 12.3.

[b, c]The ampacities for Types RHH, THHN and XHHW conductors for sizes AWG Nos. 14, 12 and 10 shall be the same as designated for 75°C conductors in this table.

Table 12.3 Characteristics of Insulated Conductors

Trade Name	Type Letter	Maximum Operating Temperature	Application Provisions
Moisture and heat-resistant rubber	RHW	75°C 167°F	Dry and wet locations.
Thermoplastic	T	60°C 140°F	Dry locations
Moisture-resistant thermoplastic	TW	60°C 140°F	Dry and wet locations
Heat-resistant thermoplastic	THHN	90°C 194°F	Dry locations
Moisture and heat-resistant thermoplastic	THW	75°C 167°F	Dry and wet locations
Moisture and heat-resistant thermoplastic	THWN	75°C 167°F	Dry and wet locations
Moisture and heat-resistant cross-linked thermosetting polyethylene	XHHW	90°C 194°F	Dry locations
		75°C 167°F	Wet locations
Silicone-asbestos	SA	90°C 194°F	Dry locations
Asbestos and Varnished cambric	AVA	110°C 230°F	Dry locations only

Source. Extracted from the National Electrical Code.

cussed at length in the section on circuitry but it is well to fix in your mind the fact that these three sizes represent ampacities of 15, 20 and 30 amperes, respectively. Check this in Table 12.2. A brief listing of some wire types is given in Table 12.3.

As explained above, the purpose of the wire's insulation is to prevent contact between electrically "hot" wires and to withstand the heat generated in the wire by the passage of current. The insulation accomplishes the latter purpose as a function of its nature—that is, asbestos withstands heat better than cloth or paper. However, the ability to insulate electrically (to withstand voltage) also depends on the insulation's thickness. Thus the same material,

Table 12.4 Dimensions of Rubber, Asbestos and Thermoplastic-Covered Conductors

Size	Type RHW[b]		Types T, THW[a], TW		Types THHN, THWN		Type AVA		Type XHHW	
AWG MCM	Approx. Diameter (Inches)	Approx. Area (Sq In.)	Approx. Diameter (Inches)	Approx. Area (Sq In.)	Approx. Diameter (Inches)	Approx. Area (Sq In.)	Approx. Diameter (Inches)	Approx. Area (Sq In.)	Approx. Diameter (Inches)	Approx. Area (Sq In.)
Col. 1	Col. 2	Col. 3	Col. 4	Col. 5	Col. 6	Col. 7	Col. 8	Col. 9	Col. 10	Col. 11
14	0.204	0.0327	—	—	—	—	—	—	—	—
14	—	—	0.162[a]	0.0206[a]	—	—	0.245	0.047	0.129	0.0131
12	0.221	0.0384	—	—	—	—	—	—	—	—
12	—	—	0.179[a]	0.0251[a]	—	—	0.265	0.055	0.146	0.0167
10	0.242	0.0460	0.168	0.0224	0.153	0.0184	—	—	—	—
10	—	—	0.199[a]	0.0311[a]	—	—	0.285	0.064	0.166	0.0216
8	0.311	0.0760	0.228	0.0408	0.201	0.0317	—	—	—	—
8	—	—	0.259[a]	0.0526[a]	—	—	0.310	0.075	0.224	0.0394
6	0.397	0.1238	0.323	0.0819	0.257	0.0519	0.395	0.122	0.282	0.0625
4	0.452	0.1605	0.372	0.1087	0.328	0.0845	0.445	0.155	0.328	0.0845
2	0.513	0.2067	0.433	0.1473	0.388	0.1182	0.505	0.200	0.388	0.1182
1	0.588	0.2715	0.508	0.2027	0.450	0.1590	0.585	0.268	0.450	0.1590
1/0	0.629	0.3107	0.549	0.2367	0.491	0.1893	0.625	0.307	0.491	0.1893
2/0	0.675	0.3578	0.595	0.2781	0.537	0.2265	0.670	0.353	0.537	0.2265
3/0	0.727	0.4151	0.647	0.3288	0.588	0.2715	0.720	0.406	0.588	0.2715
4/0	0.785	0.4840	0.705	0.3904	0.646	0.3278	0.780	0.478	0.646	0.3278
250	0.868	0.5917	0.788	0.4877	0.716	0.4026	0.885	0.616	0.716	0.4026
300	0.933	0.6837	0.843	0.5581	0.771	0.4669	0.940	0.692	0.771	0.4669
350	0.985	0.7620	0.895	0.6291	0.822	0.5307	0.995	0.778	0.822	0.5307
400	1.032	0.8365	0.942	0.6969	0.869	0.5931	1.040	0.850	0.869	0.5931
500	1.119	0.9834	1.029	0.8316	0.955	0.7163	1.125	0.995	0.955	0.7163

Source. Extracted from the National Electrical Code.
[a]Dimensions of THW in sizes Nos. 14 to 8.   No. 6 THW and larger is the same dimension as T.
[b]Dimension of RHW without outer covering is the same as THW.
No. 14 to No. 8, solid; No. 6 and larger, stranded.

for instance, PVC or rubber or even treated paper, can be used to insulate cables at 300 v or 3000 v simply by thickening the insulation. There are limitations, of course, both technical and economic, that lead to the use of certain materials and not to others, but in general the thicker the insulation the higher the cables' voltage rating, and vice versa. Table 12.4 gives dimensional data on 600 v insulated cables. Over the insulated conductor(s) the manufacturer frequently places a covering or jacket that protects the cable against all sorts of damage—physical, chemical, heat, water and the like. The interlocked armor on BX cable gives physical protection; the heavy plastic jacket on type UF direct underground burial cables gives water and chemical protection; the neoprene jacket on industrial use cables gives physical and chemical protection, and so on. Special use cables are available in literally hundreds of different types, but for our purposes the types shown in Table 12.2 are the most important.

Conductor insulation can be colored during manufacture as desired. This makes possible a standard system of color coding for branch circuits which allows us to keep wires and phases straight. Without this coding, installation would be much more time-consuming because of having to "ring-out" all the wires to identify them. The standard branch circuit color code is as follows:

Neutral   white or gray
Phase A   black
Phase B   red
Phase C   blue
Ground   green (when insulated)

The color coding can be the insulation color, a stripe in the insulation or even paint on the cable. Neutral and ground, however, must be color coded throughout their lengths. Identification of phase legs at the ends only, is permitted. An uninsulated ground is obviously not color coded. It is quite obviously the ground wire. Whether to use an insulated or uninsulated ground is the designer's choice. In BX cable, the ground is the continuous metallic strip along the armor, since the armor alone is not a suitable ground. In Romex, the ground wire is generally uninsulated. Two examples of more complex cable assemblies are shown in Figure 12.12.

12.6. Connectors

All conductor connections in modern building construction work are made with solderless connectors of various types. Branch circuit connectors are most often made with twist-on insulated connectors frequently called Wire-Nuts after Ideal Industries' trade name. The general term for all pressure-type connectors is solderless-connectors, since at one time all joints were soldered. Today, only busbars are occasionally connected by brazing, but also frequently by bolting. Figure 12.13 shows some typical branch circuit and larger cable connectors.

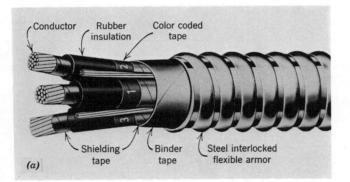

Figure 12.12 (a) High voltage 3-conductor interlocked steel armored cable. (Courtesy of General Electric)

Figure 12.12 (b) Multiconductor 300 v control cable. (Courtesy of General Electric Co.)

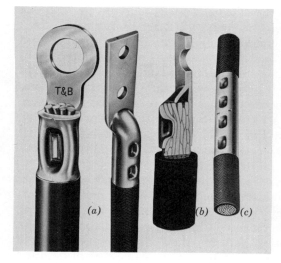

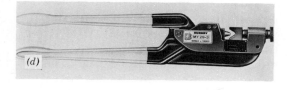

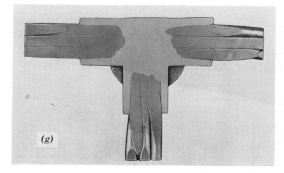

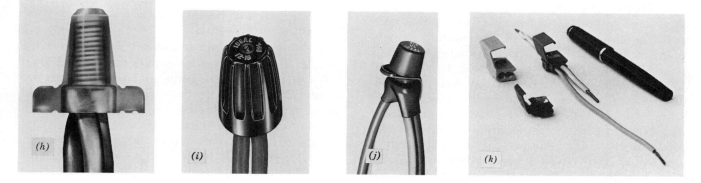

Figure 12.13 Electrical connectors: Typical lug
connector (a) and sleeve type (c) are put on with pressure
indenting tool (d). High quality of splice is shown in
cutaway (b). "Burndy Servit" screw type pressure
connector (e) can be used for straight through or "T" taps;
larger cables require special bolted connectors (f).

Thermal welding process actually welds the conductors
together (g). Branch circuit wiring is usually spliced by
use of small screw-on pressure connectors of various
design (h), (i), (j). [Photos courtesy of Thomas Betts Co. (a
to c, and h), Burndy Co. (d to f) and Ideal Industries (i to j).]

12.7. Conduit

By NEC definition, a raceway is any channel expressly designed for holding wires. Since round pipe, or conduit, is the most commonly used electrical raceway, we begin our discussion with it. The purpose of electrical conduit in addition to providing a means for running wires from one point to another, is to physically protect the wires, to provide a grounded enclosure and to protect the surroundings against the effects of faults in the wiring. This last point is often overlooked. We should, however, take note of the number of fires that are caused annually by short circuits and other electrical faults. Many of these would probably not occur if the faulted and overheated wiring were enclosed in steel conduit, since the steel pipe would contain the arcs, help dissipate the heat and tend to snuff out the fire. This same reasoning is behind the NEC tendency to require that the entire electrical system be enclosed in metal. This includes not only the wiring but also the switches, receptacles, panels, switchgear and other components of the wiring system. Note in Figure 12.14 how all of the components are steel enclosed—by pipe, metal cabinets and boxes. The purpose is, as stated, protective, and it works two ways. It acts to protect the wiring system from damage by the building and its occupants and, even more so, to protect the building and its occupants from damage by the electrical system. The power available in even the smallest electrical system in a private residence is awesome, and must be carefully controlled, limited and isolated. Insulation and conduit do the isolation job. Fuses and other devices do the control and limiting job. These will be discussed in the section on current protection.

In Figure 12.5 we show the most common types of rigid steel conduit. Most branch circuitry that does utilize conduit is run in small size conduit, namely ½ in., ¾ in., and 1 in. This is because the small branch circuit wiring, generally Nos. 14, 12 and 10 AWG, fits easily into these size conduits in the quantities normally encountered in branch circuit work. Refer to the NEC table which shows the number of wires that can be accommodated by different size conduits. Note, for instance, that a typical run of four No. 12 AWG type TW wires can fit into a ¾-in. conduit with room to spare. These small branch circuit conduits are normally installed in walls and floor slabs and no problems arise in supporting them. Although the larger size conduits are not normally encountered in branch circuit work, we discuss their installation problems at this point for convenience. First, however, some idea must be gained of the sizes and weights involved when we speak of electric conduit. Refer to Table 12.5 for a listing of conduit dimensions and weights for rigid steel, EMT, the relatively new intermediate weight conduit and aluminum conduit. Note the very considerable weight of even small conduits and, therefore, the necessity for adequate supports when these conduits are run exposed horizontally or vertically. Minimum spacing of supports is specified in the NEC, and we will not duplicate these data, except as required to illustrate use.

With exposed conduit it is frequently necessary to detail conduit supports and hanging methods. This is particularly true in areas where space is limited, such as in hung ceilings, closets and shafts, where coordination between the requirements of all the different trades is necessary. This type of detailed space coordination work is frequently given to a capable electrical draftsman or technologist to work out and therefore, he must be familiar with support methods and hardware. The simplest types of support are the one-hole conduit strap and the "C" clamp (Figure 12.15a), which can be used to clamp both vertical and horizontal runs. Where more rigid fixing of the conduit is required, a 2-hole pipe strap (Figure 12.15b) can be used. Often, horizontal runs of exposed conduit must be supported by suspension. Individual hangers (Figure 12.15c) or trapeze arrangements with individual conduit clamps at each trapeze support (Figure 12.15d), are very common. Indeed, the trapeze is frequently assembled into two or more layers of conduit, as is shown. Supports for vertical conduit can be assembled very much like a trapeze support, except of course, they are vertical instead of horizontal, or one can use units manufactured especially for this purpose. See Figure 12.5e and f. These are the basic support elements, which can be utilized to assemble a good conduit support system.

To properly lay out a conduit system with the correct dimensioning for tight fits, requires a knowledge of how conduit joints and terminations are made. We restrict ourselves at this point to rigid conduit fittings, which are threaded. Lengths of conduit are joined together by screwed steel couplings. When a conduit that carries wires larger than No. 6 AWG enters a box or cabinet, the NEC requires insulated bushings. For smaller wires (and conduits) the joint at the cabinet is usually made up with a locknut and bushing or double locknuts and bushing—insulated or uninsulated. For many applications, special grounding bushings are used. When attaching a conduit to a ceiling or floor outlet box, a chase nipple is generally used. These various

Typical Electrical Building Equipment

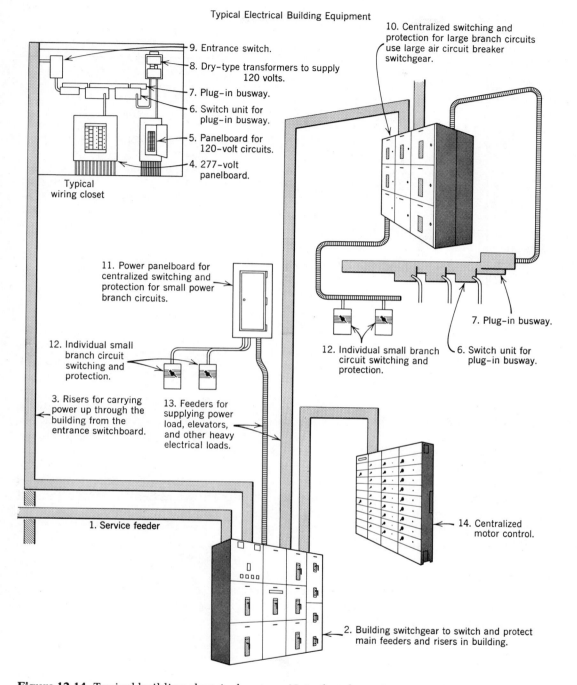

9. Entrance switch.

8. Dry-type transformers to supply 120 volts.

7. Plug-in busway.

6. Switch unit for plug-in busway.

5. Panelboard for 120-volt circuits.

4. 277-volt panelboard.

Typical wiring closet

10. Centralized switching and protection for large branch circuits use large air circuit breaker switchgear.

11. Power panelboard for centralized switching and protection for small power branch circuits.

12. Individual small branch circuit switching and protection.

3. Risers for carrying power up through the building from the entrance switchboard.

13. Feeders for supplying power load, elevators, and other heavy electrical loads.

12. Individual small branch circuit switching and protection.

7. Plug-in busway.

6. Switch unit for plug-in busway.

1. Service feeder

14. Centralized motor control.

2. Building switchgear to switch and protect main feeders and risers in building.

Figure 12.14 Typical building electrical system. Note that the entire system is, in effect, jacketed in steel. (Courtesy of General Electric Co.)

Table 12.5 Comparative Dimensions and Weights of Metallic Conduit

Nominal or Trade Size	Outside Diameter (In.)				Inside Diameter (In.)				Weight per 10 ft Length (Pounds)[a]			
	R.S.[b]	I.S.[c]	EMT[d]	AL.[e]	R.S.	I.S.	EMT	AL.	R.S.	I.S.	EMT	AL.
½	0.84	0.82	0.71	0.84	0.62	0.69	0.62	0.62	7.9	5.3	2.9	2.7
¾	1.05	1.03	0.92	1.05	0.82	0.89	0.82	0.82	10.5	7.2	4.4	3.6
1	1.32	1.29	1.16	1.32	1.05	1.13	1.05	1.05	15.3	14.4	6.4	5.3
1¼	1.66	1.64	1.51	1.66	1.38	1.47	1.38	1.38	20.1	17.7	9.5	7.0
1½	1.90	1.88	1.74	1.90	1.61	1.70	1.61	1.61	24.9	23.6	11.0	8.6
2	2.38	2.36	2.20	2.38	2.07	2.17	2.07	2.07	33.2	23.6	14.0	11.6
2½	2.86	2.86	2.88	2.88	2.47	2.61	2.47	2.47	52.7	38.2	20.5	18.3
3	3.50	3.48	3.50	3.50	3.07	3.23	3.07	3.07	68.3	46.9	25.0	23.9
3½	4.00	3.97	—	4.00	3.55	3.72	—	3.55	83.1	54.7	—	28.8

[a]Standard length including one coupling.
[b]Standard heavy-wall rigid steel conduit.
[c]Intermediate-weight steel conduit (see NEC Art. 345).
[d]Electric metallic tubing.
[e]Aluminum.

fittings are illustrated in Figures 12.16 and 12.17. Typical dimensional data for conduit couplings, chase nipples and conduit bushings are given in Table 12.6. To properly lay out rigid steel conduit where close spacing is required in runs, or at cabinet entrances, it is necessary to remember that the conduit terminations are larger than the conduits themselves, and that space must be left between locknuts for a wrench. We recommend that this space be ¼-in. minimum, and Table 12.7 furnishes spacing data based on this assumption. Table 12.8 presents dimensional data on conduit elbows. Although these data are for manufactured elbows, they are approximately the same as a good tight field bend and can be used for that, as well as EMT bends. However, the fitting data presented, as stated, are definitely not applicable to EMT, which use compression, tap-on or set-screw type couplings and fittings, a few of which are shown in Figure 12.18.

12.8. Other Raceways

Rigid conduit is used for straight runs and for connection to equipment that is noise and vibration free. However, connection to motors, which vibrate no matter how well balanced, or transformers, which hum, should be made with a loop of flexible conduit that will minimize transmission and amplification of noise and vibration. This type of flexible conduit is called by the trade name of Greenfield. When it is covered with a plastic jacket it becomes liquid-tight, and is known as Sealtite (trade name of Anaconda Company's product) for application in wet locations. See Figures 12.19 to 12.23.

In addition to conduits of various materials and degrees of rigidity, there is an entire line of raceways that are intended for surface attachment. These are generally not round, but have a flat bottom which serves as an attachment surface. A few typical sections and their wire capacities are shown in Table 12.9. These raceways are used where:

(a) Concealed raceways are not practical.
(b) Frequent access to the raceway wiring is desired.
(c) Wiring devices are required at frequent intervals.

This last application has become very popular in industrial, commercial and even residential work. A few applications are shown in Figures 12.24 to 12.28. The cost of a surface raceway with receptacle outlets at 18-in. centers is much lower than a conduit and outlet box arrangement with the same spacing of outlets. Furthermore, surface raceways are not as

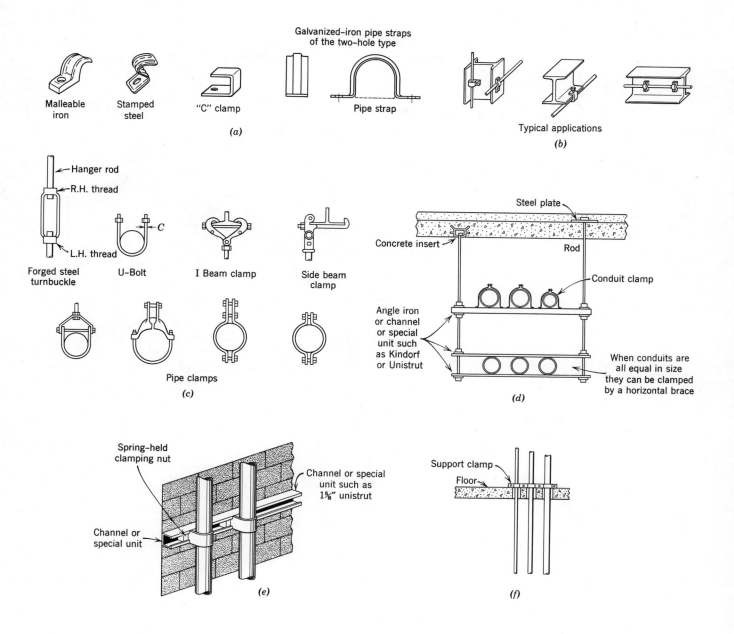

Figure 12.15 Conduit supports.

Figure 12.15 (a) and (b) One-hole conduit clamps, "C" clamp, two-hole pipe strap and typical applications.

Figure 12.15 (c) Single-pipe suspension methods. These fittings are also applicable to pipe hanging.

Figure 12.15 (d) Trapeze conduit hangers can be made of standard steel members or assembled of channels specifically intended for the purpose. Each conduit is clamped in place individually. Trapeze hanging can be in one or more layers.

Figure 12.15 (e) Vertical conduit support is similar to horizontal.

Figure 12.15 (f) When supported at a floor opening special clamps can be used.

Figure 12.16 Typical conduit fittings and their application.

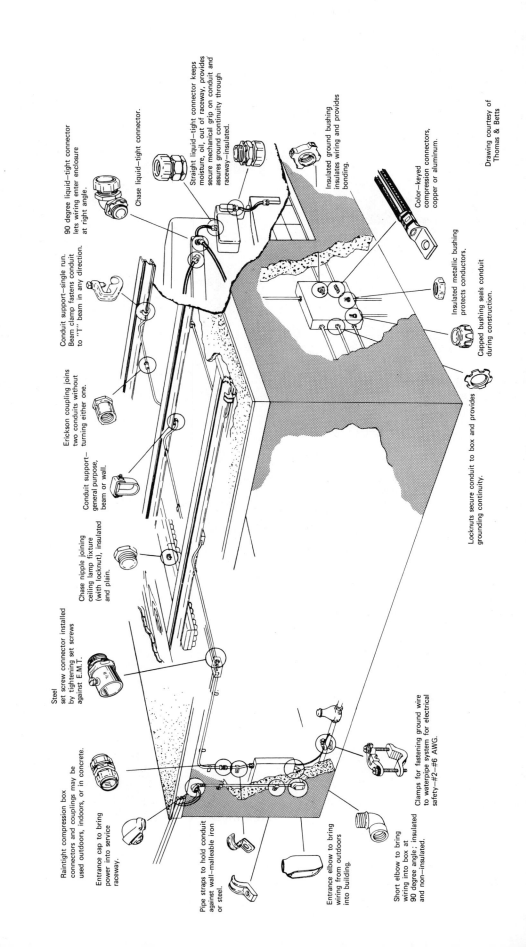

90 degree liquid–tight connector lets wiring enter enclosure at right angle.

Chase liquid–tight connector.

Straight liquid–tight connector keeps moisture, oil, out of raceway, provides secure mechanical grip on conduit and assures ground continuity through raceway–insulated.

Insulated ground bushing insulates wiring and provides bonding.

Color–keyed compression connectors, copper or aluminum.

Insulated metallic bushing protects conductors.

Capped bushing seals conduit during construction.

Locknuts secure conduit to box and provides grounding continuity.

Drawing courtesy of Thomas & Betts

Conduit support–single run. Beam clamp fastens conduit to "T" beam in any direction.

Erickson coupling joins two conduits without turning either one.

Conduit support– general purpose, beam or wall.

Chase nipple joining ceiling lamp fixture (with locknut), insulated and plain.

Steel set screw connector installed by tightening set screws against E.M.T.

Raintight compression box connectors and couplings may be used outdoors, indoors, or in concrete.

Entrance cap to bring power into service raceway.

Pipe straps to hold conduit against wall–malleable iron or steel.

Entrance elbow to bring wiring from outdoors into building.

Short elbow to bring wiring into box at 90 degree angle; insulated and non–insulated.

Clamps for fastening ground wire to waterpipe system for electrical safety–#2–#6 AWG.

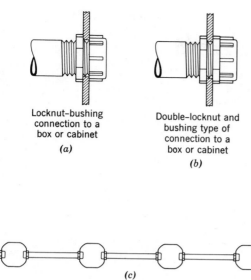

Locknut–bushing
connection to a
box or cabinet
(a)

Double–locknut and
bushing type of
connection to a
box or cabinet
(b)

Figure 12.17 *(a), (b)* Conduit can be fastened to a cabinet with a bushing and one locknut as in *(a)* or two locknuts as in *(b)*. The double locknut connection is superior. (From *American Electricians' Handbook,* by T. Croft and C. C. Carr, McGraw-Hill Book Company. Used with permission of McGraw-Hill Book Company.)

(c)

(c) Good electrical continuity is maintained along the conduit with proper box terminations.

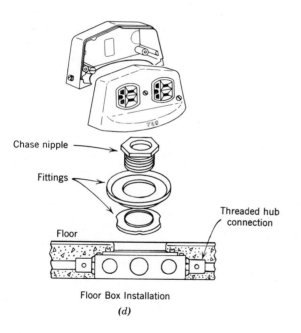

Chase nipple

Fittings

Floor

Threaded hub
connection

Floor Box Installation
(d)

(d) Floor mounted receptacles are fastened to floor boxes using chase nipples. These are also used to terminate conduits emerging from the concrete floor and connecting to floor receptacles. (Courtesy of Thomas and Betts Co.)

Table 12.6*a* Dimensional Data for Rigid Steel Conduit (Courtesy of Youngstown Sheet & Tube Company)

Conduit

Size (Inches)	Nominal Diameter (Inches) Internal	Nominal Diameter (Inches) External	Nominal Wall Thickness (Inches)	Threads per Inch	Length Without Coupling	Weight per Foot (Incl. Coupling)
½	0.622	0.840	0.109	14	9'11¼"	0.79
¾	0.824	1.050	0.113	14	9'11½"	1.05
1	1.049	1.315	0.133	11½	9'11"	1.53
1¼	1.380	1.660	0.140	11½	9'11"	2.01
1½	1.610	1.900	0.145	11½	9'11"	2.49
2	2.067	2.375	0.154	11½	9'11"	3.32
2½	2.469	2.875	0.203	8	9'10½"	5.27
3	3.068	3.500	0.216	8	9'10½"	6.83
3½	3.548	4.000	0.226	8	9'10½"	8.31
4	4.026	4.500	0.237	8	9'10¼"	9.72
5	5.047	5.563	0.258	8	9'10"	13.14
6	6.065	6.625	0.280	8	9'10"	17.45

Couplings

Size (Inches)	Outside Diameter (Inches)	Length (Inches)	Weight per Piece
½	1.010	1.562	0.12
¾	1.250	1.625	0.17
1	1.525	2.000	0.30
1¼	1.869	2.062	0.37
1½	2.155	2.062	0.52
2	2.650	2.125	0.67
2½	3.250	3.125	1.70
3	3.870	3.250	2.10
3½	4.500	3.375	3.40
4	4.875	3.500	2.90
5	6.000	3.750	4.50
6	7.200	4.000	7.30

Conduit furnished in 10 ft. lengths, threaded both ends with one coupling.

Table 12.6*b* Dimensions of Conduit Nipples and Bushings*

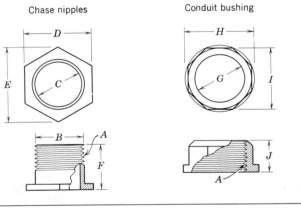

Size (Inches)	A	B	C	D	E	F	G	H	I	J
½	14	0.82	0.62	1.00	1.15	0.62	0.62	1.00	0.94	0.37
¾	14	1.02	0.82	1.25	1.44	0.81	0.75	1.25	1.12	0.44
1	11.5	1.28	1.04	1.37	1.59	0.94	1.00	1.50	1.37	0.50
1¼	11.5	1.63	1.38	1.75	2.02	1.06	1.25	1.81	1.75	0.56
1½	11.5	2.87	1.61	2.00	2.31	1.12	1.50	2.12	2.00	0.56
2	11.5	2.34	2.06	2.50	2.89	1.31	1.94	2.56	2.37	0.62
2½	8	2.82	2.46	3.00	3.46	1.41	2.37	3.06	2.87	0.75
3	8	3.44	3.06	3.75	4.33	1.50	2.87	3.75	3.50	0.81
3½	8	3.94	3.54	4.25	4.91	1.62	3.25	4.25	4.00	1.00

Note: A = threads per inch.

B = thread diameter.

*These dimensions vary slightly from one manufacturer to another.

Table 12.7 Rigid Steel Conduit Spacing at Cabinets (All Dimensions in Inches)

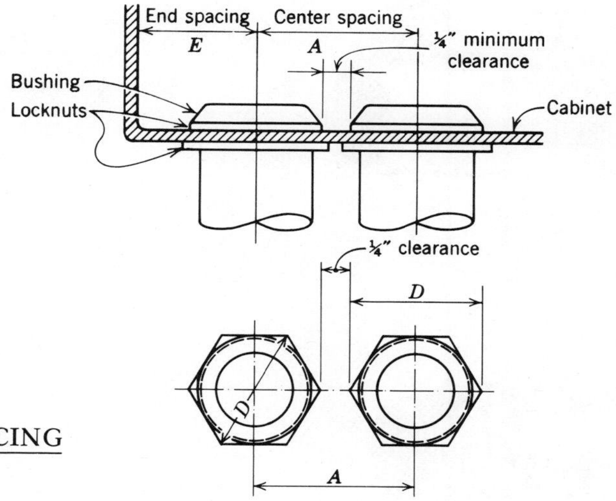

CENTER SPACING "A"

Locknut "D"	End spacing "E"	Conduit size (Inches)	4	3½	3	2½	2	1½	1¼	1	¾	½
5⅝	3¼	4	5⅞									
4⅞	2⅞	3½	5½	5⅛								
4 5/16	2⅝	3	5¼	4⅞	4 9/16							
3½	2⅛	2½	4 13/16	4 7/16	4 3/16	3¾						
2⅞	1⅞	2	4½	4⅛	3⅞	3 7/16	3⅛					
2 5/16	1⅝	1½	4¼	3⅞	3 9/16	3 3/16	2⅞	2 9/16				
2	1½	1¼	4 1/16	3 11/16	3 7/16	3	2 11/16	2 7/16	2¼			
1⅝	1⅛	1	3⅞	3½	3¼	2 13/16	2½	2¼	2 1/16	1⅞		
1 7/16	1	¾	3 13/16	3 7/16	3⅛	2¾	2 7/16	2⅛	2	1 13/16	1 11/16	
1⅛	⅞	½	3⅝	3¼	3	2 9/16	2¼	2	1 13/16	1⅝	1 9/16	1⅜

Table 12.8 Dimensions of Rigid Steel Conduit Elbows (All dimensions in inches)

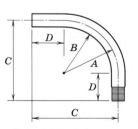

[a]Varies for different manufacturer

Nominal Trade Size	Actual Inside Diam.	Actual Outside Diam.	Radius A	Radius B	C[a]	D[a,b]	Weight (ea. Lb)
½	0.63	0.84	4.00	3.58	6.5	2.5	0.82
¾	0.83	1.05	4.50	3.98	7.25	2.75	1.09
1	1.05	1.32	5.75	5.09	8.63	2.88	2.01
1¼	1.38	1.66	7.25	6.42	10.0	2.75	3.13
1½	1.61	1.90	8.25	7.3	11.0	2.75	4.14
2	2.06	2.38	9.50	8.31	13.62	4.13	7.07
2½	2.47	2.86	10.50	9.06	15.68	5.19	14.11
3	3.06	3.50	13.00	11.25	17.75	4.75	18.50
3½	3.56	4.00	15.00	13.0	20.0	5.0	29.79
4	4.06	4.50	16.00	13.75	21.31	5.31	35.28

[a]Varies with different manufacturers.

[b]This dimension represents the straight portion of the elbow.

Figure 12.18 Various types of EMT connectors:

(a)

(a) Set screw coupling.

(e)

(e) Compression-type rain-tight connector.

(b)

(b) Set screw connector, insulated.

(f)

(f) Compression-type insulated connector in cutaway.

(c)

(c) Tap-on coupling.

(g)

(g) Compression-type rain-tight coupling.

(d)

(d) Tap-on connector, insulated.

(Courtesy of Thomas E. Betts, Inc.)

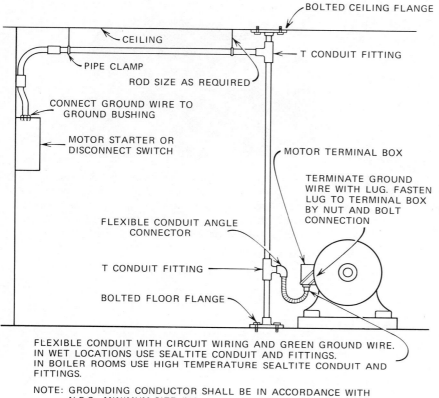

Figure 12.19 Typical indoor motor installation. Note flexible conduit connection, conduit brace to ceiling, overhead conduit supports, and method of attachment at floor to fix vertical run. In no case should the vertical run come down without a rigid floor connection. (Courtesy of New York Telephone Co.)

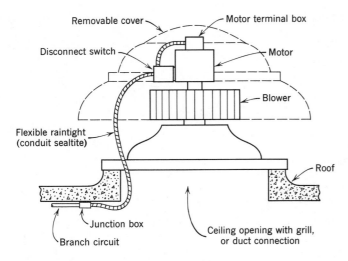

Figure 12.22 Connection of motor without any slack in the flexible conduit. If the conduit is tight, motor vibration will be transmitted to the adjacent box and to the branch circuit wiring. Result may be loose terminals and objectionable noise.

Figure 12.20 Typical detail of power roof ventilator. Note flexible conduit and disconnect switch at motor.

Figure 12.21 Actual installation of a motor in a wet location. Note connection with "Sealtite" and special waterproof switch.

Figure 12.23 Poor transformer installation. Note rigid conduit connection, obstructed top of unit, and lack of vibration pads.

Table 12.9 Abbreviated Listing of Some of WIREMOLD's Surface Raceways and Their Wire Capacities[a]

Raceway Type	Section Through Raceway	Wire Size	Number of Wires	
			Type RHW	Types T, TW
No. 200	11/32", 1/2"	14	3	3
		12	2	3
No. 500	17/32", 3/4"	14	5	6
		12	4	6
		10	2	4
No. 700	21/32", 3/4"	14	7	8
		12	6	8
		10	3	6
No. 1000	15/16", 1 5/16"	14	10	10
		12	10	10
		10	6	8
No. 1500	1 9/16", 11/32"	14	4	8
		12	4	6
		10	—	4

Raceway Type	Section Through Raceway	Wire Size	With Devices	Without Devices	With Devices	Without Devices
No. 1900	9/16", 13/16"	14	3	3	3	3
		12	3	3	3	3
No. 2000	1 9/16", 3/4"	14	3	3	3	3
		12	3	3	3	3
No. 2100	1 1/4", 7/8"	14	b	17	b	17
		12	b	14	b	14
		10	—	10	b	10
No. 2200	2 3/8", 3/4"	14	—	—	10	10
		12	—	—	10	10
		10	—	—	10	10

Table 12.9 (continued)

Raceway Type	Section Through Raceway	Wire Size	Number of Wires			
			Type RHW		Types T, TW	
No. 3000	2 3/4", 1 7/16"	14	b	44	b	56
		12	b	40	b	42
		10	b	20	b	20
No. 4000	4 3/4", 1 3/4"	14	17	28	17	68
		12	15	24	15	53
		10	11	20	11	41
		8	7	12	7	22
No. 6000	3 9/16", 4 3/4"	14	61	97	61	234
		12	54	82	54	184
		10	38	68	38	141
		8	27	41	27	77
		6	20	25	20	38

Source. Wiremold Company.

[a]For a complete listing and details of devices that can be mounted on these raceways refer to Wiremold catalog.

[b]See Wiremold catalog.

Figure 12.24 Applied communications laboratory requires use of many instruments. Plugmold 2000 assures adequate, conveniently located outlets. (Courtesy of Wiremold Co.)

Figure 12.25 Note that connection can readily be made between conduit and surface raceway systems. The second conduit emerging from the floor carries communication cable that will be fed into a second surface raceway. Note also how easily special devices, such as the partition mounted timer and the large receptacle, can be installed in any circuit in the raceway. See Figure 12.28 (Courtesy of Wiremold Co.)

Figure 12.26 In this school rewiring job, type 1000 Wiremold was used for groups of branch circuits extending from the panel box. Individual branch circuits are 500 and 700 Wiremold. Locating a fixture directly above piping is undesirable. In rewiring jobs, however, it is frequently unavoidable. (Courtesy of Wiremold Co.)

Figure 12.27 Note that the larger wireways (G-6000 illustrated) are often used to carry conductors of different voltages in addition to mounting of switches and receptacles. (Courtesy of Wiremold Co.)

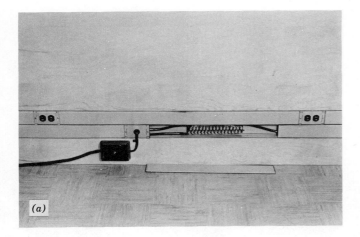

Figure 12.28 *(a)* Running parallel raceways for power and telephone wiring results in an installation that looks like this.

Figure 12.28 *(b)* By using a divided raceway with a single cover, the same results can be accomplished with a neater installation.

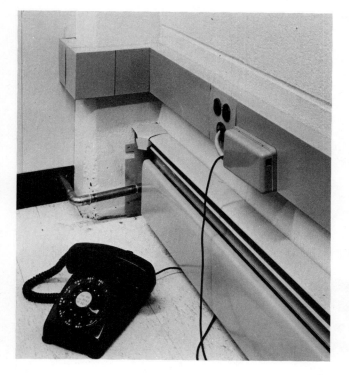

Figure 12.28 *(c)* A close-up of the combined raceway shows how easily corners are negotiated and how telephone equipment can be mounted. (Courtesy of Wiremold Co.)

unsightly as they once were, and now are used routinely in schools, labs and offices. A new system of raceways has been developed that can handle the large open areas found in "landscaped" offices, libraries and multi-use classrooms. These raceways extend *vertically* from floor to ceiling and are wired into a hung ceiling track system. This solves the problem of how to feed a desk in the middle of a room without spending the large amounts that are required for underfloor wiring systems. These Tele-Power poles (Wiremold) and their application are illustrated in Figures 12.29 to 12.31. Finally, surface raceways are very frequently employed for communication wiring. In such cases, they are normally installed empty and the wiring is put in later.

Note in Figure 12.6, our raceway symbol list, that surface raceways are represented very generally by a double parallel line and some sort of identifying letters. The multi-outlet assembly is similarly represented. There is no consensus on symbols for this type of raceway, and the technical man who prepares the drawings is free to invent any symbol he chooses, provided that it is well identified and clear. As we have mentioned, surface raceway with receptacles placed as desired, is readily applicable to residential work. Referring to our basic-plan house, there are two locations in which surface raceways could well be applied—the kitchen and the basement. See Figure 12.32. The modern housewife has an abundance of electrical devices and frequently a shortage of outlets. This could be remedied by the layout shown. Similarly, in the basement—assuming a conventional heating plant installation—the owner's worktable could probably be placed as shown, and the layout given would provide an excellent application for surface raceway.

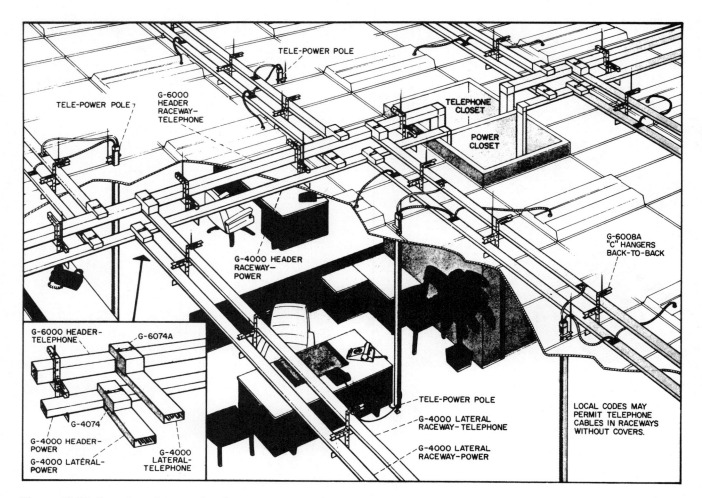

Figure 12.29 Drawing of an overhead raceway system fed from a central service core. Note the extreme versatility of this kind of feeder (header) and branch circuit raceway system. (Courtesy of Wiremold Co.)

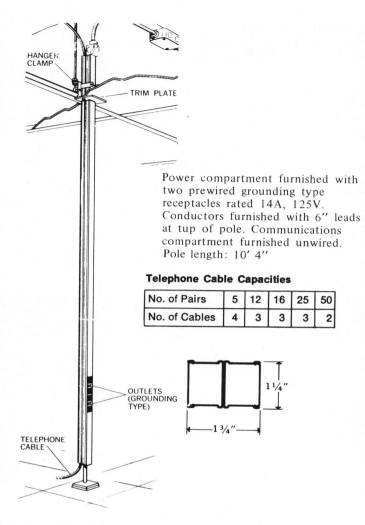

Power compartment furnished with two prewired grounding type receptacles rated 14A, 125V. Conductors furnished with 6" leads at tup of pole. Communications compartment furnished unwired.
Pole length: 10' 4"

Telephone Cable Capacities

No. of Pairs	5	12	16	25	50
No. of Cables	4	3	3	3	2

Figure 12.30 Details of Wiremold's "Tele-Power" pole. Note method of feed and the outlets provided.

Figure 12.31 Actual installation of Tele-Power poles shows how they are connected at the ceiling to the source of power. In this case, because of absence of a hung ceiling, the horizontal ceiling raceways are provided with receptacles. The poles take power from these receptacles with a simple flexible cord and cap. (Courtesy of Wiremold Co.)

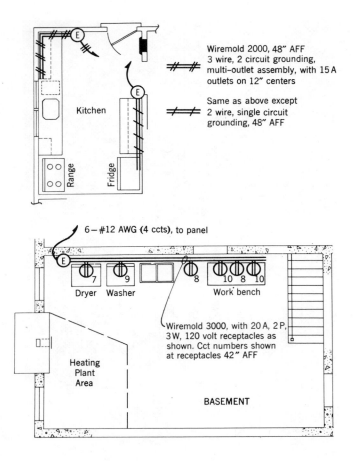

Wiremold 2000, 48" AFF
3 wire, 2 circuit grounding,
multi–outlet assembly, with 15 A
outlets on 12" centers

Same as above except
2 wire, single circuit
grounding, 48" AFF

Figure 12.32 Application of surface wireways in Basic Plan House. Note alternate ways of indicating surface raceways, and information required on the drawing.

12.9. Outlets

Referring again to the National Electrical Code for a definition, we find that an "outlet" is defined as "a point on the wiring system at which current is taken to supply utilization equipment." In simpler terms, any point that supplies an electric load is called an outlet. Therefore, a wall receptacle, a motor connection, an electric heater junction box, a ceiling lighting box and the like, are all outlets. A wall switch is *not* an outlet, although in the field one frequently hears the term "switch outlet"—so much so that the term is generally accepted and understood. When counting the numbers of outlets on a circuit, switches are *not* included. They *are* included when counting outlets for the purpose of making a rough cost estimate. Thus an electric contractor will normally quote a figure of *X* dollars per switch outlet and *Y* dollars per receptacle outlet. This same contractor, however, will understand perfectly an in-

struction to wire no more than six outlets on a circuit and will not include switches in his outlet count. What then, physically, is an outlet? Generally, it consists of a small metal box into which a raceway and/or cable extends. See Figure 12.33. Outlet boxes vary in size and shape and are specifically suited to different kinds of construction. Thus we have the "jiffy" box and the "gem" box, intended for single wall devices; octagonal and round 3½-in. boxes intended for ceiling outlets; the 4-in. square general purpose box, and the like. See Figure 12.34. Boxes intended for fastening to wall studs are furnished with mounting holes for nails and screws. Ceiling outlets boxes for construction of this kind are mounted on steel straps that, in turn, fasten to ceiling joists. Box entrances are equipped to receive cable, BX, or conduit. The important fact to remember is that every outlet has a box. (Recently a line of "boxless" wiring devices has appeared on the electrical industry market. These are actually

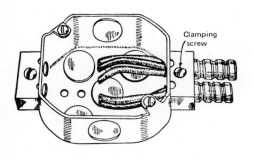

(a)

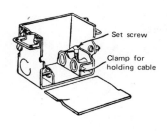

Set screw

Clamp for
holding cable

(b)

Fibre bushing

(c)

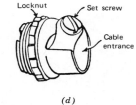

Locknut

Set screw

Cable
entrance

(d)

Figure 12.33 Typical outlet box construction for armored (BX) cable. In *(a)* and *(b)* we have boxes intended for ceiling and wall installation, respectively. Preparation of cable includes using a fiber bushing to protect cable against the sharp edge of cut armor. A typical connector is shown in *(d)*. Details of rigid conduit entrances to boxes are shown in Figure 12.17, page 319. (From *American Electricians' Handbook,* by T. Croft and C. C. Carr, McGraw-Hill Book Company. Used with permission of McGraw-Hill Book Company.)

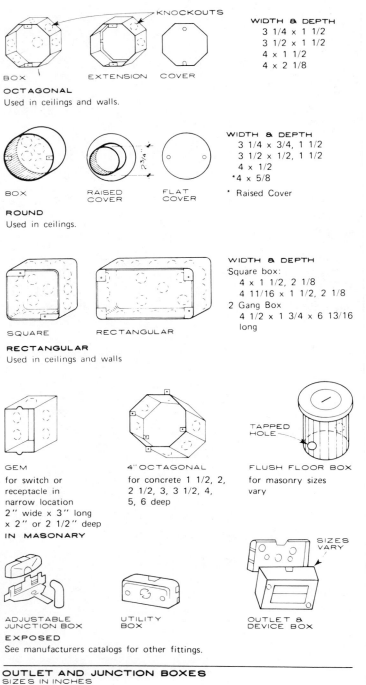

KNOCKOUTS

WIDTH & DEPTH
3 1/4 x 1 1/2
3 1/2 x 1 1/2
4 x 1 1/2
4 x 2 1/8

BOX   EXTENSION   COVER

OCTAGONAL
Used in ceilings and walls.

WIDTH & DEPTH
3 1/4 x 3/4, 1 1/2
3 1/2 x 1/2, 1 1/2
4 x 1/2
*4 x 5/8

* Raised Cover

BOX   RAISED COVER   FLAT COVER

ROUND
Used in ceilings.

WIDTH & DEPTH
Square box:
4 x 1 1/2, 2 1/8
4 11/16 x 1 1/2, 2 1/8
2 Gang Box
4 1/2 x 1 3/4 x 6 13/16 long

SQUARE   RECTANGULAR

RECTANGULAR
Used in ceilings and walls

GEM
for switch or receptacle in narrow location
2" wide x 3" long x 2" or 2 1/2" deep

4" OCTAGONAL
for concrete 1 1/2, 2, 2 1/2, 3, 3 1/2, 4, 5, 6 deep

TAPPED HOLE

FLUSH FLOOR BOX
for masonry sizes vary

IN MASONARY

ADJUSTABLE JUNCTION BOX

UTILITY BOX

SIZES VARY

OUTLET & DEVICE BOX

EXPOSED
See manufacturers catalogs for other fittings.

OUTLET AND JUNCTION BOXES
SIZES IN INCHES

Figure 12.34 Outlet boxes of different dimensions and intended for various uses, as shown. (Courtesy of John Wiley & Sons.)

self-contained box-device combinations.) The term "gang" when referring to a box indicates the number of wiring device positions supplied. Thus if three receptacles are to be mounted side by side, a three-gang box would be used. Boxes can normally be fastened to each other to form such an assembly.

Although each outlet requires a box, there is only one situation that requires a box that is not an outlet. This is the common *junction box*. When wire is being run in a building, it frequently becomes necessary to make a tap, or junction, in the wiring to take power off to another point. See Figure 12.38. That is, wiring to more than one device in a circuit requires a tap. These taps, whenever possible, are made at outlet boxes to avoid the expense of a junction box. Sometimes, however, the necessity for a junction box is unavoidable and one is installed. In it, wires enter, are spliced to each other, and leave to feed outlets. Such a box is *not* an outlet since, by definition, it does not supply current to a utilization device. Some contractors, incorrectly, count junction boxes as outlets. Care should be used to clarify this point if payment is being made by number of outlets. On a drawing, junction boxes should be properly shown and symbolized. As is shown in Figure 12.38 outlets are normally symbolized by simple circles unless there is a possibility of confusion with round columns or other circular architectural features. This confusion is possible because outlet boxes are obviously not shown to scale. This is true of most electrical work. If we were to attempt to show to scale a 4-in. box on ⅛ in. = 1 ft 0 in. scale drawings, the box would be 1/24-in. in diameter—a pencil dot. For this reason almost all electrical work is shown at a size that makes for easy drawing reading—and that is unrelated to the scale of the drawings. This creates coordination problems in close spots, but it cannot be avoided. The only possible exception to this (no scale) convention on architectural-electrical plans is fluorescent fixtures. They are normally shown to scale in *length* and occasionally in width, as well. Of course, construction details are almost always drawn to scale.

A few details of outlet boxes and their mounting are given in Figure 12.35. Observe that in each case we have basically the same material—raceways or cable, a metal box and cover, and the mounting arrangement. In recent years, nonmetallic boxes and conduits have come into use. Their application is somewhat specialized, and the interested reader should refer to the NEC for applicability and to manufacturers' literature for equipment details. One further clarification must be made. The outlet is the point at which current is taken off to feed the "utilization equipment" (NEC term for the electrical load apparatus), and *does not* include the item itself. In the case of a receptacle outlet where the receptacle is mounted inside the outlet box, the receptacle device is not an electrical load, but rather an extension of the box wiring. Its purpose is to permit easy connection of the load equipment. The outlet is to be considered as separate from the load device, even if it is included as part of this device. For instance, a recessed electric wall heater (see Figure 12.36) frequently comes with an outlet box permanently mounted in or on the unit. Nevertheless, the outlet is only the box itself. The drawing *must* show the final connection to the load device by symbol or note if there is any field work to be done. This work normally consists of final connection with a short piece of flexible conduit as for example, to motors, transformers, and unit heaters. (See Figure 12.6, the fourth symbol from the end and Figure 12.37, the last symbol.) If this connection is not shown a contractor might claim that the circuitry required by the drawings includes only the outlets and not the equipment connections. For these, he would then request additional payment. Refer to Figure 2.21, page 48. Note that the wiring chamber on the left-hand end of the baseboard electric heater is actually the outlet even though it constitutes part of the unit. The wiring to the heater elements is internal. No connection need be shown between the box and the heater, since no field work is necessary. See Figure 12.38. However, good installation practice would include an outlet box on the wall and a flexible connection to the heater. This permits the heater to be removed without disturbing the circuit wiring. Similarly, in the case of units containing a motor such as fan coil units, unit heaters and the like, it is good practice to terminate the branch circuit in a receptacle. The unit itself is connected with a cord and cap. This practice allows quick disconnection of the unit for service, since the receptacle acts as the required disconnect means (see the National Electrical Code) without any disturbance to the remainder of the circuit. This is particularly important in the wiring of fan coil units. These motors are very small, and the general practice is to wire three or more units on a single circuit. The additional cost of the receptacle device is minimal.

In showing ceiling lights, particularly incandescent lights (see Figure 12.37), it is customary to show only a circle, symbolizing both the outlet *and* the fixture. For this reason, ceiling outlets must be clearly shown as *including* the appropriate lighting fixture. The outlet itself, by definition, is only the wired outlet box. Similarly, when a lighting fixture

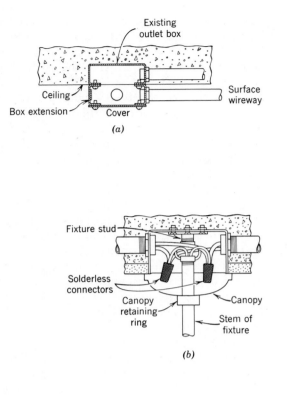

(a)

Figure 12.35 Outlet boxes and mounting arrangements.

(a) Surface extension to concealed wiring system.

(b)

(b) One method of mounting a suspended ceiling fixture.

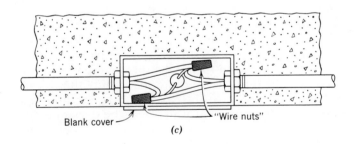

(c)

(c) Typical recessed ceiling junction box.

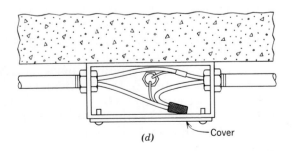

(d)

(d) Typical surface-mounted box, either ceiling or wall. Box is attached to the surface with appropriate mounting device.

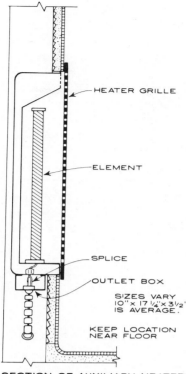

HEATER GRILLE

ELEMENT

SPLICE

OUTLET BOX

SIZES VARY
10" × 17¼" × 3½"
IS AVERAGE.

KEEP LOCATION
NEAR FLOOR

SECTION OF AUXILIARY HEATER
SCALE : 1½" = 1'–0"

Figure 12.36 Outlet box may be separate from or integral with the device being supplied. In this case the outlet is a box welded to the heater. Connection from the box to the heater is generally factory-made. (Courtesy of John Wiley & Sons.)

is built with an outlet box as part of the fixture, a separate outlet box is not shown, nor is a "pigtail" connection, since the wiring is internal. Refer to the symbol list, part II, Figure 12.37. This portion of the list is devoted to outlets. Switches and receptacles are covered in the next section and in part III of the symbol list. Your attention is called particularly to Note 1 of Figure 12.37. Only a single outlet is used in a continuous row of fluorescent fixtures. This is always the most economical method of installation. It is therefore normally true that an installation with, say, two continuous rows of 6 fixtures each, for a total of 12 fixtures, will be cheaper than an installation of 10 fixtures that are separately installed. In the first situation we have 12 fixtures and 2 outlets,

while in the second we have 10 fixtures and 10 outlets. The 8 additional outlets of the second situation will normally outweigh the added cost of 2 fixtures in the first. This type of economic consideration should be kept in mind when laying out fixtures. As a rule of thumb, a ceiling outlet separately installed costs about the same as a two-lamp fluorescent fixture or a medium quality incandescent downlight. This rule is obviously very approximate but is close enough so that in a specific case the technologist, given the choice of continuous rows or individual fixtures, will have some guidance. He can then turn to the designer or engineer for a more accurate evaluation of the case, and a final decision.

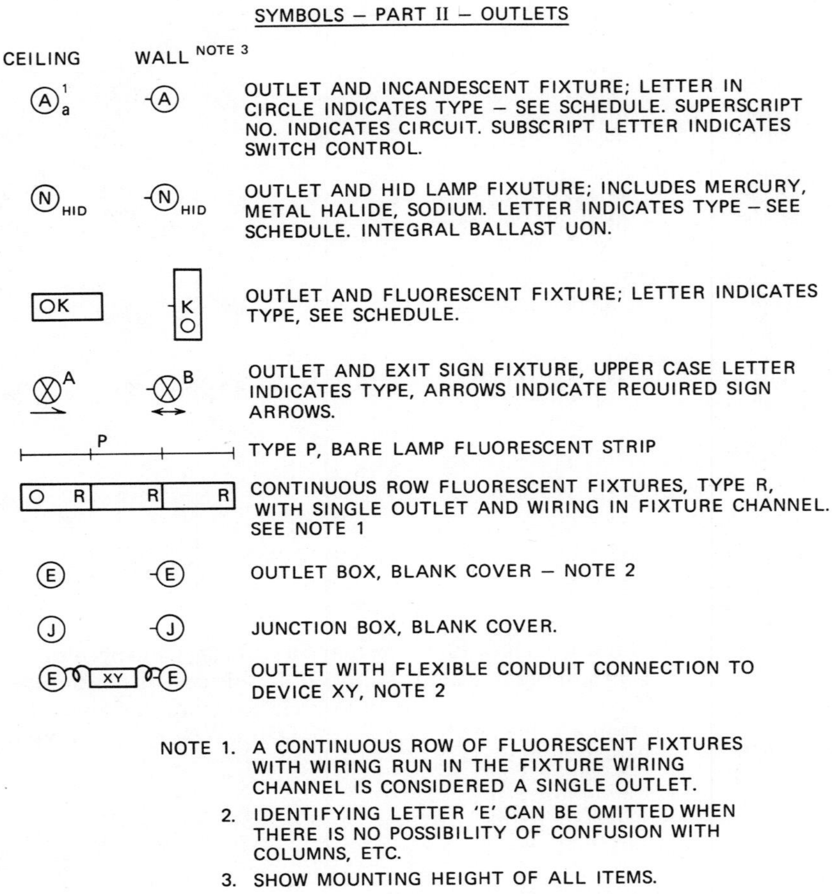

SYMBOLS — PART II — OUTLETS

CEILING   WALL NOTE 3

(A)^1_a	-(A)	OUTLET AND INCANDESCENT FIXTURE; LETTER IN CIRCLE INDICATES TYPE — SEE SCHEDULE. SUPERSCRIPT NO. INDICATES CIRCUIT. SUBSCRIPT LETTER INDICATES SWITCH CONTROL.
(N)$_{HID}$	-(N)$_{HID}$	OUTLET AND HID LAMP FIXUTURE; INCLUDES MERCURY, METAL HALIDE, SODIUM. LETTER INDICATES TYPE — SEE SCHEDULE. INTEGRAL BALLAST UON.
[OK]	[K O]	OUTLET AND FLUORESCENT FIXTURE; LETTER INDICATES TYPE, SEE SCHEDULE.
⊗A	-⊗B	OUTLET AND EXIT SIGN FIXTURE, UPPER CASE LETTER INDICATES TYPE, ARROWS INDICATE REQUIRED SIGN ARROWS.
⊢——P——⊣		TYPE P, BARE LAMP FLUORESCENT STRIP
[O R R R]		CONTINUOUS ROW FLUORESCENT FIXTURES, TYPE R, WITH SINGLE OUTLET AND WIRING IN FIXTURE CHANNEL. SEE NOTE 1
(E)	-(E)	OUTLET BOX, BLANK COVER — NOTE 2
(J)	-(J)	JUNCTION BOX, BLANK COVER.
(E)⊸[XY]⊸(E)		OUTLET WITH FLEXIBLE CONDUIT CONNECTION TO DEVICE XY, NOTE 2

NOTE 1. A CONTINUOUS ROW OF FLUORESCENT FIXTURES WITH WIRING RUN IN THE FIXTURE WIRING CHANNEL IS CONSIDERED A SINGLE OUTLET.

2. IDENTIFYING LETTER 'E' CAN BE OMITTED WHEN THERE IS NO POSSIBILITY OF CONFUSION WITH COLUMNS, ETC.

3. SHOW MOUNTING HEIGHT OF ALL ITEMS.

Figure 12.37 Architectural–electrical plan, symbol list, Part II, outlets.

For other portions of the symbol list see:
Part I, Raceways, Figure 12.6, page 304.
Part III, Wiring Devices, Figure 12.39, page 341.
Part IV, Abbreviations, Figure 12.43, page 345.
Part V, Single Line Diagrams, Figure 13.4, page 354.
Part VI, Equipment, Figure 13.14, page 369.
Part VII, Signaling Devices, Figure 15.34, page 500.
Part VIII, Motors and Motor Control, Figure 16.19, page 527.
Part IX, Control and Wiring Diagrams, Figure 16.22, page 530.

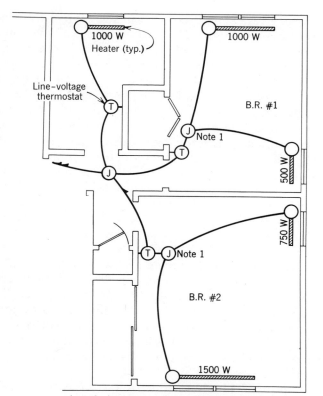

Figure 12.38 Electric heating for the Basic Plan House. Note use of junction and outlet boxes. Absence of pigtail wiring symbol between outlet box and heater indicates internal, factory connection.

12.10. Receptacles and Other Wiring Devices

The only outlet with a self-contained device is the receptacle. This usually takes the form of the common wall outlet or, as will be illustrated, much larger and more complex devices. We must comment here about terms. The common wall outlet is properly called a *convenience outlet*. The term "wall plug," which is heard so often, is really incorrect. A plug is another name for the *attachment cap* on the wire coming from a device such as a lamp or appliance. It is plugged into the wall; hence, the name "plug." However, since the term wall plug is used so often, it has been accepted in trade circles. The technologist should try to follow proper terminology.

Receptacles belong in the general classification of *wiring devices*, which includes not only all receptacles but also their matching caps (plugs), wall switches, small dimmers and outlet box mounted pilot lights. Most people restrict the term *wiring device* to any device that will fit into a 4-in.-square box, thus excluding devices larger than 30 amperes. Switches, too, are limited to 30 amperes, single pole, 220 v in this category; above that size they are separately mounted and are called disconnect switches. Essentially, wiring devices are those found in common use in lighting and receptacle circuits, which normally do not exceed 30 amperes in rating. The NEC defines a receptacle as "a contact device installed at the outlet, for the connection of a single attachment plug." Since the normal wall receptacle will take two attachment plugs, it is referred to as a duplex receptacle. However, at a single outlet more than one receptacle can be installed. Just as a row of fluorescent fixtures wired together is considered one outlet (see Note 1, Figure 12.37), so any number of receptacles mounted together in one or more coupled boxes is considered one outlet. Thus three receptacles so mounted would be a three-gang outlet. This is extremely important when one is circuiting under conditions limiting either (or both) the number of outlets or the number of receptacles on a circuit. Furthermore, as previously stated, the lower the number of outlets, the lower the cost of installation. A circuit with six duplex receptacles individually mounted is normally more than twice the cost of the same six receptacles installed in two outlet groups of three gangs each.

Receptacles are described and identified by *poles* and *wires*. The number of poles equals the number of hot contacts, thus *exluding* the grounding pole. The number of wires includes all connections to the receptacle, *including* the green ground wire. This system of description is awkward and often confusing, since a normal three-slot (two hot plus "U" ground) wall receptacle is officially a 2-pole, 3-wire device even though it has three slots and is often connected with only two wires, with the third connection—the green ground—being connected to the conduit system. For this reason, we strongly suggest that the drawing's receptacle symbol list use NEMA (National Electrical Manufacturer's Association) designations *and* a graphic representation. This is the system we follow to avoid the almost certain confusion of attempting a description by poles and wires.

Receptacles are available in ratings of 10 to 400 amperes, 2 to 4 poles and 125 to 600 v. The typical wall convenience receptacle is 2 pole, 3 wire, 125 v, 15 or 20 amperes. The quality or *grade* of the unit is specified in the job specifications and can be economy (cheap), standard (good) or specification (excellent).

In preparing drawings, many draftsmen insert a different symbol for each type of receptacle. This is acceptable practice for a job with up to three types, for example, a small residence. Most nonresidential jobs, however, have four or more receptacle types, and the drawings become confusing. For this reason we strongly recommend using one standard symbol for all receptacles other than the common wall convenience outlet, and then including a schedule of all these special items on the drawings. This will avoid confusion, save drafting time, allow standardization from job to job and show a systematic professional approach. See the symbol list, part III, Figure 12.39. The receptacle lists, Figures 12.40 and 12.41, give the physical configurations and NEMA designa-

tions. These should be used in the table of special receptacles. Figure 12.42 shows some of the common types of receptacles. Mounting height must always be specified with all wall devices, as well as configuration. Wall convenience receptacles are normally mounted vertically, between 12 and 18 in. AFF (above finished floor). Some architects prefer horizontal mounting, and if this is desired it must be so specified, preferably on the drawings, along with the mounting height. In industrial areas, shops, workrooms and the like, where the receptacle outlets must be accessible above tables, mounting height is 42 in., and horizontal mounting is preferable so that cords do not hang on top of each other. This, too, must clearly be specified on the drawings. See Figure 12.32.

A newcomer to the receptacle field is a device with built-in ground fault protection. This type of receptacle is indicated as a special type and is generally known as a GFI or GFCI (ground fault circuit interrupter) receptacle. It is used in locations where sensitivity to electric shock is high, such as in wet areas. Further details on this device and its application are given in Section 13.3.

Figure 12.43 is a continuation of the symbol list— that is, Part IV, and lists abbreviations commonly used in architectural-electrical work. The draftsman should use only widely accepted abbreviations to avoid possible misunderstanding.

A word of caution here is important with respect to 2 specific receptacle outlet types. Clock hanger outlets and fan outlets are outlets only, *without* clocks or fans. They are manufactured to support a clock or fan and to supply power from the built-in single receptacle. If a clock or fan is desired with the outlet, it must clearly be called for. See Figure 12.42h.

The second major type of wiring device is the wall switch. See Figure 12.39 for a listing and Figure 12.44 for a drawing of a few common types. Switches serve to open and close the electric circuits. The NEC defines a number of switch types such as general-use, and snap switch, but these definitions need not concern the technologist. It is sufficient to know that a switch is a one-gang size wiring device, available in ratings from 15 to 30 amperes, generally 1 or 2 pole and single or double throw. From outward appearance it is frequently difficult to distinguish between different units, since they are generally made to fit, as mentioned, a single-gang box. One easily recognized special type of switch is the low-voltage control switch (see symbol list, Figure 12.39) which is illustrated in Figure 15.14, page 471.

SYMBOLS — PART III — WIRING DEVICES

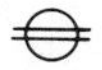

⊖ DUPLEX CONVENIENCE RECEPTACLE OUTLET 15 AMP[1] 2 P 3 W 125 VOLT, GROUNDING, WALL MTD.[2], VERTICAL, ₵ 12″ AFF.

◖A SPECIAL RECEPTACLE, LETTER DESIGNATES TYPE, SEE SCHED. DWG. NO.____ WALL MOUNTED.

🕐 CLOCK HANGER OUTLET, SEE SPEC. MTG. HT. 9′ AFF.[3]

🕐I CLOCK HANGER OUTLET WITH TYPE I CLOCK, SEE SPEC., 9′ AFF, DTL. DWG. NO.____

(F) FAN HANGER OUTLET, 8′6″ AFF.

⊙B FLOOR OUTLET TYPE B[4] SEE DWG. NO.____

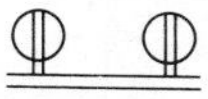

⊕ ⊕ MULTI—OUTLET ASSEMBLY[4] SEE DWG. NO.____ FOR SCHEDULE AND DETAILS (SEE SPEC.)

S_a SINGLE POLE SWITCH, 15 A[1] 125 V, 50″ AFF[3] UON. SUBSCRIPT LETTER INDICATES OUTLETS CONTROLLED.

S_L SWITCH, LOW VOLTAGE SWITCHING SYSTEM.

$_aS_3$ SWITCH, 3 WAY, 15A 125 V, SEE SPEC.[6]; CONTROLLING OUTLETS 'a' [5]

S_{DP} SWITCH, DOUBLE POLE, 15 A 125 V.

S_4 SWITCH, 4 WAY, 15 A 125 V.

S_K SWITCH, KEY OPERATED, 15A 125V.

S_D DOOR SWITCH, SEE SPEC. FOR RATING AND TYPE.

↰⊘ SWITCH/RECEPTACLE COMBINATION IN 2 GANG BOX.

S_P SWITCH, SP 15A 125V. WITH PILOT LIGHT.

S_{SA} SWITCH, SPECIAL PURPOSE, TYPE A, SEE SPEC.; SEE DWG. NO.____

———— ABBREVIATIONS RELEVANT TO SWITCHES:
SP   — SINGLE POLE
DP   — DOUBLE POLE
SPDT — SINGLE POLE DOUBLE THROW
DPDT — DOUBLE POLE DOUBLE THROW
RC   — REMOTE CONTROL

S_{WP} SWITCH, WEATHER PROOF ENCLOSURE, SEE SPEC.

S_{MC} SWITCH, MOMENTARY CONTACT.

Ⓜ A SWITCH, REMOTE CONTROL, TYPE A, MECHANICALLY HELD. SEE SPEC.

Ⓔ A SWITCH, REMOTE CONTROL, TYPE A, ELECTRICALLY HELD. SEE SPEC.

[R] OUTLET—BOX—MOUNTED RELAY

[D] OUTLET—BOX—MOUNTED DIMMER.

[S/D] OUTLET—BOX—MOUNTED SWITCH AND DIMMER.

⊖GFCI RECEPTACLE, RATED AS SHOWN, EQUIPPED WITH GROUND FAULT CIRCUIT INTERRUPTER.

NOTE 1. SPECIFY 20 AMP IF DESIRED.
     2. ALL RECEPTACLES ARE WALL MOUNTED UON
     3. ALL MOUNTING HEIGHTS ARE TO OUTLET ₵; SPECIFY MH. OF EACH OUTLET.
     4. ALSO SHOWN IN SYMBOLS, PT. I, FOR COMPLETENESS, USING ALTERNATE SYMBOL.
     5. REFER TO SPECIFICATIONS FOR DATA ON SWITCHES.

Figure 12.39 Architectural–electrical plan symbol list, Part III, wiring devices.

For other portions of the symbol list see:
Part I, Raceways, Figure 12.6, page 304.
Part II, Outlets, Figure 12.37, page 338.
Part IV, Abbreviations, Figure 12.43, page 345.

Part V, Single Line Diagrams, Figure 13.4, page 354.
Part VI, Equipment, Figure 13.14, page 369.
Part VII, Signaling Devices, Figure 15.34 page 500.
Part VIII, Motors and Motor Control, Figure 16.19, page 527.
Part IX, Control and Wiring Diagrams, Figure 16.22, page 530.

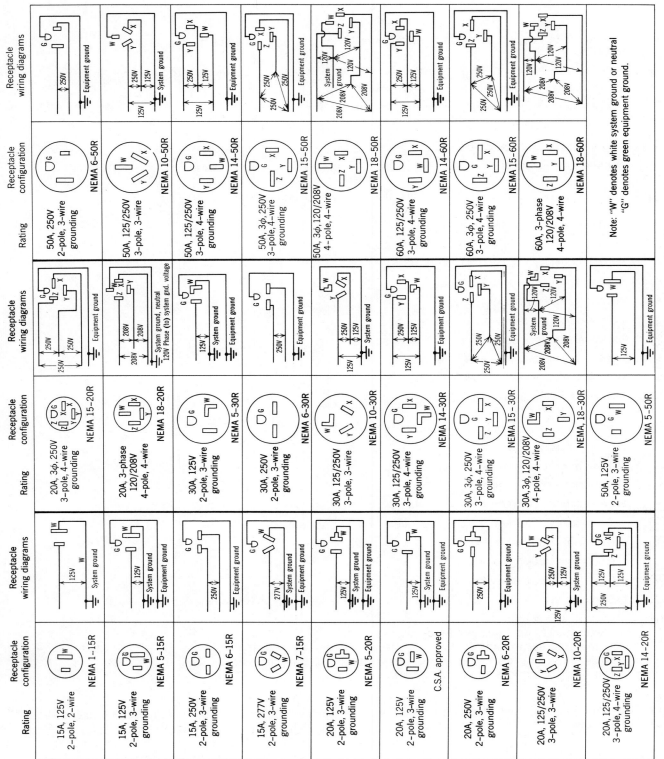

Figure 12.40 Receptacle configuration chart of general purpose, nonlocking devices, with applicable NEMA configuration numbers. (Courtesy of Pass and Seymour Inc.)

Rating	Configuration & NEMA No	Rating	Configuration & NEMA No	Rating	Configuration & NEMA No	Rating	Configuration & NEMA No
15A, 125V 2-pole, 2-wire	L1-15R	30A, 125V 2-pole, 3-wire grounding	L5-30R	30A 3φ, 480V 3-pole, 3-wire	L12-30R	20A, 3φ, Y 277/480V 4-pole, 4-wire	L19-20R
15A, 250V 2-pole, 2-wire	L2-20R	30A, 250V 2-pole, 3-wire grounding	L6-30R	30A 3φ, 600V 3-pole, 3-wire	L13-30R	20A, 3φ, Y 347/600V 4-pole, 4-wire	L20-20R
15A, 125φ, Y 2-pole, 3-wire grounding	L5-15R	30A, 277V AC 2-pole, 3-wire grounding	L7-30R	20A 125/250V 3-pole, 4-wire grounding	L14-20R	30A, 3φ, Y 120/208V 4-pole, 4-wire	L18-30R
15A, 250V 2-pole, 3-wire grounding	L6-15R	30A, 480V 2-pole, 3-wire grounding	L8-30R	20A 3φ, 250V 3-pole, 4-wire grounding	L15-20R	30A, 3φ, Y 277/480V 4-pole, 4-wire	L19-30R
15A, 277V, AC 2-pole, 3-wire grounding	L7-15R	30A, 600V 2-pole, 3-wire grounding	L9-30R	20A 3φ 480V 3-pole, 4-wire grounding	L16-20R	30A, 3φ, Y 347/600V 4-pole, 4-wire	L20-30R
20A, 125V 2-pole, 3-wire grounding	L5-20R	20A 125/250V 3-pole, 3-wire	L10-20R	30A 125/250V 3-pole, 4-wire grounding	L14-30R	20A, 3φ, Y 120/208V 3-pole, 5-wire grounding	L21-20R
20A, 250V 2-pole, 3-wire grounding	L6-20R	20A 3φ 250V 3-pole 3-wire	L11-20R	30A 3φ 250V 3-pole, 4-wire grounding	L15-30R	20A, 3φ, Y 277/480V 4-pole, 5-wire grounding	L22-20R
20A, 277V, AC 2-pole, 3-wire grounding	L7-20R	20A 3φ 480V 3-pole, 3-wire	L12-20R	30A 3φ 480V 3-pole, 4-wire grounding	L16-30R	20A, 3φ, Y 347/600V 4-pole, 5-wire grounding	L23-20R
20A, 480V 2-pole, 3-wire grounding	L8-20R	30A 125/250V 3-pole, 3-wire	L10-30R	30A 3φ, 600V 3-pole, 4-wire grounding	L17-30R	30A, 3φ, Y 120/208V 4-pole, 5-wire grounding	L21-30R
20A, 600V 2-pole, 3-wire grounding	L9-20R	30A 3φ 250V 3-pole, 3-wire	L11-30R	20A, 3φ, Y 120/208V 4-pole, 4-wire	L18-20R	30A, 3φ, Y 277/480V 4-pole, 5-wire grounding	L22-30R
						30A, 3φ, Y 347/600V 4-pole, 5-wire grounding	L23-30R

Figure 12.41 Receptacle configuration chart locking devices showing NEMA configuration numbers. (Courtesy of Pass and Seymour Inc.)

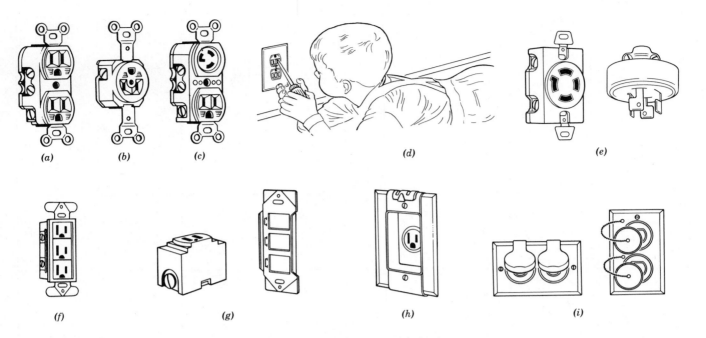

Figure 12.42 Receptacle and cap types. (a) 2-pole 3-wire 15 amp, 125 v, duplex grounding type. (b) The same as (a) except single. (c) The same as (a) except one normal and one locking device. (d) The same as (a) except safety type. (e) 3-pole, 4-wire locking receptacle, 20 amp, 125/250 v, with matching cap. (f) Triplex 15 amp grounding receptacle. (g) Miniature (interchangeable) device with mounting strap. (h) Clock outlet with hanger plate. (i) Outdoor weatherproof receptacles.

Symbols—Part IV—Abbreviations

A,a	Amperes	N	Neutral
AFF	Above finished floor	NC	Normally closed
C	Conduit	NO	Normally open
C/B	Circuit breaker	NL	Night light
CCT	Circuit	NIC	Not in contract
DF	Drinking fountain	OH	Overhead
DN	Down	OL	Overload relay
EWC	Electric water cooler	OC	On center
EM	Emergency	PB	Push-button,
EL	Elevation		Pull-box
EC	Empty conduit	PC	Pull chain
F	Fuse	PL	Pilot light
FA	Fire alarm	RHC	Reheat coil
F-3	Fan No. 3	SW	Switch
FC	Fan coil unit	TC	Telephone cabinet
GND	Ground	T	Thermostat,
GFCI	Ground fault		Transformer
	cct-interrupter	TEL	Telephone
GFI	Ground fault	TV	Television
	interrupter	TYP	Typical
HOA	Hand-off-automatic	UON	Unless otherwise noted.
	selector switch	UF	Unfused
HP	Horsepower	UG	Underground
L	Line	UH	Unit heater
LTG	Lighting	VP	Vaporproof
MCC	Motor control center	WP	Weatherproof
MH	Mounting height,	XP	Explosion proof
	Manhole	XFMR	Transformer

Figure 12.43 Architectural–electrical plan symbol list, Part IV, list of common abbreviations.

For other portions of the symbol list see:
Part I, Raceways, Figure 12.6, page 304.
Part II, Outlets, Figure 12.37, page 338.
Part III, Wiring Devices, Figure 12.39, page 341.
Part V, Single Line Diagrams, Figure 13.4, page 354.
Part VI, Equipment, Figure 13.14, page 369.
Part VII, Signaling Devices, Figure 15.34, page 500.
Part VIII, Motors and Motor Control, Figure 16.19, page 527.
Part IX, Control and Wiring Diagrams, Figure 16.22, page 530.

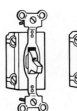

Toggle / Key
Specification Grade switches
15, 20 and 30 ampere

Momentary contact switch
15 and 20 ampere
(Double throw, center off)

Maintained contact switch
15 and 20 ampere

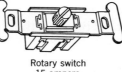

Rotary switch
15 ampere
120 V

Tap–plate switch
15, 20 ampere
120, 277 V

Special actuator plate

1 to 4 gang

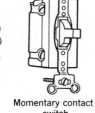

Three fit one strap in standard boxes

Screw terminals / Screwless terminals

Specification Grade Interchangable switches
15 and 20 ampere

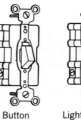
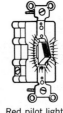

Button

Lighted button in "Off" position

Red pilot light (button lights in "On" position)

Press switches
15 and 20 ampere

Figure 12.44 Typical branch circuit switches.

The final feature of outlets that we must discuss is the device cover. On a blank outlet or a junction box, we use a plain metal cover. With switches and receptacles, the cover is made to match the wiring device and, of course, to close the outlet box. Covers are available in one or more gangs and in various materials. These should be specifically listed on drawings or in specifications. Floor outlets are mounted in special floor boxes and must be shown as a specific receptacle in a specific type of floor box. Both must be specified.

Problems

12.1. For the sleeping area of the house in which you live, lay out the rooms with all outlets, and then lay out the interconnecting raceways as if the wiring method were
(a) BX.
(b) Romex.
(c) Plastic Conduit.

12.2. Do a material take-off of the raceway systems of Problem 12.1, counting outlets, conduit, feet of cable, and the like (refer to electrical supply catalogs for fittings).

12.3. Draw the living and kitchen area of your home showing the outlets, but use surface raceways throughout. Refer to the manufacturer's catalog (we suggest Wiremold Co). Show any installation details that you think the contractor will need.

12.4. Repeat Problem 12.3 for the classroom or lecture room you are now using.

12.5. Assume that each student in your class requires a source of power and communication at his seat. (This situation may become a real problem if access to library by computer link is required.) How would you wire up the room? Draw a layout.

12.6. List four actual places where surface raceway would be preferable to concealed raceway and four more where the reverse is true.

12.7. Using Figure 12.40, make up a table of special receptacles for your school, as you would place it on the drawings.

12.8. Using Table 12.1, show how the area in circular mils is derived from the diameter of these wire sizes: No. 10, No. 1/0, 500 MCM. Is the diameter of solid or stranded wire?

12.9. (a) Using Table 12.2, list the ampacity of each of the following wires. (Remember that this table is for up to three wires, in a raceway.)
No. 12 TW, No. 12 THW, No. 12 THHN
No. 10 RHW, No. 2 THWN, No. 4/0 XHHW
250 MCM T, 250 MCM THHN
500 MCM TW, 500 MCM THW, 500 MCM XHHW
Check these values in the corresponding table in the NEC (1975 edition, Table 310-16).
(b) Adjacent to each ampacity in (a) above, place the ampacity for the same wire in *free air*. Use the appropriate NEC table (1975 edition, Table 310-17).

12.10. The NEC provides (Chapter 9, Table 1) that four or more conductors in a conduit may occupy 40% of the conduit's cross-sectional area. Using Tables 12.4 and NEC Table 4, Chapter 9, find the conduit size required for the following two groups of single conductors
(a) 6-No. 2 AWG Type THW plus 4-No. 1/0 AWG Type T.
(b) 3-No. 4/0 AWG Type THWN, 3-250 MCM Type XHHW.
and 1-No. 4 Type T equipment ground wire, 600 v insulation.

12.11. Using Table 12.7, lay out with all dimensions the end of a pull box that is receiving two layers of conduit as follows:
Top layer 4–3 in., 2–2 in., 2–1½ in.
Bottom layer 10–1½ in. conduits
Use ⅜-in. minimum locknut clearance instead of the ¼ in. shown in the table for conduits in the same layer, and 1 in. minimum locknut clearance between layers.

Additional Reading

National Electrical Code, source for all rules governing electric construction work. The Code as it is interpreted and applied in the field is the final authority on whether or not an installation is proper.

Steel Electric Raceways—Design Manual, the American Iron and Steel Institute. This soft-cover booklet gives explanatory and well illustrated examples of Code rules, and explains basic design considerations. In addition, it devotes much space to special occupancies such as hazardous areas and residential wiring.

American Electricians' Handbook, T. Croft and C. C. Carr, McGraw-Hill. An excellent source book on all facets of electrical wiring. Well illustrated.

Architectural Graphic Standards, American Institute of Architects and G. G. Ramsey and H. R. Sleeper, Sixth Edition, John Wiley. This architect's "bible" provides, in its electrical section, handy physical dimensional data and detailed drawings of many electrical items.

Mechanical and Electrical Equipment for Buildings, William J. McGuinness and Benjamin Stein, John Wiley. This book contains the same type of material as in the present book, except in greater detail. It is useful for deeper level study in all areas.

Electrical Conduit Reference Manual, Youngstown Steel Co. Contains technical material, tables, dimensional data and manufacturing information on conduit. It also provides much other technical material (some of which may be out of date).

13. Building Electric Circuits

In the preceding chapter we discuss in detail the branch circuit and its components and describe several usual applications. We emphasize that, by NEC definition, the branch circuit comprises only the wiring between the circuit protective device and the outlets. We then extended our study to cover outlets, including receptacles and other wiring devices. The third basic circuit element—the circuit protective device—is considered in this chapter where we examine it in detail. After discussing this third, and last, basic circuit element, our study expands to cover related areas. These areas are branch circuit criteria, types of branch circuits, basic motor circuits as used in residential building design and drawing application of these concepts and techniques. Finally, we study the next stage, which is the assembly of branch circuits into a complete wiring system. This latter includes the layout of actual buildings complete with circuitry, panel scheduling, switching and basic load study. Applications in this chapter will be to the same Basic Plan that you have studied previously in the HVAC sections of this book.

After studying this chapter, you will be able to:

1. Understand the functioning of overcurrent devices.
2. Determine where in a circuit to place the overcurrent devices, and what type to use.
3. Recommend the type of overcurrent device required for the circuit being considered.
4. Specify fuses and circuit breakers according to size and type.
5. Understand the use of grounding electrodes, grounding conductors and ground fault circuit interrupter (GFCI) devices.
6. Draw single and 3-phase panelboard schedules.
7. Lay out and draw a residential electrical plan, starting from the architect's plan.
8. Circuit a complete electric plan, including all wiring, devices and loads, and prepare a complete panel layout.

9. Thoroughly understand the principles and guidelines for residential layout and wiring.
10. Apply these layout and circuitry skills to any type of building.

13.1. Circuit Protection

Referring to Figure 12.2, page 302, we see that if we consider the circuit protective device from the branch circuit point of view, it represents the source of voltage. In reality, the overcurrent device *follows* the source of voltage in the circuit. However, since the overcurrent (o/c) device is always connected at its hot (line) end to the voltage source, and at its load end to the circuit wiring, it becomes the apparent source of voltage. When we study panelboards, of which overcurrent devices are a part, we shall learn that the busbars become the source of voltage as we look "upstream" from the overcurrent devices. The apparent voltage source depends on where in the electric network we happen to be looking at the moment. Glance back now at Figure 11.25, page 293. Note that at each stage of the system the preceding stage (upstream) is the source of voltage or power, whereas in reality, the only actual source of power is the generator at the beginning of the line. This same relationship in branch circuit form is shown in Figure 13.1. As you will remember, the NEC defines the branch circuit as that portion of the circuit beyond the overcurrent device. Therefore, in examining overcurrent devices, we are leaving the branch circuit and moving upstream one step in the electric network. What, then, is an overcurrent de-

vice? What is its function, appearance and method of operation? How is it shown and how much space does it occupy? These are the questions that we will now answer.

The alert reader will already have noticed that there are two principal, and different, causes of overcurrent. One is an overload in the equipment. The second is a fault of some sort, frequently a short circuit or an accidental ground. Both of these conditions result in excessive current flowing in the circuit. The overcurrent device protects both the branch circuit *and* the load device against this excess current.

a. Branch Circuit Protection

In our studies in basic electricity and materials in Chapters 11 and 12, we learn that the current in a circuit produces heating in the circuit conductors, known as line I^2R loss. We also learn that circuit wiring can safely handle a specific amount of current, and that this amount of current, expressed in amperes, is called the ampacity of the conductor. If this ampacity is exceeded, the wire overheats and a fire may result. The overcurrent device prevents this from happening. It does not matter whether the excess current is being caused by an equipment problem such as an overload, or by a circuit problem, such as an unintentional ground. The overcurrent device "sees" only excess current—and interrupts it. We will have more to say on this topic later on when we discuss fuses and circuit breakers in detail.

b. Equipment Protection

Consider, for instance, a machine tool that is drawing excess current due to overloading or other cause. This excess current in turn causes excess internal heating. If this is not relieved, permanent damage can easily result. Such damage is prevented by the action of the overcurrent device.

As its name suggests, the overcurrent device prevents an overcurrent from continuing by the simple method of opening the circuit when it senses excessive current. It, therefore, acts in the same manner as a safety valve in a mechanical system.

There, the valve opens to relieve excess pressure just as here, the overcurrent device opens to relieve excess current. In both instances, the action prevents an overload condition from becoming dangerous. But there the similarity ends. The overcurrent device opens the circuits much as a switch would,

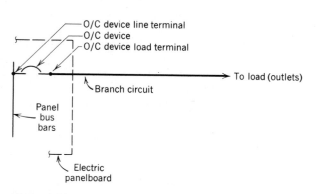

Figure 13.1 Typical electric circuit. Note that the branch circuit, which extends from the overcurrent device load terminal to the outlets, "sees" the overcurrent device as its power source.

stopping current flow entirely and deactivating the circuit. With a valve, the excess pressure is relieved by venting or bypassing, but the system is normally not shut down by the valve action. The difference in action is illustrated in Figure 13.2, where a fused circuit is shown in the three stages of operation: prior to overload, during the protective action, and after clearing the overcurrent. The action of an overcurrent device is called clearing, since it clears the circuit of the overload or the fault.

Notice that the overcurrent device is always "upstream" of the equipment being protected, that is, electrically *ahead* of the load. This is obviously where it belongs, since current flows "downstream." Therefore, to cut off excess current, the overcurrent device must be placed ahead, in an electrical sense, of the protected device. In the case of branch circuits, the overcurrent device is in the electric panel that supplies the branch circuits. The panel is the source of current. There, current must be cut off to protect the branch circuits downstream from the panel. The upstream side of any device is called the *line* side. The downstream side is called the *load* side. In the case of a switch, circuit breaker, or any other circuit interrupting device, the line side

remains hot after it is opened, but the load side is dead, or deenergized. Figure 13.3 illustrates the location of overcurrent devices. Observe from this diagram that the line side of a switch remains hot (live, energized) even with the switch open. When we have two disconnect devices in series, the load side of the upstream one is the same as the line side of the downstream one. If this seems confusing, refer again to Figure 13.3. The busbars in the main switchboard are hot from connection to the electric service wiring. The line terminals of the fused switches are, therefore, hot. Note that a switch is always connected that way—and never with the blade hot—for safety reasons. Moving downstream, we come to the lighting panel. Here the feeder supplies the panel busbars. The circuit breakers are connected with their line side to the panel and their load side facing downstream, and so on. By drawing convention, a switch is shown open. (A closed switch would simply look like a straight line.) Circuit breakers and fuses alone are shown closed, since no drawing convention exists to show an open circuit breaker (c/b). The open fuse in the third diagram of Figure 13.2 is drawn that way to illustrate a point. In actual practice, a blown fuse is never shown, either

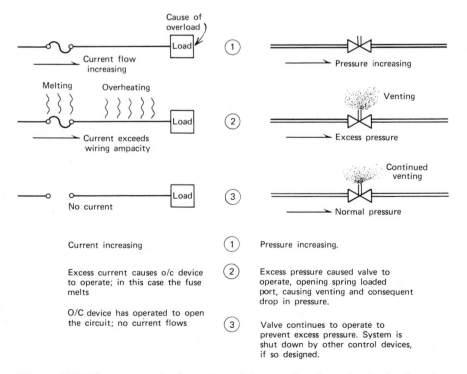

Figure 13.2 Three stages in the action of electrical and mechanical safety devices.

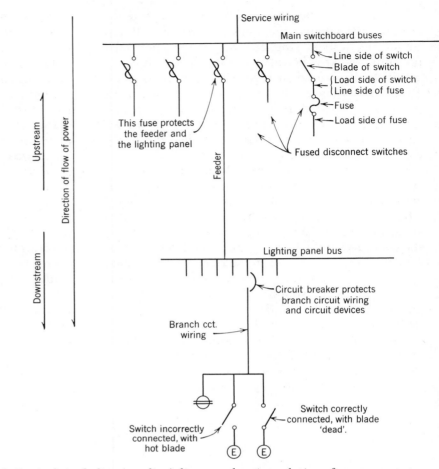

Figure 13.3 Typical single-line (one-line) diagram showing relation of components to each other, proper location of overcurrent devices, and terminology in general use.

as a break in the circuit or in any other way. This is because circuits are always shown in *normal* condition, and a blown fuse or open circuit breaker represents an *abnormal* condition. When a fuse is shown in conjunction with a switch, it is usually shown as in Figure 13.3. Refer to Figure 13.4 for the symbols involved.

13.2. Fuses and Circuit Breakers

The simplest and most common circuit protective device in use is the fuse. Although in modern technology fuses are available in literally hundreds of designs, ratings and shapes, they are all basically the same. A fuse consists of two terminals with a piece of specially designed metal, call the fusible element, in between. Power flows into the line terminal, through the fusible element and out the load terminal. When the current flow is excessive, the element melts due to excessive I^2R loss in it, and the fuse opens the circuit. As simple as that. The technical aspects of fuses—ratings, clearing time and interrupting capacity are the concern of the design engineers. As far as the electrical draftsman is concerned, the fuse is an overcurrent element in a panel, a switch or even in an individual box or cabinet, to which the circuit to be protected is connected. Several types of fuses are shown in Figure 13.5. Fuse representation in electric plans and diagrams (Section 13.1) is shown in Figures 13.3 and 13.4 Since the fuse element melts out when the fuse operates, it is obviously a one-time device and must be replaced after it clears a fault. This melting, of course, refers only to the fusible element in the fuse, and not to the whole fuse. In *one-time*, or *nonrenewable* fuses, such as are shown in Figure 13.5, the fusible element is not replaceable. The entire fuse must be discarded after operation since it has become useless. With renewable fuses, the melted fusible element can be removed and replaced with a new one. This type of fuse is not in general use because it is easy to replace the element with another of a different rating, thus defeating the purpose and usefulness of the fuse. For this reason, most of the fuses in use are nonrenewable. In part, because of this one-time operation characteristic, and the replacement and inventory problems it causes, the circuit breaker was developed.

The circuit breaker is a device that combines the functions of an overcurrent circuit protective device with that of a switch. Basically it is a switch equipped with a tripping mechanism that is activated by excessive current. An understanding of how this occurs can be gained from studying Figure 13.6. The important fact to remember is that circuit breakers, like fuses, react to excess current. The circuit breaker, however, unlike the fuse, is not self-destructive. After opening to clear (disconnect) the fault, it can be simply reclosed and reset. This means that replacement problems are eliminated. Furthermore, a circuit breaker can be manually tripped so that, in many cases, it also acts as the circuit switch.

Consider for instance, a residential circuit that feeds a kitchen electric range, wall oven or other stationary appliance. The NEC requires a disconnecting device within easy reach of the unit. This means that if the house panel is mounted in or near the kitchen, the circuit breaker in the panel may be used as the disconnect *and* the circuit protective device, saving the cost of switch or heavy receptacle. With fuse protection of the circuit this is obviously not possible and a separate disconnect is required. Such a disconnect frequently takes the form of a switch mounted adjacent to the fuse in what is known as a *switch and fuse* panel. See Figure 13.13b. Panels of this kind were once very common, but the majority of panels installed today are of the circuit breaker type.

Several other advantages of the circuit breaker as compared with the fuse are of interest to the technologist. Fuses are single pole devices; they are put into a single wire and can protect only a single electric line. Circuit breakers, on the other hand, can be multipole. This means that a single circuit breaker can be built with 1, 2, or 3 poles to simultaneously protect and switch one, two or three electric lines. An overcurrent in any line causes the circuit breaker to operate and disconnect all the lines controlled by that circuit breaker. This has an obvious advantage when protecting circuits with two or three hot legs, such as in 208 v single-phase or 208 v 3-phase lines. Another advantage of the circuit breaker over the fuse is that it is readily tripped from a remote location—a very useful control function that is impossible with fuses and quite difficult with a fused switch. A further advantage of circuit breakers over fuses is that their position—closed, trip, open—shows at the handle. A blown fuse is not easily recognized, since the melted element is inside the fuse casing. As a result of these and other features, the circuit breaker has taken precedence over the fuse for most residential and commercial use.

The single great advantage of the fuse is its stabil-

SYMBOLS — PART V — (ONE LINE DIAGRAMS)

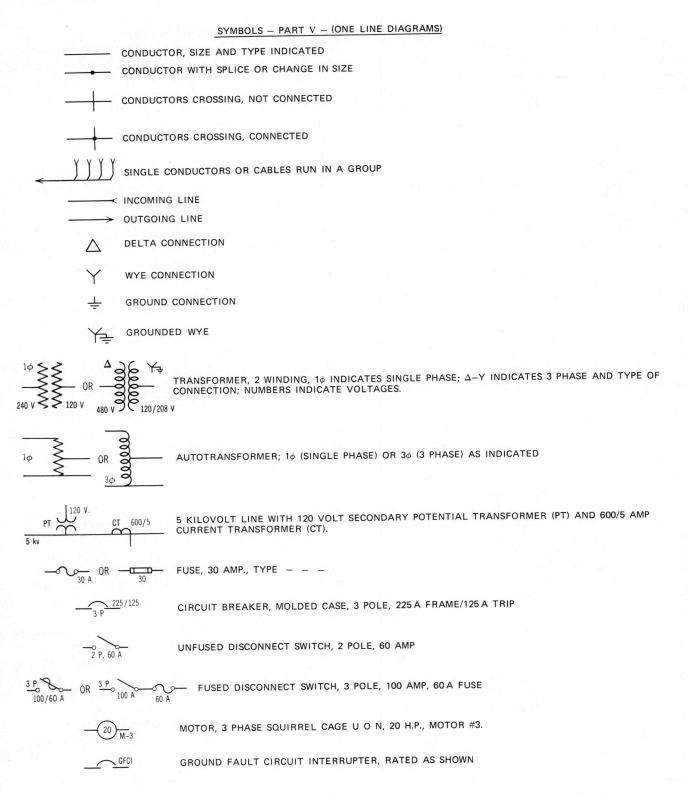

CONDUCTOR, SIZE AND TYPE INDICATED

CONDUCTOR WITH SPLICE OR CHANGE IN SIZE

CONDUCTORS CROSSING, NOT CONNECTED

CONDUCTORS CROSSING, CONNECTED

SINGLE CONDUCTORS OR CABLES RUN IN A GROUP

INCOMING LINE

OUTGOING LINE

DELTA CONNECTION

WYE CONNECTION

GROUND CONNECTION

GROUNDED WYE

TRANSFORMER, 2 WINDING, 1ϕ INDICATES SINGLE PHASE; Δ–Y INDICATES 3 PHASE AND TYPE OF CONNECTION; NUMBERS INDICATE VOLTAGES.

AUTOTRANSFORMER; 1ϕ (SINGLE PHASE) OR 3ϕ (3 PHASE) AS INDICATED

5 KILOVOLT LINE WITH 120 VOLT SECONDARY POTENTIAL TRANSFORMER (PT) AND 600/5 AMP CURRENT TRANSFORMER (CT).

FUSE, 30 AMP., TYPE — — —

CIRCUIT BREAKER, MOLDED CASE, 3 POLE, 225 A FRAME/125 A TRIP

UNFUSED DISCONNECT SWITCH, 2 POLE, 60 AMP

FUSED DISCONNECT SWITCH, 3 POLE, 100 AMP, 60 A FUSE

MOTOR, 3 PHASE SQUIRREL CAGE U O N, 20 H.P., MOTOR #3.

GROUND FAULT CIRCUIT INTERRUPTER, RATED AS SHOWN

Figure 13.4 (Facing page) Electrical drawing symbol list, Part V. Single-line diagram symbols in common use.

Part I, Raceways, Figure 12.6, page 304.
Part II, Outlets, Figure 12.37, page 338.
Part III, Wiring Devices, Figure 12.39, page 341.
Part IV, Abbreviations, Figure 12.43, page 345.
Part VI, Equipment, Figure 13.14, page 369.
Part VII, Signalling Devices, Figure 15.34, page 500.
Part VIII, Motors and Motor Control, Figure 16.19, page 527.
Part IX, Control and Wiring Diagrams, Figure 16.22, page 530.

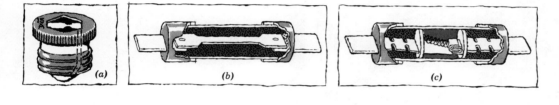

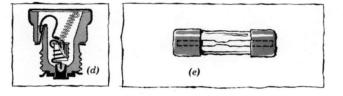

Figure 13.5 Standard types of fuses are *(a)* nonrenewable plug fuse, *(b)* nonrenewable knife-blade fuse, *(c)* nonrenewable dual-element time-delay ferrule cartridge fuse, *(d)* Buss Fustat time-delay plug fuse, *(e)* nonrenewable miniature fuse (electronic and instrument applications). Since fuses are inherently very fast-acting devices, time delay must be built into a fuse to prevent "blowing" on short-time overloads such as those caused by motor starting. A dual-element fuse such as type *(c)* or *(d)* allows the heat generated by temporary overloads to be dissipated in the large center metal element, preventing fuse blowing. If the overload reaches dangerous proportions the metal will melt, releasing the spring and opening the circuit. The time required to clear (blow) a fuse is generally inversely proportional to the amount of current.

Figure 13.6 (Facing page) 1. **Frame sizes.** Molded case circuit breakers are made in a few frame sizes that cover the ampere range from 0 to 2500 amperes. See chapter text and manufacturers' data for details of ratings. Illustrated unit is available to 1200 amperes.

2. **Molded case.** Of phenolic material, provides insulating casing for the breaker.

3. **Trip indication.** When mechanism trips, handle moves to middle, or "Trip-Indicating" position midway between ON and OFF. To reset, move handle manually to extreme OFF, then to ON.

4. **Quick-make, quick-break, trip-free mechanism.** Fast action contacting provides minimum arcing when breaker operates. Trip-free mechanism is independent of manual control so breaker can trip open under short circuit or overload conditions even though the operating handle is held in the ON position.

5. **Front-adjustable magnetic trip.** Magnetic elements in the trip circuit provide instantaneous trip action in the event of a short circuit. Any sudden current surge above the trip setting produces a magnetic field strong enough to instantly activate the trip mechanism and open the circuit.

6. **Thermal Trip.** Provides protection against sustained overloads. A bimetallic element similar to that in a home heating thermostat, reacts, time-wise, in inverse proportion to the current. When a circuit becomes overloaded, heat resulting from the passage of current affects this bimetal and causes it to bend, activating the trip mechanism and opening the circuit.

7. **Common-trip bar.** Assures instant disconnection of all conductors when an overload or short circuit occurs on any one conductor in the circuit.

8. **Interchangeable Trip Units.** Trip units on frames above 100 amperes in rating can be interchanged, to meet application requirements.

9. **Silvered contacts.** Conducting properties of silver assure good electrical contact, eliminate pitting and burning, and assure long breaker life.

10. **Arc chute.** Constructed of heat-absorbing insulating material and metal grid plates, it provides fast arc interruption and guards against damage to the breaker. As the contacts part, the arc and its heat are instantly forced into the chute. The arc is quickly "snuffed out" and the hot, gaseous arc products are quickly cooled and expelled.

11. **Line connections.** Designed to take appropriate sizes of copper or aluminum cable. This is UL listed with a full frame rating. (Courtesy of General Electric Company)

ity and reliability. Unlike the circuit breaker, the fuse can stay in position for years and, when called on to act, it will, just as designed. Circuit breakers, on the other hand, have mechanisms with many moving parts and, therefore, require maintenance, periodic testing and operation to keep them in top shape.

One-line diagrammatic representation of circuit breakers is illustrated in Figure 13.4. A typical molded-case circuit breaker and its mechanism is shown in Figure 13.6. Circuit breakers, like fuses, are rated in amperes. However, instead of having a separate design for each ampere rating, manufacturers use *frame* sizes within which a range of ampere sizes is available. Unfortunately, there is no real standardization of frame sizes between manufacturers. One major manufacturer makes frame sizes of 100, 400, 800 and 1200 amperes. Another covers this same range with frames of 100, 600, and 1200 amperes. The trip sizes, however, *are* standard. In amperes, and as divided by one major manufacturer into frame sizes, they are as follows:

100-amp frame	15, 20, 30, 40, 50, 70, 90, 100 A trips
225-amp frame	70, 90, 100, 125, 150, 175, 200, 225 A trips
400/600-amp frame	125, 150, 175, 200, 225, 250, 300, 350, 400, 500, 600 A trips
800/1200-amp frame	250, 300, 350, 400, 500, 600, 700, 800, 1000, 1200 A trips

Note that there is considerable overlapping between frame sizes, for reasons of design convenience. For exact data always consult a current manufacturer's catalog.

The NEC lists standard ratings of circuit breakers *and* fuses together, indicating that the same ratings should be commercially available in both. Actually, the NEC list represents what is commercially available in fuses, while the ratings listed above cover generally available circuit breaker (c/b) ratings. Again, the technologist is cautioned to consult manufacturer's data, which is readily available, up to date and accurate. He will find, as mentioned above, that ratings frequently vary from one manufacturer to another. When this occurs, not only the item but also the manufacturer must be specified. A typical drawing note would show: circuit breaker, NEMA I enclosure, 100-amp frame, 60-amp trip, XYZ type A-37.2.

13.3. Grounding and Ground Fault Protection

Much publicity and emphasis has been given recently to a family of devices that are intended to clear a special type of circuit fault—the ground fault. The detailed study of ground faults is a complex subject and not suited to our purposes here. But an understanding of what a ground fault (circuit) interrupter (GFI or GFCI) is, how it is applied and what it protects against, is important. For this reason, we must backtrack a little and discuss the subject of grounding. Refer again to Figure 11.26, page 294. Notice that the neutral line is shown at ground potential. This is accomplished by *physically* connecting the system neutral to the ground. There are many ways to actually accomplish this, and acceptable ground electrodes are listed in the NEC. A few of them are:

(a) Connection to a buried cold water main.
(b) Connection to a ground rod or group of rods.
(c) Connection to a buried ground plate.

The most common method is (a) above. The purpose of all this is to fix, permanently, a zero voltage point in the system. Without a fixed ground, the system voltages may drift, causing all sorts of control and protection problems. Once a ground is firmly established, it becomes the reference for all voltages in the system. You will hear persons saying that point

A is "600 volts above ground" and point B is "120 volts above ground." *Ground is zero voltage by definition.* The connection between the system and ground is never broken. The grounded line of a circuit is never fused, so that a solid, uninterrupted connection to ground is always maintained. The reason for this is essentially for safety, as we explain below. Figure 13.7 illustrates what we have explained thus far.

The essential aspect of electrical safety is illustrated in the three diagrams of Figure 13.8. In the not too distant past before grounding outlets were required in all new installations, the situation shown in Figure 13.8, Diagram 1, existed. (This situation still exists in millions of buildings, since the NEC requires all *new* installations to use grounded outlets but cannot legally require all existing installations with non-grounding, 2-pole, 2-wire outlets to change over to 2-pole, 3-wire grounding outlets. These existing installations are the most dangerous, and GFI application is advisable.) In this situation, if any contact occurs between the metal case of the electric drill (or stove, saw, broiler, washer, or similar appliance), and the hot leg of the wiring, the case also becomes hot. Then, anyone touching the drill and another grounded surface such as a pipe, or a concrete floor (damp concrete is a good conductor) gets a full 120v electric shock. If this unfortunate person happens to have wet hands, he may be electrocuted, as has actually happened. To prevent this, the grounded outlet, and the 3-wire electric cord was introduced.

In Figure 13.8, Diagram 2, the common 3-wire receptacle and cord is shown. Here, if a contact occurs between metal case and hot line, no danger should be present because the case is directly grounded by its *green* ground wire. Indeed, if the ground is a good one, no danger does occur, and the branch circuit overcurrent device will trip. Unfortunately, the ground path is frequently not a good one, and it presents a resistance so high that the overcurrent device does not trip, and a voltage appears on the case of the drill. Here again, an electric shock awaits anyone who touches it.

To eliminate this dangerous situation, which occurs anytime there is a "leak" of current to ground in an electric circuit, the ground fault circuit interrupter (GFCI or GFI) was developed. See Figure 13.8, Diagram 3. This device compares the current flowing in the hot and neutral legs of a circuit. If there is a difference, it indicates a ground fault (dangerous condition) and the device trips out. The device can be applied at the panel to replace a normal circuit breaker (see Figures 13.9 and 13.10) or at an individual outlet to replace a normal receptacle device (see

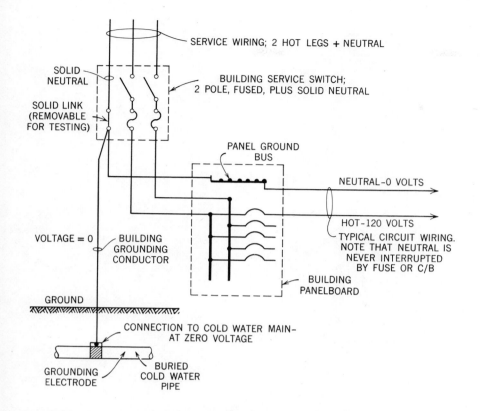

Figure 13.7 Typical service grounding. Note that the grounded conductor (neutral) is continuous throughout. Removable link in service switch is often provided to allow entire circuit to be disconnected for testing. See Appendix A.11, page 568, for complete grounding diagram.

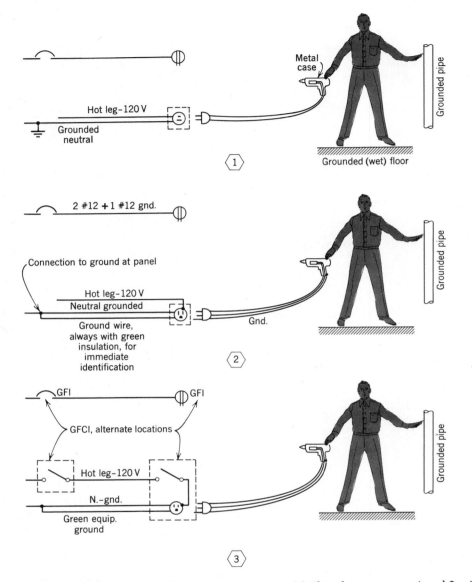

Metal case

Grounded pipe

Hot leg–120 V

Grounded neutral

Grounded (wet) floor

①

2 #12 + 1 #12 gnd.

Connection to ground at panel

Hot leg–120 V

Neutral grounded

Ground wire, always with green insulation, for immediate identification

Gnd.

Grounded pipe

②

GFI   GFI

GFCI, alternate locations

Hot leg–120 V

N.–gnd.

Green equip. ground

Grounded pipe

③

Figure 13.8 Three types of circuit arrangement. (1) This shows conventional 2-wire, grounded neutral circuit, with no means of preventing shocks from ground faults. (2) This is a similar arrangement but includes a green ground wire, which will considerably lessen the danger of ground fault shocks. (3) This illustrates the use of ground fault circuit interrupter devices, now mandatory for bathrooms, outdoor locations, swimming pool circuits and other areas sensitive to ground faults.

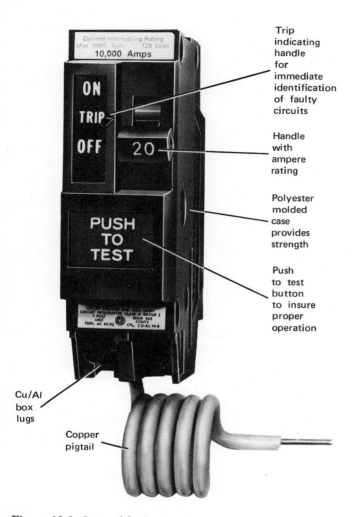

Trip indicating handle for immediate identification of faulty circuits

Handle with ampere rating

Polyester molded case provides strength

Push to test button to insure proper operation

Cu/Al box lugs

Copper pigtail

Figure 13.9 Ground fault circuit interruption feature is built into a single pole molded case circuit breaker of the type usually used to protect normal receptacle circuits. These circuit breaker units will fit into an ordinary panelboard, replacing a conventional circuit breaker. See Figure 13.10. (Courtesy of General Electric Company)

Figure 13.10 *(a)* The special GFI circuit breaker can be installed in a panel along with conventional circuit breakers, to protect circuits prone to ground faults.

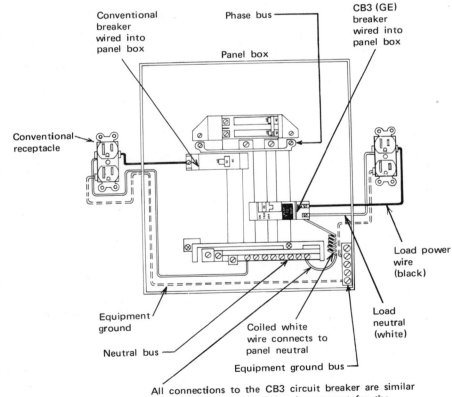

Conventional breaker wired into panel box

Phase bus

CB3 (GE) breaker wired into panel box

Panel box

Conventional receptacle

Load power wire (black)

Equipment ground

Coiled white wire connects to panel neutral

Load neutral (white)

Neutral bus

Equipment ground bus

All connections to the CB3 circuit breaker are similar to those of conventional breakers except for the addition of a neutral connection.

(b)

Figure 13.10 *(b)* A panel might look like this. Remember that conventional receptacles are used when the entire circuit is protected with a GFCI. (Courtesy of General Electric Company)

Figure 13.11 Where it is preferable to protect only a single outlet, the GFI receptacle *(a)* can replace the conventional receptacle *(b)*. Notice that the receptacle is duplex, and that it provides indication, reset and test features. (Courtesy of General Electric Company)

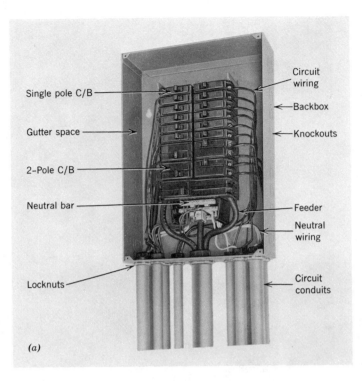

(a)

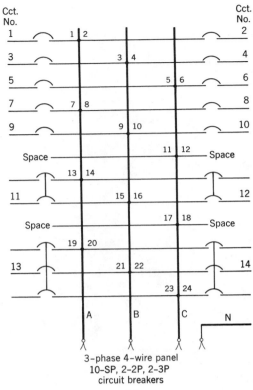

3-phase 4-wire panel
10-SP, 2-2P, 2-3P
circuit breakers

(c)

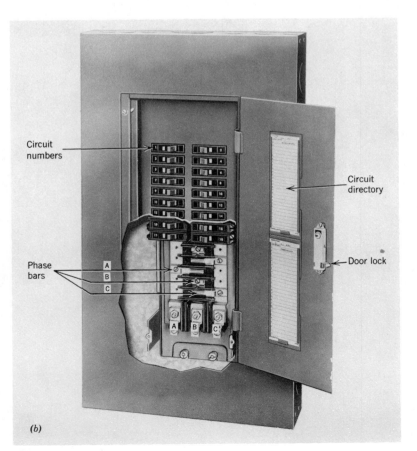

(b)

Figure 13.12 (Facing page) Panelboards may be of the circuit-breaker or fuse type. The illustrated panels contain 1, and 2-pole branch circuits. Panels are provided with a minimum 4-in. gutter space to allow routing of circuit wiring and any feed-through conductors. Lighting panels average 4½ in. deep by 16 in. to 20 in. wide; power panels ar 5 to 6 in. deep and 20 to 30 in. wide. Panels are mounted with the top circuit device no higher than 78 in. AFF and the bottom device no lower than 18 in. AFF. A wired panel and backbox are illustrated in *(a)*; the panel front has been mounted in *(b)*. A typical schematic for a panel is shown in *(c)*.

Figure 13.11). Since this device will sense a ground fault and disconnect it, the green ground of the 3-wire branch circuit or appliance cord is really not essential, although the NEC requires it. It is now clear why we said above that the GFCI finds a ready application in the old 2-wire circuits. These aging circuit components are prone to ground faults and can best be protected with a GFI. It is advisable to use GFI devices on all appliance circuits. See the NEC for locations where GFI use is mandatory, such as outdoors and in bathrooms. Application to lighting circuits is not as essential, since fixtures are generally out of reach and are switch controlled. In mixed lighting and receptacle circuits the GFCI is best applied at the outlet that is to be protected.

13.4. Panelboards

We make passing reference to panelboards, more commonly referred to as *panels* or *electric panels*. An electric panel is simply the box in which the circuit protective devices are grouped and from which they are fed. If the devices are fuses, we have a fuse panel. And if the devices are circuit breakers, we have a circuit breaker panel. Fuses and breakers are very rarely mixed in a panel, except that a circuit breaker panel occasionally has a main switch and fuse for overall protection of the panel. Basically, then, a panel consists of a set of electric busbars to which the circuit protective devices are connected. The assembly is then placed in a metal box, with knockouts to allow for entrance of the branch circuit conductors. These, as we already know, are connected to the load side of the protective devices. The components of an electric panel are shown in Figure 13.12. A close-up of the inside of a panel is also shown in Figure 13.10*a*. Observe that although the buses are shown schematically (Figure 13.12*c*) as straight lines running the length of the box, they are actually offset to allow circuit breakers to be installed in *two vertical rows*. This will become clearer when we demonstrate the making of a panel schedule. See Figure 13.13. The panels in Figure 13.12 are 3 phase whereas the one in Figure 13.10 is single phase, 3-wire (2 buses). Note that in both, the breakers are

PANEL 1 A						
CIRC. NO.	LOAD DESCRIPTION	CIRCUIT BREAKERS			LOAD DESCRIPTION	CIRC. NO.
1	LTG. RMS. 126 – 127, 1250 W	20	1 2	20	LTG. RMS. 128 – 130, 1150 W	2
3	LTG. AUDITORIUM 1800 W	20	3 4	20	LTG. LOUNGE 1400 W	4
5	12 – ⏀ RM 126, 127	20	5 6	20	6 – ⏀ RM 130	6
7	10 – ⏀ RM 128, 129	20	7 8	20	SPARE	8
9	PROJECTOR OUTLET 1500 W	20	9 10	20	SPARE	10
	SPACE		11 12		SPACE	
11	3 KW HEATER	20	13 14 15 16	20	3 KW KILN	12
	SPACE		17 18		SPACE	
13	5 HP PUMP P–2	40	19 20 21 22 23 24	20	3 HP VENT. FAN F – 1	14

100 A MAINS, 60 A GND BUS
MAIN LUGS ONLY (NO MAIN C/B)
ALL BRANCH C/B TO BE 10,000 A MIN. I.C.
FLUSH MTD.

120/208 VOLTS

Figure 13.13 *(a)* Typical circuit breaker panel schedule, corresponding to schematic diagram shown in Figure 13.12*c*. Circuits 13 and 14 feed 3-phase motors and are, therefore, 3 pole (see Appendix A.18).

PANEL 1 A								
CIRC. NO.	LOAD DESCRIPTION	SW	F		F	SW	LOAD DESCRIPTION	CIRC. NO.
1	LTG. RM. 126, 127 1250 W	30	15	1 2	15	30	LTG. RMS. 128, 130 1150 W	2
3	LTG. AUDITORIUM 1800 W	↓	20	3 4	20	↓	LTG. LOUNGE 1400 W	4
5	12 – ⏀ RM. 126, 127		20	5 6	15		6 – ⏀ RM 130	6
7	10 – ⏀ RM. 128, 129		20	7 8	20		SPACE	8
9	PROJECTOR OUTLET 1500 W		20	9 10	20		SPARE	10
	SPACE			11 12			SPARE	
11	3 KW HEATER	30 2 P	20 20	13 14 15 16	20 20	30 2 P	3 KW KILN	12
	SPACE			17 18			SPACE	
13	5 HP PUMP P–2	30 3 P	18 18 18	19 20 21 22 23 24	12 12 12	30 3 P	3 HP VENT. FAN F – 1	14

100 A MAINS, 60 A GND. BUS
FLUSH MOUNTING

MAIN LUGS ONLY

120/208 VOLTS
ALL FUSES BUSSMAN CO.
CCTS. 13, 14 FUSETRON FUSES

Figure 13.13 *(b)* Typical switch/fuse panel, corresponding to circuit breaker panel of Figure 13.13*a* (see Appendix A.19).

installed in two rows. The busbars are ingeniously designed to allow for this. A single-phase, 3-wire panel is fed with two hot lines and a neutral, connected to the two line buses and the neutral bus, respectively. A 3-phase panel has an additional hot line and bus. Panels vary in ampere rating of the buses, type of protective devices installed, whether or not the panel has a main device that completely disconnects the entire panel, and type of mounting, that is, whether the panel is to be flush or surface mounted. A typical drawing note describing a panel would be:

House panel, circuit breaker type, for surface mounting, single-phase, 120/240 v, 150-amp mains, 100/70-amp 2-pole main circuit breaker. Branch breakers—all 70-amp frame; 12–20 amp single pole, 2–30 amp 2 pole, 1–20 amp SP, GFI.

To clarify in our minds the structure of a panelboard and to give us a "feeling" for what it comprises, we study at this point the preparation of a panel schedule. Almost everyone has seen the schedule of circuits found in a small directory on the inside of a panel door. This schedule, sometimes neatly typed but more often handwritten, is a listing of the panel circuits and the loads fed by these circuits. This data should properly be available directly from the drawings, and making the door directory should simply be a matter of copying the data off the drawings. Often, however, the complete panel schedule information is not available on the drawings. Then, this very useful schedule on the panel door becomes a hit-or-miss affair hastily written by the installing electrician, or filled in after being laboriously traced out by the owner or maintenance man. As we will discuss later, the technologist doing the circuiting makes a complete panel schedule in any case and, therefore, its failure to appear on the drawings in a complete and easily usable form is simply poor practice.

For our analysis we present the schedule shown in Figure 13.13. We say "schedule of choice" because we do not recommend that a single schedule be used for all work. The schedule used must meet the needs of the job. Frequently, a single job will use more than one schedule format. Unfortunately, there are almost as many different formats of panel schedules as there are draftsmen and technologists, and it has been our experience that each person is convinced that his version is clearly best. The panel schedule formats presented in this book have been arrived at after studying literally dozens of schedules, and they are readily usable and practical. The format of

Figure 13.13 was chosen for illustration here because, in addition to being easy to use and practical, it clearly demonstrates to the novice technologist the "workings" of a panel. Also it can be transferred as it stands to the physical panel directory. There it comprises a permanent and extremely useful record for the resident or his electrician when he wishes to do any maintenance or repair work. Furthermore, and this is of great importance, it is sufficiently detailed so that changes in the wiring system or loads can be shown on it, and thus it can be kept up to date. The abbreviated panel schedule so frequently found on drawings, which simply consists of a listing of the breakers or fuses, is not useful to the owner. It does *not* tell him which outlets are on what circuit, or what the fixed loads are. As such, it is useful only at the time of layout and even there causes problems when design changes are made. It should therefore be avoided as a false economy.

Referring now to Figure 13.13, note how the schedule reflects both the physical reality as shown in Figure 13.12a and b, and the schematic drawing of Figure 13.12c. One difference is the absence in the panel schedule of the neutral, which is not often shown schematically either. A moment's thought will show that nothing is to be gained by showing the neutral bar. If a *ground* bus is called for, it need not be shown but must be specified, and a space at the bottom of the panel schedule exists specifically for that purpose. Note *carefully* the following:

(a) A *pole* is a single connection to a bus. Therefore, a multipole circuit breaker such as circuit breaker no. 11 is connected to more than 1 pole and cannot, without causing considerable confusion, carry the pole numbers. Indeed, the pole numbers do not appear on the physical panel and are only used as a convenience in circuitry. Circuit no. 11, fed through circuit breaker no. 11, is connected to poles 13 and 15. If we wish to know which poles and/or phases are connected to each circuit protective device, we must consult the panel schedule as designed and as shown on the manufacturer's shop drawings.

(b) The *pole* connections are made in pairs from the buses, that is, 1–2 from phase A, 3–4 from phase B, 5–6 from phase C, and so on. If, as in Appendix A31, we have a single-phase panel with only two busbars, then 5–6 are again taken from A phase. Therefore, to get consecutive phases, we must skip a pole each time, since phases A, B, C correspond to poles 1, 3 and 5 on the left side of the panel and poles 2, 4 and 6 on the right side of the panel. This is important in circuitry, as we shall learn. Keep in mind that these are *pole* numbers and *not* circuit numbers.

(c) The usual lighting and appliance panel (see the NEC for definition) is limited to 42 poles, *not including* the main disconnect. For this reason most panel schedules are made for 42 poles. (See Appendix A.19.) If we use less, we simply cut off the panel schedule at the desired point.

(d) Space should be provided on the panel schedule for busbar and circuit breaker data, as shown. This information is supplied by the job electrical designer.

(e) The Load Description column should contain a complete but brief listing of the loads and the design wattage or, in the case of a motor, horsepower. An entry such as "1000 watts lighting" is usually insufficient.

(f) Panel designations often indicate the panel's location in a large building, for example, panel 2C is in closet C, second floor. Since there are basically only two kinds of panels—lighting and appliance panels, and power panels, it is well to avoid misleading titles, such as lighting panel, receptacle panel and distribution panel, and stick to LP for lighting and appliance panels and PP for power panels. Even more simple is avoiding titles altogether, calling panels simply P1, P2 and the like without reference to use.

As mentioned above, panel schedules change not only during layout as a result of design changes and after panel installation as a result of field changes, but also at the manufacturing stage. The manufacturer sends to the design office a panel layout representing the panel as it actually will be built. For one reason or another, the actual arrangement shown in this *shop drawing* may differ from the panel layout on the drawings. If the designer or electrical technologist who is checking the shop drawing agrees to the manufacturer's version, he must make the corresponding changes on the drawings. This will assure that the drawing used in the field corresponds to the equipment on the job in all respects. Leaving the panel circuit arrangement to the contractor may save office time, but it creates field problems and, what is worse, maintenance problems for the owner. Symbols for panelboards as well as other equipment items that appear on electrical drawings are given in Figure 13.14. This figure is Part VI of the overall symbol list.

13.5. Procedure in Wiring Planning

Now that we have studied the basic elements that make up the electrical system of a building, namely, the branch circuit, the panel and protective devices, and the outlets and wiring devices, we can give our attention to the procedure normally followed in electrically planning a building. We refer to the Basic Plan House and apply to it the procedures normally followed. In the next chapter, we generalize the procedure as applied to residential buildings, using a large residence. We have selected single residences to begin our study for two reasons:

1. This is the type building that persons are usually most familiar with.
2. The residence presents many types of wiring and circuitry problems in a single building. Thus, through having mastered a residence, you will have handled lighting, control, kitchen equipment, motors, receptacles, outdoor lighting and signals. No other small building gives such diverse loads. You can then easily go on to other type buildings.

The procedure normally followed is:

(a) Identify all spaces on the architectural plan by use, present and future.

(b) Assemble all criteria applicable to those spaces. These criteria will be discussed in part below, and more generally in the chapters that follow.

(c) Show to scale and in position all items of equipment that require electrical connection. This includes mechanical and heating items. Items that do not require electrical power, but *do* affect outlet location must also be shown. Thus furniture placement, *if it is known*, should be shown.

(d) Locate the remaining electrical devices according to the needs of each space and the applicable criteria. At some point between (a) and (d) a decision has to be made as to the number of drawings and the scale of the drawings. This, too, will be discussed as it applies.

(e) Circuit all the equipment and also prepare a panel schedule. This step or the previous one also includes all switching and circuit control elements.

(f) Check the work.

We now apply the above steps to the Basic Plan House while also learning the techniques involved in this application.

SYMBOLS — PART VI — EQUIPMENT

3 P SN — INDIVIDUALLY MTD. DISC. SW.; 30A 3P, SN (SOLID NEUTRAL) 250V., UNFUSED, TYPE ND, NEMA I ENCLOSURE, U O N

60/30 II — SW., 3 POLE AND SN., 60A/30 AF. NEMA II ENCL., ND (NORMAL DUTY) 250 V.

CB 100/60 — ENCLOSED CIRCUIT BREAKER, 3 POLE, NEMA TYPE I ENCLOSURE U O N, 100A FRAME, 60A TRIP. SEE SPEC.

E 2 P, 30 A — CONTACTOR, ENCLOSED, ELECTRICALLY OPERATED ELECTRICALLY HELD, 2 POLE, 30 AMP.

M OR RC 3 P, 60 A — CONTACTOR, ENCLOSED, ELECTRICALLY OPERATED, MECHANICALLY HELD, 3 POLE 60 AMP.

T S 3 P, 100 A — AUTOMATIC TRANSFER SWITCH, ENCLOSED, 3 POLE, 100 AMP.

□ — ITEM OF ELECTRICAL EQUIPMENT, AS INDICATED

XFR A — TRANSFORMER; TYPE A, SEE SCHEDULE

LP–1 — ELECTRIC PANELBOARD LP–1, RECESSED, SEE SCHEDULE ON DWG. _____ .

P–2 — ELECTRIC PANEL (BOARD) P–2, SURFACE MTD.

TEL. 18 × 24 × 6 — CABINET, MOUNTING, SIZE AND PURPOSE AS SHOWN.

PBI 12 × 24 × 5″ — PULLBOX, SIZE AND IDENT. SHOWN.

Figure 13.14 Architectural–electrical plan symbols, Part VI, equipment items.

Part I, Raceways, Figure 12.6, page 304.
Part II, Outlets, Figure 12.37, page 338.
Part III, Wiring Devices, Figure 12.39, page 341.
Part IV, Abbreviations, Figure 12.43, page 345.
Part V, Single Line Diagrams, Figure 13.4, page 354.
Part VII, Signalling Devices, Figure 15.34, page 500.
Part VIII, Motors and Motor Control, Figure 16.19, page 527.
Part IX, Control and Wiring Devices, Figure 16.22, page 530.

13.6. The Architectural — Electrical Plan

Step one, as stated above, is the preparation of the architectural–background plan, which is the raw material of our work. This plan is a stripped down version of the architectural plan showing *only* walls, partitions, windows, door swings and pertinent fixed equipment. Dimensions, room titles and other architectural data are omitted. Such a plan of the Basic Plan House is shown in Figure 2.16, page 39. Since the electrical work must stand out clearly and be completely unambiguous, it is false economy to attempt to use a prepared architectural plan for electrical work. The required background should always be retraced with light line work, showing the elements listed above. It was once almost universal practice to trace the architectural background on the *back* of the sheet. This had two purposes: (1) erasures on the electrical plan would not affect the architectural outline, and (2) the print would show the architectural work as lighter than the electrical plan, which was done on the front side of the tracing. With the increasing use of pencil cloth tracings, which have a glossy back surface, this has become impossible. Hence, it is a very good idea to obtain a screened photo print of the stripped architectural, which allows the electric work drawn on it to show as boldly as is desired. Architectural changes, which always occur, can be made on such a photo print with no trouble. Spaces should be identified by room titles placed *outside the plan* if at all possible, to avoid conflicting with the electrical work. Where this is not possible or is undesirable, room titles should be penciled in *lightly* to identify spaces, and only lettered in final form when the electrical work is complete, to avoid space conflicts. This is also true for equipment. The initial drawing must be as open and uncluttered as it is possible to make it. After the electrical work is in, the required names, descriptions, titles and equipment identification can be added.

13.7. Residential Electrical Criteria

By this title we simply mean the process by which we decide what is required and where we put it. The answer to those questions is varied. The NEC gives minimum criteria, and the architect and engineer in conjunction with the owner establish additional requirements on the basis of need, convenience and common sense. As a guide to this latter group, numerous books and pamphlets have been published by trade and governmental agencies that list room-by-room recommendations for lighting and other electric outlets. Some of these are listed in the Additional Readings at the end of the chapters of this book. The student will do well to read these, while keeping in mind that they are *recommendations*. In contrast to this, the minimum requirements of the NEC are *mandatory*. Since you must be fully familiar with these NEC requirements, we review them in this chapter while putting off the optional material until the more detailed discussion of residential buildings in the next chapter.

13.8. Equipment and Device Layout

This stage of the work includes steps (c) and (d) of Section 13.5. We continue to use the Basic Plan as our reference. Since in Chapter 2 the heating layout was shown for both hot-water and electric systems, we show here the electric layout corresponding to these systems. Refer now to Figure 13.15, which corresponds to the hot water heated building of Figures 2.20, and to Figure 13.16, which corresponds to the electrically heated house of Figure 2.27.

In accordance with the step-by-step layout procedure described in Section 13.5, we would first show all electrically connected equipment. Referring to Figure 13.15, this would include the electric range, refrigerator, dishwasher and exhaust fan in the kitchen, and the laundry equipment and boiler in the basement. In the case of the electrically heated building of Figures 13.16 and 2.27, we have the electric baseboard units and room thermostats in all the spaces, but we eliminate the boiler of Figure 13.15. At this point we must decide whether showing the electric heating on the drawing, in addition to all the remaining electrical work, will cause the drawings to be cluttered and difficult to read. Here, we decide that by use of tables this clutter can be avoided. The expense of an additional drawing for electric heat is also avoided.

Since no furniture layout was provided, the remaining devices are located in accordance with the NEC requirements and what the layout man considers to be good practice. The criteria for both are discussed in Chapter 14. The resulting drawings as they stand prior to circuiting are shown in Figures 13.15 and 13.16. Note that at this point we have already made a symbol list and have shown the switching of lights. The symbol list should be prepared *at this stage* and not, as is often done, as an afterthought. This way all needed symbols are added to the list as they are put onto the drawing, along with required notes, mounting heights and the like. Similarly, light switching should be indicated as soon as each lighting outlet is drawn. This avoids any possibility of forgetting a switch. Deciding on the fixture type, location, and control or switching, are three parts of a single activity, which are best done together. The type of fixture is selected by the designer or technologist and is shown in the specification or as here, in a fixture schedule on the drawing.

13.9. Circuitry Guidelines

This step and the next one of actual drawing are the heart of the electrical work and as such will be covered slowly and in detail. The preliminary work of layout and switching has been shown; it remains to make the connections to the panelboard. But this is not as simple as it sounds. There are many ways in which the circuiting can be done, but there is no optimum way. There are, however, certain guidelines to keep in mind which, if followed, will yield a flexible, economical and convenient layout. Also NEC rules must be followed, and there are a few technical considerations that will help to yield a good layout. The larger and more complex the job, the more ways there are to do it. Yet, if guidelines are followed, all the layouts will be acceptable.

(a) The NEC (1975 edition) requires for residences sufficient circuitry to supply a load of 3 w/sq ft in the building, excluding unfinished spaces such as porches, garages and basements. This requirement of 3 w/sq ft works out to 800 sq ft/20-amp circuit (2400 w) or 600 sq ft/15-amp circuit (1800 w). Good practice, however, dictates a load of no more than 1600 w for a 20-amp circuit or 1200 w for a 15-amp circuit. Therefore, on the basis of 3 w/sq ft, we get a good practice rule of:

15-amp circuit, 1200 w max.   400 sq ft max.
20-amp circuit, 1600 w max.   530 sq ft max.

Of course, the final decision in this, as in all subsequent discussion, rests with the job engineer. If the decision is left to the technologist, as it frequently is, he should follow the above recommendation.

(b) The NEC requires a minimum of two 20-amp appliance branch circuits to feed all the small appliance outlets in the kitchen, pantry, dining room and family room, and *only* these outlets. Furthermore, all kitchen outlets must be fed from at least two of these circuits (which may also feed other appliance outlets). This NEC requirement needs some explanation. We already know what a branch circuit is. The NEC, however, goes a step further and defines three types of branch circuits: appliance, general purpose, and individual. Figure 13.17 should make the differences clear. Thus the NEC requires that at least two circuits be reserved for *appliance outlets*, but it does not specify what these appliance outlets are except to say, by inference, that all kitchen outlets are appliance outlets. In effect then, according to the NEC, all the receptacles are potential appliance outlets, and at least two circuits should be supplied to serve them. Good practice dictates that certain receptacles in each room be designated as appliance outlets even though they do not differ from the other outlets in appearance. These outlets are:

(1) All kitchen receptacles.
(2) One dining room receptacle.
(3) One receptacle in the family (or living) room.

These receptacles should be circuited with preferably two, but no more than four such outlets on a 20-amp circuit, and the circuits should be arranged so that the kitchen has part of at least two circuits feeding its outlets.

(c) Additional circuits similar to appliance circuits should be furnished to supply one outlet in each bedroom of a house that is not centrally air-conditioned. Such outlets are intended for window air conditioners. Also, on circuits of this kind we would place basement workbench outlets. These additional circuits must not be mixed with the appliance branch circuits discussed above, as they are not strictly appliance circuits by NEC definition.

(d) The NEC requires that at least one 20-amp circuit supply the laundry outlets. This requirement satisfies good practice. If an electric clothes dryer is anticipated (and it should be

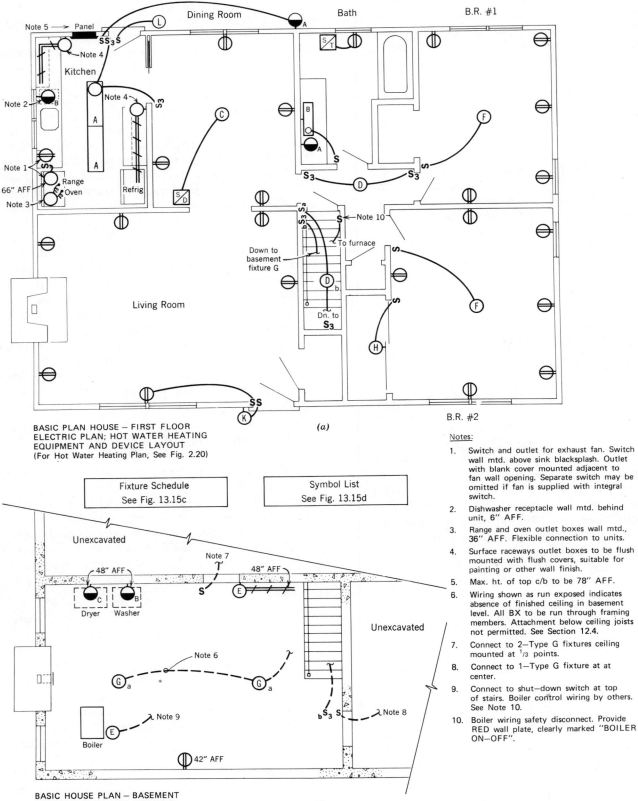

BASIC PLAN HOUSE — FIRST FLOOR
ELECTRIC PLAN; HOT WATER HEATING
EQUIPMENT AND DEVICE LAYOUT
(For Hot Water Heating Plan, See Fig. 2.20)

(a)

Fixture Schedule
See Fig. 13.15c

Symbol List
See Fig. 13.15d

BASIC HOUSE PLAN — BASEMENT
ELECTRIC PLAN; H.W. HEATING
EQUIPMENT AND DEVICE LAYOUT

(b)

Notes:

1. Switch and outlet for exhaust fan. Switch wall mtd. above sink blacksplash. Outlet with blank cover mounted adjacent to fan wall opening. Separate switch may be omitted if fan is supplied with integral switch.

2. Dishwasher receptacle wall mtd. behind unit, 6" AFF.

3. Range and oven outlet boxes wall mtd., 36" AFF. Flexible connection to units.

4. Surface raceways outlet boxes to be flush mounted with flush covers, suitable for painting or other wall finish.

5. Max. ht. of top c/b to be 78" AFF.

6. Wiring shown as run exposed indicates absence of finished ceiling in basement level. All BX to be run through framing members. Attachment below ceiling joists not permitted. See Section 12.4.

7. Connect to 2—Type G fixtures ceiling mounted at $1/3$ points.

8. Connect to 1—Type G fixture at at center.

9. Connect to shut—down switch at top of stairs. Boiler control wiring by others. See Note 10.

10. Boiler wiring safety disconnect. Provide RED wall plate, clearly marked "BOILER ON—OFF".

Figure 13.15

LIGHTING FIXTURE SCHEDULE

TYPE	DESCRIPTION	MANUFACTURER	REMARKS
A	48" L X 12" W X 4" DEEP NOMINAL, 2 LAMP/FLUORES—CENT, WRAP—AROUND ACRYLIC LENS, F 40 WW/LAMPS. SURFACE MTD.	BRITE—LITE CO. CAT. #2740/KFF OR EQUAL	4" DEPTH MAXIMUM
B	24" L, 1 LAMP 20W FLUOR. FIXTURE, WRAP—AROUND WHITE DIFFUSER, WITH SINGLE—SWITCHED RECEPTACLE. MOUNT ABOVE MEDICINE CABINET.	BRITE—LITE CO. CAT. #1/20/BFF OR EQUAL	MAX. MTG. HT. 78" TO ₵.
C	ADJUSTABLE HEIGHT PENDANT INCANDESCENT, 3—75W MAX., BUILT—IN 3—POSITION SWITCH.	HOMELAMP CO. CAT. #3/75/DRP OR EQUAL	——————
D	10" D. DRUM—TYPE FIXTURE, WHITE GLASS DIFFUSER, CENTER LOCK—UP, 2—60W INCAND. MAX., SURF. MTD.	BRITELITE CO. CAT. #2/60/HF OR EQUAL	6" MAX. DEPTH.
F	12" D. DRUM FIXTURE, CONCEALED HINGE ON OPAL GLASS DIFFUSER FOR RELAMPING WITHOUT GLASS REMOVAL, 2—75W INCAND. MAX. SURFACE MTD.	DENMARK LIGHTING SPECIAL UNIT #374821	NO SUBSTITUTION WILL BE ACCEPTED.
G	PORCELAIN LAMPHOLDER, PULL CHAIN WITH WIRE GUARD, 100W. INCAND. SURF. MTD.	——————	——————
H	SAME AS TYPE G, EXCEPT W/O GUARD.	——————	——————
K	DECORATIVE OUTDOOR LANTERN, MAX. 150W INCAND., WALL MTD. 84" AFF TO ₵.	TO BE CHOSEN BY OWNER	——————
L	UTILITY OUTDOOR LIGHT, ANODIZED ALUMINUM BODY AND CYLINDRICAL OPAL GLASS DIFFUSER. 1—100W INCAND. MAX. 84" AFF TO ₵.	UTIL—LITE CO. CAT. #1/100/BP OR EQUAL	IF VANDALISM IS OF CONCERN, SUBST. PLASTIC DIFFUSER.

(c)

SYMBOLS AND ABBREVIATIONS

—///— BX CABLES RUN CONCEALED; TICS INDICATE NUMBER OF CONDUCTORS EXCLUDING GROUND WIRES. 2 #12 + BARE GROUND, U O N

————— SAME AS ABOVE EXCEPT RUN EXPOSED.

—•—o WIRING RUN TURNING DOWN; WIRING TURNING UP

1 3 ///— HOME RUN TO PANEL; ARROWS AND NUMERALS IDENTIFY CIRCUITS; TICS INDICATE WIRING — AS NOTED ABOVE.

OUTLET BOX AND FINAL CONNECTION TO EQUIPMENT WITH FLEXIBLE CONDUIT (OR BX).

OUTLET WITH SECTION OF SURFACE RACEWAY, 2 WIRE, SINGLE CIRCUIT, AND SEPARATE GREEN GND. 15A, 2P, 3 WIRE, RECEPTACLES ON 12" CENTERS.

Ⓓa CLG. OUTLET WITH INCANDESCENT LTG. FIXTURE D, SWITCH CONTROL — 'a'.

-Ⓗ- WALL OUTLET W/INCAND. FIXT. 'H', M.HT. SHOWN.

O A CLG. OUTLET W/FLUOR. FIXT. 'A'.

B WALL OUTLET W/FLUOR. FIXT. 'B', MHT. SHOWN.

Ⓙ JUNCTION BOX

⊖ DUPLEX CONVENIENCE RECEPTACLE, 15 A, 2P, 3W, 125 V. GROUNDING, WALL MTD., VERTICAL, ₵ 12" AFF NEMA 5—15 R.

◑A DUPLEX CONVENIENCE RECEPTACLES, 15A, 2P, 3W, 125V, GROUNDING. W/INTERGAL GFCI AND GASKETED W.P. SELF—CLOSING COVER.

◑B SINGLE RECEPTACLE, 20 A, 2P, 3W, GND'G., NEMA 5—20 R.

◑C SINGLE RECEPTACLE, 30A. 125/250V. 3 POLE—4 WIRE GROUNDING NEMA 14—30 R; (NOTE 1)

Sa SINGLE POLE SWITCH, 15 A, 125 V, ₵ 50" AFF, U O N, CONTROLLING OUTLET(S) 'a'.

aS3 SWITCH, 3 WAY, 15 A, 125 V, ₵ 50" AFF, U O N, CONTROLLING OUTLETS 'a'.

S/T MANUAL TIMER SWITCH, 1 SET 15 AMP N.O. CONTACTS.

S/D OUTLET BOX MTD. SWITCH AND DIMMER, INCAND. LOAD ONLY, 600 WATTS MAXIMUM. ₵ 50" AFF.

▬ FLUSH MTD. PANELBOARD;

AFF — ABOVE FINISHED FLOOR
MHT — MOUNTING HEIGHT
T — THERMOSTAT
UON — UNLESS OTHERWISE NOTED
GFCI — GROUND FAULT CIRCUIT INTERRUPTER
WP — WEATHER—PROOF
NO — NORMALLY OPEN

NOTE 1. CONTRACTOR TO SUPPLY MATCHING CAP.

(d)

374

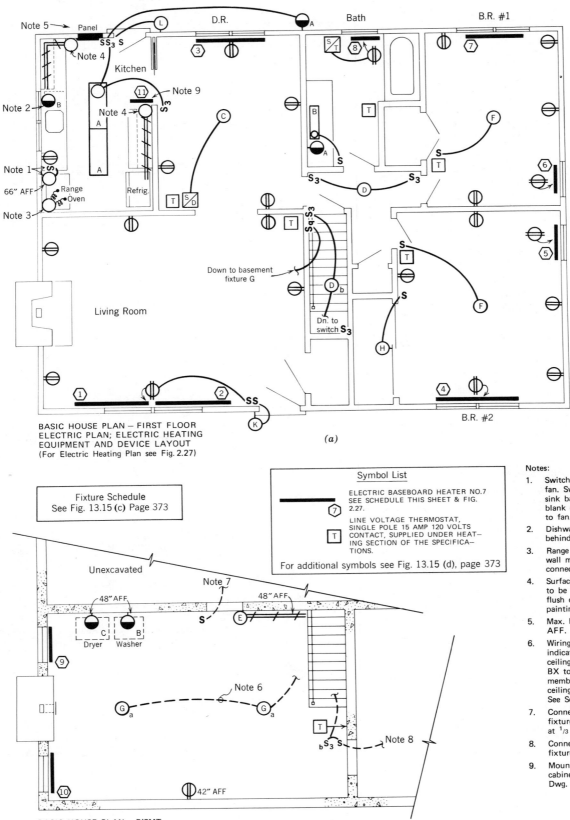

BASIC HOUSE PLAN — FIRST FLOOR
ELECTRIC PLAN; ELECTRIC HEATING
EQUIPMENT AND DEVICE LAYOUT
(For Electric Heating Plan see Fig. 2.27)

(a)

Fixture Schedule
See Fig. 13.15 (c) Page 373

Symbol List

ELECTRIC BASEBOARD HEATER NO. 7
SEE SCHEDULE THIS SHEET & FIG.
2.27.

LINE VOLTAGE THERMOSTAT,
SINGLE POLE 15 AMP 120 VOLTS
CONTACT, SUPPLIED UNDER HEAT-
ING SECTION OF THE SPECIFICA-
TIONS.

For additional symbols see Fig. 13.15 (d), page 373

Notes:

1. Switch and outlet for exhaust fan. Switch wall mtd. above sink backsplash. Outlet with blank cover mounted adjacent to fan.

2. Dishwasher receptacle wall mtd. behind unit, 6″ AFF.

3. Range and oven outlet boxes wall mtd., 36″ AFF. Flexible connection to units.

4. Surface raceways outlet boxes to be flush mounted with flush covers, suitable for painting or other wall finish.

5. Max. ht. of top c/b to be 78″ AFF.

6. Wiring shown as run exposed indicates absence of finished ceiling in basement level. All BX to be run <u>through</u> framing members. Attachment below ceiling joists not permitted. See Section 12.4

7. Connect to 2—Type G fixtures ceiling mounted at ⅓ points.

8. Connect to 1—Type G fixture at center.

9. Mount heater in end of wall cabinet. See detail, on Dwg. E

BASIC HOUSE PLAN — B'SMT.
ELECT. PLAN. — ELECT. HTG.
EQUIP'T. AND DEVICE LAYOUT

(b)

Figure 13.16

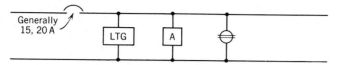

(a) General purpose branch circuit. Supplies outlets for lighting and appliances, including convenience receptacles.

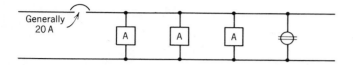

(b) Appliance branch circuit. Supplies outlets intended for feeding appliances. Fixed lighting not supplied.

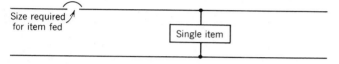

(c) Individual branch circuit, designed to supply a single, specific item.

Figure 13.17 Branch circuit types.

unless it is definitely known that a gas dryer will be used), an individual branch circuit should be supplied to serve this load, via a heavy duty receptacle.

Other good practice circuiting guidelines are as follows.

1. Avoid combining receptacles and lighting on a single circuit.
2. Avoid placing all the lighting in a building on a single circuit.
3. Circuit the lighting and receptacles so that each space contains parts of at least two circuits. Thus if a single circuit is out, the entire space is not deprived of power.
4. Do not use combination switch and receptacle outlets except where convenience of use dictates high mounting, for example, above counters. This is often done in economy construction to save a few pennies in wiring. It makes an inconvenient and unsightly receptacle.
5. Supply at least one receptacle in the bathroom and one outside the house. Both must be GFCI types. This is an NEC requirement. An additional convenience is switch control of the outside receptacle from *inside* the house. Also, a timer-controlled outlet for a plug-in bathroom heater is a welcome convenience, but obviously means additional expense.
6. In rooms without overhead lights, provide switch control of one half of a strategically located receptacle that is intended to supply a lamp. See Figure 13.18 for the wiring arrangement in such a case.
7. Provide switch control for closet lights. Pull chains are a nuisance.
8. Although convenience outlets are counted as part of the general lighting load and hence there is theoretically no limit to the number that can be placed on a general purpose branch circuit, good

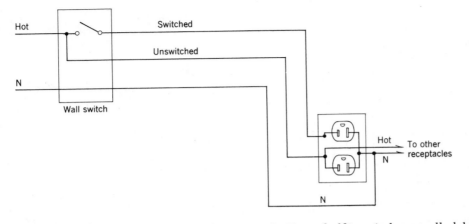

Hot

Switched

Unswitched

N

Wall switch

Hot

To other
receptacles

N

N

Figure 13.18 Split wiring of a duplex receptacle. Upper half is switch controlled; lower half is hot all the time. This allows wall switch control of a lamp or other device while maintaining part of receptacle for independent use.

practice limits them to six on a 15-amp circuit and eight on a 20-amp circuit.

Keeping in mind these guidelines and the convenience we ourselves would appreciate if we were living in the house, we can proceed to the actual circuitry.

13.10. Drawing Circuitry

At this point we apply step by step, the guidelines listed above, to the Basic Plan House in its two versions as shown in Figures 13.15 and 13.16. The difference between them is the heating method: the former is heated by a hot water system and the latter is heated by electric baseboards. We have deliberately kept all other aspects the same to show the changes in the electric layout due to electric heating.

The first thing we need is a panel schedule. We use the one shown in Appendix A 31, as previously discussed in Section 13.4. It is an accepted convention to circuit all single-pole circuits first, followed by 2-pole circuits and then, if any, 3-pole circuits. In our Basic Plan House as in most small-to-medium size residences, there are no 3-phase loads and hence no 3-pole breakers. A single-phase panel is therefore

illustrated in Figure 13.19c. The choice of panel type also depends on power available from the utility, total building load, and other factors. This decision is therefore generally made by the electrical designer and passed on to the draftsman who is doing the job circuiting.

It is also customary that all lighting outlets are circuited first, followed by appliance circuits and individual branch circuits. Since most wiring is a minimum of No. 12 AWG, we use 20-amp circuits. In accordance with the guideline stated in Section 13.9a, we should allow 500+ sq ft per 20 amp circuit.

The main floor of our Basic Plan is 1430 sq ft gross. We therefore need a minimum of three circuits for general lighting and receptacles, assuming that we can reasonably circuit this plan with about 1600 w per circuit. Refer now to Figure 13.19a. Because of the relatively small size of the house, we find it impossible to follow all the guidelines of Section 13.9. We say there in paragraph (d) that it is best to avoid combining lighting and receptacles on a single circuit. Also, that it is inadvisable to place *all* the lighting on one circuit. This is so that the failure of one circuit will not black out the entire house. In this building, to place a reasonable load on a circuit, we will have to mix lighting and receptacles, since the main floor lighting totals only about 1000 watts. Note that although convenience outlets (duplex receptacles) do not add to the general lighting load of 3 w/sq ft as noted above, in figuring individual circuit capacity they are counted at 180 w (1.5 amp) each. Thus, six receptacles are 1080 w and eight are 1440 w. These receptacles are the ones *not* connected to appliance circuits. To show clearly which outlets are intended as appliance outlets, we have placed a dot next to the symbol. This dot should be erased when the circuiting is finished. It simply serves as a reminder during circuiting.

Convenience outlets do *not* count in the building load total but *do* count in the circuit load. Therefore, it is necessary to keep a double load record—one for the circuit and one for the building. This is explained in detail in Section 15.8. To avoid confusion with this double load record, it has been our custom to show the panel load as a combination of fixed outlet wattage plus the number of receptacles. For instance, circuit no. 1, detailed below, has 820 w (actually volt-amperes) of fixed lighting, plus three receptacles. Instead of showing 820 + 3 (180) or 1360 w we show in the panel schedule 820 + 3*R*. See Figure 13.19*c*. This kind of notation gives more data because it separates fixed lighting from convenience outlets.

We made an exception to this in circuit no. 3 below when we showed the living room lighting as 225 w even though it is supplied by a receptacle. This is because there is no ceiling outlet in the living room. This receptacle therefore becomes the fixed lighting outlet, and we show it as such.

One further item must be clarified. The NEC requires that [see Section 13.9b above] at least two circuits be provided to feed small appliances in the kitchen, pantry, dining room and family room. The intention is to provide circuits for small appliances, but the language is restrictive. For this reason we suggest in Section 13.9c the supplying of additional appliance–type circuits for room air conditioners. Also, although the Code states "family room," we interpret this to mean the room where the family meets. Therefore, in houses without a "family room," it is our practice to supply one appliance outlet in the living room, since in that house the *living* room is the *family* room. You will therefore see that outlets indicated with a dot have been provided in the living room and bedrooms. Now let us follow the circuiting as it appears in Figure 13.19, keeping in mind that this is only one of many ways that it could be done, and we make no claim to presenting a single best solution. Following the suggestions listed above we start with lighting and with circuit no. 1, connecting the ceiling lighting outlets in the kitchen, dining room, hall, bedroom and bath. From the kitchen switch we also feed through to the back

patio light fixture L and to the outside receptacle. This gives us a load of:

Kitchen ceiling lights	200 w
Dining room lights	225 w
Bath fixture B	25 w
Hall fixture D	120 w max
Bedroom fixture F	150 w max
Back porch	100 w max
	820 w lighting
Outside receptacle	180 w
	1000 w

Since this is too light a load for a 20-amp circuit and also, since the two receptacles near bathroom fixture B can be picked up with very little wiring, we add them to the circuit. This gives us a total of 1000 + 2 (180) = 1360 w. This information is then placed on the panel schedule in the space reserved for circuit no. 1. The schedule should be ruled up with enough space (⅜ in.) so that 2 lines of lettering can be used if needed.

Next we consider the remaining lighting in the house, including the basement and unexcavated areas. These are circuited to circuit no. 3, for reasons we explain below. We start with outlet D in the hall outside bedroom no. 2, and run to fixtures F and H in bedroom no. 2. A second line goes to the switched outlet under the living room window that serves as the lighting outlet in the living room. Then we go to the front lantern, and finally down to the basement lights and wall receptacle. This basement wall receptacle was picked up on the lighting circuit because at least two other receptacle circuits appear in the basement (washer and workbench), so that there is no possiblity that a single fault will cause the area to be without power. This is an extremely important factor in general, and all the more so in an isolated area with little or no daylight such as the basement. An initial load check for circuit no. 3 shows:

Hall fixture D	120 w max
Bedroom fixture F	150 w max
Closet light H	60 w (assume 60 w lamp)
Living room floor lamp	225 w (estimated for receptacle)
Basement ceiling lights	200 w max
Basement unexcavated areas	180 w (assume 60 w lamps)
Maximum total lighting	935 w

Since it is obvious that the probability of all of these lights being on at the same time is just about zero, we can readily add receptacles to this circuit. This

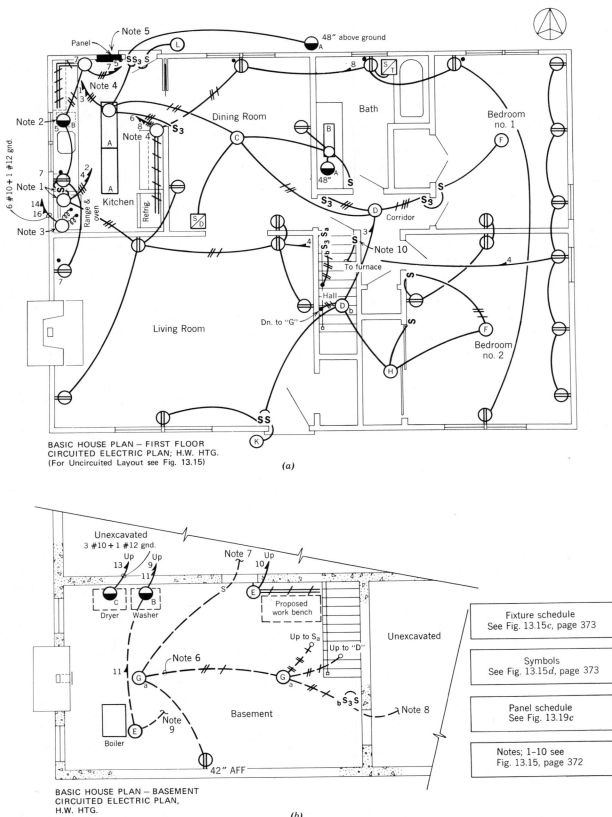

BASIC HOUSE PLAN — FIRST FLOOR
CIRCUITED ELECTRIC PLAN; H.W. HTG.
(For Uncircuited Layout see Fig. 13.15)

(a)

BASIC HOUSE PLAN — BASEMENT
CIRCUITED ELECTRIC PLAN,
H.W. HTG.

(b)

	**PANEL SCHEDULE FOR BASIC PLAN OF FIG. 13.19**

CIRC. NO.	DESCRIPTION	LOAD VA.	CIRCUIT BREAKERS		DESCRIPTION	LOAD VA.	CIRC. NO.
1	LTG — { KIT. DR., BR. #1 / OUTSIDE, BATH + ⏀	820 3R	20 1│2	20	OUTLETS — LR. & DR. + EXH. FAN	30 6R	2
3	LTG — { OUTSIDE / LR., HALL, BR. #2, BSMT. + ⏀	935 4R		3│4	OUTLETS — BR. 1 & 2	6R	4
5	DISHWASHER	1500	5│6		APPLIANCE OUTLETS — KIT., DR.	——	6
7	APPLIANCE OUTLETS — KITCHEN, LR.	——		7│8	OUTLETS — BATH, BR's.	——	8
9	LAUNDRY OUTLET — BSMT.	——	9│10		OUTLETS — BSMT.	——	10
11	HOT WATER BOILER	1300		11│12	SPARE		12
13	ELECTRIC CLOTHES DRYER	5000 {	30 A 13│14 / 2P 15│16	30 A / 2P	RANGE {	6000 {	14
15	SPACE FOR 2—1P	——	17│18	30 A	OVEN —— {	4800 {	16
	OR 1—2P	——	19│20	2P			

PANEL DATA
MAINS, GND. BUS: 150 A MNS., 60 A GND. BUS   VOLTAGE   120/240   1 PH.
MAIN C/B ~~OR SW/F~~   150/100
BRANCH C/B INT. CAP.   5000 AMP.
MOUNTING — ~~SURF~~/RECESS
REMARKS: FRONT SUITABLE FOR PAINTING

(c)

Figure 13.19

we have done by picking up two receptacles in bedroom no. 2 and one in bedroom no. 1. We chose these receptacles first, because the wiring required to reach them is short and, secondly, because they present little demand load. By this we mean that it is hard to imagine a situation when the basement, the crawl spaces, the living room and the bedroom lights *and* receptacles are all in use together. Even so, the maximum load would only be:

Lighting (above)	935 watts
Basement receptacle	180 w
Bedroom receptacles	540 w
Maximum load	1655 w

This is *well* within the maximum capacity of a 20-amp circuit and no. 12 AWG wiring, which according to the NEC should not exceed 80% of rating, or 1920 w. This information for circuit no. 3 is placed on the panel schedule, as was the circuit no. 1 data.

At this point we pause in the actual circuiting to study the reasons for connecting circuits nos. 1 and 3 as we did, in order to explain in detail the technique of circuiting. We began circuit no. 1 at the kitchen outlet, since it is nearest to the panel and thus gives the shortest *home-run*. A home-run is defined as the wiring between the first outlet and the panel. Obviously then, in the interest of economy, the home-run should be taken from an outlet as near as possible to the panel. Since we were picking up lighting on this first circuit, we drew a looped line to the dining room ceiling outlet. From the outlet box at dining room fixture C several lines emerge. One extends to corridor fixture D and another extends to a wall switch and dimmer. Refer to Figure 13.20 for the basic wiring of a standard single pole switch. Note carefully that circuiting lines are *always* drawn curved, to avoid any possibility of confusion with architectural work or equipment. The type of curve used is unimportant. Corridor fixture D, unlike fixture C, is switched from two points by a method known as *3-way* switching. The name is somewhat misleading, since the switches are 2 position, and two of them are used. Refer to Figure 13.21 which shows how 3-way switching works.

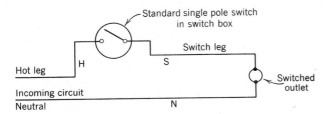

Figure 13.20 Typical standard single–point switching arrangement. A single pole switch is used. Note the terms and designations: hot leg, H;, switch leg, S; neutral, N.

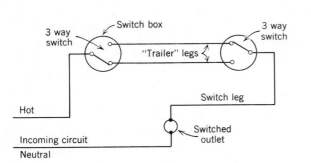

Figure 13.21 Typical 3-way switching arrangement. Note that two "trailer" legs are required between the two 3-way switches. The circuit is shown with the outlet OFF.

Figure 13.22 shows the three alternative methods by which the wiring could be connected between outlets C, D and F in the dining room, corridor and bedroom no. 1, plus their associated switching. We considered all three methods and decided on that shown in Figure 13.22*a* because it is the simplest and cheapest, requiring minimum wiring. Note in diagram (a) and on the plan, Figure 13.19*a*, that we feed through from S_3 at the end of the corridor to fixture F, via its wall switch. Feedthrough between switches is *not* usually a good idea because it clutters the small switch box with wiring that is not related to the switch. Here, however, it is an excellent arrangement because the hot leg is already at the switch box and only a run-through neutral leg is required. The same is true for the three switches grouped at the back door in the kitchen. To understand that feedthrough is not usually desirable, study the diagram in Figure 13.22*b*. There we ran the wiring from dining room outlet C into S_3 at the corridor. Observe how this wiring arrangement requires more wiring than plan (a) and clutters the switch boxes so that a larger box would be required. Plan (c) shows a system whereby all feedthrough in switch boxes is eliminated. This is also an acceptable plan, but it is somewhat more expensive than plan (a) and was therefore not used. It does have the advantage of not running any wiring through switch boxes that is not related to the switches. This is important because switches need periodic replacement which is made difficult by a cluttered box. Also, when doing such switch replacement in a light box with old wiring and dried out insulation, a fault may develop. This would not be true if all the wiring not related to the switch is in the ceiling box, where it is normally left untouched for the life of the wiring system.

The alert student who has been following the discussion and studying the diagrams may well ask a good question at this point. Why didn't we run the wiring from fixture C to bathroom fixture B and then to D in the corridor? Indeed, that appears good at first glance. Look at Figure 13.22*d* to see what it involves. Fixture B is wall mounted, and its box already contains wiring to feed a switch and two receptacles. To understand why it is not advisable to add any additional wiring to the box at fixture outlet B, we reproduce in Table 13.1 an excerpt from the NEC that deals with the wiring capacity of outlet boxes. In general, the purpose of restricting the number of wires in these boxes is simply to avoid crowding, which may cause overheating and insulation damage followed by faults. Note, for instance, the capacity allowed for a $3 \times 2 \times 2\frac{1}{2}$-in. device box.

Table 13.1 Wiring Capacity of Outlet Boxes

Box Dimensions, Inches, Trade Size		Maximum Number of No. 12 AWG Conductors
4 × 1¼	round, octagonal	5
4 × 1½	round, octagonal	6
4 × 2⅛	round, octagonal	9
4 × 1¼	square	8
4 × 1½	square	9
4 × 2⅛	square	13
3 × 2 × 1½	device	3[a]
3 × 2 × 2	device	4[a]
3 × 2 × 2½	device	5[a]
3 × 2 × 2¾	device	6[a]

Note. This table is extracted from Table 370-6(a) of the NEC, 1975 edition. In using this table, note that a deduction must be made where the box contains fixture studs, cable clamps, wiring devices, grounding conductors, and other devices. For wiring capacity in these situations, see the NEC. Connection to a fixture is not counted. A wire running through the box *without* a splice is counted as a single wire. All other wires entering the box are counted individually.

[a]Box capacity with no device in the box.

In such a box containing a wiring device and cable clamps for **BX**, the allowable capacity is reduced to three wires. This is exactly what is required when using such a box to switch a hot leg and also to feed through a neutral. However, if the device were a 3-way switch or we were feeding through a hot leg *and* a neutral (as we did, for instance, at the kitchen door switches) larger single boxes or gang assemblies would be used. In the case at hand, the box behind fixture B, which is normally a 4 × 1½ in. round, has a nominal capacity of six wires. Deducting one from this for cable clamps leaves a capacity of five wires. But, as you can plainly see from the diagram, we already have eight wires in the box, requiring us to use at least a 2⅛-in. deep box. Any additional wiring would require a collar on the box to deepen it. This is unusual in a wall box, although it is frequently done in ceiling boxes to give them additional capacity. Therefore, we would need a ceiling junction box as shown in Figure 13.22d. The cost of installing this box is greater than the cost of the additional 4 or 5 ft. of BX cable in layout (a). Note that to use layout (d) without ceiling box J, we would have to go down to outlet B *and up again*, thus losing some of the advantage we are attempting to gain in shortening the runs.

One further point must be noted. By means of small diagrams of the type shown in Figure 13.22, the technologist can arrive at the number of wires required in each run. He then shows by hatch marks, or tics, any run with more than two wires. See Figure 13.19. After some practice, the draftsman can use the single line type sketches of Figure 13.22c, d and e which show the wiring by abbreviation, without actually drawing the wires as in a and b. Finally, with enough experience, the layout man will be able to work out the choices and wiring mentally, without the necessity of making these little scratch paper diagrams at all, except in the most complex cases.

Summing up, thus far we have learned that:

(a) Small interconnection wiring sketches help us determine best circuit routing and the number of wires in each run.
(b) Feeding circuits through switch boxes is generally not advisable, but can be used in specific cases.
(c) It is good practice to limit the number of cables entering a wall box to three and in ceiling boxes to four. When using more than that number, a count of wires should be made to see whether the additional box capacity that will be needed will fit without creating space or construction problems.

Continuing with the analysis, let us examine the wiring of circuit no. 3. First note that we skipped circuit no. 2 temporarily and passed on to circuit no. 3, in order to reach another phase. Notice on the panel schedule, Figure 13.19c, that circuit no. 2 is on the same phase as circuit no. 1, phase A. If we had used circuit no. 2 for the second lighting circuit, we would end up with the two lighting circuits on the same phase that from the NEC point of view carry continuous load. This is undesirable for two reasons. First, it is good practice to balance the loads on the panel to the extent possible. This means that not only must the sum of the *connected loads* on the individual phases be approximately equal but the actual *demand loads*, that is, the loads actually being drawn, must also be balanced to the extent possible. Since lighting loads are by nature much more "continuous" than receptacle loads, we attempt to spread the lighting evenly over the phases. Thus we place the second lighting circuit on the other phase. The first available circuit on this second phase, Phase B, is circuit no. 3. The second reason for choosing circuit no. 3 is also a technical one and is mentioned above. We are permitted to carry two circuits

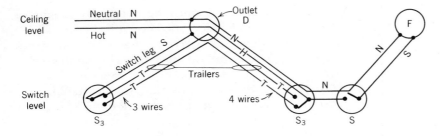

Figure 13.22 *(a)* Wiring between outlets, corresponding to ceiling lights in corridor and bedroom no. 1 of Figure 13.19. Heavy dot indicates an electrical connection or junction. Refer to Figure 13.21 for basic 3-way switching.

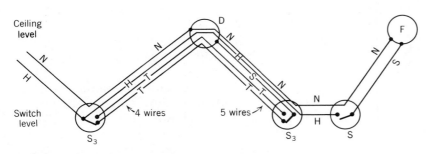

Figure 13.22 *(b)* Wiring for the same outlets as shown in *(a)* above, except with circuit entering via a switch outlet. The arrangement of *(a)* is clearly preferable.

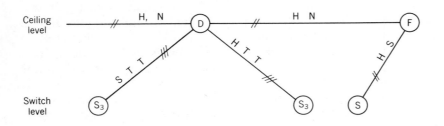

Figure 13.22 *(c)* Simplified version of wiring runs, identifying wires by abbreviation (H, N, S, T) and showing wire count by hatch marks (tics).

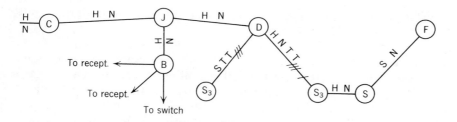

Figure 13.22 *(d)* Wiring arrangement including fixture B in bath, fed from ceiling junction box J. See chapter text. Note that fixture outlets are shown here larger than switch and junction boxes, for clarity and ease of identification. This was not done in *(a)* and *(b)* above because the switch configurations, with contact points, were shown there. Note that where the wiring consists of only two wires, no tics are shown. Also, where hatch marks are shown the neutral is separated from the other marks as between *D* and *S₃*. This is for ease of counting neutrals. The tics for the two trailer legs can also be separated, as in *(e)*. See the chapter discussion for further explanation.

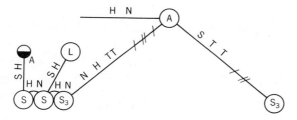

Figure 13.22 *(e)* Wiring arrangement in kitchen of Basic Plan. Note that feedthrough at the bank of switches is the most practical wiring method. Switches are mounted in a 3-gang box facilitating feedthrough.

on a *single* neutral wire, *provided that the circuits are on different phases*. If circuits nos. 1 and 3 are connected on phases A and B, respectively, we are permitted to use three wires—two hots and a *single, common* neutral. (See the hatch marks on the home-run from fixture A.) On the other hand, if we had taken the second lighting circuit from A phase as well, that is, if we had used circuit no. 2, we would need *two* neutral wires, or four wires in all. This obviously is more expensive and is to be avoided as an unnecessary expense and, therefore, as poor practice. The reason that we are allowed this economy in running neutrals is highly technical and will not be discussed in detail except to comment that the common neutral carries no more current than *one* of the two circuits and therefore is not overloaded.

As we will discuss in more detail later, this same principal is followed when wiring 3-phase panels. Three circuits, all on different phases, can be run with only a single neutral, without any danger of overload. In fact, if the loads are balanced (equal), the neutral current is zero. We have now learned two more principles of circuiting:

(d) Circuit loading should be balanced between phases to the extent possible.
(e) A single neutral will carry up to three circuits provided that these circuits are taken from different phases or panel buses.

Now let us return to Figure 13.19*a*, circuit no. 3. The loading of this circuit was explained above. Here again as in circuit no. 1, we use a feedthrough technique at the front door switches to feed the ouside lantern and the wall outlet. Feedthrough of this type to serve an outside light is fairly standard. The wiring coming from hall fixture D, to bedroom no. 2 fixtures F and H, plus switches and receptacles, can be arranged in a number of ways. Three of these are illustrated in Figure 13.23, along with the corresponding wiring–routing diagrams with which we have already become familiar. Let us analyze those diagrams and find out which presents the best arrangement and what special considerations are involved.

Wiring as in Figure 13.23*a* is a straightforward arrangement using single runs of 2 conductor BX. The only disadvantage is the amount of cable used because of the duplication of runs between feeds and switch legs. Look at the runs to fixtures F and H and their switch legs to see what we mean by this comment. This somewhat objectionable condition is corrected in the arrangement of Figure 13.23*b*. Note how the addition of a junction box (also called a J box) shortens the runs and avoids the back-and-forth wiring of Figure 13.23*a*. But there is a catch here also. The J box has 10 wires in it, which if we add the cable clamp as one wire, requires a 4-in. box with extension. Also, all boxes require access. Therefore, unless the house has a crawl space or unfinished attic, we cannot use this arrangement, unless we are willing to put up with a capped outlet stuck into the ceiling. Most people will not. If we ask why we do not feed across from one switch to another and avoid running back up to the box, the answer lies in the type of construction. We are assuming here a wood stud wall, which requires drilling the studs to make a horizontal run. Vertical runs are much easier and cheaper, and runs in an unfinished

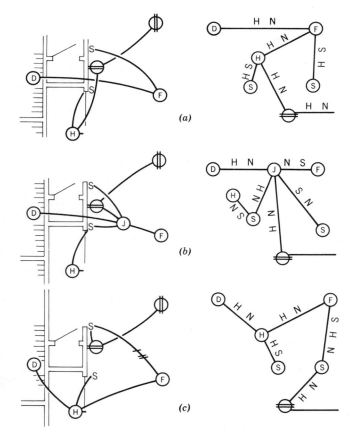

Figure 13.23 Alternate wiring schemes for part of Basic Plan, Figure 13.19 (see chapter discussion).

ceiling space can be run on top of the ceiling beams (see Figure 12.8).

The arrangement of Figure 13.23c is the system we have chosen, since in our opinion it offers the shortest runs, does not require junction boxes in the ceiling and has only a short horizontal run which is more than offset in cost by the cable saved. You are invited to try other arrangements to see if you can come up with a better one than any of those illustrated.

The last area that will be considered here in detail in this Basic Plan House is the wiring to stair fixture D. In Figure 13.24 we have drawn three different wiring layouts, each with merits and drawbacks. The layout of Figure 13.24a has the advantage that there is no backtracking of wiring. That is, we drop down from the first (main) floor ceiling fixture D to the switches, from there to the ceiling of the basement and, then, to the switches in the basement. We never go back up and down again and, therefore, we are being economical on cable. The disadvantage here is the heavy wiring runs between the two banks of switches and fixture G. In one case we have six wires and in the other, five. This means two 3-conductor cables. Also we have some feedthrough at switches, but this is only a neutral wire and is not objectionable. In layout (b), on the plus side, we have a straight short drop from fixture D to the basement switches. Looking at the plan, we see that this is almost a vertical run, with a turn at the ceiling—an excellent way to wire. This run unloads fixture G, so that we can feed the basement receptacle and the far side switch directly from here without running to the second G fixture as in layouts (a) and (c). On the minus side, we have a double box at fixture D to accommodate the 13 wires in it. (The fixture stud and cable clamps count as another, fourteenth, wire.) The two connecting legs of a 3-way switching arrangement count only once, since they run through without a splice. These wires are called trailers. Fixture D is in the ceiling of a stairwell—an extremely difficult place to reach if any repairs are required. Therefore, the less wiring in the outlet box at D, the better, unless access from an attic is provided. Finally in layout (c), on the plus side, we have no feedthrough at switches, minimal wiring at D, and heavy wiring at fixture G which is highly accessible. The drawback is some duplication of runs, in that we feed up from G to both D and to the switches. It would appear that no layout is obviously superior to the other two. We leave it to the student, as an exercise, to evaluate all three layouts in detail and to make a recommendation. We have chosen layout (c) to show on the drawing of the Basic Plan House, (Figure 13.19).

13.11. Circuitry of the Basic Plan House

We now describe the remainder of the circuitry for the Basic Plan House. Make any wiring sketches that you think necessary to follow the description and wiring as shown.

(a) Circuit no. 1 is fully described in the preceding section, covering approximately one half the lighting in the house.

(b) Circuit no. 3, like circuit no. 1 described in the previous section, is essentially a lighting circuit, covering the remainder of the Basic Plan House. It is connected at its closest point to circuit no. 1, that is, from fixture D in the hall to fixture D in the corridor. The single arrowhead on this run indicates that circuit no. 3 joins the wiring at this point and is carried with circuit no. 1 through the dining room and kitchen to the combined home-run from the A fixture outlet box. Being on separate phases, the two circuits can be carried with a single neutral as explained above. Hence, the home-run is three wires—a hot leg each for circuit nos. 1 and 3, and a neutral.

(c) Circuits nos. 2 and 4 are convenience outlet circuits arranged to cover a number of rooms. Both circuits carry six receptacles, although up to eight can be safely connected to a 20-amp circuit in accordance with the guidelines above. Circuit no. 2 also carries the kitchen exhaust fan, rated 30 watts. Each room of the house has parts of at least two, and normally three, circuits. Thus no room can be easily blacked out, even if one of the two phases goes out. Although such an occurrence is relatively rare, it happens occasionally as a result of power company outages. Note that the panel schedule shows circuits nos. 2 and 4 as receptacle circuits with a number of receptacles rather than showing a wattage. This is because, as stated, convenience outlets are part of the general lighting load of 3 w/sq ft. Load calculation is not part of our study here. Consult the NEC for information on this point. Refer to Sections 13.10 and 15.8 for a brief explanation of receptacle loads.

(d) Circuit no. 7 feeds the strip of kitchen outlets where most of the load will occur plus the living room appliance outlet. The NEC specifies that this is to be an appliance circuit. The dishwasher, circuit no. 5, is on an individual branch circuit. See Figure 13.17.

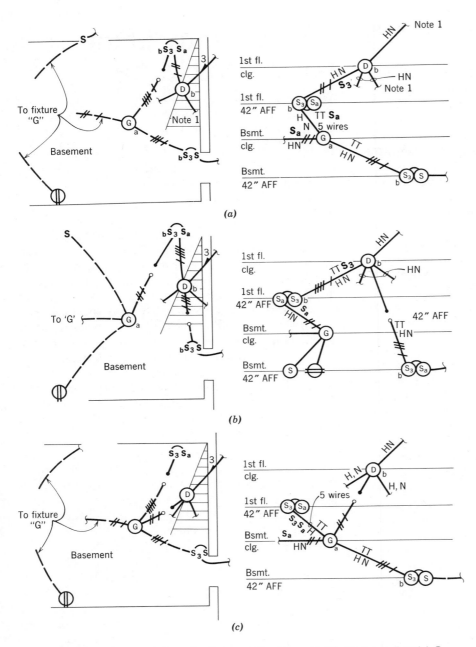

(a)

(b)

(c)

Note 1. Outlet D has 3 cable entries (from corridor fixture D, BR #2 fixture F and L.R. switches) which are shown for the sake of completeness. Since they do not affect, nor are they affected by, the various arrangements (a), (b) and (c), they would ordinarily be omitted.

Figure 13.24 Three wiring arrangements for the same outlets appearing on the Basic Plan of Figure 13.19. See chapter discussion for a comparison of the three layouts. Observe that both the physical layout (left) and the wiring connections (right) are necessary to be in a position to decide on the best layout. The physical plan gives distances and physical arrangement, and the wiring sketch shows connections and number of wires.

(e) Circuit no. 8 picks up appliance-type outlets in the bedrooms and a special heater outlet in the bath. Notice that this heater outlet will not be used at the same time that air conditioners are in use in the bedrooms, so it is safe to put it on circuit no. 8 along with the bedroom air conditioner outlets. Circuit no. 6 picks up the refrigerator, another wiring strip and the dining room appliance outlet. Here again a high diversity exists; that is, there is little chance of all the loads being on at the same time, and overload is very unlikely. If it occurs, the occupant will adapt his usange to avoid the problem.

(f) Circuits nos. 9 and 11 are special circuits as noted on the panel schedule, devoted to the laundry and the hot water boiler, respectively. Circuit no. 10 picks up the multi-outlet assembly in the basement. This latter may carry some rather heavy machine tool loads, but a 20-amp circuit is sufficient for all normal home-craftsmen tools. If a heavy piece of machine shop equipment is installed, such as a drill press or a lathe, a special circuit would be required. For this purpose the spare 20A circuit breaker in circuit no. 10 could be changed to a 30A unit and heavy wiring would be run to the unit.

(g) Circuits nos. 13, 14 and 16 are individual branch circuits for single items of electrical equipment. We have assumed here a four-burner electric range top plus an electric oven, since this combination is very frequently found in actual use. Obviously, if gas cooking were used in either of these places, the circuit would be eliminated and the panel would be changed to suit. The wiring shown on the drawings for these circuits would be obtained by the draftsman from the electrical designer.

(h) To finish off the panel, circuit no. 15 is actually a space for a future circuit breaker(s). We leave a double space here which can be filled by one 2-pole or two single-pole circuit breakers, 50-amp frame, from 15-amp to 50-amp trip. It is customary to leave about 20% space in the panel for spares. Here we have 17 poles, so we are leaving 3 spare poles—1 at circuit no. 12 and 2 at circuit no. 15. The remaining panel data relating to buses, circuit breaker interrupting capacity, mounting, main circuit breaker and panel voltage is filled in by the technologist from data supplied by the designer.

13.12. Basic Plan House —Electric Heat

The drawing in Figure 13.25 represents the same house plan except that electric heat has been added. The designer has decided to add a separate panel to the house to feed the electric heaters—a common practice. He therefore splits the incoming line at the service entrance point in the kitchen wall and runs a feeder to the electric heat panel, EH, which is centrally located in the house. The single-line diagram shows this arrangement. The switch and feeder sizes have been calculated by the designer and have been given to the draftsman as a rough sketch. This becomes the small riser drawing shown. To avoid cluttering the drawing, the designer and draftsman together decide to handle the circuitry of the electric heaters by using the panel schedule and a small wiring diagram on the drawing. See Figure 13.25d. The heater wiring is straightforward and simple. The branch circuit feeds throgh the thermostat to the heater, as the diagram shows. Actually showing the runs would add little if anything to the information required to wire the house. The house panel remains almost the same except for dropping the boiler circuit, which becomes a spare. This led to a slight rearrangement of circuiting which is left to the student to pick up when he traces out the circuits in panel A, Figure 13.25c, as compared with the house panel, Figure 13.19c.

With this we complete our analysis of the Basic Plan House electric plan, which was presented as a study example and not as a proposed house design. If you have followed the discussion carefully, you should now be in a position to do residential-type circuitry independently from the job engineer. In the chapters that follow we discuss more specialized wiring and other types of buildings that present particular drawing problems and situations. Also, we consider some of the design factors of residential buildings, to give the technologist a firm background in this subject.

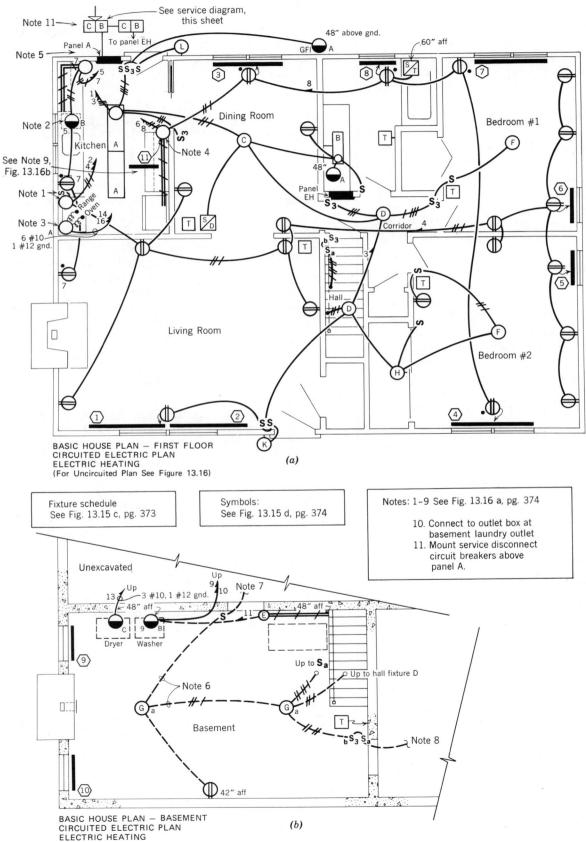

Note 11→

See service diagram, this sheet

To panel EH

48" above gnd.

60" aff

Note 5→

Panel A

GFI A

Note 2→

Dining Room

Bedroom #1

Kitchen

See Note 9, Fig. 13.16b

Note 4

Note 1→

Panel EH

Corridor

Note 3→

6 #10
1 #12 gnd.

Range
Oven

Hall

Living Room

Bedroom #2

BASIC HOUSE PLAN — FIRST FLOOR
CIRCUITED ELECTRIC PLAN
ELECTRIC HEATING
(For Uncircuited Plan See Figure 13.16)

(a)

Fixture schedule
See Fig. 13.15 c, pg. 373

Symbols:
See Fig. 13.15 d, pg. 374

Notes: 1–9 See Fig. 13.16 a, pg. 374

10. Connect to outlet box at basement laundry outlet
11. Mount service disconnect circuit breakers above panel A.

Unexcavated

Up

3 #10, 1 #12 gnd.

48" aff

Note 7

48" aff

Dryer

Washer

Up to S_a

Up to hall fixture D

Note 6

Basement

Note 8

42" aff

BASIC HOUSE PLAN — BASEMENT
CIRCUITED ELECTRIC PLAN
ELECTRIC HEATING

(b)

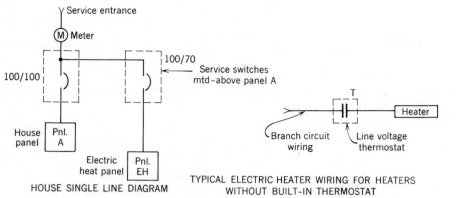

HOUSE SINGLE LINE DIAGRAM

TYPICAL ELECTRIC HEATER WIRING FOR HEATERS
WITHOUT BUILT-IN THERMOSTAT

CIRC. NO.	DESCRIPTION		LOAD VA	CIRCUIT BREAKERS		DESCRIPTION	LOAD VA	CIRC. NO.
	PANEL A FOR BASIC PLAN, FIG. 13.25							
1	LTG.	KIT., DR., BR. 1 OUTSIDE, BATH + ⊕	820 3R	20 1 ǀ 2	20	OUTLETS — LR. & DR. & FAN	30 6R	2
3	LTG.	OUTSIDE LR., HALL, BR. 2, BSMT. + ⊕	935 4R	3 ǀ 4		OUTLETS — BR. 1 & 2	6R	4
5	DISHWASHER		1500	5 ǀ 6		APPLIANCE OUTLETS — KIT., DR.	—	6
7	APPLIANCE OUTLETS — KIT, LR.		—	7 ǀ 8		OUTLETS — BATH, BR'S.	—	8
9	LAUNDRY OUTLET — BSMT.		—	9 ǀ 10		SPARE	—	10
11	BASEMENT MULTI — OUTLET STRIPS		1300	11 ǀ 12		SPARE	—	12
13	ELECTRIC CLOTHES DRYER		5000	30 A 13 ǀ 14 2P	30 A 15 ǀ 16 2P	RANGE	6000	14
15	SPACE FOR 2–1P OR 1–2P		—	17 ǀ 18	30 A	OVEN	4800	16
			—	19 ǀ 20	2P			
				21 ǀ 22				

PANEL DATA

MAINS AND GND BUS: 150 A. MNS. + 60 A GND BUS   VOLTS 120/240

~~MAIN C/B OR SW/FUSE~~ LUG IN MAINS ONLY

BRANCH C/B INT. CAP. 5000 AMP.

MOUNTING—~~SURF~~/RECESS

REMARKS: FRONT SUITABLE FOR PAINTING

(c)

CIRC.	DESCRIPTION		BUSES		DESCRIPTION		CIRC.
	PANEL EH — ELECTRICAL HEAT						
1	HTR. #1, L.R.,	1500W	1 ǀ 2		HTR. #11, KITCHEN	1500W	2
3	HTR. #2 L.R.,	1500W	3 ǀ 4		HTR. #3, DIN. RM.	1250W	4
5	HTRS. 5 & 6, BR. 1 & 2	1250W	5 ǀ 6		HTR. #7, BR. #1	1000W	6
7	HTR. #4 BR. #2	1500W	7 ǀ 8		HTR. #8, BATH	1000W	8
9	SPARE		9 ǀ 10		HTRS. 9 & 10, BSMT.	1500W	10
11	SPACE		11 ǀ 12		SPACE		12

PANEL EH DATA   120/240 VOLTS

100 A MNS, 60A GND. BUS'   12 CIRCUITS

LUGS IN MAINS ONLY.

ALL BRANCH C/B 50 AF/20 AT. 5000 A.I.C.

FLUSH MOUNT; COVER SUITABLE FOR PAINTING

(d)

Figure 13.25.

Problems

13.1. Draw a one-line diagram of a branch circuit consisting of a 20-amp circuit breaker, a No. 12 feeder and a 1500 w load. If you feel any additional data is necessary, add it.

13.2. Draw in single line or block diagram form a circuit showing the following items properly, with respect to each other:

Service entrance feeder, a wall receptacle, house panel, main service switch, overhead room light, panel branch circuit switch and fuse.

13.3. (a) List all the fuse sizes and circuit breaker trip sizes that would protect a wire of 50-amp capacity from overload. Use the NEC or the list in Section 13.2 for sizes.

(b) Which of these sizes would also be useful if this 50-amp feeder were carrying a 40-amp load?

13.4. Draw schematically a single-phase 120/240 v 16 circuit panelboard. Show 100A mains, a 100/70A main circuit breaker, and these circuit breakers: 8-SP, 20A, 4-SP, 30A, 2-2P, 20A and 2-2P, 30A.

13.5. In a single phase 120/240V panel on what phase are these poles: 1, 4, 8, 10? Answer the same question for a 3-phase panel. Use phases A and B for single phase and A-B-C for 3 phase.

13.6. In residential wiring

(a) What is the square foot load allowance for general lighting?

(b) What is the minimum number of appliance circuits?

(c) What is the minimum number of laundry circuits?

(d) Define an appliance circuit, a general purpose branch circuit and an individual branch circuit.

13.7. What is the purpose of appliance outlets? When are they *not* used?

13.8. Where are GFCI receptacles required by the NEC? Where else would you recommend them?

13.9. In figuring circuit loads, what is used for a duplex receptacle? An appliance circuit? (see the NEC).

13.10. What is the maximum to which a panel circuit breaker can be loaded according to the NEC: 50%, 60%, 80%, or 100%? What does this mean in watts on a 120 v, 20-amp circuit?

13.11. Draw a 3-way switching arrangement controlling two outlets together. Show all wiring.

13.12. In figure 13.24 select the wiring arrangement you think is best and explain why. If you can improve on it, do so, also explaining why.

13.13. Find a method for switching an outlet from three locations (called 4-way switching) and draw it. Note any special equipment required.

13.14. A client has examined the electric plan of Figure 13.15 and requests that certain changes be made to the electric layout, before construction begins. These are:

1. Kitchen—add a light over the sink and an outlet under the sink for a garbage disposal unit.
2. Dining room—arrange the ceiling outlet for 3-way switching from the two doorways.
3. Living room—arrange 3-way switching for the switch controlled receptacle, from the two doorways.
4. Bathroom—add a night-light.

Make all the necessary alterations to Figure 13.19 to accommodate these changes and add devices, such as switches, made necessary by these changes. Make all appropriate chages in the panel schedule, fixture schedule, notes, and the like.

Additional Reading

Handbook of Residential Wiring Design, Edison Electric Institute.

Simplified Electric Wiring Handbook, Sears Roebuck & Co.

Residential Wiring Handbook, Committee on Interior Wiring Design

American Standard Requirements for Residential Wiring.

All of the above volumes contain recommendations for wiring residential spaces, and as such provide supplementary material.

Architectural Graphic Standards, C. G. Ramsey and H. R. Sleeper, Sixth Edition, John Wiley.

Mechanical and Electrical Equipment for Buildings, William J. McGuinness and Benjamin Stein, John Wiley.

14. Lighting Fundamentals

There is obviously more to lighting than locating a ceiling lighting outlet. So much more, in fact, that lighting design has become a specialty. Most building lighting design work is done at present by the building electrical designer with the assistance of a technologist, and much of this design work is very well done. Once the technologist has mastered the fundamentals of lighting, he can pursue its technical and artistic aspects, to the extent of his ability. This chapter is devoted to a study of the basics of lighting. It is divided into three parts: how light behaves, how light is produced and how light is used. In the course of this study we will learn about light sources, illumination levels and lighting fixtures. This information coupled with the knowledge the technologist has already obtained about building circuits will give him the necessary background to approach an overall building electrical layout.

After completing study of this chapter you will be able to:

1. Understand fundamental behavior of light including reflection, transmission and diffusion.
2. Distinguish between the factors that affect both the quantity and the quality of light.
3. Calculate illumination in both conventional and SI (metric) units, and convert between the two systems, as desired.
4. Understand the effect of brightness ratio, contrast and glare on the quality of a lighting installation.
5. Know how to lay out a lighting system with minimum direct and reflected glare.
6. Select a light source for an installation that will give proper quantity and quality of light, along with operating economy.
7. Perform illumination and reflectance measurements using simple meters.
8. Understand the operating and illumination characteristics of all the major light sources.
9. Design and lay out lighting for interior spaces, given illumination requirements.
10. Draw details of lighting fixtures and architectural lighting elements.

14.1. Reflection of Light

You are able to read this book because light is reflected from the page and enters your eye. This process is illustrated in Figure 14.1. *Reflection* is one aspect of the behavior of light that is of particular interest to us. Other factors are *absorption* and *transmission*, and the particular way in which these processes occur. But, we are not physicists studying light as such. We are principally interested in how to apply light—in other words, we are interested in *illumination*. Therefore, we discuss the five factors that affect our ability to see clearly and well—*brightness, contrast, glare, diffuseness* and *color*, plus the usual terms related to these factors. When light falls on an opaque object some of it is reflected and some of it is absorbed. The ratio between the amount reflected and the original amount is called the *reflection factor*, or simply *reflectance*. The reflection factor of an ordinary mirror is quite high—90% or more. The paper on which this book is printed has a reflectance of about 75%. The light that is not reflected is absorbed by the material and is lost. Therefore, we can see that in lighting fixtures the reflecting surfaces must be treated to give high reflectance and, therefore, minimum light loss. Actually, the white paint found on the inside of a typical fluorescent fixture has a reflectance of about 88%. That is, 12% of the light is lost and 88% of the light from the lamps is reflected as useful light. A glance at Figure 14.2 will make this concept clear.

Although light reflection is obviously necessary to the act of seeing, it can also be quite annoying if it is mirrorlike, or what is technically termed *specular* reflection. In specular reflection the source of light is reflected in the object at which we are looking, causing glare. That is why reading a "slick" magazine can be very troublesome if the light is not placed properly. We will have more to say on this subject when we study glare and glare control.

If the surface of the object being viewed is not specular, or "shiny," we get a type of reflection that is much more pleasant to the eye—*diffuse reflection*. This is the type of reflection given by a dull, flat finish surface. The difference between diffuse and specular reflection can readily be seen by comparing the appearance of matte finish and glossy finish photographs. Most materials give both diffuse and specular reflection but one kind of reflection is more pronounced than the other. See Figure 14.3 for a diagrammatic illustration of these two types of reflection.

14.2. Light Transmission

We are probably as familiar with this characteristic of light as with reflection. Sunlight comes in through the window; light comes from the frosted incandescent lamp; light comes through the plastic lens of the fluorescent fixture, and so on. Just as in reflection, the ratio between the incident and transmitted light is called the *transmission factor*, or

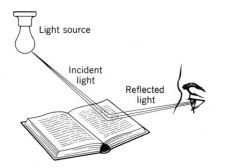

Figure 14.1 The ability to see the words in the book is a result of light being reflected from the book into the eye. Thus we see most objects by reflected light.

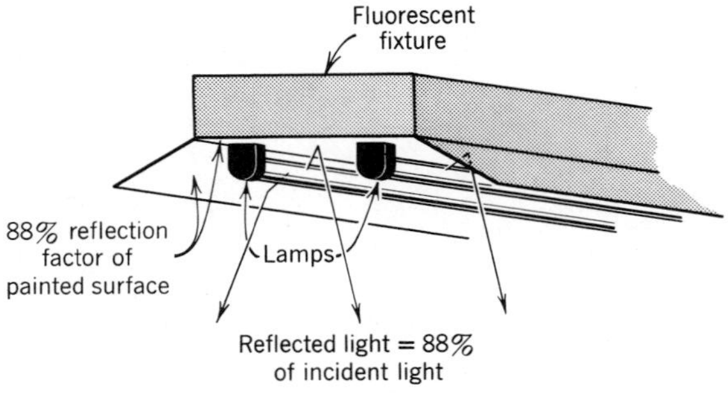

Figure 14.2 The light output of the lamps in a fixture is reflected from the inside fixture surfaces as shown. A 12% loss results if the reflectance of the surfaces is 88%.

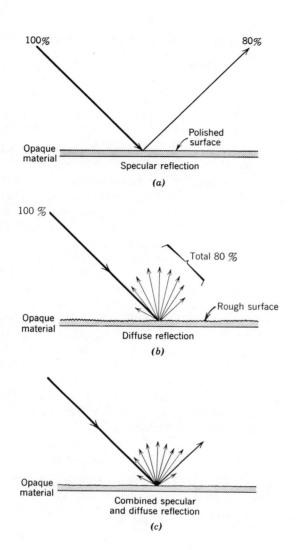

Figure 14.3 Types of reflection.

(a) Mirrorlike or specular reflection on a glossy, polished surface. This type of reflection allows the source of light to be seen by reflection and frequently causes glare.

(b) In diffuse reflection light is spread in many directions. Note that here also, the reflectance is 80%, with 20% absorbed. This type of reflection, which is found in all flat or matte fininshes, is easy on the eye and reduces glare.

(c) Most surfaces show a combination of specular and diffuse reflection. Control of surface finish is often used to control glare problems.

transmittance. Also, as with reflection, we have diffuse and nondiffuse transmission of light. See Figure 14.4 for simple ray diagrams that illustrate these two types of transmission. A piece of clear glass or plastic shows almost complete nondiffuse transmission with very little absorption or reflection. A piece of translucent material, however, such as frosted or white glass, milky plexiglass or tissue paper, gives diffuse transmission, low reflection, but relatively high absorption. Although, frequently, high absorption is the price paid for good diffusion, inside frosted incandescent lamps have no higher loss than a clear glass lamp, but give almost perfect diffusion.

Transmission characteristics of materials are of particular importance in concealing the light source of a lighting fixture. We generally select a material that has high transmission so that losses are low and efficiency is high. However, we don't want to "read" (see) the lamps and, therefore, we want to high diffusion or some sort of lens action. And, finally, we want minimum reflection because this causes re-reflections inside the fixture, and lowering of efficiency. The materials generally used are various types of glass and plastic that fulfill these requirements fairly well. Interestingly, these fixture closures are called diffusers, after their original purpose and design. This term is used even if the material is not a diffusing material, such as a glass or plastic lens. This topic will be studied further in our discussion of lighting fixtures.

14.3. Light and Vision

In Section 14.1 we mention some of the illumination factors that affect how well we see. We say illumination factors because since the other principal factor is the eye itself, and that is not our concern. Our concern is to provide the best possible illumination within our budget limits. Also, when we speak of illumination, or simply lighting, we are referring to man-made lighting. Daylighting, which is excellent, is not included in our study here. We must assume a nighttime condition.

When lighting designers talk about lighting, they refer to two things—the quantity and the *quality* of lighting. The first, quantity, can be calculated, measured, and is fairly easy to handle. The second item, quality, is a mixture of all of the items related to illumination other than quantity of light. This mixture includes brightness, brightness ratio, contrast, glare, diffuseness and color. In addition, most de-

signers include in *quality* items such as psychological reactions to color and fixture patterns.

14.4. Quantity of Light

It is somewhat difficult to speak of the "quantity" of light as if it were an item that can be boxed or bottled. However, we have already overcome this type of difficulty in our study of heat. There, we speak of the amount of heat generated or lost, and measured in BTU per hour. Here, we speak of the light, generated continuously, and give it the unit of *lumens* (abbreviated lm). A specific source, say, a 60 w incandescent lamp, produces 850 lm of light, continuously. A 40 w, 48-in. cool white, rapid-start, fluorescent lamp produces 3200 lm of light, continuously. When this light *flux*, or continuously generated light energy, falls on an object, the object is illuminated, and we see it by reflection, as explained above. We are interested in the amount of light that falls on the areas we want to illuminate. Therefore, we want to know the lumens per square foot in a space. This quantity, *light flux density*, is the familiar *footcandle* (abbreviated fc).

$$\text{footcandles} = \frac{\text{lumens}}{\text{square feet}}$$

Let us illustrate this simple relation. Suppose that we have a light fixture in a room that causes 1000 lm to strike the floor. The room is 10 ft sq. What is the illumination on the floor?

$$\text{footcandles} = \frac{1000 \text{ lm}}{10 \times 10 \text{ sq ft}} = 10 \text{ fc}$$

(This is poor lighting for close work.) We shall learn later how to calculate the room illumination if given the lighting fixture data and the room dimensions. At this point we want to emphasize that the footcandle is the important unit of light to the technologist. In practical design and layout work the technologist will be given the required illumination in footcandles and will be asked to calculate the lighting required and to lay out the fixtures. Undoubtedly you will encounter tables of footcandle recommendations. Such tables have been issued by many sources and are generally based on IES (Illuminating Engineering Society) recommendations. Recently, with emphasis on energy conservation, most authorities have reduced their recommendations. Also, research into glare problems has resulted in some new concepts. They include a quantity called *ESI footcandles*, which indicates the usable footcan-

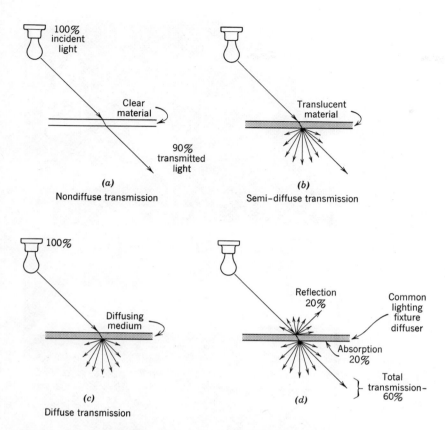

Figure 14.4 Light transmission through different materials.

(a) Clear, transparent materials transmit about 90% of incident light.

(b) Instead of the clear glass or plastic of (a), the diffusing medium here is a translucent glass or plastic. It is called semidiffusing, since the source can be seen somewhat—that is, there is some nondiffuse transmission. The light source can be "read" through the diffuser.

(c) Complete diffusion results in an even brightness on the diffuser and complete hiding of the light source. Milk white glass has this effect.

(d) With most diffusing materials there is some reflection from the top surface, some absorption in the material, some direct transmission, and some diffuse transmission. For common, commercial light diffusers, a figure of 60% for total transmission of light is reasonable.

dle illumination at a given position in a room as opposed to the "raw" footcandles obtained from the previous calculation. For these reasons no table of recommended footcandles is included here. A popular rule of thumb for levels today is the 10-30-50 rule. This states that 10 fc is adequate for halls and corridors, 30 fc is adequate for areas between work stations, such as those in an office other than the desk areas, and 50 fc is adequate at desks where office work is done. These levels are much lower than the recommendations of just a year or two ago. Hence, one must look closely at the source and *date* of any footcandle requirement tables. Anything prior to 1973 probably is prior to our energy-conscious time, and may be much higher than is desirable.

14.5. Brightness and Brightness Ratios

As we have stated several times, we see by reflection. If we place a piece of black velvet and a piece of white paper on a table, obviously the paper will look brighter than the velvet even though both have equal illumination. The paper reflects more light and is therefore brighter. In lighting terms, the paper has a higher *brightness* than the velvet.

It has been found by experience that if a light object is placed on a dark background, eye discomfort can result. The IES recommends a maximum brightness ratio of 3 to 1 between the object being viewed and the background. Therefore, a piece of white paper on a dark desk top can easily cause discomfort. See Figure 14.5. It is for this reason that modern office furniture is generally light colored—tan or light green being most comfortable to the eye.

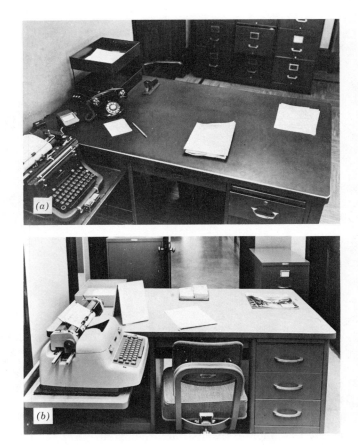

Figure 14.5 White paper on a dark desk as in *(a)* gives a brightness ratio of as much as 15 to 1 which causes eye discomfort. A much more desirable condition is shown in *(b)* where the light color of the furniture gives a brightness ratio between work and background of less than the recommended 3 to 1 maximum. (Courtesy of Illuminating Engineering Society)

14.6. Contrast

In the previous section we explain that a large difference between the brightness of what we are looking at and the background can be annoying. This is true when the background is dark and the object is light as in Figure 14.5. It is also true in reverse. When we pass a person in the street and the light is behind him (bright background), we have difficulty seeing his face clearly. All we can really see is his outline, or silhouette. This effect can also be helpful by providing what is called *contrast*. See Figure 14.6. You are

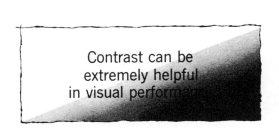

Contrast can be extremely helpful in visual performance

Figure 14.6 The importance of contrast is clearly shown here. Look at this drawing in poor light and notice that the black on white can still be easily read. Compare this with the amount of light you need to read the end of the word "performance" where there is almost no contrast.

now reading black print on white paper. You see the print clearly because you are reading the shape or outline of dark letters on a bright background. If this book were printed in yellow or white letters on a white paper it would be very hard to read. The outline of the letters would fade into the background because of lack of contrast.

14.7. Glare

More effort and money has been spent in attempting to reduce glare effects than on any other lighting problem. There are two types of glare: *direct* and *indirect*. Direct glare is the annoying brightness of light sources in a person's normal field of vision. A draftsman sitting at his table with his head up can see the ceiling lighting fixtures, his desk and the area in front of him. See Figure 14.7. Therefore, to control direct glare, ceiling fixtures are shielded and designed in such a way that the bulbs or their reflections in the fixture are not seen. Also, in a well-designed fixture, the luminous areas such as the sides and bottom are not too bright, so that direct glare is not a real problem in a modern office. Recently, we have seen and heard of users in commercial and institutional buildings removing diffusers from fixtures and also some of the lamps in an effort

to reduce energy consumption. The result is a bare bulb installation that brings back the problem of direct glare—a problem that had long since been solved. Worse yet, it aggravates the problem of reflected glare.

Reflected glare is much more serious and difficult to control. Reflected glare, which is called in technical lighting language, *veiling reflection*, is just what the term says—glare due to reflection. Look at Figure 14.8. The technologist or draftsman sitting at his table generally has his eyes down, looking at the drawings. Glossy pencil-cloth and his plastic triangles reflect the ceiling lights into his eyes, causing reflected glare. You have undoubtedly experienced this type of glare and know how bad it can be. It causes the line work and lettering on a drawing to wash out in some places and to shine in others (see Figure 14.9), generally making it impossible to see properly. A draftsman faced with a serious glare problem will try to do one or more of these things to reduce the glare:

(a) Move his entire table.
(b) Change the angle of the table top.
(c) Reposition his desk lamp (if he doesn't have one he'll see about getting one).
(d) *Reduce* the brightness of the ceiling fixture by removing lamps, or changing the diffuser.
(e) Change the type of paper being used for drawing.

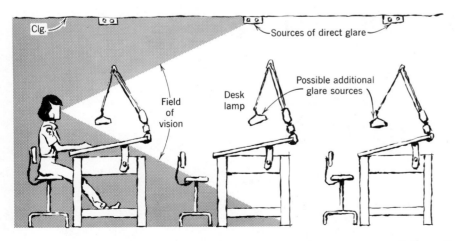

Figure 14.7 The draftsman sitting at his table with head up can see all the ceiling fixtures in front of him and all the desk lamps. Each one is a source of direct glare.

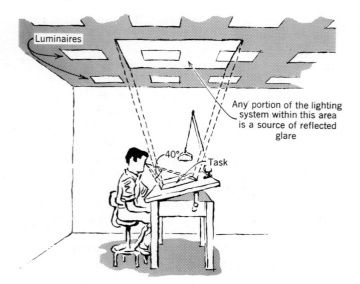

Figure 14.8 The ceiling fixtures in the outlined area reflect onto the draftsman's work, causing reflected glare on his drawing, triangles, parallel straight edge and instruments.

Figure 14.9 Notice the effect of an increasing amount of light in the area of the ceiling which causes reflected glare. The black print washes out. The effect is even worse for a draftsman working on glossy pencil-cloth with reflective instruments and triangles. Normal viewing angle varies between 20 and 40 deg. from the vertical. (Courtesy of Illuminating Engineering Society)

Why do these "cures" work? Let's examine them individually, while we look at Figure 14.8.

(a-1) Moving his table will not help in this room, since the entire ceiling is uniformly covered with lighting fixtures. This solution only helps when there is a single fixture or row of fixtures as in Figure 14.10. There, moving the desk so that the fixture is at position (b) will eliminate the reflected glare. This is the origin of the rule that for best lighting the light should come from over your left shoulder. (If it comes over your right shoulder, your right hand casts a shadow on your work.)

(b-1) Since the desk top is acting like a mirror, changing its angle will change what is reflected on it. Notice that as the desk top gets higher and higher, the area of ceiling that can create this problem gets smaller and smaller. (As an exercise, redraw Figure 14.8 with the desk at 60 deg from the horizontal and note how small the "offending" area of ceiling becomes.) This is why many draftsmen and artists work with tables that are almost vertical. Reflected glare is almost completely eliminated.

(c-1) Positioning the desk lamp so that the entire work area becomes bright eliminates the glare. This is the same as eliminating the glare of a flashlight in a dark room by opening the blinds and letting in the sun. The overall light level becomes so high that we no longer see the glare. This method is often the only one the draftsman can use.

(d-1) Referring again to Figure 14.8, notice that if we shut off the fixture(s) causing the glare, the glare will disappear. Obviously the illumination or footcandle level will drop. Despite this drop, we can frequently see better.

(e-1) Reflected glare is caused by a light reflected in a glossy object. In (d-1) above we remove one end of the problem by reducing the light. The other end of the problem can also be handled. Remove the glossy objects. Use diffuse white paper, matte finish triangles, parallel straight edge and the like.

With the recent emphasis on energy conservation, drafting rooms are now being designed with lower level general lighting plus desk lamps at each desk for concentrated local light. Since the overhead lighting is reduced and the desk lamp can be positioned at will, the reflected glare problem in such an office should be minimal. It is important, however, that we do not go back to the bare bulb era in the name of energy savings. It is a poor economy measure. A good low-brightness diffuser is very efficient and goes a long way toward reducing glare. With reduced glare comes increased production and real economy.

Figure 14.10 If the desk is moved so that the fixture (or row of fixtures) is at position *(b)* with respect to the desk instead of position *(a)*, reflected glare will be completely eliminated.

14.8. Diffuseness

This quality of light is a measure of its directivity. A single lamp produces sharp deep shadows and little diffusion. A luminous ceiling produces a completely diffuse illumination and no shadows. Usually, neither extreme is desirable.

14.9. Color

Volumes have been written on color: its definition, effects, how to produce it and other characteristics. To the technologist, however, color of lighting is generally a secondary factor. We normally assume all lighting to be white. The fact that incandescent lamps produce a yellowish light while cool white fluorescents produce a blue-white is normally given little consideration. And with good reason. The eye adapts quickly to the color of the light provided. After a short while in the room, the light produced by all the major lamp sources looks white. We will, however, give some specific recommendations for lamp choice based on color when we discuss the different types of lamps.

14.10. Footcandle Measurement

As we learned in Section 14.4, the unit of illumination is the footcandle. This is the unit most frequently used when describing the amount of light in a room. For instance, the lighting designer specifies that general office space be illuminated at 50 fc and that corridors be illuminated at 10 fc. Not only must the technologist know how to calculate these levels in new areas, he must also know how to measure them in existing spaces. To measure footcandle levels, one of the types of footcandle meters shown in Figure 14.11 would most probably be used. These meters are generally direct reading in footcandles. When one is measuring illumination levels, the meter should be held with its sensitive surface horizontal, and at least 12 in. from the body. If possible, the meter should be placed on a stable surface and read from a distance. Care must be taken that the person doing the survey does not block any of the light. When doing a general illumination check in a room, the meter should be held about 30 in. above the floor, which is desk height. Readings should be taken throughout the room and the results should be recorded on a plan of the room. Detailed instructions for conducting field surveys are provided in the IES publication, "How to Make a Lighting Survey."

14.11. Reflectance Measurements

We often want to know the reflectance of a material that we are thinking of using in a lighting installation, or of a surface in an existing space. Two methods of calculating reflectance are shown in Figure 14.12, using a simple, readily available footcandle meter. Results are accurate to about 5%, which is close enought for normal lighting design work.

14.12. Artificial Light Sources

In the sections that follow we briefly survey the common types of light sources in use today. These include incandescent, fluorescent, and high intensity discharge (HID) types, of which mercury lamps are the oldest example. The type of lamp and the fixture to be used in a particular application are chosen by the engineer or lighting designer. The technologist is responsible for proper wiring, switching, circuiting and detailing. He therefore should know the lamp sources, their electrical characteristics, physical shapes and dimensions, and something about their application. Note that we are dealing only with artificial light sources here. Daylighting is a specialty that is not handled by the electrical design team.

14.13. Incandescent Lamps

Construction of a typical general service incandescent lamp is shown in Figure 14.13. Incandescent lamps are made in a wide variety of shapes and sizes with different types of bases. See Figures 14.14 to

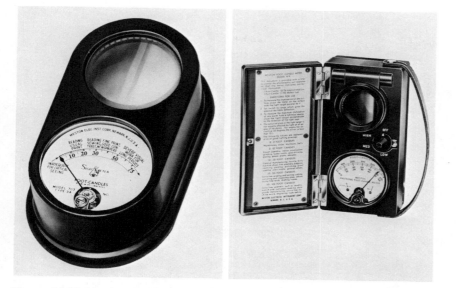

Figure 14.11 Two common types of direct-reading footcandle meters. A third pocket-type is illustrated in Figure 14.12, being used for reflectance measurements. (Courtesy of Weston Electric Instrument Corporation)

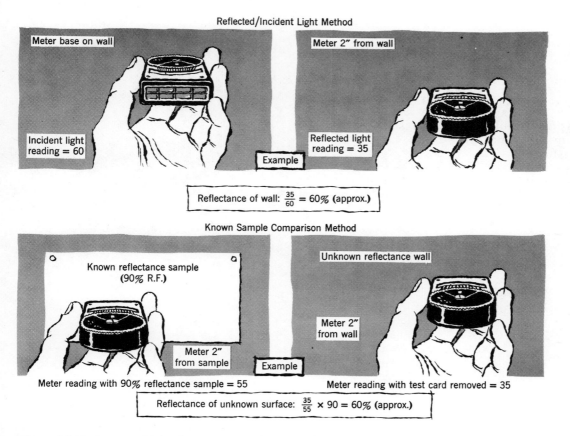

Figure 14.12 Two methods of measuring reflectance factors.

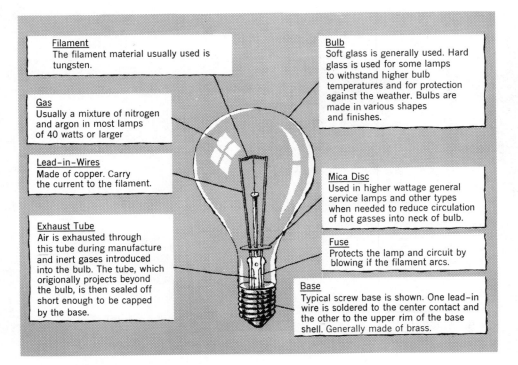

Figure 14.13 Typical incandescent lamp.

Table 14.1 Some Typical Incandescent Lamp Data (Listing a Few of Many Sizes and Types of 115, 120, and 125 v Lamps)

Watts and Life		Lumens		Physical data	
Lamp Watts[a]	Av'g Rated Life (Hours)	Initial Lumens	Lumens per Watt	Shape of Bulb[b]	Base
60	1000	855	14.2	A-19	Med
60	1100	855	14.2	A-19	Med
75	750	1180	15.7	A-19	Med
100	750	1750	17.5	A-19	Med
100	750	1710	17.1	A-19	Med
100 SB	1000	1450	14.5	A-21	Med
150	750	2760	18.4	A-21	Med
200	750	4000	20.4	A-23	Med
200/SBIF	1000	3300	16.5	PS-30	Med
300/SBIF	1000	5250	17.5	PS-35	Mog

[a]All inside frosted unless otherwise noted. Other letters have these meanings: W, white; SB, silver bowl; SBIF, silver bowl, inside frosted.

[b]See Figures 14.14 and 14.15.

BULB SHAPES

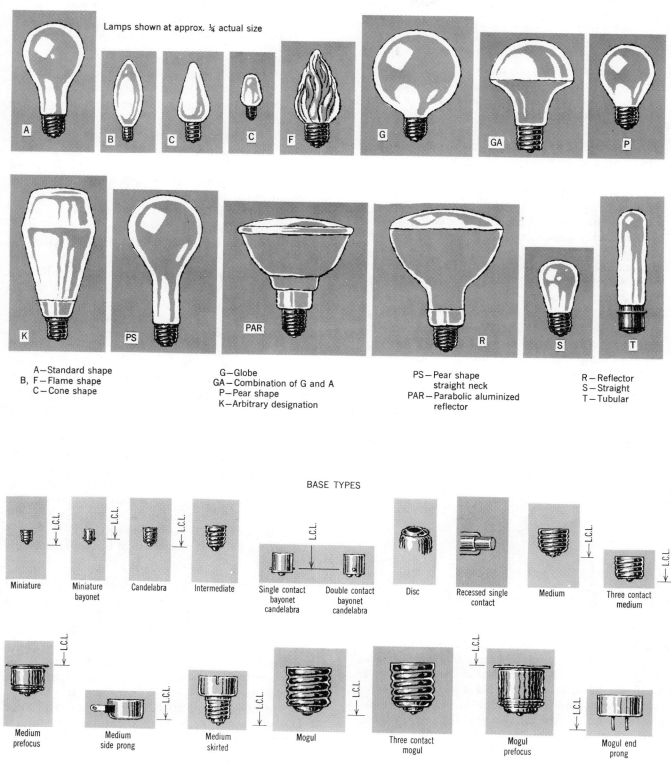

Lamps shown at approx. ¼ actual size

A—Standard shape
B, F—Flame shape
C—Cone shape

G—Globe
GA—Combination of G and A
P—Pear shape
K—Arbitrary designation

PS—Pear shape
       straight neck
PAR—Parabolic aluminized
       reflector

R—Reflector
S—Straight
T—Tubular

BASE TYPES

Miniature   Miniature bayonet   Candelabra   Intermediate   Single contact bayonet candelabra   Double contact bayonet candelabra   Disc   Recessed single contact   Medium   Three contact medium

Medium prefocus   Medium side prong   Medium skirted   Mogul   Three contact mogul   Mogul prefocus   Mogul end prong

L.C.L.—Light center length

Bases shown at approx. ⅓ actual size

Figure 14.14 Incandescent lamp shapes and base types.

14.16 and Table 14.1. The important characteristics of incandescent lamps are briefly discussed below.

(a) Incandescent lamps are very inefficient producers of light. On the average, less than 10% of the wattage goes to produce light; the remainder is heat. Therefore, incandescent sources are a poor choice from an energy conservation point of view. Efficiency increases with larger sizes, varying from about 8% for a 25 w lamp to 13% for a 1000 w lamp.

(b) Principal advantages of incandescent lamps are low cost, instant starting, cheap dimming, good warm color which is flattering to the skin, and small size. This last item allows the incandescent lamp to be used as almost a point source in fixtures that focus the light. This point will be illustrated in our discussion of fixtures. See Figure 14.35.

(c) Incandescent sources have a relatively short useful life, and the life is very voltage sensitive. At 10% undervoltage, life is increased about 250%. At 10% overvoltage, life is reduced about 75%. This means that for a nominal 1000-hr-life lamp, a swing of 10% in voltage either way can change lamp life from 3500 to 250 hr. Lamps operated at rated voltage give maximum efficiency. Voltage effects are shown in Figure 14.17.

Reflector and projector lamps (see Figure 14.16) have built-in beam control and really require only a lampholder and not a fixture. Refer to the catalog of any major lamp manufacturer for complete details of sizes and ratings of all incandescent lamps.

14.14. Quartz Lamps

The quartz lamp, or what is technically called *a tungsten-halogen lamp*, is a special type of incandescent lamp. It uses a quartz tube, generally mounted inside an R or PAR shaped bulb. See Figure 14.18 and Table 14.2. Its great advantage over normal incandescent lamps is that it maintains its light output at an almost constant level throughout its life. A normal incandescent gradually blackens during its life. Also, the quartz lamp has about three to four times the life of a normal incandescent. Its disadvantage is higher price. It is also slightly more efficient than the incandescent, producing on the average about 13% light and 87% heat.

14.15. Fluorescent Lamps —General Characteristics

The fluorescent lamp is in extremely common use, second only to the incandescent lamp. Like the incandescent lamp, the fluorescent comes in literally hundreds of sizes, wattages, colors, voltages and specific application designs. The typical fluorescent lamp is a hot cathode type, consisting of a sealed glass tube containing a mixture of inert gas and mercury vapor. See Figure 14.19. The cathode causes a mercury arc to form inside the tube. This arc produces ultraviolet (UV) light which is not visible to the naked eye. The UV light strikes the phosphors coating the inside of the tube which then *fluoresce*, producing visible light. By changing the type of phosphors the lamp color and output can be controlled. Like all arc discharge lamps the fluorescent lamp requires a ballast in its circuit. Refer to Figure 14.20a. As this figure shows, the ballast is basically a coil. Its purpose is to limit the current in the circuit. Without the ballast the lamp would draw excessive current, and the fuse or circuit breaker would open. Do not confuse the *ballast* with the *starter*. Notice that in the above illustration the starter shows as a switch, because that is what the starter is—a switch. Its purpose is to start the lamp by a switching action. This is explained in detail in Section 14.16(a) below. The starter is a small cylinder about 1 in. in diameter and 2 in. long that snaps into the fixture, from the outside. The ballast is much larger (see Table 14.3) and is mounted inside the fixture and thus is concealed from view. The absence of a starter indicates a rapid-start or other starter-less circuit, as explained below.

The principal characteristics of fluorescent lamps are:

1. The lamps are large and, therefore, a large and relatively expensive fixture is required to hold them. The fixture also houses the ballast which is heavy and large. See Table 14.3. The fixture must provide the required light control, since the source is a long tube emitting light along its entire length. Focusing and accurate light beam control are not possible; therefore, the fluorescent lamp is best applied to general area lighting.

2. The efficiency of a fluorescent lamp is much higher than that of an incandescent lamp. Between 16 and 20% of the input energy becomes visible light, with the remainder being converted to heat. This *includes* the energy loss in the ballast, which is all heat energy. Indeed, getting rid

(continued on page 415)

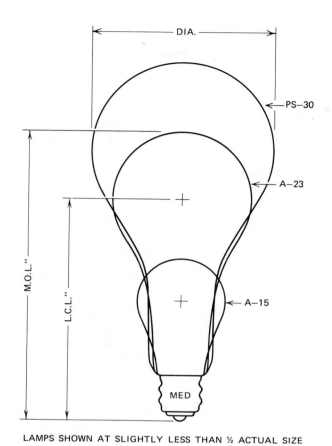

BULB DIAMETER IS GIVEN IN
$1/8$ INCH. EXAMPLE: AN A–19
BULB HAS A DIAMETER OF
$19/8$ INCH OR $2\,3/8''$.

M.O.L.—MAXIMUM OVERALL
LENGTH: THIS FIGURE REFERS
TO THE MAXIMUM LENGTH OF
THE BULB.

L.C.L.—LIGHT CENTER LENGTH:
THIS DIMENSION, IMPORTANT
WHEN DESIGNING REFLECTORS,
IS MEASURED FROM THE RILA—
MENT TO A POINT THAT VARIES
WITH BASE TYPE. SEE FIG' 14.14.

LAMPS SHOWN AT SLIGHTLY LESS THAN ½ ACTUAL SIZE

	A — STANDARD SHAPE									PS — PEAR SHAPE							
WATTS	15	25	40	60	75	100	100	150	150	150	200	300	300	500	750	1000	1500
BULB	A–15	A–19$_1$	A–19$_2$	A–19$_3$	A–19$_3$	A–19$_3$	A–21$_1$	A–21$_2$	A–23	PS–25	PS–30	PS–30	PS–35	PS–40	PS–52	PS–52	PS–52
DIAMETER"	$1\,7/8$	$2\,3/8$	$2\,3/8$	$2\,3/8$	$2\,3/8$	$2\,3/8$	$2\,5/8$	$2\,5/8$	$2\,7/8$	$3\,1/8$	$3\,3/4$	$3\,3/4$	$4\,3/8$	5	$6\,1/2$	$6\,1/2$	$6\,1/2$
M.O.L."	$3\,1/2$	$3\,7/8$	$4\,1/4$	$4\,7/16$	$4\,7/16$	$4\,7/16$	$5\,1/4$	$5\,1/2$	$6\,3/16$	$6\,15/16$	$8\,1/16$	$8\,1/16$	$9\,3/8$	$9\,3/4$	13	13	13
L.C.L."	$2\,3/8$	$2\,1/2$	$2\,15/16$	$3\,1/8$	$3\,1/8$	$3\,1/8$	$3\,7/8$	4	$4\,5/8$	$5\,3/16$	6	6	7	7	$9\,1/2$	$9\,1/2$	$9\,1/2$
BASE	MED	MED	MED	MED	MED	MED	MED	MED	MED	MED	MED	MED	MOG	MOG	MOG	MOG	MOG
STANDARD FINISH	IF	IF	IF W	IF	IF	IF	IF	IF	CL IF	CL IF	IF	IF	IF	IF	CL IF	CL IF	CL IF

CL — CLEAR   IF — INSIDE FORSTED

Figure 14.15 Typical dimensional data for general service incandescent lamps.

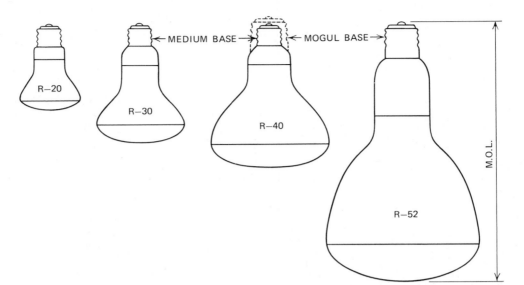

REFLECTOR	R–20		R–30		R–40						R–52	
WATTS	30	50	75	75	150	150	300	300	300	500	500	750
BEAM TYPE	FL	FL	SP	FL	SP	FL	SP	FL	SP		FL	
DIAMETER"	$2^1/_2$		$3^3/_4$		5		5		5		$6^1/_2$	
M.O.L."	$3^{15}/_{16}$		$5^3/_8$		$6^9/_{16}$		$6^9/_{16}$		$7^1/_4$		$11^3/_4$	
BASE	MED		MED		MED		MED		MOG		MOG	

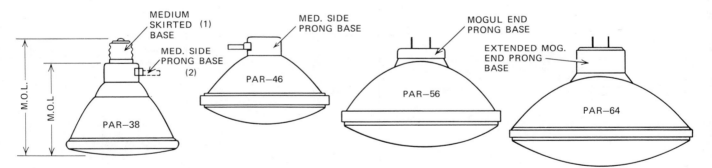

PROJECTOR	PAR–38				PAR–46	PAR–56	PAR–64
WATTS	75	100	150	150	200	300	500
BEAM TYPE	SP	FL	SP	FL	SP	MED. FL	MED. FL
DIAMETER"	$4^3/_4$	$4^3/_4$	$4^3/_4$	$4^3/_4$	5.6	7	8
M.O.L."	$5^5/_{16}$	$5^5/_{16}$	$5^5/_{16}$ (1) $4^5/_{16}$ (2)		4	5	6

SP = SPOTLIGHT
FL = FLOODLIGHT

Figure 14.16 Typical dimensional data for incandescent lamps, reflector and projector types. (This figure, and material in figures 14.27, 14.29, and Table 14.3 are from *Architectural Lighting Graphics,* by J. Flynn and S. Mills, © 1962 by Litton Educational Publishing, Inc. Reprinted by permission of Van Nostrand Reinhold Co.)

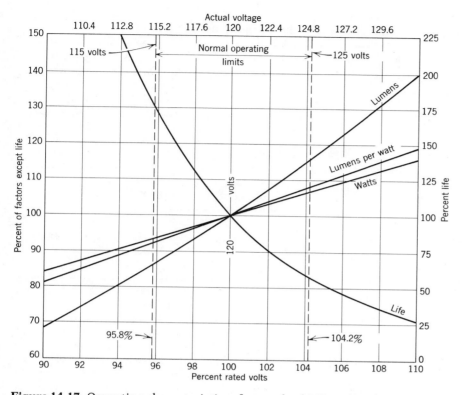

Figure 14.17 Operating characteristics of a standard 120 v incandescent lamp as they vary with voltage.

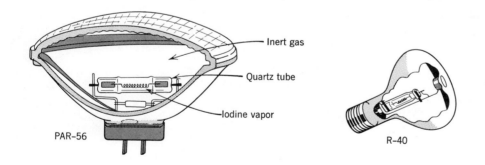

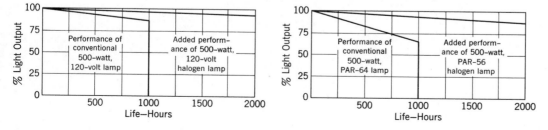

Figure 14.18 The quartz (tungsten-halogen) lamp gives low light depreciation whether the lamp is bare or in an enclosure.

Table 14.2 Typical Data for Quartz Tungsten-Halogen Lamps
Par, Reflector and Tubular 120-v Lamps for Spot, Flood, and General Lighting[a]

Watts	Bulb	Maximum Overall Length (Inches)	Base	Rated Life (Hours)	Beam Type	Approximate Initial Total Lumens	Mean Lumens Through Life (Percentage)
250	PAR-38	5⁵/₁₆	Medium skirted	4000	Spot Flood	3220 3220	94 94
500	PAR-56	5	Mogul end prong	4000	Narrow spot Medium flood	8000 8000	94 94
1000	PAR-64	6	Extended mogul end prong	4000	Narrow spot Medium flood	19,400 19,400	94 94
1000	R-60	10⅛	Mogul	3000	Spot Flood	17,000 17,000	95 95
250	T-4	3	DC bay	2000	—	4850	95
300	T-4	3⅛	RSC	2000	—	5650	95
400	T-4	3⅝	Mini-can	2000	—	7970	95
500	T-4	6	Med.-PF	2000	—	10,450	95
750	T-6	6	Med.-PF	2000	—	15,750	95
1000	T-6	5⅝	RSC	4000	—	19,800	95

[a]For a complete listing, see manufacturer's catalog.

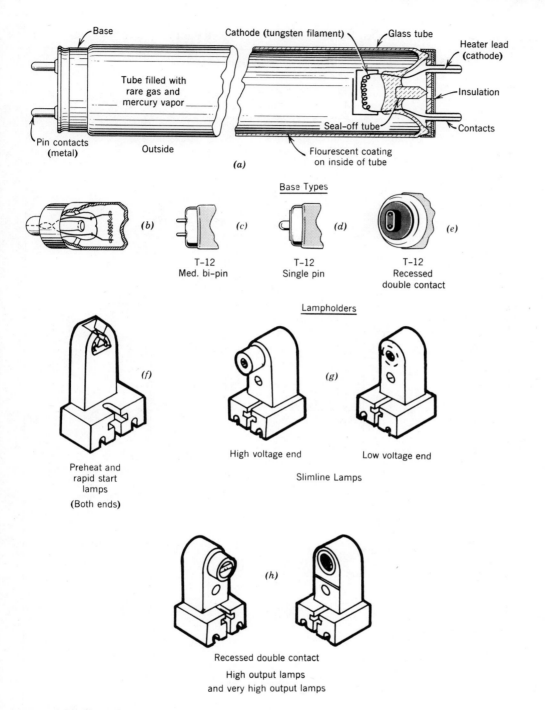

Base Types

T–12
Med. bi-pin

T–12
Single pin

T–12
Recessed
double contact

Lampholders

Preheat and
rapid start
lamps

(Both ends)

High voltage end

Low voltage end

Slimline Lamps

Recessed double contact

High output lamps
and very high output lamps

Figure 14.19 Details of typical fluorescent lamps and associated lampholders. *(a)* Construction of preheat/rapid-start bipin base lamp. This type lamp has type *(c)* base and is held in type *(f)* lampholder.

Instant start lamps *(b)* have a single pin base *(d)* and use single pin lampholders *(g)*, which are different for each end.

High output HO and VHO rapid-start lamps use recessed dc base *(e)* and lampholders *(h)*.

Table 14.3 Typical Fluorescent Lamp Ballast Data (All Ballasts Are High Power Factor, 120 v Service)[a]

Lamp		Ballast Characteristics				
Type	Lamp Watts	Case Type	Case Length	Ballast Weight	Sound Rating	Cross Section
Preheat (starters required)						
F15 T12	15	Single ①	8¼ in.	2.1 lb	D	
F20 T12	20	Single ①	8¼ in.	2.0 lb	D	
Rapid Start						
F30 T12 RS F40 T12 RS	30, 40 (425 ma)	Single ①	9½ in.	4.0 lb	A	
F40 T12 RS	40 (425 ma)	2-lamp ①	9½ in.	4.0 lb	A	
F48 T12 HO	60 (800 ma)	Single ③	11¾ in.	9.75 lb	C	
		2-lamp ③	11¾ in.	10.0 lb	C	
F96 T12 HO	105 (800 ma)	Single ④	11¾ in.	12.0 lb	C	
		2-lamp ④	11¾ in.	13.0 lb	B	
F48 T12 VHO	110 (1500 ma)	Single ④	14¹⁵⁄₁₆ in.	15.0 lb	C	
		2-lamp ④	14¹⁵⁄₁₆ in.	16.0 lb	C	
F96 T12 VHO	215 (1500 ma)	Single ④	14¹⁵⁄₁₆ in.	16.0 lb	D	
		2-lamp ⑤	14¹⁵⁄₁₆ in.	18.0 lb	D	
Instant Start						
F42 T6	25 (200 ma)	Single ②	9½ in.	5.0 lb	B	
F48 T12	39 (425 ma)	Single ③	9½ in.	6.1 lb	C	
F64 T6	38 (200 ma)	Single ②	9½ in.	5.0 lb	B	
F72 T12	57 (425 ma)	Single ③	11¾ in.	8.0 lb	C	
F96 T12	74 (425 ma)	Single ③	11¾ in.	10.0 lb	C	
		2-lamp ④	14⁵⁄₁₆ in.	15.0 lb	D	

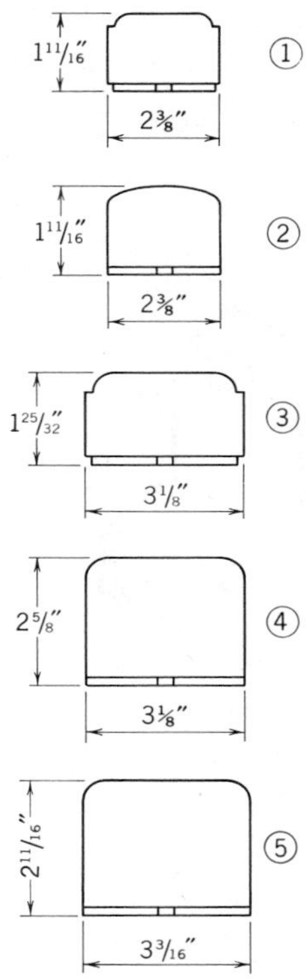

Cross Section diagrams:

① 1¹¹⁄₁₆″ height, 2⅜″ width

② 1¹¹⁄₁₆″ height, 2⅜″ width

③ 1²⁵⁄₃₂″ height, 3⅛″ width

④ 2⅝″ height, 3⅛″ width

⑤ 2¹¹⁄₁₆″ height, 3³⁄₁₆″ width

[a]Consult current manufacturers' catalogs for accurate data.

Figure 14.20 Simplified preheat lamp circuits for one lamp *(a)* and *(b)* and two lamps *(c)*. The circuit does not show compensators and other detail elements, for the sake of clarity. Most lamps are T-12 and operate at 425 ma.

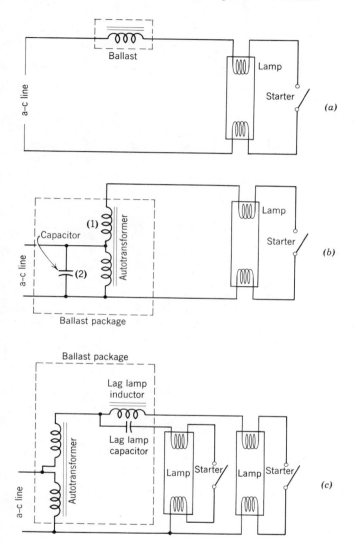

(a) Basic preheat circuit. Starter may be any of several types, manual or automatic.

(b) Preheat circuit modified with (1) autotransformer to adjust line voltage to ballast voltage, and (2) capacitor to make the entire device high power factor.

(c) Two lamp preheat circuit—also known as a lead-lag circuit because of the phasing of the lamps. This arrangement gives high power factor and minimizes flicker effects, since the two lamps are exactly out of phase.

Table 14.4 Typical Fluorescent Lamp Data

Lamp Abbreviation[a]	Watts	Lamp Data Diameter (Inches)	Length (Inches)	Lamp Current (milliamperes)	Ballast Watts[b,c]	Total Watts	Lamp Life (hr)[d]	Initial Output Lumens[e]	Actual Efficacy (lm/w)[f]	Lamp Efficacy (lm/w)[g]
Preheat lamps										
F15 T8 CW	15	8/8	18	425	8	23	7500	870	38	58
F20 T12 CW	20	12/8	24	425	10	30	9000	1300	43	65
Rapid start—preheat lamps[h]										
F40 T12 CW	40	12/8	48	425	6	46	18000	3150	68	79
F40 T12 WW	40	12/8	48	425	6	46	18000	3200	70	80
F40 T12 CWX	40	12/8	48	425	6	46	18000	2200	48	55
F40 T12 D	40	12/8	48	425	6	46	18000	2600	57	65
Rapid start—high output										
F48 T12 CW/HO	60	12/8	48	800	15	75	12000	4300	57	72
F60 T12 CW/HO	75	12/8	60	800	15	90	12000	5400	60	72
F72 T12 CW/HO	85	12/8	72	800	15	100	12000	6650	67	78
F96 T12 CW/HO	105	12/8	96	800	15	121	12000	9200	76	88
Rapid start—very high output										
F48 T12 CW/VHO	110	12/8	48	1500	8	118	9000	6250	53	57
F72 T12 CW/VHO	165	12/8	72	1500	8	173	9000	9900	57	60
F96 T12 CW/VHO	215	12/8	96	1500	13	228	9000	14500	64	67
Instant start (Slimline) lamps										
F42 T6 CW	25	6/8	42	200	15	40	7500	1750	44	70
F64 T6 CW	40	6/8	64	200	10	50	7500	2800	56	70
F48 T12 CW	40	12/8	48	430	16	56	9000	3000	54	75
F64 T12 CW	55	12/8	72	430	16	71	12000	3600	51	65
F96 T12 CW	75	12/8	96	430	17	92	12000	6300	68	84

[a]Standard ordering abbreviation: CW, cool white; WW, warm white; CWX, cool white deluxe; D, daylight.
[b]Figures are for a two-lamp circuit.
[c]See Table 14.3 for typical ballast data.
[d]Life figures are for 3 hr burning per start.
[e]After 100 hr burning.
[f]Includes ballast loss.
[g]Excludes ballast loss.
[h]Data given for lamps in a rapid-start circuit.

of this ballast heat, which amounts to about 25% of the rated lamp wattage, is an important function of the fixture. When we speak of *efficiency* here, we are referring to the amount of energy converted to visible light. The remainder is heat, with a small amount of invisible ultraviolet light. There is, however, another term that is frequently confused with efficiency, which is a measure of the *lumens per watt* produced by the lamp. This is called *efficacy*. It, too, is a measure of how much light energy is produced per watt but is expressed in lighting terms. Efficacy, that is, lumens per watt, is tabulated in Table 14.4, with and without ballast loss. Most published figures usually *do not* show ballast loss and can be very misleading. For fluorescent lamps, lumen output at 100 hr burning is used rather than initial lumens, since at 100 hr the lamp output has dropped to its stable point.

3. Fluorescent lamps have outstandingly long life. This life, however, is affected by the number of times the lamp is turned on and off, since switching tends to wear out the cathode. An average fluorescent lamp burned continuously will last about 30,000 hr; with 3 burning hours per start, it will last about 12,000 hr. These figures are constantly being increased by new developments in fluorescent lamps. This long lamp life, when the lamp is not switched on and off, is the reason that it was general practice to leave fluorescents burning continuously—it turned out to be cheaper. But today, with the increased cost of electric power and, more important, with the need to conserve energy, this is not true. Fluorescents should be switched off when not in use, unless it is only a very short time until they will be needed again. Life figures can also be deceiving, since they are generally given to burnout. Most users replace lamps when they reach about 75% of burnout life because the light output has dropped at that point to about two thirds.

4. Fluorescents have the advantage of being cheap, readily available in a very wide range of sizes and colors and relatively insensitive to changes in voltage. This is particularly important in areas where "brownouts" are common.

5. The color of the light produced by fluorescent lamps was originally very high in blues and greens. As such, it was unkind to human skin color, making people look pale and ill. However, the phosphors have long since been improved so that this problem no longer exists. Lamps are available in cool white and warm white for gen-eral use, plus a host of other shades and colors including daylight. Recently, fluorescent lamps have been developed whose color closely duplicates the color of incandescent lamps. In effect, the "color barrier" no longer exists for fluorescent lamps.

6. The noise that is always associated with fluorescent lamp installations is created by the ballast. All ballasts hum. Recently ballast manufacturers established noise ratings for ballasts ranging from A to D, with A being the quietest. Selection of a proper noise level ballast can now be made that is suitable to the area where the fixture will be used. A class A ballast is appropriate for residences and other quiet areas. A class D is entirely adequate for a machine shop or foundry. Frequently the fixture and not the ballast is to blame for a noisy installation. A poorly designed fixture acts as a noise amplifier for the ballast. This should be checked when examining a fixture for suitability in a particular installation.

7. Other characteristics of fluorescent lamps that should be kept in mind are as follows.
   (a) Possible difficulties in starting at low temperatures, which limits outdoor use. In addition, output drops with temperature. Special ballasts are available for low temperature application.
   (b) Rapid-start lamps require a piece of grounded metal adjacent to the lamp. This is important where using lamps in architectural coves, valances, and the like with remote ballasts.
   (c) Dimming of fluorescent lamps is possible but requires special expensive dimming ballasts. An existing fluorescent installation cannot be dimmed without changing the ballasts.
   (d) As a rule of thumb, ballasts lose one half their life with each 10° C rise in their operating temperature. It is therefore important to provide for adequate heat radiation from the fixture. Ballast manufacturers now make a ballast called a class P unit which contains a thermal protector. When the ballast overheats, the protector opens the circuit to prevent the ballast from melting. Some of these devices reset after the ballast cools.

As mentioned above, there are many types of fluorescent lamps in addition to a huge array of sizes, wattages, and special purpose lamps. The types differ basically in starting techniques and amount of current drawn by the lamp. We now briefly describe the major types of lamps and their special characteristics. Keep in mind that all the descriptions and tables presented here are brief extracts chosen to illustrate the typical types. Complete listings of types and ratings are available in manufacturers' literature which is up to date and available for the asking. The three major types of fluorescent lamps and their circuits are preheat, instant-start and rapid-start.

14.16. Fluorescent Lamp Types

a. Preheat Lamps

This is the original (1937) type lamp, which requires a separate starter. The starter allows the cathode to preheat and then opens the circuit, causing an arc to flash across the lamp, starting it. See Figure 14.20. Most of these starters are automatic, although in desk lamps the preheating is done by pressing the start button for a few seconds and then releasing it. This closes the circuit and allows the heating current to flow. All preheat lamps have bipin bases. They range in power from 4 to 90 w and in length from 6 to 90 in. A typical ordering abbreviation for a preheat lamp would be F15T12WW. This translates: fluorescent lamp, 15 w, tubular-shaped bulb, 12/8-in. diameter (number represents diameter in one-eighths of an inch), warm white color. See Table 14.4. In large measure preheat lamps have been superseded by the rapid-start and instant-start types.

b. Instant Start (Slimline) Lamps

This type was the second one developed (1944), and operates without starters. The ballast provides a high enough voltage to strike the arc directly. Since no preheating is required, Slimline, instant-start lamps have only a single pin at each end. A typical ordering description for such a lamp would be: F42T6CW Slimline, which means: fluorescent, 42-in. length, tubular, 6/8-in. diameter, cool white, instant-start. The T-6 narrow tube indicates a low current, 200 milliampere (ma) lamp, in lieu of the usual 425 ma lamp. Note also that in instant-start lamps the number following F indicates length not wattage. This is true with all lamps that operate at other than 425 ma, which is the normal current. To find wattage a catalog must be consulted. See Figure 14.21 for a typical lamp circuit.

c. Rapid-Start Lamps

The third type of lamp became available in 1952. These lamps, also called rapid-start/preheat, operate like the preheat type except that the cathodes are always energized. The delay in starting the preheat lamp results from the time required to heat the cathode. In rapid-start circuits the cathode is heated constantly by a special winding in the ballast. Therefore, no heat-up delay is required and the lamp can be started rapidly. See circuit diagram Figure 14.22. Because of this similarity of operation, rapid-start lamps will operate satisfactorily in a preheat circuit. The reverse is not true because the preheat requires more current to heat the cathode than the rapid-start ballast provides. See Table 14.5 for interchangeability of lamps in the various circuits. By far the most popular lamp is the 40 w T-12 lamp. A standard ordering abbreviation for a lamp would be F40T12WW/RS which indicates fluorescent, 40 w, T-12 bulb, warm white color, rapid-start. Understood is 425 ma operation. It must always be kept in mind that the ballast has a considerable wattage (heat) loss. Therefore, a two-lamp fixture using 40 w tubes gives a nominal load of 100 w (96 actual) due to the 16 w ballast loss. Special lower wattage "energy-saving" lamps now being produced reduce this total to about 85 w.

d. High-Output Rapid-Start Lamps

As mentioned, all preheat, most instant-start and most rapid-start lamps operate at 425 ma. If this current is increased, the output of the lamp also increases. Two special types of higher output rapid-start lamps are available. One operates at 800 ma and is called simply high output (HO). See Table 14.4. The second, which operates at 1500 ma (1.5 ampere), is called by different manufacturers—very high output (VHO), super-high output or simply 1500 ma rapid-start. There is also a 1500 ma special lamp that uses what looks like a dented or grooved glass tube. This lamp, called Power Groove by General Electric, has somewhat higher output than the standard VHO tube. All of these high-output lamps have recessed double contact bases, require special circuits and ballasts and are *not* interchangeable with any other type lamp. They are used in application where high output is required from a limited size source such as outdoor sign lighting, street lighting and merchandise displays. Because of the serious heat problems involved, VHO lamps are frequently operated without enclosing

Figure 14.21 Basic instant-start lamp circuits. Notice that lamps are single pin, unlike the rapid-start and preheat types, which have a separate circuit to heat the lamp filament. T-6 and T-8 lamps normally operate at 200 ma; T-12 lamps operate at 425 ma.

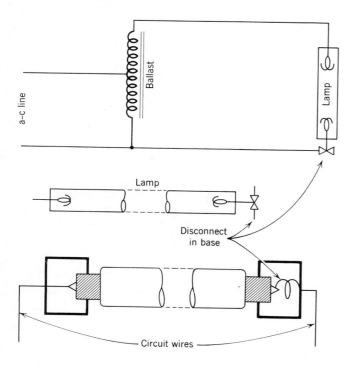

Basic instant-start lamp circuit. Cathodes are not preheated. Voltage from ballast-transformer causes arc to strike directly. Bases are single pin.

Due to the high voltage involved, the lampholder at one end of the lamp is a disconnecting device that opens the circuit when the lamp is removed.

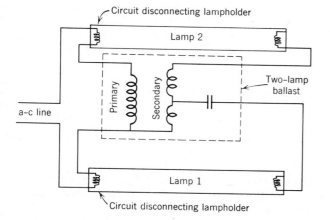

Typical two lamp instant-start circuit. Note disconnecting lampholders and autotransformer in the ballast. The capacitor provides for a phase shift that assists operation. The circuit is called series-sequence, since the lamps start in sequence, in series.

Figure 14.22 Typical basic rapid start circuits. To assure proper starting all standard RS lamps *must* be mounted within ½ in. of a grounded metal strip, extending the full length of the lamp (1 in. for HO and VHO lamps). Normal output lamps are 425 ma T-12. High-output lamps operate at 800 ma; very high-output lamps at 1500 ma.

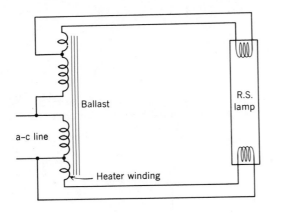

Basic rapid-start circuit. Note the special small end windings used to supply voltage to heat the cathodes. Cathodes (filaments) are heated *constantly*.

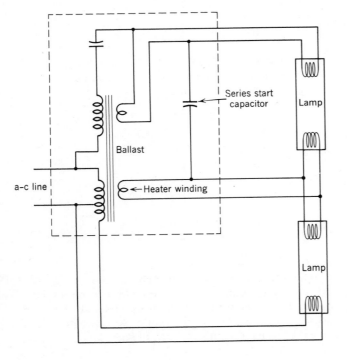

Typical series-sequence two-lamp rapid-start circuit. The series start capacitor assists in lamp starting; the other capacitor improves power factor.

Table 14.5 Fluorescent Lamp Interchangeability

Lamp Type	Ballast/Circuit Type		
	Preheat	Rapid-start	Instant-start
Preheat	OK	NG, poor starting	NG, poor starting short life[a]
Instant-start (Slimline)	Won't start, NG[b]	Won't start, NG[b]	OK
Rapid-start	OK	OK	NG, poor starting short life[a]
Preheat/ rapid-start	OK	OK	NG, poor starting short life[a]

[a]Normally no possibility of interchange because of incompatible base and lampholder: preheat has bipin base, instant-start lampholder is single pin.

[b]Normally no possibility of interchange. Instant-start lamp is single pin base; preheat/rapid-start lampholders are for bipin bases.

fixtures. This, however, creates a glare problem, limiting the application of these very bright sources. Also, the HO and VHO lamps are generally less efficient than the standard 425 ma rapid-start lamp and have considerably shorter life. Typical ordering abbreviations for these high-output lamps are similar to the standard rapid-start except that the number indicates length, not wattage. For instance, F72T12/CW/HO is fluorescent, 72-in. long, T12 bulb, cool white, high output (800 ma). Similarly, F72T12/CW/VHO is fluorescent, 72-in. long T-12 bulb, cool white, very high output (1500 ma). By consulting the catalog we find that these two lamps are rated 85 w and 165 w, respectively, *without* ballast loss.

e. Special Fluorescent Lamp Types

In addition to the standard tubular fluorescent lamp there are U -shaped lamps, square panel lamps and tubular types with built-in reflectors and apertures. All these types plus newer developments and their technical data can be found in any current manufacturer's catalog.

14.17. The HID Source

The general name high intensity discharge (HID) source is now used to cover a number of lamp sources. Just a few years ago the only lamps in this field were the low pressure sodium lamp and the mercury vapor lamp. Today, HID includes mercury, metal halide and sodium sources. Here we briefly discuss these three sources and their physical and operating characteristics. These HID sources have become very important because of their high efficiency, and they are now used indoors. The poor color that had restricted mercury use to parking lot lighting (and incidentally making it difficult to find a red car) has been corrected. The metal halide source whose color is inherently good was initially used for both interior and exterior applications. The sodium lamp whose yellow color was initially thought to be unsuitable for indoor use has also moved inside. Sodium lamps are finding wide acceptance in high-ceilinged spaces like banks, lobbies and auditoriums. In combination with other sources, sodium is also used in indoor sports arenas and in industrial spaces.

14.18. The Mercury Lamp

This lamp combines the compact focusable shape of an incandescent with the arc discharge character of the fluorescent. The result is an efficient, long-life source with many applications. Figure 14.23 shows the construction of a typical mercury lamp while Figure 14.24 gives some dimensional data. A mercury lamp differs from a fluorescent in that it is a high-pressure mercury vapor lamp. The fluorescent is a low-pressure lamp. The mercury lamp produces most of its output in *visible* light, in the blue-green range. This accounts for the typical color of a clear glass mercury. Such clear lamps are used only in outdoor application where the color distortion produced its not important. The efficiency and output is high, however, and this accounts for its use. Efficacy of a clear mercury lamp runs about the same as a fluorescent lamp. Wattages run from 50 to 1500 w in standard commercially available lamps. Life of a mercury lamp runs more than 20,000 hr.

To improve the color of mercury lamps, phosphors are added that act the same way as in fluorescent lamps. Other devices are also employed, so that today these lamps are available with excellent color—almost as good as that of fluorescent lamps.

These color-improved lamps, however, have lower efficacy than the clear lamp. They are now in wide use in many interior applications such as schools, stores, banks, and institutions. The advantage over fluorescents is the shape of the lamp which permits the use of a small fixture. Some important facts to remember about mercury lamps are:

(a) Mercury lamps are available in a wide variety of ratings, colors, sizes and shapes. A small sampling is shown in Table 14.6. Consult a current catalog for complete technical and physical data.

(b) In common with all discharge lamps, mercury lamps require a ballast. This ballast can be mounted remotely from the lamp—and indeed, frequently is. Ballasts are of several different types. Some typical ballast data are given in Table 14.7.

(c) Mercury lamps require a warm-up period of *up to 5 minutes* before giving full output, depending on type of lamp, temperature and type of ballast. Therefore, these lamps can only be used where such a delay is acceptable. In mercury installations, it is customary to include some incandescent sources to provide emergency light after a blackout. Even if the power failure is only a few seconds, the lamps will not restrike their arcs. They must be allowed to cool somewhat before they will relight, and then take more time to reach full output.

(d) In the wiring of mercury lamps care must be taken to keep voltage within ±5% unless voltage-compensating ballasts are used.

(e) Since most mercury lamp ballasts have a high inrush current, the technologist should check with the engineer or specifier before circuiting. The inrush current, not the wattage, may limit the number of fixtures on a circuit.

(f) Until very recently mercury lamps were not dimmable, since below a certain voltage, depending on the ballast, the lamp extinguishes. A dimming ballast has now been developed.

(g) Mercury lamp ballasts are noisy. Where this might be a problem, one should consider remote mounting.

(h) The lamp designation system for mercury lamps is not as simple as the one for fluorescents. Using the ANSI (American National Standards Institute) system, an H-33-GL-400/DX lamp is a mercury, 400 w, BT-37 bulb, mogul base phosphor-coated, deluxe white lamp. Some manufacturers use their own abbreviation system in

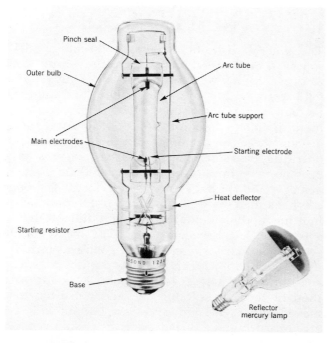

Figure 14.23 Construction details of a typical clear mercury lamp.

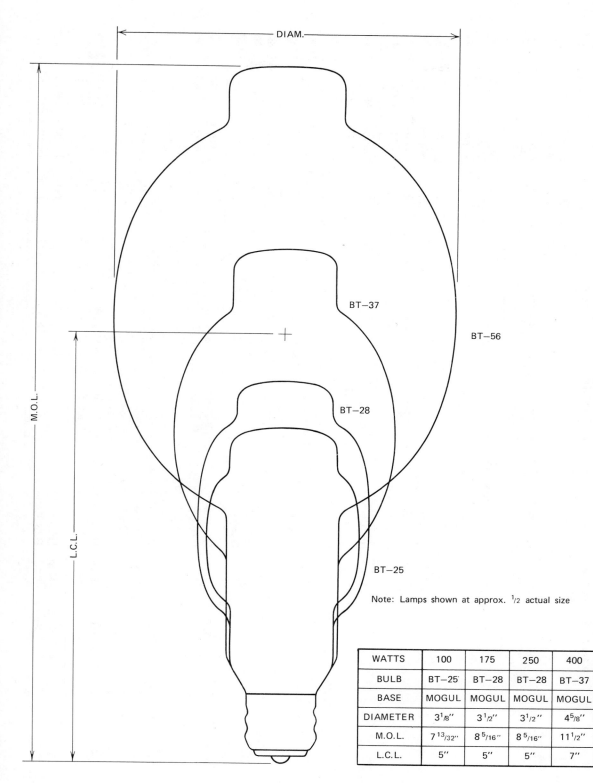

Note: Lamps shown at approx. $^1/_2$ actual size

WATTS	100	175	250	400	700	1000
BULB	BT–25	BT–28	BT–28	BT–37	BT–46	BT–56
BASE	MOGUL	MOGUL	MOGUL	MOGUL	MOGUL	MOGUL
DIAMETER	$3^1/_8''$	$3^1/_2''$	$3^1/_2''$	$4^5/_8''$	$5^3/_4''$	$7''$
M.O.L.	$7^{13}/_{32}''$	$8^5/_{16}''$	$8^5/_{16}''$	$11^1/_2''$	$14^1/_2''$	$15^3/_8''$
L.C.L.	$5''$	$5''$	$5''$	$7''$	$9^1/_2''$	$9^1/_2''$

Figure 14.24 Typical dimensional data for BT shape mercury lamps. Refer to Table 14.6 for typical technical data.

Table 14.6 Typical Data for Mercury Vapor Lamps[a]

Lamp Watts	Bulb	Base	ANSI Ordering Abbreviation	Former Abbreviation	Description (See below)	Light Center Length (in.)	Max. Overall Length (in.)	Rated Aver. Life (hrs)	Approximate Lumens Initial	Mean
40	B-17	Med.	H46DL-40-50/DX	H46DL/DX	*G	3⅛	5⅛	16000+	1100	800
50	B-17	Med.	H46DL-40-50/DX	H46DL/DX	*G	3⅛	5⅛	16000+	1550	1150
75	B-21	Med.	H43AZ-75	H43AZ	*G, S	3¾	6½	16000+	2800	2350
			H43AY-75/DX	H43AY/DX	*G, S	3¾	6½	16000+	2800	2200
			H43AY-75/N		*G, S	3¾	6½	16000+	2050	1600
			H43AY-75/R		*G, S	3¾	6½	16000+	2800	2200
100	A-23	Med.	H38LL-100	H38-4LL	*G	3½	5 7/16	24000+	4100	3450
			H38MP-100/DX	H38-4MP/DX	*G	3½	5 7/16	24000+	4300	3200
			H38MP-100/N		*G	3½	5 7/16	24000+	3600	2650
	BT-25	Mog.	H38HT-100	H38-4HT	*G, S, B	5	7½	24000+	4100	3450
			H38JA-100/R		*G, S	5	7½	24000+	4400	3300
	R-40	Med.	H38BP-100/DX	H38BP/DX	*RF, FF, VW		7½	24000+	2850	2280
			H38BP-100/N		*RF, FF, VW		7½	24000+	2450	1950
			H38BP-100/R		*RF, FF, VW		7½	24000+	2850	2280
175	BT-28	Mog.	H39KB-175	H39-22KB	*G, S, B	5	8 5/16	24000+	7700	6600
			H39KC-175/DX	H39-22KC/DX	*G, S	5	8 5/16	24000+	8500	6800
			H39KC-175/N		*G, S	5	8 5/16	24000+	7000	5600
			H39KC-175/R	H39KC/R	*G, S	5	8 5/16	24000+	8500	6800
	R-40	Med.	H39BM-175	H39-22BM	*RF, FF, W		7½	24000+	6100	5150
			H39BP-175/DX	H39-22BP/DX	*RF, FF, VW		7½	24000+	5750	4600
250	BT-28	Mog.	H37KB-250	H37-5KB	*G, S, B	5	8 5/16	24000+	12100	9850
			H37KC-250/R	H37KC/R	*G, S	5	8 5/16	24000+	13000	9750
300	BT-37 (Econ-o-watt)	Mog.	H33CD-300		*G, S	7	11½	16000+	14000	
			H33GL-300/DX		*G, S	7	11½	16000+	15700	

400	BT-37	Mog.	H33CD-400	H33-1-CD	*G, S, B	7	11½	24000+	21000	18300
			H33GL-400/DX	H33-1-GL/DX	*G, S	7	11½	24000+	23000	18400
	R-57	Mog.	H33FY-400	H33-1-FY	*G, B, RF, FF, W		12¾	24000+	18500	16400
			H33DN-400/DX	H33-1-DN/DX	*G		12¾	24000+	23000	18400
700	BT-46	Mog.	H35NA-700	H35-18NA	*G, S	9½	14½	24000+	41000	35700
1000	BT-56	Mog.	H34GV-1000	H34-12GV	*G, B	9½	15⅜	16000+	55000	44000
			H34GW-1000/DX	H34-12GW/DX	*G	9½	15⅜	16000+	56000	36400
			H36GV-1000	H36-15GV	*G, S, B	9½	15⅜	24000+	57500	47100
			H36GW-1000/DX	H36-15GW/DX	*G, S	9½	15⅜	24000+	63000	44700

Source. Westinghouse Corporation.
aFor accurate, current data consult the manufacturers' catalogs.

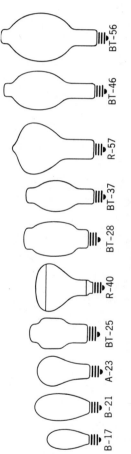

B-17 B-21 A-23 BT-25 R-40 BT-28 BT-37 R-57 BT-46 BT-56

• Explanation of color suffix in ordering abbreviation:

/DX Deluxe White	/R Beauty Lite
/N Style-Tone	No suffix—Clear
	(non-phosphor coated)

*Descriptive Symbols For color identification see ordering abbreviation.

B Black Light	RF Reflector Flood	W Wide Beam
FF Frosted Face	S Street Lighting	
G General Lighting	VW Very Wide Beam	

Table 14.7 Typical Mercury Vapor Lamp Ballast Data. Constant Wattage Ballasts—Indoor Service[a,b]

Lamp type and Wattage	Nominal Line Voltage	Line Voltage Range	Watts Loss	Line Current	Min. Start. Temp. (°F)	Ship. Wgt. (lbs.)	Dimension in Inches				
							Over. Lgth. A	Case Width B	Case Hgt. C	Case Lgth. D	Mtg. Lgth. E
1-175 w H-39	120 × 240	108-132 × 216-264	35	1.83 0.94	0	17	13⅜	6⅛	3⅜	12	12⅝
	240 × 277	216-264 × 250-300		0.94 0.79							
1-250 w H-37	120 × 240	108-132 × 216-264	35	2.56 1.27	0	20	13⅜	6⅛	3⅜	12	12⅝
	240 × 277	216-264 × 250-300		1.27 1.08							
1-400 w H-33	120	108-132	55	4.0	0	22	15¾	6⅛	3⅜	14¹¹⁄₃₂	15
	240 × 277	216-264 × 250-300		2.0 1.7							
	208	190-230		2.3							
	480	432-528		1.0							
1-1000 w H-36	120 × 240	108-132 × 216-264	90	9.30 4.65	−20	37	16¹⁄₁₆	6⅛	5¹⁵⁄₁₆	14½	15¼
	277	250-300		4.05							

Source. Sola Electric Co.
[a]All constant-wattage ballasts have minimum power factor of 90%.
[b]For accurate, current data refer to the manufacturers' catalogs.

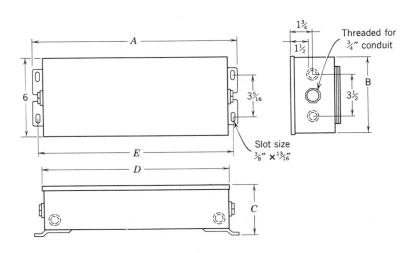

addition, which is sometimes clearer, depending on the manufacturer. The only sure way to know is to consult a catalog.

(i) The energy situation created a demand for replacement of low efficiency incandescents with high efficiency mercury lamps. In response to this, manufacturers made available direct replacement kits including screw-in ballasts and low wattage mercury lamps. These good color, phosphor-coated small lamps are only somewhat more efficient than the incandescent lamp *when ballast loss is included*. They have the advantage of long life but the disadvantage of not being instant start. Careful analysis should be given to any such proposed replacement. The use of a standard fluorescent lamp giving high output, long life and instant start should also be considered in these cases.

(j) Self-ballasted mercury lamps are available with a ballast built into the lamp. These lamps show lowered output and are relatively expensive. Their efficacy is only marginally better than incandescents, and their use is only advisable where long life is the deciding factor.

14.19. The Metal Halide Lamps

The long life, small size and high output of the mercury lamp led to research to overcome its color problem. One result of that research is the metal-halide lamp. This is basically a mercury lamp with alterations to its arc tube. These lamps have very good color making them applicable to almost all indoor use. Their efficacy in lumens per watt is higher than mercury, making them a very efficient source. Their life, however, is shorter. A brief comparison shows the following.

	Mercury	Metal-Halide
Color	Poor to good	Good to excellent
Life	20,000 + hours	10,000 + hours
Avg. efficacy (with ballasts)	50 lm/w ±	70 lm/w ±

Common trade names used for these lamps by two leading manufacturers are "Metal-Arc" and "Multi-Vapor."

14.20. Sodium Lamps

The most recent development in HID lamps is the *high pressure* sodium lamp. This lamp, illustrated in Figure 14.25, was developed by General Electric and is marketed by them under the trade name "Lucalox." Other trade names for this lamp when produced by other companies are "Ceramalux" and "Unalux." The outstanding fact about this lamp is its high output. Note from Table 14.8 that output including ballast losses approaches 100 lm/w (also abbreviated lpw). This is double the efficacy of a color corrected mercury lamp and is, at least, 50% better than a standard fluorescent. The HPS lamp has a yellowish color similar to low wattage incandescent and warm white fluorescent. See Table 14.9. Its color makes it suitable for indoor use, where its small size, high output and long life make it highly desirable and economical. Like all HID sources, it does not start instantly but its start and restrike time are shorter than for a mercury lamp. The *low pressure* sodium lamp has such poor color that despite its high efficacy it is only used for road lighing.

14.21. Control of Light

In the preceding sections we have studied some of the important facts about how light is produced, and how it acts. We have also discussed in some detail how light and vision act together. In other words, we now know how artificial light is produced

Table 14.8 Efficacy of Various Light Sources

Source	Efficacy (Lumens per Watt)
Candle	0.1
Oil lamp	0.3
Original Edison lamp	1.4
1910 Edison lamp	4.5
Modern incandescent lamp	14–20
Tungsten halogen lamp	16–20
Fluorescent lamp[a]	50–80
Mercury lamp[a]	40–70
Metal-halide lamp[a]	60–80
High-pressure sodium[a]	90–100

[a]Including ballast losses.

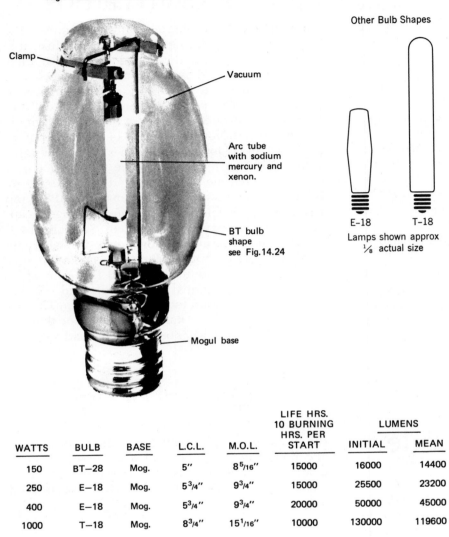

High Pressure Sodium Lamps

Clamp

Vacuum

Arc tube
with sodium
mercury and
xenon.

BT bulb
shape
see Fig.14.24

Mogul base

Other Bulb Shapes

E-18   T-18

Lamps shown approx
⅛ actual size

WATTS	BULB	BASE	L.C.L.	M.O.L.	LIFE HRS. 10 BURNING HRS. PER START	LUMENS	
						INITIAL	MEAN
150	BT—28	Mog.	5″	8⁵⁄₁₆″	15000	16000	14400
250	E—18	Mog.	5³⁄₄″	9³⁄₄″	15000	25500	23200
400	E—18	Mog.	5³⁄₄″	9³⁄₄″	20000	50000	45000
1000	T—18	Mog.	8³⁄₄″	15¹⁄₁₆″	10000	130000	119600

Figure 14.25 Construction details and typical data for high pressure sodium (HPS) lamps. (Courtesy of Sylvania, General Electric Companies)

Table 14.9 Color Properties of Light Sources

Type of Lamp	Lamp Appearance Effect on Neutral Surfaces	Effect on "Atmosphere"	Colors Strengthened	Colors Grayed	Effect on Complexions	Remarks
FLUORESCENT LAMPS						
Cool white CW	White	Neutral to moderately cool	Orange, yellow, blue	Red	Pale pink	Blends with natural daylight
Deluxe cool white CWX	White	Neutral to moderately cool	All nearly equal	None appreciably	Most natural	Best overall color rendition; simulates natural daylight
Warm white WW	Yellowish white	Warm	Orange, yellow	Red, green, blue	Sallow	Blends with incandescent light
Deluxe warm white WWX	Yellowish white	Warm	Red, orange, yellow, green	Blue	Ruddy	Good color rendition; simulates incandescent light
Daylight	Bluish white	Very cool	Green, blue	Red, orange	Grayed	Usually replaceable with CW
INCANDESCENT LAMPS						
Incandescent filament	Yellowish white	Warm	Red, orange, yellow	Blue	Ruddiest	Good color rendering
HIGH INTENSITY DISCHARGE LAMPS						
Clear mercury	Greenish blue-white	Very cool, greenish	Yellow, blue, green	Red, orange	Greenish	Very poor color rendering
White mercury	Greenish white	Moderately cool, greenish	Yellow, green, blue	Red, orange	Very pale	Moderate color rendering
Deluxe white mercury	Purplish white	Warm, purplish	Red, blue, yellow	Green	Ruddy	Color acceptability similar to CW fluorescent
Metal halide	Greenish white	Moderately cool, greenish	Yellow, green, blue	Red	Grayed	Color acceptability similar to CW fluorescent
High pressure sodium	Yellowish	Warm, yellowish	Yellow, green, orange	Red, blue	Yellowish	Color acceptability approaches that of WW fluorescent

Source. Lamp Department, General Electric Co.

and how we see. The next subject to be studied is how this light is controlled. The lamp sources produce the light, generally radiating in all directions. What do we do with this light to make it useful? How do we redirect the light energy produced, so that it provides room illumination? Like most technical questions, the answer is simple in principle but more complex in practice. What we do is to build enclosures for the lamp sources. These enclosures, which are generally lighting fixtures, are designed to:

(a) Hold and energize the lamp(s).
(b) Direct the light.
(c) Change the quality of the light produced.
(d) Provide shielding (cut off) to prevent direct glare.

Not all fixtures do all of these things. For instance, a simple incandescent lampholder does only (a) above, yet most people would call it a lighting fixture, although a very simple one. In the trade, a lampholder is listed as a lighting fixture.

Sometimes, not all four characteristics are required. Reflector types lamps (see Section 14.13 and Figure 14.16) have their light-directed mechanism built into the lamp, and produce a spotlight or

floodlight beam when they are installed in a simple lampholder. These lamps are their own fixture and, therefore, function (b) above is not necessary. Also, we frequently combine the fixture with the building structure to get the desired lighting. We build coves, coffers, hung ceilings and other arrangements where the structure acts as a part of the lighting control system. This type of lighting, referred to as *architectural lighting elements*, will be discussed separately from lighting fixtures. These elements are of great importance to electrical draftsmen because they are always detailed on the drawings.

14.22. Lampholders

As mentioned above, the most elementary lighting fixture is a simple lampholder. Lampholders can be cord- or box-mounted sockets for incandescent lamps as in Figure 14.26 or wiring "strips" for fluorescent lamps as in Figure 14.27. The fluorescent lamp-wiring channel also provides mounting for the lamp ballast. This is a requirement for all fixtures handling discharge type lamps. HID lamps are almost never used in a simple lampholder. The HID lighting fixture must provide for mounting the ballast and, because the lamps are normally very bright, also provide some shielding. A precision type lampholder is shown in Figure 14.28.

14.23. Reflectors and Shields

The lampholders described above can be readily modified by the addition of various types of reflectors. These reflectors are finished with a high reflectance paint or metallizing process, so as to reflect most of the lamp's light. The reflectors provide two of the functions listed above—(b) and (d). Their primary purpose is reflection, but cutoff and some shielding are automatically provided as well. The reflectors can be simple units that fasten on to lampholders (see Figure 14.29) or part of the fixture.

Fixtures consisting of a lampholder channel and a reflector, with no diffuser, are generally called industrial fixtures. This is because that type of unit was, until a decade ago, the common lighting unit in industrial areas. The best quality units of this type carry an RLM (Reflector Luminaire Manufacturer)

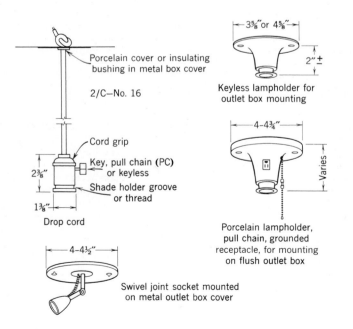

Figure 14.26 Typical basic, bare-lamp, incandescent lampholders. For other details see Appendixes A.2 and A.4.

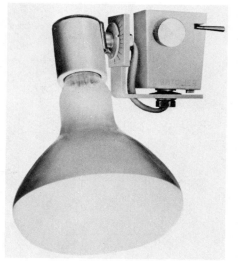

Single-Lamp Units

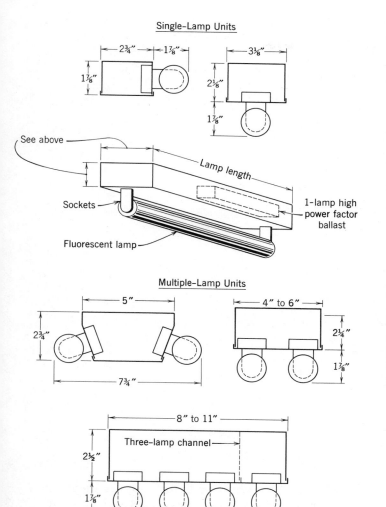

Multiple-Lamp Units

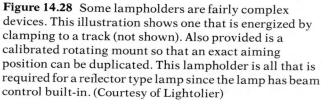

Figure 14.28 Some lampholders are fairly complex devices. This illustration shows one that is energized by clamping to a track (not shown). Also provided is a calibrated rotating mount so that an exact aiming position can be duplicated. This lampholder is all that is required for a reflector type lamp since the lamp has beam control built-in. (Courtesy of Lightolier)

Figure 14.27 Typical bare fluorescent lamp, surface-mounting wiring channel units, also known as fluorescent strips. All dimensions are typical and vary with the manufacturer. See Appendix for additional drawings.

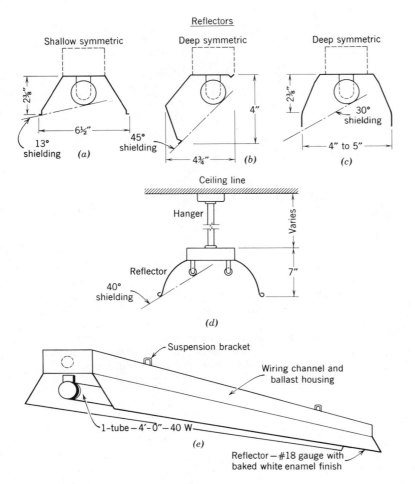

Figure 14.29 Accessory reflectors (a), (b), and (c) provide varying degrees of shielding. Commercial grade industrial fixture (d) normally provides minimum of 35 deg shielding. Typical pendant single lamp fluorescent strip with type (a) reflectors, is shown in detail (e). For additional details of reflector type fluorescent fixtures, see Appendix A.5.

label to indicate that the fixture quality meets the industry standard. Today these RLM and RLM-type open reflector units are used in nonindustrial installations as well. Figure 14.30 shows a number of open reflector design fixtures. Others are found in the Appendices.

Simple reflectors like those in Figures 14.29 and 14.30 provide shielding, although their principle purpose is to reflect light. In the case of fluorescent units, the shielding provided is lengthwise on the fixture, that is, it shields the length of the lamp. This shielding is the important one for fluorescent lamps. If crosswise shielding is also desired, baffles (also called shields or louvers) must be added to the unit as in Figure 14.31. This fixture gives shielding in both directions. Above the shielding angle the lamp is not visible. This angle varies with the design of the fixture. The shielding provided by the reflector is usually quite low, because the reflector is normally shallow. A deep reflector provides good shielding, but this cutoff also reduces the spread of light from the fixture. See Figure 14.32. To provide good shielding in both directions a two-way louver is used.

RLM standard dome

RLM symmetrical angle

Shallow dome

Deep bowl

Elliptical angle

Diameter	Lamp
12″	60–100 W
14″	150 W
16″	200 W
18″	300–500 W

Figure 14.30 Standard shapes of open industrial reflectors for incandescent lamps. See Appendix A.4 for additional details and description.

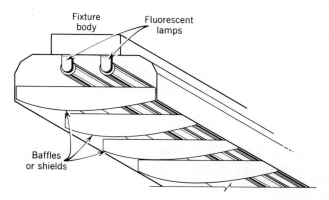

Figure 14.31 Basic fluorescent fixture with crosswise shielding provided by reflector and lengthwise shielding provided by baffles.

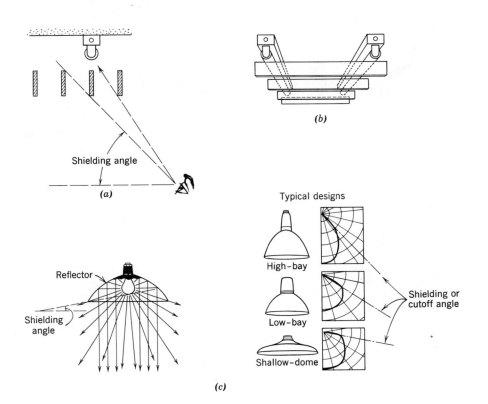

Figure 14.32 One-way shielding is provided in fluorescent fixtures by a simple slat louver *(a)*, a baffle *(b)*, or the fixture reflector (see Figure 14.29). For open-bottom incandescent reflectors, the shielding or cutoff is symmetrical around the lamp. The angle varies with the type of reflector *(c)* and Figure 14.30.

This consists of slats of metal or plastic in both directions, with height and spacing designed to give the shielding angle wanted. Such a louver is called an "egg-crate" louver, because it looks like one. See Figure 14.33. This type of two-way louver provides a relatively low brightness surface over the bottom of the fixture instead of the very bright bare fluorescent lamps. The same louvering technique can be used with an incandescent fixture, by making the louvers round. Usually these louvers are concentric circles mounted on the fixture bottom. See Figure 14.34 for an interesting example of this design.

One other aspect of reflectors should be mentioned here, but since it is very complex, it will only be touched on. This is the use of reflectors to focus the light of a point source. You will remember that we stated that one excellent characteristic of incandescent (and HID) lamps was their ability to be focused. This results from the source being almost a point source, which can be focused by designing a proper reflector. This involves the complex and detailed subject of reflector design, which is far beyond the scope of this book. But, since the electrical draftsman will have many occasions to draw and examine fixture details, he should be able to recognize a reflector type. Figure 14.35 shows two of the principal types in common use.

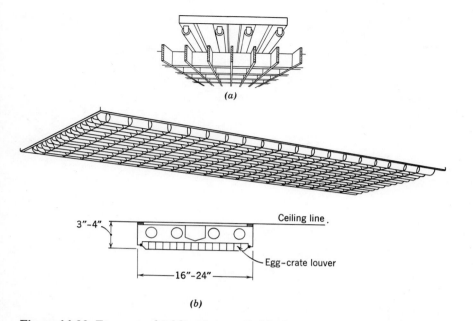

Figure 14.33 Two-way shielding is provided by louvering in both directions. Principle of the design is shown in (a) and an actual fixture with an egg-crate louver is shown in (b). Louver material can be metal or plastic.

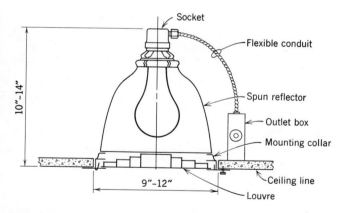

Figure 14.34 Recessed incandescent downlight for general service lamp. Note that the circular louvers increase in depth as they near the center. This increases the shielding angle so that the source remains shielded as the viewer approaches the fixture. Dimensions vary with the wattage of the lamp.

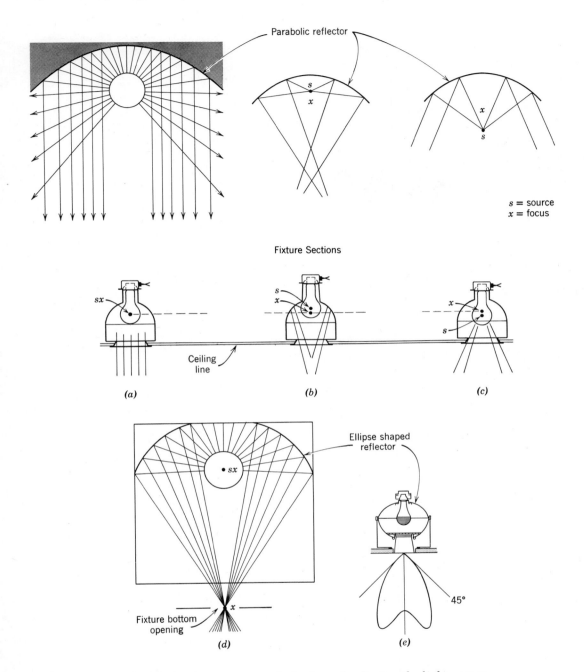

Parabolic reflector

s = source
x = focus

Fixture Sections

sx

Ceiling line

(a)

s
x

(b)

x
s

(c)

sx

Fixture bottom opening
x

(d)

Ellipse shaped reflector

45°

(e)

Figure 14.35 When a light shines on a parabola-shaped reflector, the light comes straight back *(a)*, concentrates *(b)* or spreads *(c)*, depending on where in the parabola the light source is. This is the basis of the common fixture section shown and shows the need to know the exact light center length (LCL) of the lamps. By using the light concentrating properties of an ellipse-shaped reflector *(d)*, light can be focused through a very narrow hole *(e)*, with very little loss.

14.24. Diffusers

Of the four items listed as the functions of a fixture in Section 14.20, we have discussed (a), (b) and (d) in detail. This leaves (c), the item concerned with change of light quality. You will remember that in *quality* we included items such as contrast, glare, diffuseness and color. Adding a *diffuser* to a fixture affects the quality of the light produced by the fixture. The diffuser, whether it is a piece of glass or plastic or a complex lens, affects the quality. The item is called a diffuser, because the original diffuser was a white or frosted glass bowl below the incandescent lamp. See Figures 14.36 and 14.37. Its function was to *diffuse* the light. In so doing it also decreased glare, increased diffuseness, decreased sharp shadows and contrast, and generally reduced the quantity of light. Such diffusers are still very much in use, as, for instance, the currently used fixtures shown in Figure 14.38. In industrial use, a white glass diffusing element is frequently added to a simple open RLM dome, to reduce direct and reflected glare.

The use of lenses, either glass or plastic, as fixture diffusers is also widespread. These lenses, which are of many different designs, redirect the light, reduce the fixture brightness and glare and, to an extent, diffuse the light. A lens acts much more efficiently than white glass or plastic and is generally more costly. A few examples of fixtures and diffusers are shown in Figures 14.39 and 14.40 for incandescents and in Figure 14.41 for fluorescents.

14.25. Lighting Systems

Most of the illustrated fixtures in the above sections give light directly downward. Exceptions are the diffusers of Figures 14.37 and 14.38, which give light in many directions. Many reflector-type fixtures have slots on top to give up-light. Up-light bounces off the ceiling, making the ceiling a low-brightness secondary source of light. The same is true of light from the sides of fixtures, which hits the wall and ceiling. Fixtures are classified by their

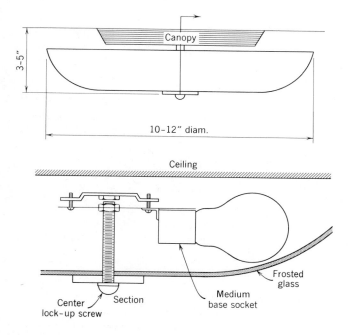

Figure 14.36 Elementary ceiling-type diffuser for incandescent lamps. Frosted or white glass diffuses light and reduces glare. Open glass dish quickly fills with dust, bugs, and the like, sharply reducing light output.

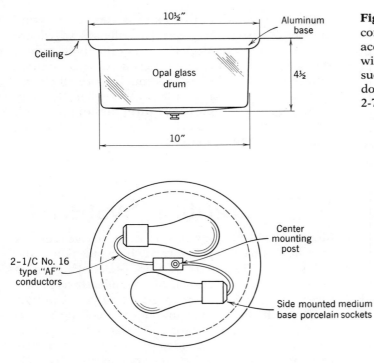

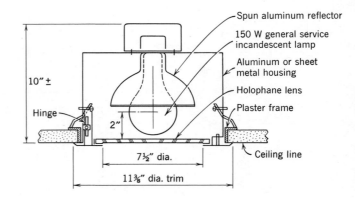

Figure 14.37 Center lockup glass drum type fixture in common use. Enclosing glass largely eliminates dirt accumulation inside fixture. This type fixture is also made with spring attached drum, to make relamping easier. In such fixtures the drum pulls down to expose lamps but does not come off. Illustrated unit dimensions suitable for 2-75 watt lamps.

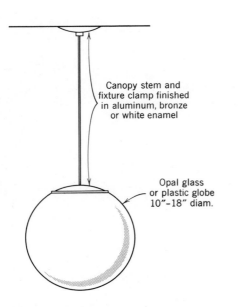

Figure 14.38 Globe-type diffuser for incandescent fixture is identical in operation to fixtures of Figures 14.36 and 14.37 except for shape and size. To achieve good diffusion and lamp hiding, these fixtures sacrifice efficiency.

Figure 14.39 Typical recessed downlight with a glass lens-type diffuser. The Holophane Company developed these lenses which are known in the fixture trade simply as Holophane lenses. There are many types of such lenses, each having its own particular optical characteristics.

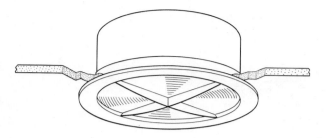

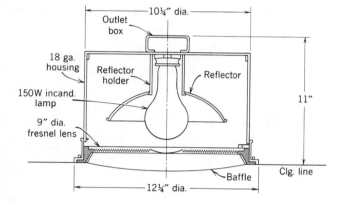

Figure 14.40 An unusual downlight that combines the light-directing effect of the Fresnel lens with the shielding of a decorative baffle.

light patterns, or distribution. Look at Figure 14.42. Note that the seven types of distribution are different, and give different effects:

(a, b) Direct lighting is most frequently used in hung ceiling areas. With proper spacing and diffusers, even illumination over the area can be readily obtained. The concentrating type of down light is used for special effects and highlighting.

(c) Semidirect lighting is used with pendant fixtures to provide some light on the ceiling. Otherwise, stem hung direct lighting units produce a very dark ceiling.

(d) Direct-indirect lighting uses the ceiling as a reflector and results in soft diffuse room light.

(e) General diffuse is similar to direct-indirect with the addition of side light. Room effect is very similar.

(f, g) Semi-indirect and indirect lighting shines the light on to the ceiling and makes the ceiling into part of the lighting fixture. The effect is very soft, low intensity, diffuse light.

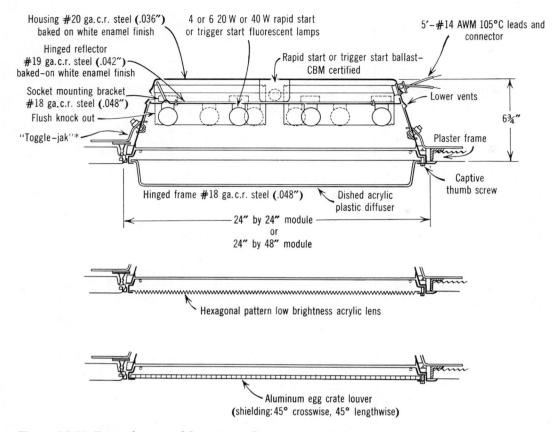

Figure 14.41 Typical recessed fluorescent fixture detail with various types of diffusers. The aluminum egg-crate louver acts both as a shielding mechanism and, by means of reflections on the louvers, as a diffuser. (Courtesy of Gotham Lighting)

14.26. Lighting Methods

All of the above lighting systems are designed to give general room illumination. For many types of close work general room illumination is insufficient. For these areas, such as drafting tables, jewelers benches and display tables, we add local supplementary light. This brings the level up to that needed at the work site without making the rest of the room too bright. Also, from the energy viewpoint this is certainly the right approach. The typical drafting room, therefore, uses direct-spread recessed ceiling fixtures (troffers) for general lighting and direct-concentrating desk lamps for supplementary local lighting.

14.27. Lighting Uniformity

Footcandle lighting calculations are made to determine the *average* illumination in a room, at the *working plane* level. This level is taken as 30 in. above the floor, which is approximately desk height. It is assumed that enough fixtures are used to give even room lighting. To check this, look up the manufacturer's recommended maximum spacing-to-mounting-height ratio (S/MH) and check, as follows:

Example 14.1 A room is to be lighted with direct lighting fluorescent troffers with a S/MH of 1.2. Ceiling height is 8 ft. What is the maximum fixture spacing?

Solution:

$$\frac{S}{MH} = 1.2 = \frac{S}{8\,ft}$$

therefore,

$$S = 8 \times 1.2 = 9.6\,ft$$

Maximum *side-to-side* spacing is 9 ft, 7 in.

Example 14.2 A warehouse is to be lighted with pendant RLM dome incandescents with an S/MH of 1.3. Spacing, because of architectural considerations, must be on a grid of 16 ft by 16 ft. What is the minimum mounting height?

Solution:

$$MH = \frac{Spacing}{Ratio} = \frac{16\,ft}{1.3} = 12.3\,ft = 12\,ft\,4\,in.$$

Note that this gives the *minimum* mounting height.

If manufacturers' data for the particular fixture involved is not available use the following approximate figures:

System	S/MH Ratio
Direct concentrating	0.4
Direct spread	1.2
Direct indirect, diffusing	1.3
Semi-indirect, indirect	1.5

14.28. Fixture Efficiency and Coefficient of Utilization

A lighting fixture emits less light than is generated by its lamp sources. Light is lost inside by internal reflection and absorption. The ratio of fixture output lumens to lamp output lumens is the efficiency of the fixture. Although this figure is important, it is *not* useful in lighting calculations. What is needed is a number that expresses how well a *particular fixture lights a particular room*. In other words, what we really need is a number indicating the efficiency of the fixture-room combination. This figure, normally expressed as a decimal, is called the *coefficient of utilization*, or CU.

$$CU = \frac{lumens\ utilized}{lamp\ lumens}$$

We cannot go into the derivation of this figure here. What is important to remember is that every manufacturer publishes tables of CU for each fixture he makes. With the help of these tables the lighting designer or engineer can find the CU for each lighting fixture in each space. These he will give to the technologist who in turn will do the final lumen calculation and fixture layout. Most designers will use a method called the Zonal Cavity Method to

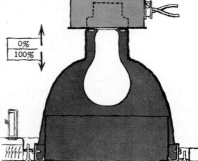

(a) Direct Light — Concentrating

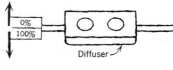

(b) Direct Light — Spread

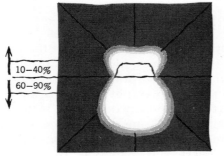

(c) Semidirect Light

(d) Direct–Indirect Light

Figure 14.42 Lighting system patterns, and typical fixtures for the different types of lighting systems. The room surfaces that are lighted directly (not by reflection) are shown white. Other surfaces are shown black.

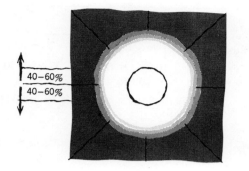

(e) General Diffuse Light

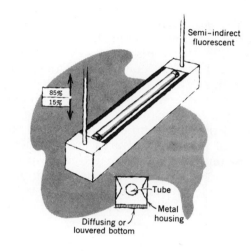

(f) Semi-Indirect

(g) Indirect

arrive at the CU. Consult the sources listed in the Additional Reading at the end of the chapter for details of this method. We also suggest careful study of the literature of major lighting manufacturers to gain familiarity with the CU table.

14.29. Metric Lighting Units

Before proceeding with illumination calculations, it is important that you become familiar with metric lighting units. Up to this point we have been using the system that is in common use in the United States. In this system, distance is measured in feet, area in square feet, luminous flux in lumens and illumination in footcandles. In the metric system, also called the SI system or International System, the units of distance and area are the meter and square meter (abbreviated m and m² respectively). Luminous flux remains in lumens, but illumination or light flux density is expressed in lux as follows:

$$\text{lux} = \frac{\text{lumens}}{\text{square meters}}$$

Placing the units of the two systems side by side we have

	American	SI
Length	Feet	Meters
Area	Square feet	Square meters
Luminous flux	Lumens	Lumens
Illumination (flux density)	Footcandles	Lux

The conversion factors between feet and meters are well known and are given in the table of conversion factors on page 556. To convert between lux and footcandles remember that 1 footcandle = 10.76 lux. That is, multiply lux by 0.0929 to get footcandles and multiply footcandles by 10.76 to get lux.

Let's derive these factors to show how simple they are. In Section 14.4 we considered this example: find the illumination on the floor of a room 10 ft square, produced by 1000 lm falling evenly on that surface. We will now do parallel calculation in the American and SI systems. The basic formula that is applicable to both systems is

$$\text{illumination} = \frac{\text{light flux}}{\text{area}}$$

	American	SI
Light flux	1000 lm	1000 lm
Area	$10\,\text{ft} \times 10\,\text{ft} = 100\,\text{sq ft}$	$\dfrac{10 \times 10}{3.28^2}\text{m}^2$
Illumination	$\dfrac{1000\,\text{lm}}{100\,\text{sq ft}} = 10\,\text{fc}$	$\dfrac{1000\,\text{lm}}{\dfrac{100}{3.28^2}\text{m}^2}$

$$= 10 \times 3.28^2 \text{ lux}$$
$$= 10 \times 10.76 \text{ lux}$$
$$= 107.6 \text{ lux}$$

Thus we see that the relation between footcandles and lux is the same as the relation between square meters and square feet, that is, a factor of 10.76. We could have worked out the answer in footcandles and have multiplied by 10.76 to get lux. However, it is better to become accustomed to using metric units throughout. The day is not too far off when this same room will not be called off as 10 ft sq, but as 3.05 m² or, more probably, 305 cm sq. (In the section that follows we use American units for the most part.)

14.30. Illumination Calculations

The *average* footcandles in a space can be simply calculated if we remember that by definition, one footcandle equals one lumen per square foot. Or

$$\text{footcandles (fc)} = \frac{\text{lumens (lm)}}{\text{area in square feet (sq ft)}}$$

Therefore, all we need do is calculate the usable lumens in a room, divide by area, and we have the average footcandle level. The usable *initial* footcandles is equal to the footcandles produced, reduced by the coefficient of utilization:

initial footcandles = produced footcandles × CU

We emphasize "initial" because the output of a fixture becomes less with time as the fixture becomes dirty and the lamp output drops. The factor assigned to describe this output drop is called the *maintenance factor*, MF. Therefore, to find the maintained, average illumination, we reduce the initial illumination by the maintenance factor.

$$\text{initial illumination} = \frac{\text{produced (lamp) lumens} \times \text{CU}}{\text{area}}$$

$$\text{maintained illumination} = \frac{\text{lamp lumens} \times \text{CU} \times \text{MF}}{\text{area}}$$

Lamp lumens is simply the rated output of the lamps. To show how easily these calculations work, let us try a few examples.

Example 14.3 A classroom 22 ft by 25 ft is lighted with 12 fluorescent fixtures, each containing 4–F40WW/RS lamps. Calculate the initial and maintained illumination in footcandles. Assume a CU of 0.38 and an MF of 0.75.

Solution

$$\text{lamp lumens} = \begin{array}{l} 12 \text{ fixtures} \\ \times \ 4 \text{ lamps per fixture} \\ \times \ 3200 \text{ lumens} \\ \qquad\qquad \text{per lamp} \end{array}$$

$$\text{lamp lumens} = 12 \times 4 \times 3200$$
$$= 153{,}600$$

$$\text{initial footcandles} = \frac{153{,}600 \text{ lm} \times 0.38}{25 \times 22 \text{ sq ft}}$$
$$= 106 \text{ fc}$$

$$\text{maintained footcandles} = 106 \times 0.75 = 80 \text{ fc}$$

Or, the entire calculation can be done in one step:

$$\text{initial footcandles} = \frac{12 \times 4 \times 3200}{25 \times 22} \times 0.38$$
$$= 106 \text{ fc}$$

$$\text{maintained footcandles} = \frac{12 \times 4 \times 3200}{25 \times 22} \times \begin{array}{l} 0.38 \\ \times \ 0.75 \end{array}$$
$$= 80 \text{ fc}$$

As a practice exercise we do this problem in SI units also. Remembering that maintained illumination equals lamp lumens × CU × MF/area, we have

$$\text{lux} = \frac{\text{lumens} \times \text{CU} \times \text{MF}}{\text{m}^2}$$

or

$$\text{lux} = \frac{(12 \times 4 \times 3200) \text{ lumens} \times 0.38 \text{ CU} \times 0.75 \text{ MF}}{\dfrac{22}{3.28} \times \dfrac{25}{3.28} \text{ m}^2}$$

Therefore,

$$\text{maintained lux} = \frac{153{,}600 \times 0.38 \times 0.75}{6.71 \text{ m} \times 7.62 \text{ m}}$$
$$= 856.2 \text{ lux}$$

To check the accuracy of this answer, we use the conversion factor listed above:

  multiply lux by 0.0929 to get footcandles
Therefore,
$$856.2 \text{ lux} \times 0.0929 = 80 \text{ fc}$$
Which checks the above result.

Most of the time the problem is presented the other way around. That is, given a room and required footcandles, find the number of fixtures. This is readily done
  Since

$$\text{footcandles} = \frac{\text{lumens}}{\text{area in square feet}}$$

we have
$$\text{lumens} = \text{footcandles} \times \text{square feet}$$
or
$$\text{maintained lumens} = \begin{array}{l} \text{maintained footcandles} \\ \qquad\qquad \times \text{ square feet} \end{array}$$
but
$$\text{maintained lumens} = \text{lamp lumens} \times \text{CU} \times \text{MF}$$
so we have
$$\text{lamp lumens} = \frac{\text{maintained lumens}}{\text{CU} \times \text{MF}}$$
or
$$\text{lamp lumens} = \frac{\text{maint. fc} \times \text{sq ft}}{\text{CU} \times \text{MF}}$$

This same relation expressed in SI (metric) units is very similar:

$$\text{lamp lumens} = \frac{\text{maint. lux} \times \text{m}^2}{\text{CU} \times \text{MF}}$$

Applying these equations to the same problem, we have the following.

Example 14.4 Given a classroom 22 by 25 ft to be lighted to an average maintained footcandle level of 75 fc. Find the number of four-lamp 48-in RS fixtures required. Assume CU = 0.38, MF = 0.75.

Solution

$$\text{lamp lumens} = \frac{75 \text{ fc} \times (22 \times 25) \text{ sq ft}}{0.38 \text{ CU} \times 0.75 \text{ MF}}$$

$$\text{lamp lumens} = 144{,}737 \text{ lm}$$

Since each F40WW lamp has an output of 3200 lm

$$\text{number of lamps} = \frac{144{,}737}{3200} = 45$$

Since we have four lamps per fixture, we have

$$\text{number of fixtures} = \frac{45}{4} = 11.25 \text{ fixtures}$$

Obviously, we can't have one-fourth of a fixture. We would most probably raise the number to 12 fixtures for two reasons. First, we want a minimum of 75 fc and, secondly, an even number of fixtures (12) is much easier to lay out than an odd number such as 11. We discuss layout below.

This same problem in metric units would be: given a classroom 6.71 m by 7.62 m, to be lighted to an average maintained level of 807 lux (75 fc × 10.76), find the number of four-lamp 48-in. RS fixtures required. The solution would be:

number of fixtures =

$$\frac{\text{illumination} \times \text{area}}{\text{CU} \times \text{MF} \times \text{lumens per lamp} \times \text{lamps per fixture}}$$

$$= \frac{807 \text{ lux} \times 6.71 \text{ m} \times 7.62 \text{ m}}{0.38 \times 0.75 \times 3200 \text{ lm} \times 4 \text{ lamps}}$$

$$= 11.3 \text{ fixtures}$$

which checks with the above answer. The slight difference is due to the rounding off of the dimensions of the room when they are expressed in meters.

One further technique in illumination calculation should be mastered. Frequently, we must calculate illumination for a very large space such as an office floor. Instead of doing this for the entire floor, it is easier and more meaningful to calculate the number of fixtures required per bay. This can be done directly or by calculating the area covered by a single fixture as follows.

Since

number of fixtures =

$$\frac{\text{illumination} \times \text{area}}{\text{lamps per fixture} \times \text{lumens per lamp} \times \text{CU} \times \text{MF}}$$

it follws that the area lighted by a single fixture is

area per fixture =

$$\frac{\text{lamps per fixture} \times \text{lumens per lamp} \times \text{CU} \times \text{MF}}{\text{illumination}}$$

An example would help here.

Example 14.5 An entire office floor is to be lighted to an average maintained footcandle of 50. The floor measures 320 × 150 ft and is divided into bays measuring 40 × 25 ft. Using two-lamp recessed troffers with 40 w RS lamps, find the number of fixtures required. Assume an economy grade fixture with a low CU of 0.35 and MF of 0.7.

Solution A

fixtures per bay =

$$= \frac{50 \text{ fc} \times (40 \times 25) \text{ sq ft}}{2 \text{ lamps} \times 3200 \text{ lm} \times 0.35 \times 0.7}$$

$$= 31.88$$

This would be either 30 units arranged in 3 continuous rows of 10 the long way or 5 rows of 6 the short way. In either case, center-to-center spacing is 8 ft ±. (We suggest that you make a layout sketch now.)

Solution B

The same result can be obtained by calculating the square feet per fixture.

square feet per fixture =

$$= \frac{2 \text{ lamps} \times 3200 \text{ lm} \times 0.35 \times 0.7}{50 \text{ fc}}$$

$$= 31.36 \text{ sq ft per fixture}$$

Since the fixture is 4 ft long, centerline spacing of fixtures is

$$\text{spacing} = \frac{31.36}{4} = 8 \text{ ft}\pm$$

$$\text{fixtures per bay} = \frac{40 \times 25 \text{ sq ft}}{32 \text{ sq ft per fixture}} = 31,$$

as above.

$$\frac{\text{actual}}{\text{footcandles}} = \frac{30 \text{ fixtures}}{31 \text{ calculated}} \times 50 = 48.5 \text{ fc}$$

This is close enough. See Figure 14.43 for a diagram of this problem.

Now, to do the same problem in SI (metric) units, we first restate it as follows:

Given an office floor measuring 97.56 m × 45.73 m, with bays of 12.2 m × 7.62 m, find the number of fixtures required to produce an average maintained illumination of 538 lux. All other data remain the same.

Solution A1

fixtures per bay =

$$= \frac{538 \text{ lux} \times (12.2 \times 7.62) \text{ m}^2}{2 \text{ lamps} \times 3200 \text{ lm} \times 0.35 \times 0.7}$$

$$= 31.88$$

Solution B1

square meters per fixture =

$$= \frac{2 \text{ lamps} \times 3200 \text{ lm} \times 0.35 \times 0.7}{538 \text{ lux}}$$

$$= 2.91 \text{ m}^2 \text{ per fixture}$$

Therefore,

$$\text{fixtures per bay} = \frac{12.2 \times 7.62 \text{ m}^2}{2.91 \text{ m}^2} = 31.94$$

Which checks with Solution A1 and with Solutions A and B above.

The differences in the above solutions are due to the rounding off of sq ft per fixture. Obviously, when the change to metric units becomes general, dimensions will be in even metric units and not the odd sizes as above. That is, a bay may measure 8 × 12 m, lamps may be 125 cm long, ceiling tiles may be 75 × 150 cm, and so on. For the time being, we strongly

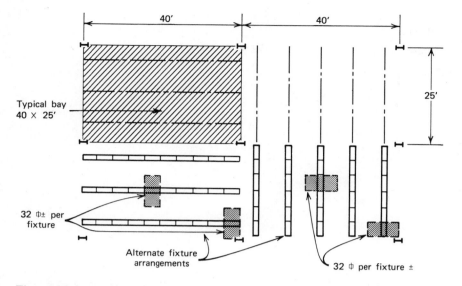

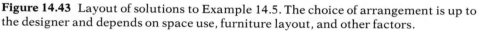

Figure 14.43 Layout of solutions to Example 14.5. The choice of arrangement is up to the designer and depends on space use, furniture layout, and other factors.

advise using the parallel system, that is, for the next few years, working in *both* systems. This will accustom you to the metric units and will make the full change much easier when it comes.

14.31. Lighting Calculation Estimates

Frequently it is necessary to make estimates and to assume figures when exact data are not available. When this is necessary use the estimates below.

Coefficient of Utilization

Efficient fixture, large light colored room	0.45
Average fixture, medium-sized room	0.35
Inefficient fixture, small or dark room	0.25

Maintenance Factor

Enclosed fixture, clean room	0.8
Average conditions	0.7
Open fixture or dirty room	0.6

It is often very helpful to be able to take an educated guess at the fixture requirements of a space. Figure 14.44 will be very helpful in this respect. To use it, simply remember that

square feet per fixture =

$$\frac{\text{fixture wattage}}{\text{chart figure} - \text{watts per square foot}}$$

Since the chart is for an average to large room, increase the wattage up to 10% for small rooms and decrease it up to 10% for very large rooms. Applying it to the two examples we considered above, we find that the results are quite good for a first estimate. That is,

Example 14.6 Using Figure 14.44, we estimate the fixture requirements of Examples 14.3 and 14.5.

Solution: Both examples use direct lighting troffers. The graph gives

$$75 \text{ fc} = 4.1 \text{ w/sq ft}$$

and

$$50 \text{ fc} = 2.7 \text{ w/sq ft}$$

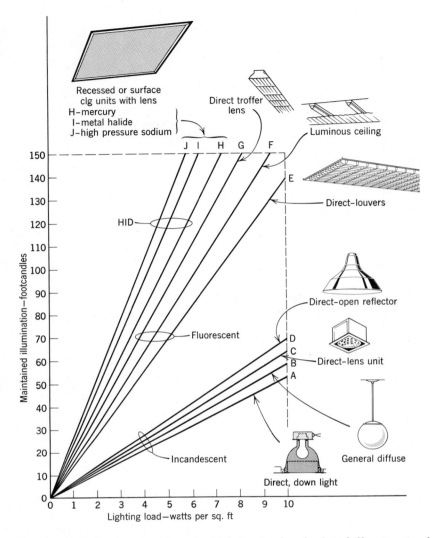

Figure 14.44 Estimating chart for lighting load and related illumination levels for different lighting sources. The chart was calculated for a fairly large room—approximately classroom size. Although all the figures are necessarily approximate because of the variables involved, the chart gives figures close enough for a first approximation. Notice the increase in output as the sources change from incandescent to fluorescent to HID.

For the 22 × 25 classroom at 75 fc, using four-lamp (200 w) fixtures, we have

square feet per fixture =

$$= \frac{200 \text{ w per fixture}}{4.1 \text{ w/sq ft}}$$

$$= 48.8 \text{ sq ft per fixture}$$

This gives

$$\frac{25 \times 22 \text{ sq ft}}{48.8 \text{ sq ft per fixture}} = 11.3 \text{ fixtures}$$

This is exactly the calculated result.

For the large office space, we must decrease the required watts per square foot by 5%. That is, 95% of

$$2.7 \text{ w/sq ft} = 2.57 \text{ w/sq ft}$$

Then,

square feet per fixture =

$$= \frac{100 \text{ w per fixture}}{2.57} = 38.9 \text{ sq ft}$$

This figure is 24% higher than the calculated result and represents the difference between a good grade unit with a CU of 0.4 and MF of 0.75 as used in the chart, and the cheaper, less efficient unit used in the calculated example. This can easily be demonstrated by correcting for CU and MF as follows:

sq ft per fixture = chart figure × corrections

$$= 38.9 \times \frac{.35}{.4}(CU) \times \frac{.70}{.75}(MF)$$

$$= 31.76, \text{ which is the calculated result.}$$

A little practice will make the technologist good at these rapid lighting estimates. Such estimates are very helpful at the beginning of a job when the decision on what type of lighting unit and what number of lamps is to be made.

14.32. Conclusion

In this chapter we examine the action of light, the production of artificial light and its control and, finally, lighting calculation. The ideas presented here will be applied in the next chapters to specific residential and industrial occupancies to demonstrate practical lighting layout technique.

Problems

14.1. (a) What is the reflection factor of an aluminum fixture reflector, which reflects 87 lm of every 100 lm falling on it?

   (b) What percent is absorbed?

   (c) This same fixture radiates 70 lm through its plastic diffuser for every 100 lamp lm. What is the transmission factor of the diffuser? What percentage does it absorb?

14.2. The fixture of Problem 14.1 above produces 12800 lamp lm. Assuming all the lumens radiating from the fixture strike the floor, calculate the average footcandle illumination level on the floor of a 10 ft × 10 ft room.

14.3. A classroom is to be lighted with two continuous rows of plastic lens fluorescent fixtures. The manufacturer gives a maximum spacing to mounting height ratio of 1.1. The room is 22 ft wide by 30 ft long. Pendant rows of fixtures are to be run the length of the room. What is the minimum mounting height? Draw a sketch of the room showing the fixtures. (*Hint:* spacing between fixture and wall is one half of the fixture-to-fixture spacing.)

14.4. A supermarket is lighted with continuous rows of two-lamp 96 in. high output rapid-start lamp fixtures mounted on 10 ft centers. Assuming a CU of 0.5 and an MF of 0.6, find the average maintained footcandle illumination in the store.

14.5. A 10 × 12 ft private office is lighted with 3 four-lamp recessed troffers using F40CW/RS lamps.

(a) With CU = 0.4 and MF = 0.7, find the maintained footcandles.

(b) After redecorating the room with dark wood paneling, the CU has dropped to 0.28. What is the new footcandles in the room? How many of the same type fixtures must be added to restore the former footcandles?

(c) Show a plan of the room with the original three fixtures, and with the new fixture layout required to restore the lighting level.

(d) Assuming that the office has a hung ceiling using 1 ft sq tiles, would the layout of (c) above change? How? Show a dimensioned ceiling plan, with tiles.

14.6. (a) Redraw Figure 14.44 on a separate sheet of paper, using SI units.

(b) Using this new set of curves, recalculate illustrative example 14.6. Do the results check out?

14.7. An office building has a 2-way, coffer-type ceiling with module dimensions of 5 ft 6 in. by 3 ft 8 in. It is desired to have an average overall maintained illumination of 30 fc. Make a recommendation for lighting the area using fluorescent fixtures. Show the calculations that led to this recommendation. Make any necessary assumptions, but state them. Draw your proposed layout for a bay that measures 28 ft × 22 ft.

14.8. A classroom is 8 m wide and 10 m long. Make at least two lighting layouts of the room that will give an average maintained illumination of 350 lux. Use two- or three-lamp, 40 w RS fixtures.

Additional Reading

IES Lighting Handbook, Fifth edition, Illuminating Engineering Society. This book is the handbook of the lighting industry and contains a wealth of technical material on all facets of lighting design and lighting equipment.

Architectural Lighting Graphics, J. E. Flynn and S. M. Mills, Van Nostrand Reinhold. This book presents lighting concepts and equipment detail graphically, making it of particular interest to draftsmen and technologists.

Westinghouse Lighting Handbook, A pocket-size booklet with a good deal of handy data on sources, fixture and design.

IES Lighting Fundamentals Course, Section ED-2, published by the Illuminating Engineering Society. This booklet contains essential technical material intended for use with a basic course in illuminating engineering. The material is presented as a technical supplement and not as an individual private study course.

Architectural Graphic Standards, American Institute of Architects and C. G. Ramsey and H. R. Sleeper.

Mechanical and Electrical Equipment for Buildings, William J. McGuinness and Benjamin Stein. These two books are discussed in preceding chapters.

15. Residential Electric Work

We now apply the techniques already learned to a large and elegant private residence. In the process we carefully examine each area in the house. For this detailed study we consider each space both as a particular case and as an example of the general case. For instance, the living and family room will be studied as rooms, and also as an example of living spaces. Our purpose is to develop a basic method for handling different kinds of spaces. This will enable the technologist to do the electrical layout for almost any type of residential space and for related spaces in non-residential buildings. In this chapter we extend our study of the electrical system beyond circuitry to load studies, riser diagrams and electric service considerations. Finally, we include a brief review of signal and communication work applicable to residences.

Study of this chapter will enable you to:

1. Analyze spaces by usage—present and potential.
2. Understand the electrical design guidelines applicable to residential spaces and apply them to drawing layout.
3. Apply the design guidelines learned to nonresidential spaces with similar requirements and uses. These include utility, circulation, food preparation, dining, storage and outside areas.
4. Select, lay out and circuit the lighting for residential and nonresidential spaces, using the applicable provisions of the NEC.
5. Draw plans showing signal and communications devices found in residential buildings, including riser diagrams.
6. Make up panel schedules, complete with loads, wire sizes and other branch circuit data.
7. Do basic load calculations, including electric heat.
8. Be familiar with electric service equipment for single and multiple residences and other buildings, including metering provisions.
9. Draw an electric plot plan, riser diagram and

one-line diagram for basic electric service.

10. Know the characteristics of underground and overhead service and be familiar with service entrance details.

15.1. General

As we stated in Chapter 3, we use for our study, with the architect's permission, the architectural plans of the Merker residence. The electrical layout that we do is not the one that was actually installed in the house. Every designer produces a different layout. Our layout expresses our approach. It is not the best possible layout, since there is no such thing, but it is a good arrangement. It meets not only the minimum requirements but also what we consider the special needs of the space. In addition, the layout attempts to anticipate problems and to make the electrical aspects of living in the house easy and comfortable. To accomplish this requires guidelines, experience and common sense. The guidelines are given here; the necessary experience must be gained in the field. We also refer back to the Basic Plan House, so that the reader can see the difference between treatment of large and small rooms that have the same use.

In doing a layout of any space, a furniture arrangement is a great help. This is particularly true in residences where the location of a bed, for instance, makes all the difference in location of outlets. If a furniture layout is not available, one must be assumed based on a reasonable arrangement. If that, too, is impossible, because the room can have more than one *good* furniture arrangement, the layout should be made to accommodate several arrangements, even at slight additional cost. Otherwise the floor will soon be covered with long extension cords, and the accessible wall outlets will be overused. This is not only unsightly but dangerous. A large percentage of fires in homes are caused by improper extension cords and overloaded outlets.

Furthermore, living areas are multi-purpose and should be treated as such. Houses with family rooms *and* living rooms are different from those with living rooms only. In the latter, all social functions take place in the living room and the electric layout should be flexible enough to meet this need. Finally, in addition to coordinating the electric layout with the furniture, it is necessary also to coordinate it with the heating/cooling equipment. This problem is much more serious in industrial and commercial buildings than in residences, and is discussed in detail in the next chapter. In houses, it is relatively simple to avoid conflicts in equipment location; we show how, in the layout description.

We have divided the spaces in a residence into five types by use and location, as follows:

(a) Living areas including living rooms and family rooms plus adjoining balconies and studies, sun rooms, television rooms, dens and libraries

(b) Kitchen and dining areas including food preparation areas, breakfast rooms, and dining rooms

(c) Sleeping areas including bedrooms, dressing rooms, closets, and adjoining balconies

(d) Circulation, storage and utility and wash areas. These include halls, corridors and stairs, garage, closets, basement, attic, laundry and other work areas, and bath- and washrooms.

(e) Outside areas attached to or adjacent to the house.

15.2. Living Areas

a. Living Room

As stated in Chapter 13, the wall receptacles in a residence are considered to be part of the general *lighting* load. This is because wall convenience outlets are frequently used to feed the lighting sources in the room. Refer to Figure 13.15a. This living room, like most small to medium size modern living rooms, has no ceiling outlet. Instead a switched receptacle is used to provide the lighting. The resident will plug a table or floor lamp into this outlet. The reason a ceiling outlet is not used is that a living room normally has a low to medium lighting level, which is easily provided by lamps. Since lamps are always a part of the furnishings, a ceiling outlet would only be used rarely, even if it were provided. Also a single ceiling outlet tends to attract the eye away from the furnishings. On the other hand, multiple ceiling outlets, wall wash units or lighted coves, valances or cornices are acceptable and desirable. Refer to the living room in the Merker residence in Figure 15.1b. This room is quite large and its unusual architecture calls for special lighting solutions, including the type of detailing that a technologist is called on to do. The important architectural features of this space are the sloping ceiling with its breakdown to the low ceiling level, the all-glass front wall leading out to the deck and the brick wall and fireplace. Let us consider these features one by one. Taking the sloping ceiling first, the break line is an ideal location for applying a lighted cove. See Figures 15.2 to 15.4 and Appendix A.6. This lighting is controlled by a switch and dimmer appropriately rated (300 w minimum), which permits level adjustment. In addition to this soft and pleasant cove lighting, the lighting designer has decided on a 3 × 3 pattern of ceiling downlights, type Z. These units are also dimmer controlled but by a standard incandescent lamp dimmer, rated 1500 w minimum. The type Z fixture has a black alzac cone so that no fixture brightness is reflected in the upper-level glass. The fixture must be obtained with a bottom slope to match the ceiling slope within 10 deg. See Figure 15.5. The nine type Z fixtures are arranged in a symmetrical pattern in the ceiling, with fixture-to-wall spacing one half of the fixture-to-fixture spacing. Figure 15.6 demonstrates to the draftsman a simple and reliable way of laying out any desired fixture pattern without detailed measurement. This method is based on the above universally accepted arrangement—fixture-to-wall spacing equal to one-half of fixture-to-fixture spacing.

Fixtures W and Y are similar types (see Figure 15.7) except that type Y is gasketed, being installed outdoors. The designer has deliberately provided lighting on both sides of the glass. This prevents the glass from acting like a mirror, which happens when the space on the far side of the glass is much darker than on the near side. This can be very annoying when you are trying to see *out* at night. The solution is, as explained above, to provide lighting on the far side of the glass as well. The fixtures have been selected so as not to show any brightness that will reflect in the glass and be a source of annoyance.

Within the living room, the point of interest is the fireplace and adjoining brick wall. To accent this, three wall wash lighting units, type V have been supplied. These units are special because they also must be made to fit into the sloping ceiling. See Figure 15.8. Control of these units is local, since they are accent lighting and will not be used all the time. Similarly, control of fixtures W and Y is separate, since they also are special purpose. Control of the general illumination furnished by the cove and the Z fixtures is provided at the entrance to the living room. It is always advisable to use a switch together with a dimmer, so that a preset dimming level can be obtained by simply flicking the switch. Otherwise, one must experiment each time, and the dimmer becomes a nuisance instead of a convenience.

The deck outside the living room is really the outdoor part of the living room. Lighting for this area is, of course, day lighting most of the time. Night lighting should provide enough light for activities on the deck such as eating and recreation. This is easily furnished by the type L adjustable floodlights similar to Appendix A.8) and the low voltage post lights, type AA. Low voltage units were chosen to enable the use of exposed wiring with minimum electrical maintenance problems. These low voltage fixtures are popular for landscape, security and general outdoor use because of the ease of wiring and minimum hazard. Control of the floodlights and the post lights is conveniently placed at the exit to the deck. At the exit to the side deck a 3-way switch is furnished for control of the post lights only. See Figure 15.9 for details of the low voltage units and their wiring.

Receptacles are placed about the room in accordance with the NEC rule that no point shall be further than 6 ft from a convience outlet. This rule is stretched a bit with respect to the all-glass, 26 ft long front wall. Permission to do this should be confirmed in the field with the local electrical inspection authorities. All wall outlets will be mounted vertically and at 12 in. AFF to the centerline unless specifically shown otherwise. The outlet in the bar area is shown at 48 in. for convenient use of electric mixers, ice crushers and the like on the bar. This completes the description of the electrical layout of the living room and connected areas. After a similar description of the Merker family room and study, we review the layout principles that have been followed.

b. Family Room

This room in many ways is the most important room in the house. Aside from eating, sleeping and formal entertaining, this is where the family spends its time. Here the family relaxes, watches television, entertains informally, plays games and so on. The room, therefore, has many functions. The electrical layout must be flexible enough to serve all these functions.

First, let us consider the lighting. Obviously, many different levels and types of lighting are required to fit all of this room's activities. Recreation activities assume a range of levels varying between high for a children's party to soft and subdued for a teenager's evening get-together. Similarly, television watching, card playing and billiards all have completely different lighting requirements. We, therefore, decide to supply adjustable level lighting, and selectable quality. The type H decorative wall bracket supplies both general room lighting

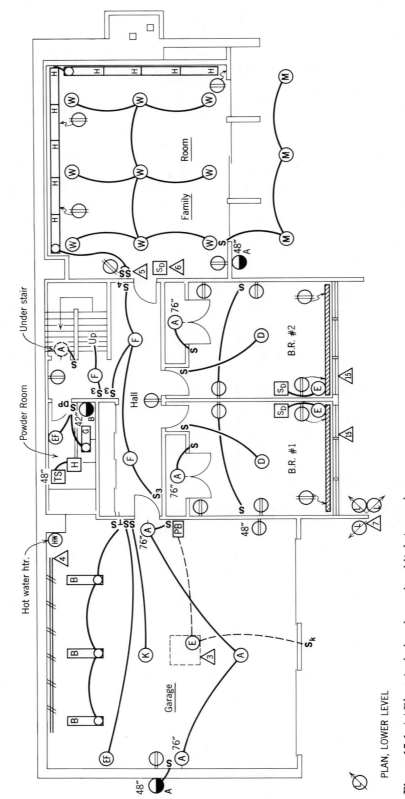

PLAN, LOWER LEVEL

Figure 15.1 (a) Electrical plan, lower level lighting and device layout.

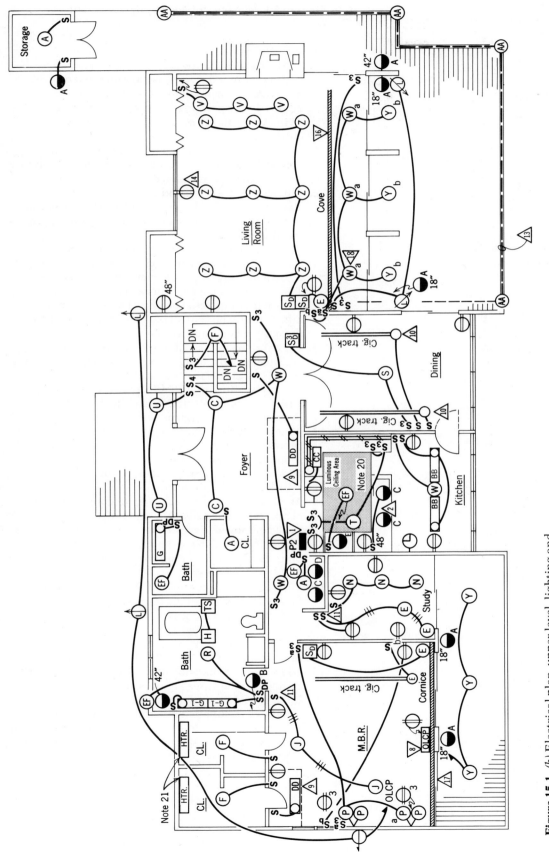

Figure 15.1 (b) Electrical plan, upper level, lighting and device layout.

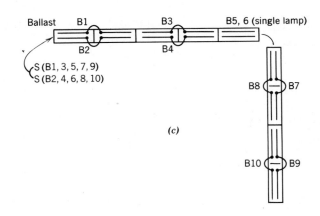

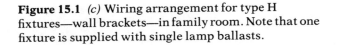

(c)

Figure 15.1 *(c)* Wiring arrangement for type H fixtures—wall brackets—in family room. Note that one fixture is supplied with single lamp ballasts.

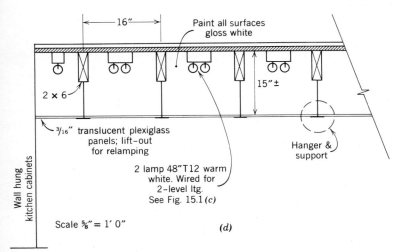

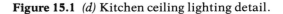

Figure 15.1 *(d)* Kitchen ceiling lighting detail.

Figure 15.1 *(e)* Notes for electric device layout

1. Panel face to be suitable for same finish as wall.
2. Outlets under counter.
3. Ceiling-mounted electric garage door opener.
4. Two-circuit wiring channel with 10 single 20-amp, 2-pole, 3-wire grounded receptacles on 24-in. centers, 48 in. AFF. Total 10 outlets; 5 per circuit. Start first outlet 12 in. from one end.
5. Wire fixtures per Figure 15.1*(c)*, so that two-lamp ballast in each fixture controls one lamp in its fixture and one lamp in the adjacent fixture.
6. Single 1500 w switch and dimmer. Install flush in closet wall and enclose rear protruding into closet.
7. Extend floodlight controls to outside lighting control panel in master bedroom.
8. The contractor is invited to submit a proposal for performing all the switching with low voltage control as manufactured by the General Electric Company. The proposal must contain a wiring diagram, complete equipment specifications and a price differential. The control panel would be mounted in the master bedroom, as shown.
9. For detail of recessed daylight lamp fixture see architectural drawings. Note on Drawing 6.4*(b)* that there is an access to the attic in this area.
10. Ceiling lighting track, field located, 15-amp capacity, with fixtures as selected by architect.
11. Provide a spare switch position.
12. Mount 2 circuit strip on wall 12 in. above base cabinet. Provide single 20-amp, 2-pole, 3-wire grounded receptacle every 18 in.
13. Run low voltage wiring exposed, under deck. Mount transformer at switch location.
14. At all locations where receptacles conflict with a heating element register or grill the receptacle can be raised or moved to either side, as directed in the field.
15. Lighted valance above the window; with 48-in. single lamp strips and dimming ballasts. See detail Figure 15.12. Approx. 15 w/ft.
16. Lighted cove approximately 9 ft. AFF. See detail Figure 15.4. 48-in. single lamp strips, with dimming ballasts. Approx. 15 w/ft.
17. Lighted cornice above the drapes. See Figure 15.3. 48-in. single lamp strips with dimming ballasts. Approx. 15 w/ft.
18. Coordinate the location of the 3-wire fixtures with ceiling registers. See Figure 6.4*(a)*.
19. Up to appliance outlet circuit 26.
20. Hood switch supplied integral with hood.
21. Each unit 750 w, 120 v, with integral thermostat and ON/OFF switch.

Figure 15.1 *(f)* Lighting fixture schedule.

Type	Mtg	Max. Watts	Description
A	—	100	Porcelain lampholder, pull chain, wire guard, mtg as shown
B	Clg	100	Recessed fluorescent, two-lamp 40 w RS industrial type, steel louver, baked white enamel
D	Clg	120	Surface-mounted drum fixture, opal glass, torsion springs, two 60 w lamps
C	7 ft	75	Decorative wall bracket
F	Clg	100	Recessed square incandescent unit, dropped–dish opal glass diffuser, white enamel frame
G	Wall	50	Fluorescent wall bracket, wraparound white plexiglass diffuser, one 48-in. T12 RS lamp, mount above wall mirror
G-1	Wall	38	Same as G except one 36-in. T12 RS lamp
H	7 ft	100	Fluorescent, wall bracket with decorative plastic front diffuser; See detail, Appendix A.6. Each unit with two 48-in. T12 RS lamps
K	Clg	100	Reel-light, 100 w portable, 25 ft retractable cable, 360 deg swivel mount.
L	—	150	WP lampholder for R-40, 150 w bracket mounted, as field directed
M	Clg	150	WP, surface-mounted cylindrical downlights 150 w, R-40 lamp
N	Clg	150	Recessed, adjustable wall wash fixture, 150 w R-40 lamp, see Appendix A.7
P	4 ft	100	Swivel mount, spherical chrome wall bracket fixtures, mounted in pairs, for single 100 w A lamp, with built-in On/Off switch.

Type	Mtg	Max. Watts	Description
R	Clg	150	Recessed, gasketed, square, incandescent unit, dropped dish frosted glass diffuser, cast aluminum frame, for single 150 w A lamp
S	Clg	1000 Max	Decorative chandelier, selected by owner
T	Clg	600	Kitchen illuminated ceiling; see Figure 15.1*d*
U	—	100	Decorative wall bracket, WP, 100 w A lamp
V	Clg	150	Special unit; see Figure 15.8
W	Clg	150	Recessed downlight; see Figure 15.7
Y	Clg	150	Same as W except WP; see Figure 15.7
Z	Clg	150	Downlight, one 150 w R-40 lamp; see Figure 15.5
AA	—	25/50	Decorative, WP, post-top lantern, 6 v A-21 lamp, 25 or 50 w, 10-in. white plexiglass globe, with 36-in. redwood post; see Figure 15.9
BB	Clg	100	Fluorescent fixture, surface, wraparound plexiglass lens. Two-lamp F40 RS WW. Shallow construction. Maximum X-section; 12-in. wide × 3½ in. deep. Diecast end with white baked enamel finish
CC	—	30	Incandescent under-counter fixture. One 17-in. T8 lumiline lamp, disc base. Maximum fixture depth 1¾ in. White plastic diffuser
DD	—	100	Fluorescent strip, 2-lamp 40 W RS, daylight lamp, built into skylight. See detail on architectural drawing, A.00

SYMBOL LIST

—///—   WIRING RUN CONCEALED. HATCH MARKS INDICATE NUMBER OF #12 WIRES U O N GROUND CONDUCTORS NOT SHOWN.

———   CONDUCTORS RUN EXPOSED, NOTES AS ABOVE.

—•—²⁴P–2   HOME RUN OF CCTS 2 & 4 TO PANEL P–2 DOT INDICATES TURNING DOWN

(E)ᴍᴍᴍ   FINAL CONNECTION FROM OUTLET TO EQUIPMENT.

—#—#—#—   MULTI–OUTLET ASSEMBLY, 2 CIRCUIT

—/—/—/—   MULTI–OUTLET ASSEMBLY, SINGLE CIRCUIT

(A)ₐ –(A)   OUTLET AND INCANDESCENT FIXTURE CLG. /WALL MOUNTED. INSCRIBED LETTER INDICATES TYPE. SUBSCRIPT LETTER INDICATES SWITCH CONTROL.

[○ B]ₐ / [○ B]   OUTLET AND FLUORESCENT FIXTURE CLG. /WALL MOUNTED. SAME NOTES AS ABOVE.

(E)   OUTLET BOX WITH BLANK COVER.

(J)   JUNCTION BOX WITH BLANK COVER.

⊖•   DUPLEX CONVENIENCE RECEPTACLE OUTLET. 15 AMP 2P 3W 125 VOLT GROUNDING. WALL MT. VERTICAL, ₵ 12″ AFF, NEMA 5–15R DOT INDICATES APPLIANCE OUTLET – SEE TEXT.

⊖₃   TRIPLE OUTLET, AS ABOVE

◖ₐ   15A 2P 3W GFCI DUPLEX OUTLET, WP.

◖ᵦ   15A 2P 3W GFCI DUPLEX OUTLET.

◖c   20A 2P 3W SINGLE OUTLET, NEMA 5–20R.

◖ᴅ   30A 125/250V 3P 4W GND. OUTLET, NEMA 14–30R.

◖ᴇ   60A 125/250V 3P 4W GND. OUTLET, NEMA 14–60R.

🕐   CLOCK HANGER OUTLET, SEE SPEC. 7′6′ AFF TO ₵.

Sₐ   SINGLE POLE SWITCH, 15A 125V, 50″ AFF U O N, LETTER SHOWS OUTLETS CONTROLLED.

S₃   THREE WAY SWITCH 15 A, 125V, 50″ AFF U O N.

S₄   FOUR WAY SWITCH, AS ABOVE.

Sᴅᴾ   DOUBLE POLE SWITCH, AS ABOVE.

Sₖ   KEY OPERATED SWITCH, AS ABOVE.

Sᴛ   SWITCH 15A 125V WITH THERMAL ELEMENT SUITED TO MOTOR.

    COMBINATION SWITCH AND RECEPTACLE IN A 2 GANG BOX.

[Sᴅ]   COMBINED SWITCH AND DIMMER.

[D]   DIMMER, RATING AS NOTED, 600W U O N

[TS]   TIMER CONTACTS, MANUAL SET, 15 AMP, 125 VOLTS, 0–20 MINUTES.

[PB]   PUSHBUTTON, 10 AMP MOM. CONTACT U O N

[H]   HEATER, RATING AND DETAILS ON DWG., WITH OUTLET.

(EF)   EXHAUST FAN, ¹⁄₁₂ HP U O N

■   RECESSED PANELBOARD, SEE SCHEDULE.

>—[ ] (M)   INCOMING ELECTRIC SERVICE, METER CABINET AND METER.

▨▨▨▨   ARCHITECTURAL LIGHTING ELEMENT; SEE PLANS.

△   REFERENCE TO NOTE 1. SEE NOTES.

ABBREVIATIONS:

A   AMPERES

AFF   ABOVE FINISHED FLOOR

C/B   CIRCUIT BREAKER

CCT   CIRCUIT

F   FUSE

GFCI   GROUND FAULT CCT INTERRUPTER

GND   GROUND

HP   HOURSEPOWER

LTG   LIGHTING

MH   MTG. HEIGHT

N   NEUTRAL

PC   PULL CHAIN

T   THERMOSTAT

TYP   TYPICAL

UON   UNLESS OTHERWISE NOTED

UF   UNFUSED

WP   WEATHERPROOF

Figure 15.1 *(g)* Symbol list.

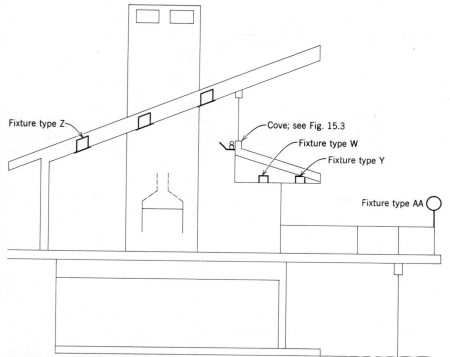

SECTION AT LIVING ROOM LOOKING SOUTH

Figure 15.2 Section through living room of Merker house showing cove lighting element (Figure 15.4), sloping ceiling downlights (fixture type Z, Figure 15.5) and recessed downlights, types W and Y, on both sides of the sliding glass doors. In addition, low voltage outdoor post light, type AA, is shown.

LIGHTED COVES

Coves direct all light to the ceiling. Should be used only with white or near-white ceilings. Cove lighting is soft and uniform but lacks punch or emphasis. Best used to supplement other lighting. Suitable for high-ceilinged rooms and for places where ceiling heights abruptly change. *Illuminating Engineering Society*

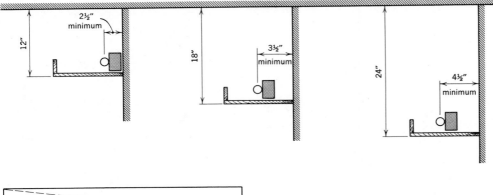

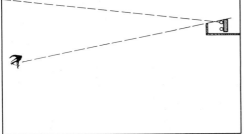

COVE INSTALLATIONS

Proper cove proportions: Height of front lip of cove should shield cove from the eye yet expose entire ceiling to the lamp. Orientation of fluorescent strip as shown is an alternate to upright arrangement. *Westinghouse Lighting Handbook.*

Figure 15.3 Lighted coves.

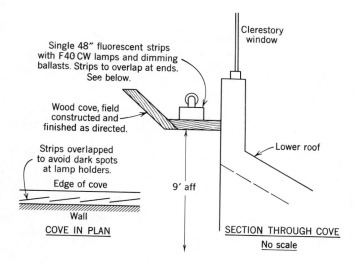

Single 48″ fluorescent strips
with F40 CW lamps and dimming
ballasts. Strips to overlap at ends.
See below.

Wood cove, field
constructed and
finished as directed.

Clerestory
window

Lower roof

Strips overlapped
to avoid dark spots
at lamp holders.

Edge of cove

Wall

COVE IN PLAN

9′ aff

SECTION THROUGH COVE
No scale

Figure 15.4 Cove detail. Refer to Figure 15.1 for living room plan and to Figure 15.2 for section showing cove location.

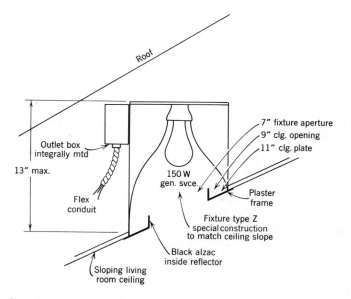

Roof

Outlet box
integrally mtd

13″ max.

Flex
conduit

150 W
gen. svce.

7″ fixture aperture
9″ clg. opening
11″ clg. plate

Plaster
frame

Fixture type Z
special construction
to match ceiling slope

Black alzac
inside reflector

Sloping living
room ceiling

Figure 15.5 Fixture type Z, of type suitable for low sloping ceiling.

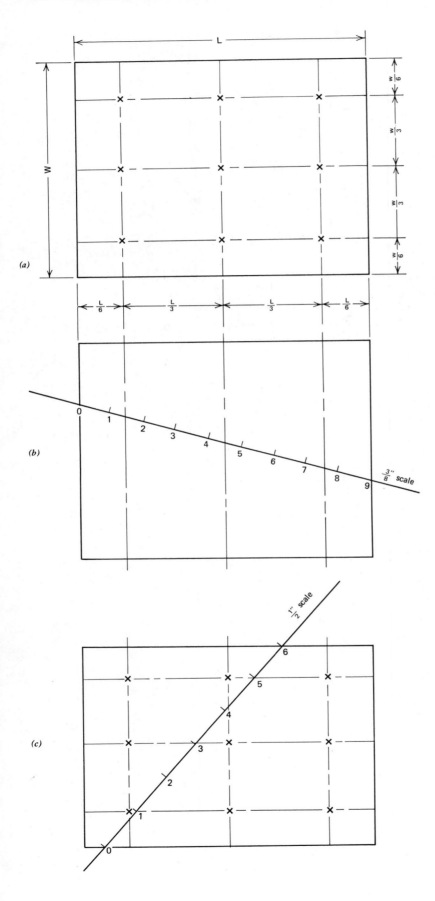

(a)

(b)

(c)

<u>REQUIRED</u>: TO LOCATE 9 LIGHTING FIXTURES IN A SPACE, WITH FIXTURE TO WALL SPACING $\frac{W}{6}$, $\frac{L}{6}$, EQUAL TO HALF THE FIXTURE–TO–FIXTURE SPACING $\frac{W}{3}$ $\frac{L}{3}$.

<u>TECHNIQUE</u> — IN LENGTH; LAY AN ARCH. SCALE OVER THE AREA, SELECTING A SCALE DIVISIBLE BY 3. HERE (b), $^3/_8$" SCALE WORKS WELL. HAVING DRAWN THE SKEW LINE, MARK $1^1/_2$, $4^1/_2$, $7^1/_2$ AS SHOWN. EXTEND LINES VERTICALLY THRU THESE POINTS.

<u>IN WIDTH</u>

REPEAT THE ABOVE PROCEDURE IN THE OTHER DIMENSION. HERE (c) ½ SCALE FITS WELL, FROM 0 TO 6". DRAW THE SKEW LINE AND MARK 1, 3, 5. EXTEND THESE POINTS HORIZONTALLY TO COM—PLETE THE GRID.

THE INTERSECTION OF THE VERTICAL AND HORIZONTAL LINES FORM THE DESIRED CENTERPOINTS OF THE FIXTURES.

Figure 15.6 (Opposite) Drawing technique for locating fixtures.

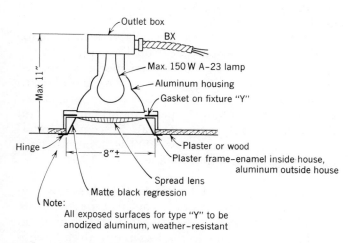

Figure 15.7 Detail of fixtures type W and Y.

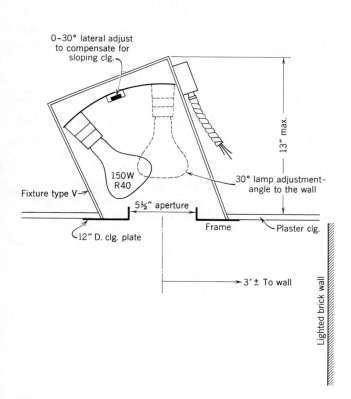

Figure 15.8 Fixture type V. Recessed adjustable wall wash unit, with sloped-ceiling compensation.

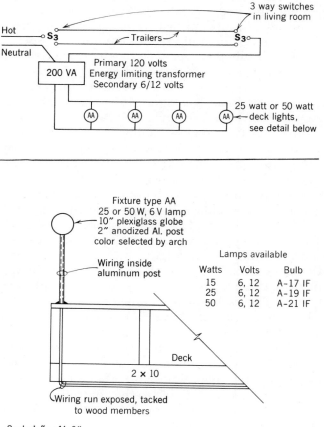

Scale ¼" = 1'-0"

Figure 15.9 Low voltage deck lights.
(a) Wiring diagram for low voltage outdoor lighting.
(b) Fixture detail.

and wall lighting. The unit can be made in the field as in Appendix A.6 or purchased commercially. Because the light is semi-indirect, the quality is soft and restful, even with a fairly high level. A two-level switching arrangement is supplied (see Figure 15.1c) to provide two lighting levels at minimum cost. Full adjustability—which is dimming—requires the use of expensive single lamp dimming ballasts and an expensive controller. A translucent plastic front is recommended for this fixture rather than a solid opaque one. This will keep the brightness ratios (Section 14.5) low for eye comfort. Also important to keep in mind are the reflections of the room lighting in the extensive glass walls. As we mention above, the glass wall acts as a mirror, and all the bright areas in the room will reflect in the glass. Thus the wall bracket will show up, but the effect will be reduced if the brightness is uniform. Therefore, we recommend a lighted front in fixture H.

For the same reason, the ceiling units are chosen with a regressed spread lens and a black, nonreflecting bottom cone. Of course, all the room's lighting could have been supplied by using more of these downlights. This would reduce the mirroring problem. The result, however, would be a ceiling covered with black holes, and no wall lighting. The layout presented, with two level fluorescent brackets is a good compromise. See Figure 15.7. These lights are supplied through a standard incandescent lamp dimmer for full-range dimming. Wall thickness at the point of dimmer installation should be checked against dimmer depth. It is very possible that the depth will have to be increased. For this reason we place the dimmer on a wall backing into a closet. In general, downlights will not provide good general lighting unless a wide spread distribution is used. This can be obtained with lens units. These also have the advantage of low brightness. We are assuming that no curtains are placed on the front wall. We furnish three cylindrical decorative downlights type M. They will act to cut down the inside mirror effect, as explained above. The outside lights are controlled by a local switch. Note that wherever possible the wiring is run along the ceiling members rather than across them, to avoid the labor and expense of drilling. Note, also, that switches are placed as close as possible to the controlled lights. This avoids large banks of switches at the room entrances. Such banks are unsightly, require nameplates to avoid confusion, and are inconvenient to use. Local switch placement is especially important in controlling special purpose or accent lighting that is not constantly in use. It is a nuisance

to have to go to the opposite end of the room when you wish to turn on the wall wash lights at this end of the room. The specification should call for identification on all switch banks of three or more switches. This labeling can vary in quality from DYMO type to etched wall plates, depending on the quality of the installation.

Receptacles have been spotted about the room for convenience. The receptacle on the wall with the hall entrance will be placed on a special appliance outlet circuit, since this is the most logical place to plug in a slide or movie projector. A pull-down screen would probably be permanently mounted on the ceiling near the opposite wall. Since many modern projectors are rated at 1000 w, a circuit with limited outlets is required. See Figure 15.16 for circuiting.

The Basic Plan (see Figure 13.15b) does not contain a family room. In houses of this type, the basement is frequently finished to provide such an area. In Figure 13.15b, since the basement already contains the heating plant and laundry, such a use is doubtful. If done, however, by partitioning off the boiler and laundry, sufficient circuits have been provided to provide lighting and outlets. High-level wall brackets, similar to type H could be used to provide general lighting and wood paneling accent. A few downlights, separately switched, could supply atmosphere and subdued lights for television.

c. Study

In the list of living areas in Section 15.1 we included libraries, studies and television rooms. The Merker residence includes such a room—a study, adjacent to the master bedroom. This room will be used for reading, family bookkeeping, correspondence and general household-related "office" work. The probable location of the family library is on the east wall of this room. The desk will most probably go on the west wall. Here, also, as in the family room, we are assuming no drapes on the outside wall. If drapes were planned, a lighted cornice similar to the one furnished in the master bedroom would be advisable.

Wall lighting and, to an extent, room lighting, is provided by wide angle, wall wash units type N, similar to those shown in Appendix A.7. Here again, this unit will avoid producing annoying reflections. Receptacles should provide amply for desk and floor lamps. In addition, two switched ceiling outlets with blank covers have been provided in the event it is desired to add any sort of ceiling fixture, track, cornice or valance. This type of provision for the future is the cheapest and best way of avoiding the problems, mess and expense that are involved with even a simple alteration. One receptacle is also switched, in the event that a desk lamp is used as the principal lighting source. Note that the wall wash switch is installed separately to avoid confusion. Also, as stated above, it controls specialized lighting that should be controlled locally.

d. Guidelines

Layout for living areas should consider the following.

1. Switched ceiling outlets in large rooms and switched receptacles in small rooms or rooms known to use only lamps.
2. Capability to change lighting levels and quality. This can be done with dimming or switching, and with different types of sources.
3. Built-in accent lighting for walls, drapes and paintings, *locally* switched.
4. Use of architectural elements in lighting such as valances, cornices, and the like.
5. Avoiding downlights for general lighting except in high ceilinged rooms.
6. Avoiding any glare sources, either direct or reflected.
7. Location of sufficient convenience receptacles to meet NEC minimum plus additional ones where useful.
8. Avoiding large banks of switches, unless grouped into a centralized house switch center. See Section 15.4a.
9. Detailing of all special fixtures, architectural elements or unusual construction or wiring.
10. Providing for the future. This provision can take the form of capped outlets, empty switch positions, empty conduits in the ceiling or wall, lightly loaded circuits, and the like.

15.3. Kitchen and Dining Areas

a. Kitchens

Looking at the kitchen of the Basic Plan House first, observe that the lighting level provided is quite high for such a small room. This is because the kitchen is basically a work room, and usually requires the highest lighting level in the house. A kitchen with a single 100 w incandescent or 32 w circline fluorescent is a depressing and badly lighted place. A general illumination level of 50 fc maintained is recommended, plus supplementary lighting on work surfaces in dark corners, under counters or where the occupant shadows the work. This kitchen is not large, and is shaped so that no supplementary lighting is needed. However, a light is placed over the sink to provide a low level of lighting for traffic that does not require the main lights. A secondary purpose for this unit is supplementary light. Switching for the general lighting is at the door. Since we have two doors, we use 3-way switches.

The second major item to attend to is the appliances. In this kitchen the fixed and stationary appliances are the range, dishwasher, and refrigerator. Refer to the NEC for the definition of appliance types.) The designer will generally furnish the technologist with the necessary location and outlet data. When such data are not furnished, the information given in Table 15.1 and 15.2 should be a considerable help. At the layout stage only the receptacle type need be chosen. At the circuiting stage (see Section 15.7), the circuit data in Table 15.2 will be very useful. The portable appliances for a typical kitchen might include a toaster, mixer, grinder, blender, coffeepot, timer and waffle iron, among others. Many of these devices stand exposed on the counter top, and having them constantly plugged in makes for ease of use. Since they are generally not used simultaneously, the load is low but the outlet requirement is large. The ideal solution is a plug-in strip that supplies all the outlets needed. The appliances need then not be moved and cords can be kept very short, for neatness. Thus the total appliance load can be handled with a few special outlets and lengths of plug-in strip under the upper wall cabinets. For kitchens where the number of appliances is large, it should be expected that several will be used at the same time. There, as in the Merker house, two-circuit strips can be used. In the Basic Plan kitchen, single-circuit strips are sufficient.

Referring now to the Merker kitchen, we note that the lighting design is a luminous ceiling. See Figure 15.1d. Here also, 3-way switching is provded—one at the entrance and one at the entrance to the eating area, and not at the other door. Occupants will use the switches when going from kitchen to eating area and back, and not from the dining room to the kitchen. In locating switches always check traffic flow. It is *not* sufficient to say flatly that switches go

Table 15.1 Average Volt-Ampere Ratings for Residential Appliances[a]

Item	Volt-Amperes	Item	Volt-Amperes	Item	Volt-Amperes
Air conditioners (room)		Projector	up to 1000	Sandwich grill	960
½ ton	880	Radio	30	Sewing machine	75
¾ ton	1200	Recorder	95	Serving tray	600
1 ton	1540	Record player	50	Shaver	11
Blanket	175	Dehumidifier	185	Sun lamp	275
Blender	275	Door chime	15	Tea kettle	550
Bottle warmer	440	Egg cooker	440	Toaster	1130
Casserole	510	Fans		Trivet	50
Clock	2	Floor circulator	120	Vacuum cleaners	
Corn popper	440	Attic	345	Bag type	340
Heating equipment		Kitchen exhaust	75	Canister type	725
Warm air furnace fan	320	Portable	50	Tank type	555
Oil burner	230	Floor polisher	475	Hand type	310
Humidifier	185	Food warmer	310	Heat lamp	250
Ice cream freezer	115	Frying pan	1085	Heating pad	60
Knife sharpener	50	Food mixer	130	Heater	up to 1650
Odorizer	11	Hair dryer	415	Vaporizer	385
Power tools	up to 1000	Roaster	1320	Waffle baker	960

[a]All these items utilize a 5–20 R receptacle.

Table 15.2 Load, Circuit and Receptacle Chart for Residential Electrical Equipment

Appliance	NEC Type[a]	Typical Connected Volt-Amperes	Volts	Wires[b]	Circuit Breaker or Fuse	Outlets on Circuit	NEMA Device[c] and Configuration (See Figure 12.40)	Notes
					KITCHEN			
Range	(F)	12000	115/230	3#6	60A	1	14–60R	Use of more than one outlet is not recommended.
Oven (Built in)	(F)	4500	115/230	3#10	30A	1	14–30R	May be direct connected.
Range top	(F)	6000	115/230	3#10	30A	1	14–30R	May be direct connected.
Range top	(F)	3300	115/230	3#12	20A	1	14–30R	May be direct connected.
Dishwasher	(F)	1200	115	2#12	20A	1	5–20R	May be direct connected.
Waste disposer	(F)	300	115	2#12	20A	1	5–20R	May be direct connected.
Broiler	(P)	1500	115	2#12	20A	1 or more	5–20R	Heavy duty appliances regularly used at one location should have a separate circuit. Only one such unit should be attached to a single circuit at the same time.
Fryer	(P)	1300	115	2#12	20A	1 or more	5–20R	Heavy duty appliances regularly used at one location should have a separate circuit. Only one such unit should be attached to a single circuit at the same time.
Coffeemaker	(P)	1000	115	2#12	20A	1 or more	5–20R	Heavy duty appliances regularly used at one location should have a separate circuit. Only one such unit should be attached to a single circuit at the same time.

Table 15.2 Load, Circuit and Receptacle Chart for Residential Electrical Equipment (continued)

Appliance	NEC Type[a]	Typical Connected Volt-Amperes	Volts	Wires[b]	Circuit Breaker or Fuse	Outlets on Circuit	NEMA Device[c] and Configuration (See Figure 12.40)	Notes
Refrigerator	(S)	300	115	2#12	20A	1 or more	5–20R	Separate circuit serving only refrigerator and freezer is recommended.
Freezer	(S)	350	115	2#12	20A	1 or more	5–20R	Separate circuit serving only refrigerator and freezer is recommended.
LAUNDRIES								
Washing machine	(S)	1200	115	2#12	20A	1 or more	5–20R	Grounding of appliance is required.
Dryer	(S)	5000	115/230	3#10	30A	1	14–30R	Appliance may be directly connected—must be grounded.
Hand iron; Ironer	(P)	1650	115	2#12	20A	1 or more	5–20R	
Water heater	(F)	4500	115/230	2#10	30A	1	—	May be direct connected.
LIVING AREAS								
Workshop	(P)	1500	115	2#12	20A	1 or more	5–20R	Separate circuits recommended.
Portable heater	(P)	1300	115	2#12	20A	1	5–20R	Should not be connected to circuit serving heavy duty loads.
Television	(S)	300	115	2#12	20A	1 or more	5–20R	Should not be connected to circuit serving appliances.
Portable lighting	(P)	1200	115	2#12	20A	1 or more	5–20R	
FIXED UTILITIES								
Fixed lighting	(F)	1200	115	2#12	20A	1 or more	—	
Air conditioner ¾ hp	(F)	1200	115	2#12	20A or 30A	1	5–20R	Separate circuit recommended.
Air conditioner 1½ hp	(F)	2400	115/230	3#12	20A or 30A	1	—	Connect through disconnect switch.[d]
Central air conditioner	(F)	5000	115/230	3#10	40A	1	—	Connect through disconnect switch.[d]
Sump pump	(F)	300	115	2#12	20A	1 or more	5–20R	Use of 1 pole thermal disconnect recommended.[d]
Heating plant (fossil fuel)	(F)	600	115	2#12	20A	1	—	Direct connected. Some local codes require separate circuit.[e]
Fixed bathroom heater	(F)	1000 to 1500	115	2#12	20A	1	—	Direct connected.
Attic fan	(F)	300	115	2#12	20A	1 or more	5–20R	Connect via thermal element 1 pole switch.

[a]Appliance types: (F) Fixed; (S) Stationary; (P) Portable.

[b]Number of wires does not include equipment grounding wires. Ground wire is #12 AWG for 20A circuit and #10 AWG for 30A and 50A circuits.

[c]Equipment ground is provided in each receptacle.

[d]For a discussion of disconnect requirements, see NEC Article 422.

[e]Provide shutdown switch at entrance to the space housing the heating plant.

at the strike side of the door. Sometimes that is an inconvenient location. An under-counter light, type CC, is provided in one location to provide supplemental lighting plus low-level midnight-snack lighting. As with the Basic Plan kitchen, special outlets are supplied for fixed and stationary appliances. Two-circuit multi-outlet assemblies are installed to handle the expected multitude of small portable appliances. Also, as in the Basic Plan, a wall switch is provided for an exhaust fan. See Figure 6.17 for duct work for this fan.

b. Eating Areas

The dining areas of a house must be examined individually and also with respect to each other. For instance, in the Basic Plan, the kitchen has very little eating space—enough perhaps for a small round table and two chairs. Therefore, the dining room becomes the three-meal-a-day eating area. In addition it is the formal dining area. Lighting for everyday meals should be fairly high level. On the other hand, formal dining calls for more subdued lighting. In the Basic Plan, in the interest of economy, we use a single dimmable pull-down pendant unit. The light is controlled from two locations but is dimmable only from the formal entrance. Wiring is similar to that of Figure 15.9, substituting the dimmer for the transformer shown. Obviously in this setup we cannot use the common combination dimmer and switch. Instead, we mount a 3-way switch in one gang of a wall box and a dimmer rated at leat 300 w, in the adjacent gang position.

In the Merker residence, everyday meals are taken in the eating area adjoining the kitchen. That leaves the dining room for meals with guests and holiday-style dining. This is also obvious from the double doors that open into the dining room. The dining room is meant to be a show place and is treated as such. A central ceiling outlet with dimming control is provided for an appropriate chandelier. Since the walls will almost surely be used to display art or other objects, ceiling tracks are provided for illuminating both walls. The number and type of fixtures to be mounted on these tracks can be selected to match the decor and function of the room. See Figure 15.10. The dining room tracks are recessed but they are also available for surface mounting. (It should be mentioned that if this dining room were very formal, tracks would probably not be used because of conflict in decor. Instead, recessed wall wash units would be installed. We are using this house as a vehicle for our study. Since we have already used wall wash units in the study and

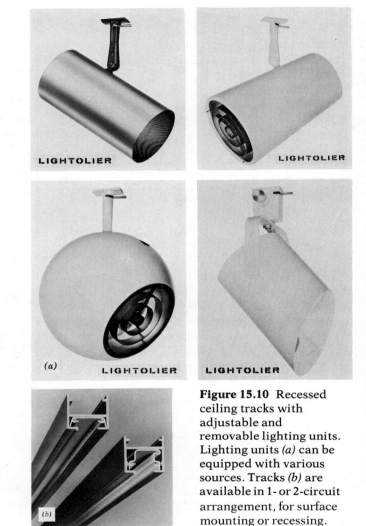

(a)

(b)

Figure 15.10 Recessed ceiling tracks with adjustable and removable lighting units. Lighting units (a) can be equipped with various sources. Tracks (b) are available in 1- or 2-circuit arrangement, for surface mounting or recessing.

(Courtesy of Lightolier)

living room, we are using ceiling tracks in this room as an example of what can be done to provide flexible room and wall lighting.) The track fixtures can also be turned around to provide room lighting, although this can create a direct glare problem. If desired, dimming can be provided for the tracks also. This, however, is not usually done, since track lighting is spot or highlighting, and as such is not generally dimmed.

The adjoining eating area is brightly fluorescent lighted as is desirable for a day-to-day eating place. Warm or cool white lamps are usable. Our recommendation is warm white to match the cooking area and to make the food look more inviting. For nighttime use of this area and when a view outside is desired, the fluorescents can be switched off and the downlight provided for this purpose can be used instead.

Receptacles are spotted around the eating areas in accordance with the NEC six-foot rule, as well as additional outlets that we think are necessary. In circuiting, these outlets will be connected to appliance branch circuits as explained in Section 13.9.

c. Guidelines

To summarize, we have established the following guidelines for kitchen and eating areas.

1. Lighting for kitchen work areas should be at least 50 fc. A combination of general and supplemental lighting is good practice.
2. Lighting for eating areas should be at least 30 fc (preferably more) for daily routine eating. More formal or holiday-style eating requires less light but a more subdued quality. Combinations of fixtures, switching and dimming can be used to achieve these results.
3. Fixed and stationary appliances are supplied by specific-use outlets. Portable appliances are supplied through individual outlets or multi-outlet strips on appliance circuits.
4. Switching should be located in accordance with traffic flow. Most often switches at the door strike side are satisfactory.
5. If, in the dining areas, a counter, sideboard or table is placed against a wall, an outlet should be supplied 4 in. above it, at the center. This applies to eat-in kitchens as well.
6. If plumbing provision is made for a food disposal unit or dishwasher, a wired and capped outlet should also be provided. The food disposal unit's control switch should be placed at such a loca-

tion that it is *impossible* to stand at the sink and turn on the unit.
7. A wall-mounted clock outlet at 90 in. AFF is a good idea in kitchens, although battery units have made this outlet less vital than it was some years ago.

15.4. Sleeping and Related Areas

a. Bedrooms

A wide range of electrical layouts is possible for bedrooms, depending on the size of the room and the uses intended. Some bedrooms are designed to be small, and only for the purpose of dressing and sleeping. Another approach uses this area for resting, reading, and television watching in addition to sleeping. These rooms are larger and the electrical layout must satisfy the requirements. In either case, a furniture layout will tell a great deal. In its absence, one must be assumed. Refer, for instance, to Figure 15.11, which is the lighting and receptacle layout for the two bedrooms of the Basic Plan, and compare it with the layout of Figure 13.15. Notice that in bedroom no. 2 the layout is the same. There, the east wall is the only logical place beds can be located, and it makes no difference whether twin beds, a double bed or a king size is used. With a double, the outlet intended for a lamp between the twin beds is blocked. However, enough outlets remain at good locations.

In bedroom no. 1, which is smaller, the layout of Figure 13.15 assumes twin beds on the east wall. A different arrangement is shown in Figure 15.11. We believe that this second layout is better because it is more flexible. It allows for use of twin or double-sized beds and leaves room for bedroom furniture. The double outlets on the north wall are needed for the twin bed's layout and only one outlet is blocked.

In bedroom no. 1 the overhead light provides room and closet lighting. In bedroom no. 2 the sliding doors and the larger closet call for a separate closet light. Notice that the closet light is wall mounted above the door, inside the closet. In that position, it lights the clothes and the shelves above. A ceiling light is less useful because of the shadow cast by the shelves, unless it is placed just inside the doors. In all cases such lights should have physical protection for the lamp to prevent breakage, in the form of a guard.

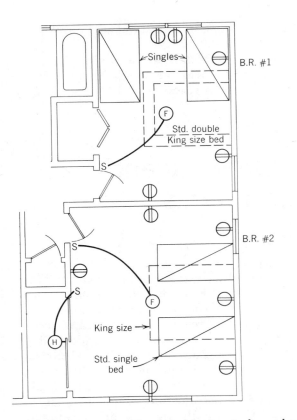

Figure 15.11 Proper layout of devices is dependent on furniture arrangement. When furniture arrangement is unknown, devices should be placed to accommodate the most logical layout(s). Note that the layouts shown will satisfy several furniture arrangements with minimum waste of devices and good availablility.

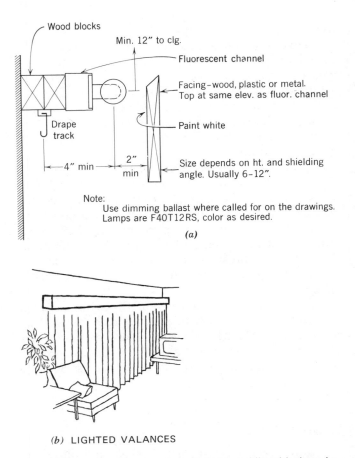

(b) LIGHTED VALANCES

Valances are always used at windows, usually with draperies. They provide up-light which reflects off ceiling for general room lighting and down-light for drapery accent. When closer to ceiling than 10 inches use closed top to eliminate annoying ceiling brightness.

Figure 15.12 Construction detail and application of lighted valances. (Courtesy of Illuminating Engineering Society)

In studying the bedrooms of the Merker residence, we find that the rooms are generally larger than conventional bedrooms. In the two bedrooms on the lower level, the bed(s) will probably be on the side wall opposite the door, and outlets have been arranged that way. Placement under the high windows is also possible, but less likely. These rooms are large and will probably be used as guest rooms. Therefore, in addition to the practical overhead light, a dimmable lighted valance is installed above the window, to provide a pleasant "sitting room" atmosphere. See Figure 15.12. Outlets are placed about the room for convenience. The front window wall has high windows and a piece of furniture or chairs will be placed there. Outlets at both ends of

this wall will handle any lamps. The outlet on the wall adjacent to the door is intended for a television set which can be watched while in bed. For this reason, it is switched from the probable bed wall.

Looking at the master bedroom on the upper level, we see that here also, despite the large size of the room, a likely bed location can be found. It is the far, or west wall. The other walls are ruled out because of doors, windows and traffic flow. This room serves as a master bedroom, sitting room and lighting control center for the house. We consider the lighting aspects first: the bed wall is provided with two pairs of wall-mounted lighting brackets—one for reading, and one for general lighting, relaxing or television watching. Each fixture has a built-in switch. A

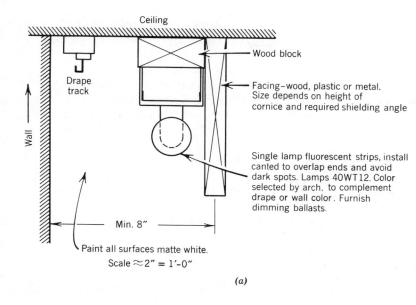

Ceiling

Drape track

Wall

Wood block

Facing–wood, plastic or metal. Size depends on height of cornice and required shielding angle

Single lamp fluorescent strips, install canted to overlap ends and avoid dark spots. Lamps 40WT12. Color selected by arch. to complement drape or wall color. Furnish dimming ballasts.

Min. 8″

Paint all surfaces matte white.
Scale ≈ 2″ = 1′-0″

(a)

LIGHTED CORNICES

Cornices direct all their light downward to give dramatic interest to wall coverings, draperies, murals, etc. May also be used over windows where space above window does not permit valance lighting. Good for low-ceilinged rooms. Courtesy of IES

(b)

Figure 15.13 Construction detail and application of lighted cornices. (Courtesy of Illuminating Engineering Society)

dimmable lighted cornice above the drapes (see Figure 15.13) extends the entire length of the room, giving the room a sitting-room atmosphere. The large wall opposite the beds (west wall) is illuminated by surface-mounted ceiling track lights. See Figure 15.10. The fixtures on this track can be aimed to highlight paintings, books or other objects. The owner has gone to the expense of providing a skylight which will give daylight when the blinds or drapes are closed. To take advantage of this desire for daylight, the lighting design has provided an artificial daylight source in this space, that is locally switched. Finally, two capped ceiling outlets are furnished with wiring for switch control. A single switch is provided, but space is left for another

switch in case separate control is desired in the future.

The deck outside the master bedroom and study is roofed, unlike the deck outside the living room. We have provided overhead lighting for this area. Notice on the front wall an outside lighting control panel. This panel controls all the outside lighting, including the deck lighting off the living room. Local control of these lights is also provided. The reasons for centralizing the outside lighting control in the master bedroom are convenience and security. Convenience, so that the owners can shut all the outside lights as they retire, without running around this large house. Security, so that outside lighting, which acts as security lighting, is in the hands of the owner at his nighttime location. For these same reasons—convenience and security—owners of large houses such as this frequently desire centralized control of *all* the lighting in the house. To do this with normal, full voltage switching is extremely expensive and clumsy because of the heavy wiring, full-size switches and pilot lights required. For this reason, a system of switch control that uses low voltage relays was developed. This allows the use of very small low voltage (24 v) control wiring and makes centralized control a relatively simple matter. Of course, local control also remains. A typical wiring diagram and some equipment photos are shown in Figure 15.14. In Figure 15.1e, note 8, the contractor is requested to furnish a proposal to perform all the switching in this manner. In a house of this size the price of low voltage would be competitive with conventional full voltage switching. If central control of *all* lighting is desired, it can only reasonably be accomplished with low voltage switching. If, on the other hand, this arrangement were desired in the Basic Plan House, it could reasonably be done both ways. Receptacles have been spotted around the master bedrom for convenience and in accordance with the NEC, including two GFI types on the outside deck.

b. Dressing Rooms and Closets

Lighting of reach-in closets is discussed above. A guarded, wall-mounted, switch-controlled light is adequate. Walk-in closets and dressing rooms must be treated as small rooms. Switches can be inside or outside, depending on shelf and pole locations. If the room is to be used for dressing and makeup, appropriate mirror lighting and outlets must be provided. The two walk-in closets in the Merker master bedroom do not function as dressing rooms and thus only a recessed ceiling light is provided.

c. Guidelines

Reviewing the guideline for sleeping areas, we have learned that we should:

1. Furnish overhead lighting and/or architectural lighting elements such as coves, valances and cornices. Lighting is important near and in closets. Mirrors will probably be placed there.
2. Provide over-bed lights with control for reading and relaxing, if desired.
3. Provide master control of all or parts of the house lighting in this room, if desired.
4. Provide interest lighting if the room is to function as a sitting room.

5. Establish a furniture layout and provide two or more duplex or triplex outlets on the bed wall.
6. Provide a strategically placed outlet for a television set (see Figure 15.32).
7. In a house without central air conditioning, provide an appliance-type outlet near a selected window, for a window A/C unit (see Basic Plan, Figure 13.15).

15.5. Circulation, Storage, Utility and Washing Areas

a. Circulation Halls, Foyers, Corridors and Stairs

Entrance halls and foyers are generally lighted with decorative fixtures. Inside corridors and stairs use simpler types of fixtures, surface mounted or recessed. Recessed fixtures should have spread distribution—lens or dropped dish-type diffuser—to give dispersed lighting. Concentrating type downlights are not desirable. Note type D fixture in Basic Plan and type F in Merker as compared with types E and W in Merker. Stair lighting must be so placed that the front edge of each step is lighted. Depending on how the stair is constructed, this can be done with a light over the stairs (see Basic Plan) or lights at the foot and at a center landing (see Merker). Lighting at the top is desirable when there is no center landing, or when the landing lighting is blocked.

Switching of circulation lighting is critically important. The technologist must take a mental trip through the building, turning lights on and off as required. Obviously a 3-way switch is required at the top and bottom of each stair. Also plain to see is the need for 3-way switches at the two ends of any corridor more than 6 ft long. See stair lighting in both plans and corridor lighting in the Basic Plan. Much less obvious is the switching of lights in the two T-shaped halls in Merker. There, because of the choices involved, 4-way switching has been provided. See Figure 15.15 for wiring of 4-way switching. Receptacles should be supplied every 15 ft of wall space, for use with vacuum cleaners, floor polishers, electric brooms and the like. If a central vacuum system is provided, this requirement can be reduced somewhat. All receptacles should be standard 20-amp, 120 v; 220 v outlets are not required.

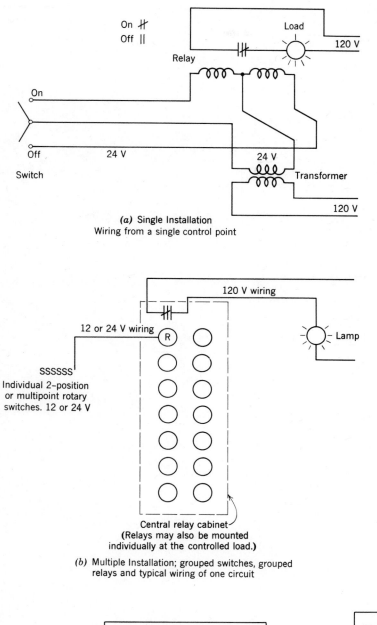

(a) Single Installation
Wiring from a single control point

(b) Multiple Installation; grouped switches, grouped
relays and typical wiring of one circuit

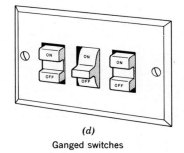

(c)
Basic rocker switch

(d)
Ganged switches

(e)
Single switch and rotary selector

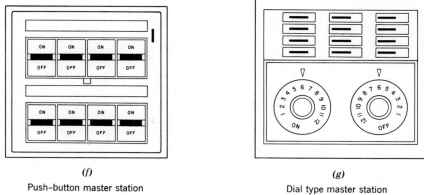

(f)
Push-button master station

(g)
Dial type master station

Figure 15.14 Low voltage switching; equipment and circuitry. (Courtesy of General
Electric)

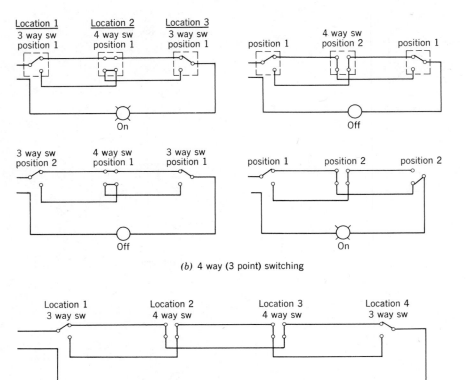

(a) 3 way (2 point) switching

(b) 4 way (3 point) switching

(c) Switching from any number of points

Figure 15.15 Multiple point switching. *(a)* Shows control from two points, using 3-way switches. *(b)* Switching from three locations, using a 4-way switch in addition to the two 3-way switches of *(a)* above. Note that complete control is accomplished from each location. *(c)* Switching from any number locations can be done by adding 4-way switches at each new location. Illustrated is switching from 4 locations.

b. Storage and Utility Areas

1. Shallow reach-in closets may be adequately lighted from the adjoining space. If the closet is deep, has recesses on the sides or top or is walk-in, it should have its own illumination. A switched porcelain lampholder is normally sufficient, if it is provided with a metal guard to protect the lamp from breakage. Convenience outlets are generally not required in closets and storage rooms. Dead storage areas such as attics and basement crawl spaces take the same type of basic lighting, switched at the entrance. In these areas a single convenience outlet is a good idea; receptacles in the lampholders are generally not advisable.

2. *Basements* serve as storage, utility, recreation and work areas. They start as empty spaces containing only the heating plant and possibly the laundry, and expand in usage later on. For this reason basement areas should start with basic utility lighting and a few outlets. However, provision must be made for easy expansion of lighting and easy addition of receptacles—both 120 and 240 v. Minimum *initial* requirements are one ceiling outlet for every 200 sq ft of area, receptacles for a spare refrigerator and a freezer, plus one wall outlet for every 20 ft of wall space. To allow for future expansion, either spare cables or empty conduits should be run to the basement from the panel. Normally, an emergency cutoff switch must be placed at the entrance to the space containing the heating plant. This switch must completely shut down the heating system. If the heating system is all electric, this requirement does not exist. However, a main disconnect for the heating system panel is a good idea. See riser diagram (Figure 15.19) and panel schedule (Figure 15.17c).

3. *Laundry areas* can be situated in the basement (Basic Plan), can be a separate enclosed space (Merker), or a porch, kitchen or balcony. In any case, the electrical requirements are appropriate outlets (see Tables 15.1 and 15.2) and adequate lighting. Laundry areas must have adequate ventilation, either natural or forced. If space permits, an outlet for a hand iron should be provided.

4. *Enclosed garages* serve many functions in addition to car storage and basic auto maintenance. In houses with no basement or utility room, the garage is also the work and repair area and a storage area for bicycles and garden equipment. Lighting outlets should be placed to illuminate the auto engine compartment with the hood open. Lights over the center of the car are useless. Notice in the Merker garage the wall brackets for general illumination and the reel-light (fixture type C), which greatly assists in repair work. The entire back of the garage is treated as work and utility space with industrial fluorescent lighting fixtures. These heavy duty steel louver fixtures fit nicely between the 12-in. on-center ceiling beams. Note that recessed fixtures are better than surface units, since the fixtures are subject to physical damage. For this reason, also, a steel louver unit has been used. The garage should be amply supplied with receptacles mounted high (48 in.) with a minimum of one on each wall. Many modern houses use an electronically operated garage door, and interior and exterior switches for control are required (see Figure 15.16). Exterior switches are key operated.

c. Bathrooms and Lavatories

1. Lighting should be over the mirror to illuminate the face head-on. If lights are placed at the sides of the mirror, they should be at both sides so that both sides of the face are illuminated. Overhead lighting fixtures in large bathrooms are acceptable for general lighting, provided that over-the-mirror, full-face lighting is furnished also. Over-the-mirror lighting alone is sufficient in a small room. See bath in Basic Plan, and Merker bath no. 2 and guest bath. In the Merker master bath the overhead fixture provides light for bathing and showering while the two mirror lights provide the facial illumination required. Notice that a separate fixture is provided over each basin. Generally, face lighting should be furnished at each mirror location. Normally, although not always, this corresponds to the basin location. Although not absolutely necessary, gasketed fixtures in rooms with baths and showers are advisable. All light should be controlled by wall switches; *never* local switches or pull chains. Switches should be located conveniently to operate the light. The switch at the door is intended to turn on the general illumination; supplemental lighting is separately switched.

A built-in night-light and switch is a convenience in houses with small children. Devices are available with this combination that fit into a standard outlet box. See Basic Plan (Figure 13.15). A light *inside* a stall shower is not a particularly good idea, even though the unit is vapor-proof. Enough light is obtained through

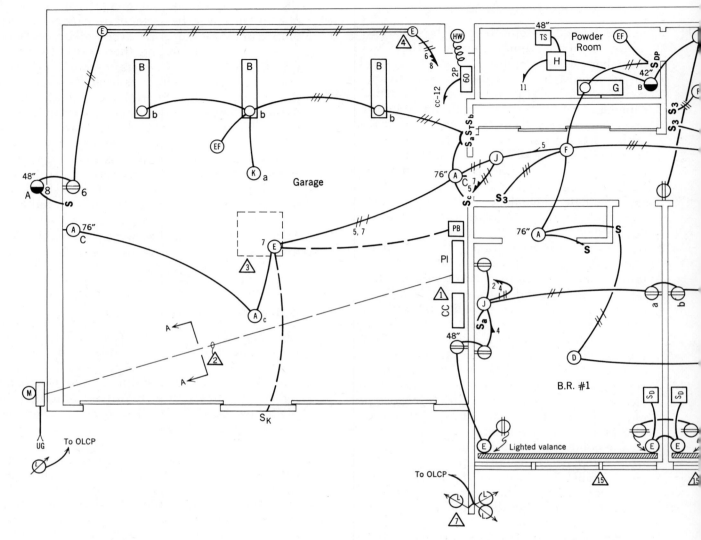

Figure 15.16 (a) Lower level electrical plan circuitry.

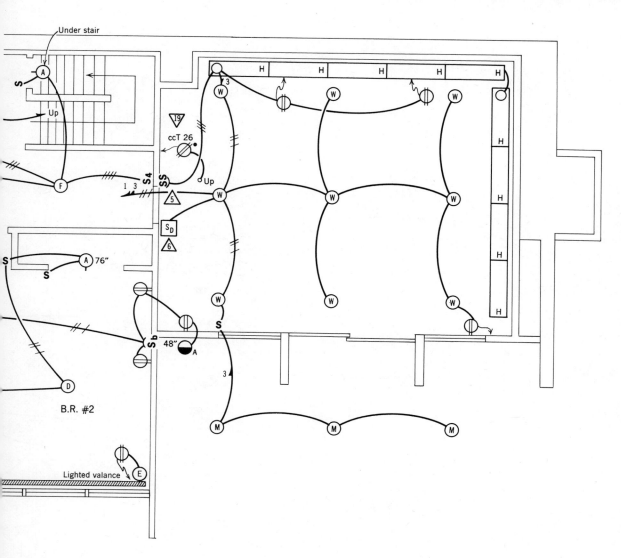

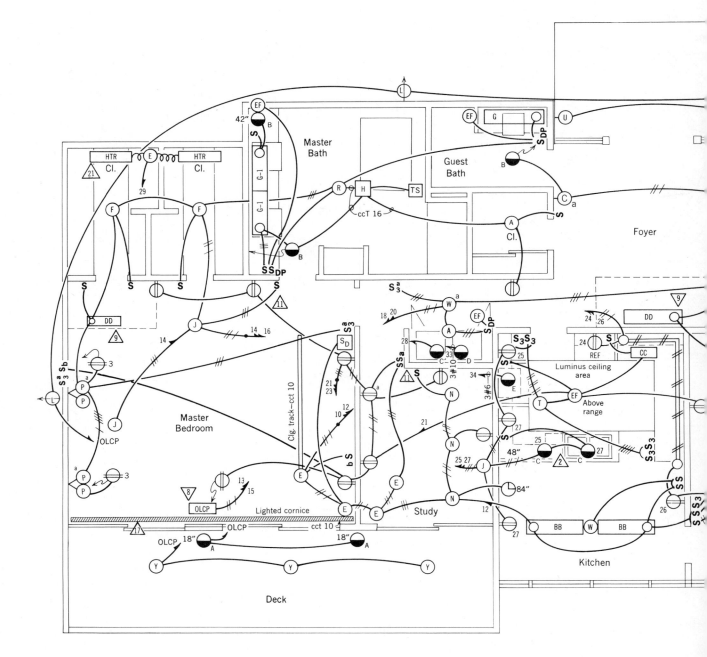

Figure 15.16 *(b)* Upper level electrical plan circuitry.

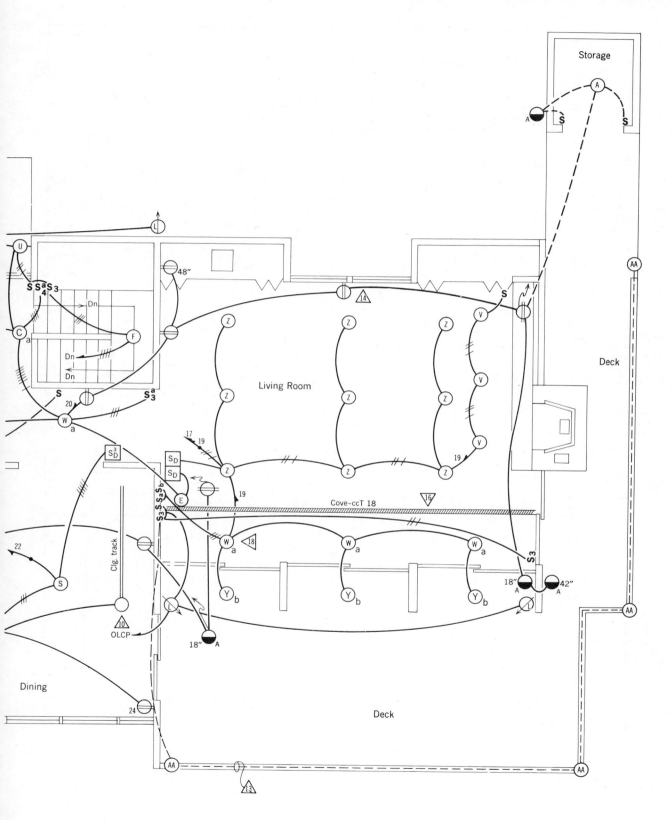

PANEL P

CIRC. NO.	DESCRIPTION	VA	CIRCUIT BREAKERS	DESCRIPTION	VA	CIRC. NO.
1	LTG. — FAM. RM.	1350 + 1R	20 · 1│2 · 20	7 REC.	7R	2
3	LTG. — FAM. RM.	1350 + 2R	3│4	BR. LTG: 2 VALANCE + Φ	360 + 6R	4
5	LOWER LEVEL LTG. + 2 E.F.[1]	1294	5│6	GARAGE OUTLETS	6R	6
7	GAR. DOOR MOTORS 2 @ ¼ HP[2]	1400	7│8	GARAGE OUTLETS	6R	8
9	SPARE		9│10	LTG. UPPER LEVEL [3][5]	1550	10
11	BATH HTR + 3 REC	1000 + 3R	11│12	LTG. UPPER LEVEL [5]	1450	12
13	OLCP PANEL	1000	13│14	LTG. + REC. UPPER LEVEL [1][3]	1480 + 3R	14
15	OLCP PANEL	1000	15│16	HTR. [4] + LTG. + 2 REC.	1100 + 2R	16
17	LIV. RM. LTG.	1350	17│18	LTG. + COVE + E.F.	1312	18
19	LIV. RM. LTG.	1350	19│20	LIV. RM. REC. + LTG.	100 + 8R	20
21	KITCHEN LTG., E.F.[1], + REC.	732 + 4R	21│22	DIN. RM. CLG. FIXTURE	1200	22
23	RECEPTACLES	8R	23│24	APPLIANCE OUTLET CCT.	1500*	24
25	APPLIANCE OUTLET CCT	1500*	25│26	APPLIANCE OUTLET CCT	1500*	26
27	APPLIANCE OUTLET CCT.	1500*	27│28	LAUNDRY WASHER	1500*	28
29	CLOSET HEAT	1500	29│30	SPARE		30
31	SPARE		31│32	SPARE		32
33	} LAUNDRY DRYER	5000	30 33│34 60	} RANGE	8KW*	34
			2P 35│36 2P			
35	SPARE		20 37│38 30	} SPARE		36
			2P 39│40 2P			
	SPACE		41│42	SPACE		

PANEL DATA
MAINS AND GND. BUS.
MAIN C/B OR SW/FUSE
BRANCH C/B INT. CAP.
SURF
REMARKS

*SEE TEXT

VOLTS

(1) Exhaust fans are rated 0.85 amp. or 102 v–a at 120 volts.
(2) Garage door motors are arranged to prevent both being started simultaneously.
(3) Assume load of each capped ceiling outlet to be 150 v–a
(4) Bath heater is 1000 watts.
(5) For all ceiling tracks a load of 50 w/ft is assumed.
    See NEC. '75, ART. 364–13

Figure 15.16 (c) Panel P, schedule.

the shower door to make this an unnecessary expense and a potential safety hazard.

2. A duplex receptacle should be installed adjacent to each mirror or lavatory basin. By NEC requirement, these must be GFCI types. Receptacles in lighting fixtures are not to be used. Because of the high cost of these GFCI receptacles, the number should be limited. A time switch controlled heater is a welcome addition to any bathroom, to prevent chills when stepping from a bath or shower. The timer switch is usually spring driven, 0-10 min. The outlet may be a receptacle (GFCI type), an outlet box intended for a fixed heater (see Basic Plan) or a fixed heater (Merker).

An exhaust fan must be provided for all interior bathrooms. Such fans should be controlled by the room light switch. This switch is double pole; one pole for the fan and the second for the light (see Merker guest bath).

15.6. Exterior Electrical Work

This section covers all exterior electrical work *not* connected with the electric service equipment. That topic will be discussed in Section 15.10. All exterior lighting fixtures must be weatherproof. This also applies to units installed in exterior soffits. Exterior lighting must be switch controlled from a nearby location inside the house. Adding master control is good if it is not over the budget. When lights are on an automatic time switch, that switch must have an override feature, preferably one that allows override from a remote location. Porches, breezeways, exterior walks, decks, patios and similar areas should be well lighted, for use *and* for security. As in the Merker residence, use of low voltage fixtures simplifies wiring and minimizes faults.

At least one weatherproof duplex receptacle should be installed on an outside wall of the house. This receptacle is a GFCI type by NEC requirement, and is therefore fairly expensive. It should be located for convenience for use with mowers, grills, and tools. Switch control from inside will prevent vandalism and will improve security. A covered carport adjacent to the house should have a duplex receptacle within easy reach.

15.7. Circuitry

The next stage in the work is the circuiting of the plans of Figure 15.1a and b. The ground rules for circuiting have already been discussed in detail in Section 13.9. The technique of circuiting has been discussed in Section 13.10. We have followed these guidelines and procedures in the circuitry shown in Figure 15.16a and b, with certain changes. It is very important that you place Figure 15.16 before you and trace out all the circuitry. Particular attention should be given to the wiring of switches.

(a) We recommend in Section 13.9 that lighting and receptacles be separately circuited wherever possible. This is good practice but is less important in residences than in other types of buildings. In residences, receptacles count as part of the lighting load. Refer to Section 13.9c (8) above and the NEC Table 220-2b for this rule. Therefore, ceiling lighting outlets and convenience outlets can be combined on a circuit if necessary. In a room with no ceiling outlet the switched wall receptacle is the lighting outlet and belongs on a lighting circuit. Note locations where we have combined lighting and convenience outlets. Economy in wiring is the main reason.

(b) Note also that in the interest of economy, we did wire through some switch outlets. Care was taken, however, to avoid carrying a second circuit through a switch outlet. Also, in the circuited drawings (Figure 15.16) we have combined switch leg runs that were shown separated in the layout drawing of Figure 15.1. A good example of this appears in the garage where the switch legs for the exhaust fan and lighting outlets B and K have been combined. This is proper, since they are all on the same circuit. Also notice that we have been careful to provide part of at least two circuits in each space, including bathrooms and halls. This is a rule that should be strictly followed, to avoid blackouts.

(c) Circuiting has taken account of the construction members, and wiring has been run along them whenever possible. Across-the-members wiring has been limited, to avoid the expense of drilling wood members. On the upper level, much of the cross wiring can be done in the attic space, thus avoiding drilling. On the lower level, wiring can be placed in the concrete slab by using conduit. The engineer selects the wiring method required. Home-runs have generally been taken from the outlet closest to the panel. Home-runs from ceiling outlets are preferable to those from wall outlets. This is so that where a deep box is required, it can be easily accommodated. The limited depth of walls does not permit this.

(d) In a few places, to avoid confusion, circuit numbers have been placed next to outlets. This is not generally done, since the numbered arrowheads and wiring hatch marks are normally enough to indicate circuit routing. The Merker residence, however, is so heavily wired that this type of identification is occasionally desirable. We urge you to follow out each circuit, using the panel schedule, Figure 15.16c, along with the circuited floor plans of Figure 15.16a and b. While doing so, you should test alternate circuiting routes mentally or by sketching them. By doing this, you will see that there are any number of ways of circuiting, each of which has its good and bad points. Also we emphasize the importance of understanding how the number of wires shown was arrived at. This can best be done with the aid of little sketches of the type shown in Figure 13.22, until you have enough experience to do the counting mentally.

15.8. Load Calculation

a. General Lighting Load

On the basis of square footage (see Section 13.9a) the house should have a *minimum* of 10 circuits for general lighting load.

$$\text{minimum number of 20A circuits} =$$
$$\frac{\text{area}}{520 \text{ sq ft per circuit}} \approx \frac{5000}{520} \approx 10$$

Actually (see panel schedule P, Figure 15.16c) it has 19, almost double the minimum. This excludes the appliance, laundry and other special circuits not supplying "general lighting load" as defined by the NEC. These 19 circuits also feed exhaust fans (cct nos. 5, 14, 18 and 21) and bathroom heaters (ccts nos. 11 and 16). If these were removed from the total and circuited separately, the circuit count for general lighting would drop to 17. This total of 7 circuits above the minimum is to be expected in a house of this size, complexity and cost.

b. Convenience Outlets

The ordinary duplex convenience outlets in a residence are counted as part of the general lighting load *when figuring the total building load.* [See Section 13.9c(8) and the NEC Table 220-2b with footnote.] This creates a situation that requires keeping a double set of loads—one for the circuit and one for the building. For the building load the receptacles count as zero load. However, with respect to an individual circuit, a normal duplex convenience receptacle is counted as a load of 180 volt-amperes (1.5 amp at 120 v). Look, for instance, at circuit no. 1 of panel schedule P. This circuit feeds 1350 volt-amperes of family room lighting plus one receptacle. This load is shown on the schedule as 1350 + 1R. This means 1350 volt-amperes of lighting plus one receptacle. It is not shown as 1530 volt-amperes. This is because with respect to the total building the circuit load is 1350 volt-amperes (zero for the receptacle) but the individual circuit load is 1530 volt-amperes (1350 + 180). The circuit load (1530) determines the size of the circuit's protection and wiring. The building load (1350 volt-amp) determines the size of the panel mains, and influences the service size. These load calculations, that is, the building load, feeder size, panel mains and main protection are done by the job engineer. We describe them in our discussion of service equipment below. Here, it is important to remember that in residential work the circuit loads should be shown split, as explained above and as they appear on panel schedule P. When a circuit consists of all receptacles, such as circuit no. 2, the load is shown as 7R. This indicates zero for the building load calculation and $7 \times 180 = 1260$ volt-amperes for the circuit load.

c. Volt-Amperes and Watts

Note particularly that all loads are figured in volt-amperes and not in watts. The difference was explained in Section 11.17. For purely resistive loads, such as most lighting and electric heating, the volt-amperes and watts are almost identical. However, small motors have a low power factor and there is a large difference between volt-amperes and watts. An example will help clarify this. Circuit no. 7 feeds 2–¼ hp garage door motors. Referring to the NEC table 430–148, we observe that full load current for a 115 v, ¼-hp motor is 5.8 amperes. Therefore

$$120 \times 5.8 = 696 \text{ volt-amperes}$$

whereas the *wattage* is approximately 250. The load shown in the panel is, therefore, 1400 volt-amperes, not *500* w. This very large difference is due to the extra-safe figures in the NEC. Most ¼-hp motors draw much less than 5.8 amperes. The NEC, however, takes the worst case to make certain that the electrical circuitry is adequate.

d. Appliance Circuits

Circuits nos. 24 to 27 are appliance circuits feeding the required appliance outlets in the kitchen, dining room and family room. The minimum number of such appliance circuits is two. Here, though, four are used to supply the large number of outlets. Each appliance circuit is figured at 1500 w. The laundry circuit, circuit no. 28, is also figured at 1500 w. These load figures are in accordance with NEC requirements, and are shown on the panel schedule.

e. Large Appliances

The electric clothes dryer, circuit no. 33 is shown as 5000 w. This is the minimum load permitted by the NEC for such an appliance. See Table 15.2 and the NEC, 1975 edition, paragraph 220-18. The electric range load is shown as 8 kw even though the unit itself is rated 12 kw. This reduction, which corresponds to a demand factor of 67%, is permitted by the NEC.

f. Panel Schedule

The panel data at the bottom of the panel schedule (Figure 15.16c) has deliberately been left blank. These data are supplied by the engineer to the technologist, and are therefore omitted here. The remaining panel information, however, is the work of the draftsman/technologist, and is shown filled in.

15.9. Climate Control System

The heating/cooling system for the Merker residence is described in Chapter 6. The corresponding electrical work is shown in Figure 15.17a and b. The panel schedule for the climate control panel CC is given in Figure 15.17c. The system utilizes electric heating on the lower level (Figure 15.17a) and heat pump heating/cooling for the upper level (Figure 15.17b). The NEC specifies certain derating factors which are applied to circuits that feed electric heating equipment. These factors and the resulting circuit capacities are given in Table 15.3. Look at Figure 15.17a and c. Note that all the electric heating (except that in the ducts) is handled by two 20-amp, 240-v, 2-pole circuits, and one 30-amp, 240-v, 2-pole circuit. Refer to Tables 15.3 and 15.4 and check the wire and circuit capacities. Circuit no. 7, which carries 4000 w, requires No. 10 wires—capacity 4720 w, and a 30-amp breaker—capacity 5760 w. These 2-pole circuits are placed after the single-pole circuits on the panel. Panels are normally made that way. The thermostats are line voltage, full-capacity type and serve to control and disconnect the heaters. Basic branch circuit data are provided in Table 15.5. The upper level heating/cooling system is more complicated. Refer to Figure 15.7b. This diagram corresponds to Figure 6.9 but in electrical terms. The NEC gives special rules for sizing circuits that feed compressors such as those in the heat pumps. They are shown as 3 hp. The electrical data of Figure 6.5c give this horsepower rating using the rules in

Table 15.3 Wire Capacity—Electric Heat Circuits

| Wire Size | 66° Wire—TW, UF | | | | 75° Wire—RHW, THW, XHHW, USE | | | |
| | Ampacity | | Power Capacity | | Ampacity | | Power Capacity | |
	Normal[a]	Derated[b]	120 v	240 v	Normal	Derated[c]	120 v	240 v
12	20	13	1575 w	3.15 kw	20	14	1.69 kw	3.38 kw
10	30	20	2360	4.72	30	21	2.53	5.07
8	40	26	3150	6.30	45	32	3.8	7.6
6	55	36	—	8.6	65	46	—	11
4	70	46	—	11.0	85	60	—	14.4
2	95	62	—	15.0	115	81	—	19.4

[a]Ampacity of wire in conduit or direct burial.

[b]Ampacity of wire used for fixed electric heating and run in wall cavities. See 1975 NEC, Art. 424.3b and 424.37b.

[c]Same as b above. Derating factor, however, changes with change in wire type.

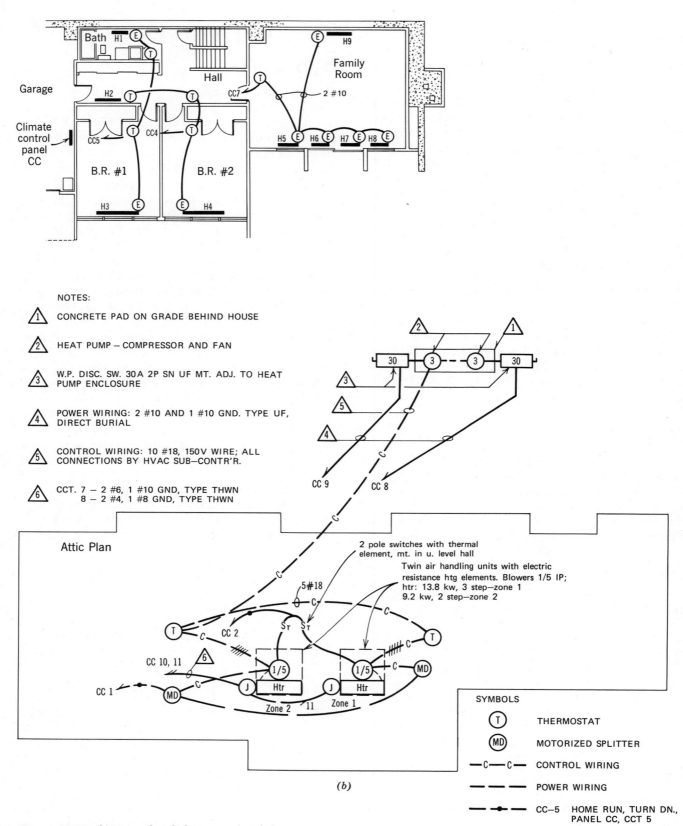

NOTES:

△1 CONCRETE PAD ON GRADE BEHIND HOUSE

△2 HEAT PUMP — COMPRESSOR AND FAN

△3 W.P. DISC. SW. 30A 2P SN UF MT. ADJ. TO HEAT PUMP ENCLOSURE

△4 POWER WIRING: 2 #10 AND 1 #10 GND. TYPE UF, DIRECT BURIAL

△5 CONTROL WIRING: 10 #18, 150V WIRE; ALL CONNECTIONS BY HVAC SUB—CONTR'R.

△6 CCT. 7 — 2 #6, 1 #10 GND, TYPE THWN
8 — 2 #4, 1 #8 GND, TYPE THWN

2 pole switches with thermal element, mt. in u. level hall

Twin air handling units with electric resistance htg elements. Blowers 1/5 IP; htr: 13.8 kw, 3 step—zone 1
9.2 kw, 2 step—zone 2

SYMBOLS

Ⓣ THERMOSTAT

ⓂⒹ MOTORIZED SPLITTER

—C—C— CONTROL WIRING

— — — POWER WIRING

— •— CC—5 HOME RUN, TURN DN., PANEL CC, CCT 5

(b)

Figure 15.17 *(b)* Upper level climate control. (see Figure 6.9).

CIRC. NO.	DESCRIPTION	VA	CIRCUIT BREAKERS		DESCRIPTION	VA	CIRC. NO.
			PANEL CC				
1	MOTORIZED DAMPERS	NEGLIGIBLE	15 ₁│₂	15	2 – ¹/₅ HP BLOWERS	1200	2
3	SPARE		20	₃│₄ 20	SPARE		
5	HTRS, LOWER LEVEL BATH, BR #1	2000 {	20 ₅│₆ 2P	20 ₇│₈ 2P	HTRS, LOWER LEVEL, HALL, BR #2	2750 {	4
7	HTRS, FAMILY ROOM	4000 {	30 ₉│₁₀ 2P	30 ₁₁│₁₂ 2P	SPARE	{	6
9	HEAT PUMP – 3HP	{	40 ₁₃│₁₄ 2P	40 ₁₅│₁₆ 2P	HEAT PUMP – 3 HP	{	8
11	13.8 KW HEATER ZONE 1	{	60 ₁₇│₁₈ 2P	40 ₁₉│₂₀ 2P	9.2 KW HEATER ZONE 2	{	10
13	SPARE	{	20 ₂₁│₂₂ 2P	40 ₂₃│₂₄ 2P	9 KW ELEC HOT WATER HEATER	{	12

PANEL DATA

MAINS AND GND. BUS.	150 A, /SN; 100A GND. BUS		
MAIN C/B ~~OR SW/FUSE~~	150A 2P, 100 AT	VOLTS	120/240
BRANCH C/B INT. CAP.	7500		
SURF /~~RECESSED~~			
REMARKS			

Figure 15.17 (c) Panel schedule for climate control Panel CC (see Figure 15.17 a and b).

Table 15.4 Branch Circuit Capacity—Electric Heat

Circuit Breaker Size	Maximum Watts[a]	
	120 v	240 v
15A	1440	2880
20A	1920	3840
30A	2880	5760

[a]Based on 80% derating of circuit breaker ie maximum allowable current in 20A circuit breaker is 16 amperes.

Table 15.5 Branch Circuit Requirements

	Branch Circuit Size				
	15 amperes	20 amperes	30 amperes	40 amperes	50 amperes
Minimum size conductors[h]	No. 14	12	10	8	6
Minimum size taps[h]	No. 14	14	14	12	12
Overcurrent device rating	15 amperes	20	30	40	50
Lampholders permitted	Any type	Any type	Heavy duty	Heavy duty	Heavy duty
Receptacle rating permitted (note g)	15 amperes	15 or 20	30	40 or 50	50 amperes
Maximum load (see note f)	15	20	30	40	50

[a]Wiring shall be types RHW, T, THW, TW, THWN, THHN, XHHW in raceway or cable.

[b]On 15-amp circuit maximum single appliance shall draw 12 amperes. On 20-amp circuit maximum single appliance shall draw 16 amperes. If combined with lighting or portable appliances, any fixed appliance shall not draw more than 7.5 amperes on a 15-amp circuit, and 10 amperes on a 20-amp circuit.

[c]On a 30-amp circuit maximum single appliance draw shall be 24 amperes.

[d]Heavy duty lamp holders are units rated not less than 750 w.

[e]30, 40, and 50-amp circuits shall not be used for fixed lighting in residences.

[f]When loads are connected for long periods, actual load shall not exceed 80% of the branch circuit rating. Conversely, continuous type loads shall be figures at 125% of actual load in all load calculations.

[g]A single receptacle on an individual branch circuit shall have a rating not less than the circuit, for example, 15 amperes on a 15-amp circuit, etc. Receptacles feeding portable and/or stationary appliances shall be limited to a load of 80% of their rating; that is, 15 amp receptacle—12 amps, 20 amp receptacle—16 amps, 30 amp receptacle—24 amps.

[h]All wire sizes in AWG. Sizes are applicable when derating is not required.

Extracted from NEC Article 210 and Table 210-24.

Section 440 of the NEC. The blowers are fed from a 2 pole, 240 v circuit through 2-pole thermal switches mounted in the upper level foyer. The electric heating elements in these air handling units are separately fed from 2-pole power circuits in panel CC. Notice that we specify the use of type THWN conductors. Referring to Table 15.3, we see that with THWN wire (75 deg rise) we can use one size smaller wire than the TW (60 deg rise) that is assumed in the remainder of the job. The smaller size wire more than pays for the more expensive insulation.

Every motor in a project must be provided with motor overload protection and with a means of disconnection. The rules covering these items are many and varied. Refer to the NEC for complete coverage of this complex subject. In this project an additional factor must be considered. The outside heat pumps are remote from the power source. In the interest of safety, the NEC requires that motors of this kind be provided with a local means of disconnection. We have furnished a 2-pole, 30-amp weatherproof switch adjacent to each heat pump. This switch meets the need for a safety disconnect.

Panel CC (climate control) supplies all the power for the heating and cooling equipment. Most of the panel data have been left blank, since this is not part of the technologist's layout work. This information is furnished to the technologist at a late stage of the work.

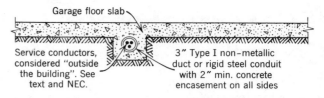

Figure 15.18 Service conductors and conduit encased in 2-in. concrete envelope below garage floor.

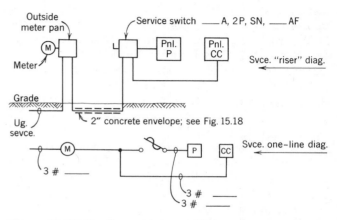

Figure 15.19 Service equipment diagrams.

15.10. Electric Service Equipment

The first item to be located on a plan is the electric panelboard. We are considering it last because it is part of the electric service equipment. When we look at Figure 15.16a, we see that the panel is located on the inside wall of the garage. It is placed there for these reasons:

(a) *Convenience to approach.* It is located near the door leading inside the house.
(b) *Central location.* This shortens all home-runs. If the outside wall were used, some of the branch circuit runs would be very long, more than 100 ft, giving excessive voltage drop.
(c) *Located near the load center.* The heaviest load is the kitchen and the laundry. Both are just above the panel.

Location on the inside wall requires that the service cable be run *under the slab, encased in concrete*, from the meter location to the service switch at the panel. See Figure 15.18. Such a run is inexpensive when poured together with the garage concrete floor.

The climate control panel is placed alongside the house panel. In an all-electric house, it is a convenience to separate the electric heating and feed it from its own panel. Figure 15.19 is the one-line diagram for the house service equipment. This diagram, and all the cable sizes are worked up after the house is laid out and circuited, and the panel loads are totalled.

The Merker residence, like most construction today, takes underground service. This means that the service cable from the utility company line to the building is run underground. Layout of the service run, detailing of service equipment and, sometimes, of the service takeoff is the work of the electrical technologist. For this reason the paragraphs that follow are devoted to a discussion of low voltage electric service to buildings.

15.11. Electric Service — General

The majority of buildings take service at low voltage, that is, below 600 v. This electric service is provided by the utility company at a service entrance point. Service can consist of 2, 3 or 4 wires including a grounded neutral wire. The service provided may be 2 wire, 120 v for a very small house, 3-wire, 120/240 v for a house like the Basic, or Merker, or 4-wire, 120/208 or 277/480 v for a medium-sized industrial facility. In each case the size of the service, in amperes, for instance, 60, 100, or 200 amperes, varies, depending on the building's load. (Generally, 2-wire, 120 v service does not exceed 60 amperes). Some very large buildings and heavy industrial plants take service at high voltage. Such arrangements are specially designed for each building and cannot be dealt with here. Other buildings that take low voltage have the utility company's transformers in or adjacent to the building. The service cables between the building property line and the utility company supply point are generally the property of the utility company in overhead service runs and the property of the *owner* in underground runs. Therefore, the Merker owner pays for the service run.

15.12. Electric Service — Overhead

The most common form of electric distribution is overhead lines. At a building requiring overhead electric service, a service drop is run from the nearest utility pole. This is connected to the building service cables at the service entrance point. Study Figures 15.20 to 15.22 to see how this is done. Figure 15.23 shows the splicing at the service pole. This work is done by the utility company. From here, the overhead service wires are extended to the building. At the building they are terminated either individually on wire holders (Figure 15.21), on a single wire holder for an entrance cable assembly (Figure 15.20) or occasionally with very heavy drop wires on a building-mounted secondary rack, as in Figure 15.22. Service entrance wires are brought out of the building from the building panel. They are spliced outside to the service drop cable with solderless connectors, and the joints are taped. (See Figures 15.20 to 15.22.) The service entrance conductors enter the building through the weatherhead and exterior electric power meter. Observe that the service wires can enter the building in pipe (conduit), or simply as exposed cable.

The weatherhead is a porcelain and steel device used to bring cables inside without allowing in the rain. The weatherhead varies in number and size of holes, and in size of conduit fitting. The detail of Figure 15.21 is for a multiple dwelling. For a single family residence the service equipment is usually placed on the street level, in the garage or utility room. Note in Figure 15.21 the mounting heights of equipment and the use of a table of materials. In Figure 15.24 the arrangement of Figure 15.21 is shown as it would appear on an electric plan. Note that all three drawings are necessary: the single line, the plan and the detail. The single line (Figure 15.24b) shows the electrical situation at a glance. Frequently the cable and equipment sizes are shown here. The location sketch (Figure 15.24a) shows the physical arrangement and should be to scale. Finally, the detail (Figure 15.21) shows the exact materials and the required construction. Construction details of this type are one of the technologist's most important tasks. Without them, the contractor is left to his own choice and the results may not be what was intended.

15.13. Electric Service — Underground

In most new construction the choice is made to bring the service into the building underground. In general, the utility company (electric power company) stops at the building's property line. Therefore, with underground service, the customer must run the service cables from the building to the property line. At that point, the service connection is made. The service work on private property must be done by a private contractor and at the owner's expense. The work is subject to the requirements and inspection of the utility company. This is to assure a proper grade of work and equipment. The type of equipment used for the underground service cable varies. Some utility companies allow direct burial of cable of acceptable design (type USE, Underground Service Entrance cable) between the service connection and the building. See Figures 15.25 and 15.26. Other companies require greater physical protection for the cable. This is provided by heavy wall (Type II) fiber conduit or rigid steel con-

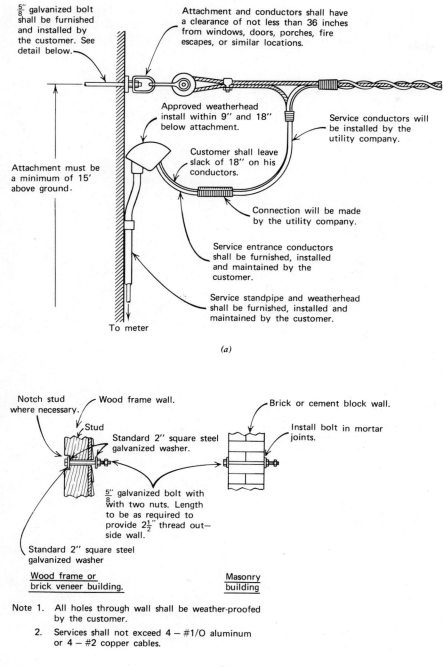

$\frac{5''}{8}$ galvanized bolt shall be furnished and installed by the customer. See detail below.

Attachment and conductors shall have a clearance of not less than 36 inches from windows, doors, porches, fire escapes, or similar locations.

Approved weatherhead install within 9'' and 18'' below attachment.

Service conductors will be installed by the utility company.

Attachment must be a minimum of 15' above ground.

Customer shall leave slack of 18'' on his conductors.

Connection will be made by the utility company.

Service entrance conductors shall be furnished, installed and maintained by the customer.

Service standpipe and weatherhead shall be furnished, installed and maintained by the customer.

To meter

(a)

Notch stud where necessary.

Wood frame wall.

Stud

Standard 2'' square steel galvanized washer.

Brick or cement block wall.

Install bolt in mortar joints.

$\frac{5''}{8}$ galvanized bolt with with two nuts. Length to be as required to provide $2\frac{1}{2}''$ thread out—side wall.

Standard 2'' square steel galvanized washer

<u>Wood frame or</u>
<u>brick veneer building.</u>

<u>Masonry</u>
<u>building</u>

Note 1.   All holes through wall shall be weather-proofed by the customer.

2.   Services shall not exceed 4 – #1/O aluminum or 4 – #2 copper cables.

(b)

Figure 15.20 Typical residential overhead electric service detail. *(a)* Service drop equipment. *(b)* Attachment bolt detail for different wall construction.

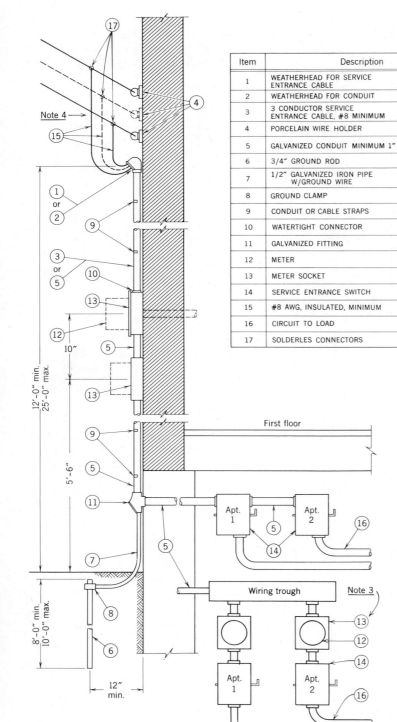

Figure 15.21 Typical electrical service, for multiple residence.

Item	Description
1	WEATHERHEAD FOR SERVICE ENTRANCE CABLE
2	WEATHERHEAD FOR CONDUIT
3	3 CONDUCTOR SERVICE ENTRANCE CABLE, #8 MINIMUM
4	PORCELAIN WIRE HOLDER
5	GALVANIZED CONDUIT MINIMUM 1"
6	3/4" GROUND ROD
7	1/2" GALVANIZED IRON PIPE W/GROUND WIRE
8	GROUND CLAMP
9	CONDUIT OR CABLE STRAPS
10	WATERTIGHT CONNECTOR
11	GALVANIZED FITTING
12	METER
13	METER SOCKET
14	SERVICE ENTRANCE SWITCH
15	#8 AWG, INSULATED, MINIMUM
16	CIRCUIT TO LOAD
17	SOLDERLES CONNECTORS

Notes:
1. Omit item #10 if conduit is used.
2. Cold water pipe ground may be used in lieu of ground rod.
3. Meters may alternatively be placed inside the building.
4. See Fig. 15.20 for arrangement with incoming multi–conductor aerial cable instead of individual wires shown here.

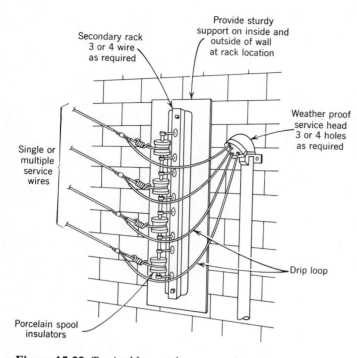

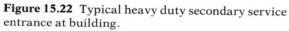

Figure 15.22 Typical heavy duty secondary service entrance at building.

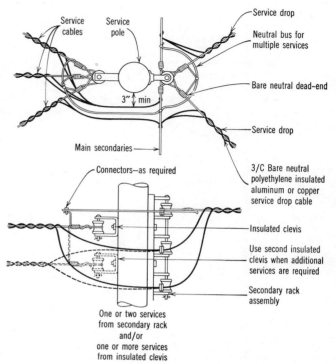

Figure 15.23 Arrangement of secondary cable connections at the service pole. These connections are used with three-conductor bare-neutral polyethylene insulated service-drop cable. Note the other end of the drop termination in Figures 15.20 and 15.21. This work is generally the responsibility of the utility company and this drawing is for information only.

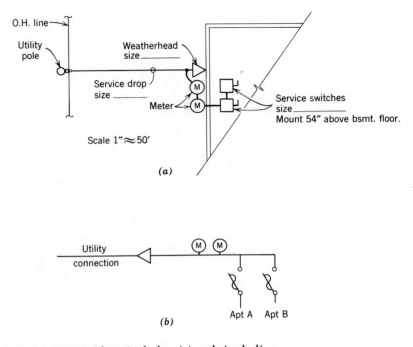

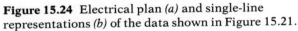

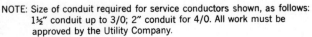

Figure 15.24 Electrical plan *(a)* and single-line representations *(b)* of the data shown in Figure 15.21.

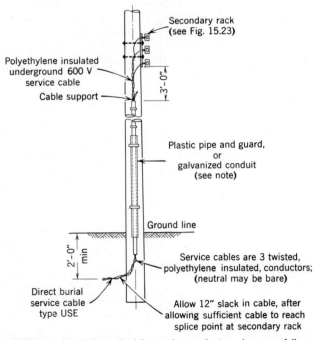

NOTE: Size of conduit required for service conductors shown, as follows: 1½" conduit up to 3/0; 2" conduit for 4/0. All work must be approved by the Utility Company.

Figure 15.25 Typical service (riser) pole feeding underground secondary service.

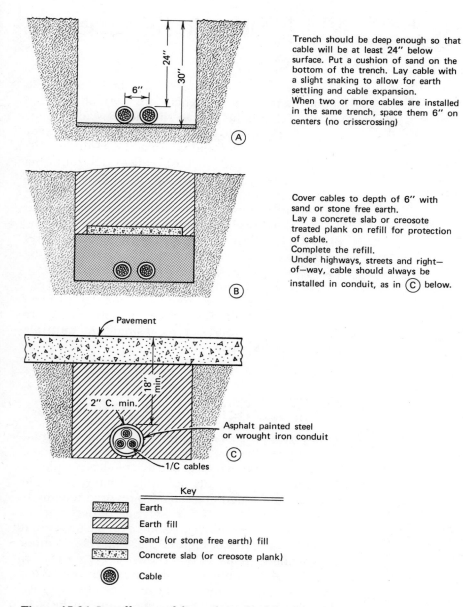

Trench should be deep enough so that cable will be at least 24" below surface. Put a cushion of sand on the bottom of the trench. Lay cable with a slight snaking to allow for earth settling and cable expansion.
When two or more cables are installed in the same trench, space them 6" on centers (no crisscrossing)

Cover cables to depth of 6" with sand or stone free earth.
Lay a concrete slab or creosote treated plank on refill for protection of cable.
Complete the refill.
Under highways, streets and right—of—way, cable should always be installed in conduit, as in Ⓒ below.

Key

	Earth
	Earth fill
	Sand (or stone free earth) fill
	Concrete slab (or creosote plank)
	Cable

Figure 15.26 Installation of direct-burial cable. Note use of separate key (symbol list).

duit. See Figure 15.27. Low voltage cable is almost never installed in a concrete enveloped raceway. An exception occurs when the service equipment is not at the point at which the underground run meets the building. See the NEC paragraph 230-44, "Conductor Considered Outside Building," and Figure 15.18, which shows this situation with respect to Merker. In some cases the utility installs cables in empty raceways supplied by the owner. In others, the entire installation is done by the owner. The drawing or specification must clearly state the division of responsibility between the utility and the customer. A typical service detail corresponding to the kind of service taken at the Merker residence is shown in Figure 15.28.

Installing Type II Nonmetallic
Underground Duct

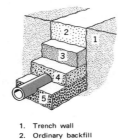

Trench should be graded true and free from stones or soft spots. Backfill should also be free of stones and be firmly tamped around the sides of the duct, to develop maximum supporting strength.

In rocky soil selected backfill (not tamped) at least 6" over the top of the conduit is recommended. After final backfill is placed, tamping may be used to finish the grade. Minimum burial depth where subject to traffic is 36".

1. Trench wall
2. Ordinary backfill
3. Selected backfill
4. Selected backfill
5. Bedding

Type II Fibre Conduit

Usage

Type II fibre conduit is an extra heavy wall conduit designed for installation *without* concrete. While the method of manufacture is basically the same as for Type I, it has the extra physical characteristics required for the service conditions of unprotected installations.

Type II is used to protect cables in power, telephone, and signal circuits wherever conditions do not necessitate concrete encasement. It is especially adapted to private property, underground services.

Strength

Type II fibre conduit strength is such that it will withstand any normal stresses when properly installed, under streets and highways, without a concrete envelope.

Longevity

Because Type II fibre conduit is composed of pitch, which saturates a framework of felted cellulose wood fibres, its life is practically unlimited.

Fibre conduit is resistant to erosion and immune to corrosion. Ground alkalis or acids have no effect on Type II fibre conduit. Termites, vegetable growth, or tree roots do not attack this pitch structure.

Figure 15.27 Specification and underground installation data for Type II heavy-wall fiber duct. (Courtesy of Carlon and Bermico Companies)

15.14. Electric Service — Metering

As can be seen in Figures 15.20, 15.24 and 15.28 single meters for residences are normally placed outside. This is helpful to the meter reader, since access to the inside of the house is not needed. For multiple residences and commercial buildings the metering is normally inside because (a) the building is open, and (b) the metering installation itself is large. With multiple residences that have individual apartment metering (many modern multiple rental residential buildings use master metering), the practice has been to install the meters in central meter rooms. The advantage of this is that reading the meters is a one-stop affair. To make an installation of this type, multiple meter pans are used. They can be assembled in groups or modules to meet almost any requirement. Examples are shown in Figures 15.29 to 15.31 of typical single and multiple meter installations.

15.15. Signal Equipment

Every residence is equipped with some signal equipment. The simplest house has a doorbell and usually a telephone. An expensive modern residence like the Merker's will most likely have intrusion alarms, smoke and fire alarms, prewired telephone jacks, amplified television antenna outlets and an intercom. These outlets are shown on the upper and lower house plans of Figure 15.32a and b. A symbol list for the systems used is shown in Figure 15.32c, and the relevant notes are shown in Figure 15.32d.

When signal systems were first installed in residences, it was usual for each system to be self-contained. The fire alarm system had its control unit and devices, the smoke detection system its equipment, the intercom its equipment, and so on. That is still the situation with small "package" systems. However, in large residences, a custom system can be furnished with a single control panel and annunciator. This connects to all the devices, and indicates visually and audibly the action of any device. The control unit is normally placed in the master bedroom. A riser diagram for such a combined, custom-designed system is shown in Figure 15.33. Note that this system is arranged to do the following when any of the remote units operate:

(a) Sound a bell in the master bedroom and a buzzer in the lower level.

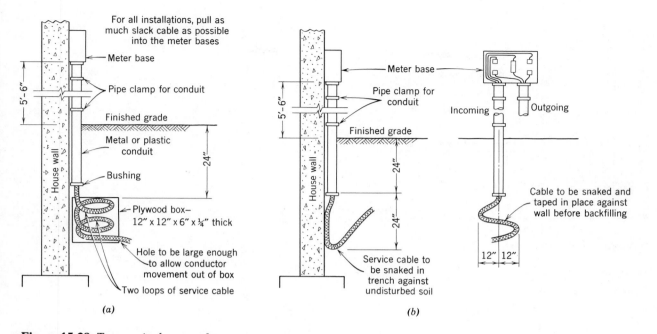

Figure 15.28 Two typical types of termination of service cable at a building. Some utility companies require a similar coil of underground cable at the base of the utility pole. See Figure 15.25.

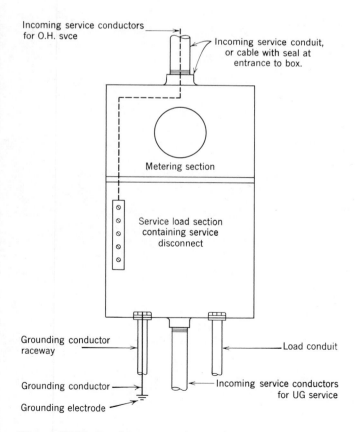

Figure 15.29 Combination meter and service cabinet for overhead or underground service.

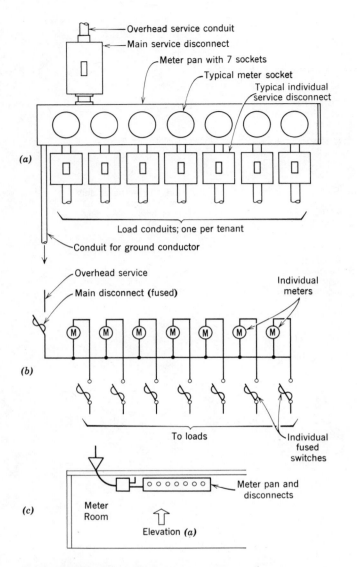

Figure 15.30 Typical arrangement of metering for multiple-occupancy building (seven tenants). *(a)* Physical arrangement in elevation. *(b)* Schematic representation, one-line diagram. *(c)* Plan representation.

Figure 15.31 Typical modular metering equipment. The cabinet at the left contains the main service disconnect—in this case a circuit breaker. Adjacent, and fed from this, are two sections of meter pan—one 8 and one 4. Below each meter socket is the circuit breaker that is the main protection for the apartment involved. More sections can be added, as needed. (Note the incoming service conduits and the outgoing load conduits.) Complete electrical and physical data are available in manufacturers' catalogs. (Courtesy of Square D Company)

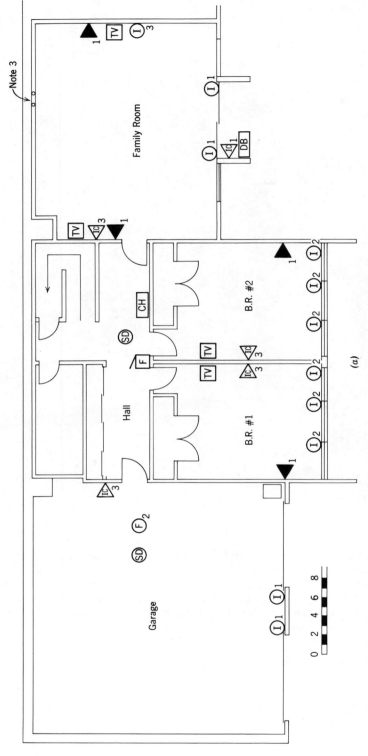

Figure 15.32 (a) Electrical plan, lower level, signal devices.

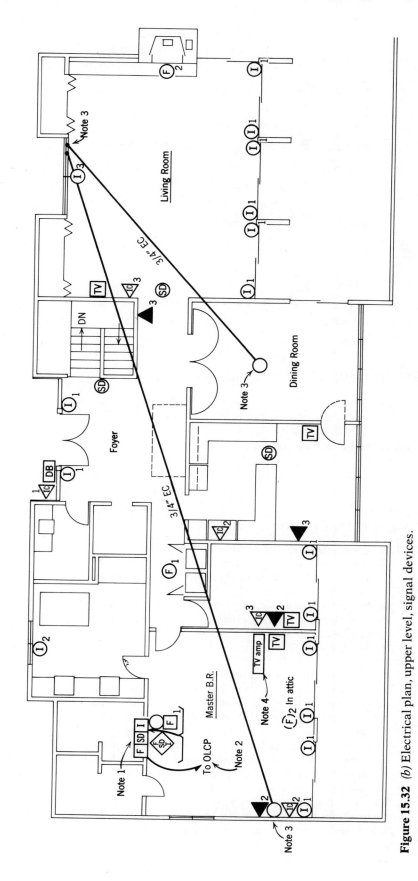

Figure 15.32 (b) Electrical plan, upper level, signal devices.

SYMBOLS FOR SIGNAL EQUIPMENT

○ F — 6" AC VIBRATING BELL, CONCEALED IN RECESSED BOX, WITH GRILL CLOTH COVER, 84" AFF.

F — BUZZER, AC, SIMILAR INSTALLATION TO ABOVE.

(F)₁ — TEMP. DETECTOR; RATE–OF–RISE & FIXED TEMP., RESETTABLE.

2 — TEMP. DETECTOR; FIXED TEMP., 185°C.

(SD) — SMOKE DETECTOR WITH RESETTABLE FIXED TEMP. DETECTOR.

(I)₁ — INTRUSION DETECTOR; MAGNETIC DOOR SWITCH.

2 — INTRUSION DETECTOR; MAGNETIC WINDOW SWITCH.

3 — INTRUSION DETECTOR; ELECTRONIC, MOTION DETECTOR.

◇ — ANNUCIATOR, CUSTOM DESIGN.

CP — CENTRAL PANEL FOR F.A., S.D. & INTRUSTION.

DB — DOOR BELL.

CH — CHIMES SIGNAL.

◀₁ — PREWIRED PHONE OUTLET; JACK 12" AFF.

2 — PREWIRED PHONE OUTLET; FIXED, 12" AFF.

3 — PREWIRED PHONE OUTLET; FIXED WALL OUTLET 60" AFF.

◁IC₁ — INTERCOM OUTLET, OUTDOOR, W.P. 60" AFF.

2 — INTERCOM OUTLET, MASTER STATION 60" AFF.

3 — INTERCOM OUTLET, REMOTE STATION 60" AFF.

TV — PREWIRED TV ANTENNA OUTLET, 12" AFF.

Figure 15.32 *(c)* Symbols for signal equipment.

Figure 15.32*(d)* Notes for Signal Systems

1. The fire detection, smoke detection and intrusion alarm devices all operate from a single control panel. The alarm bell is common. The annunciator indicates the device operated.
2. Connection between signal control panel and OLCP (outside lighting control panel) activates all outside lights when a signal device trips. Selected lights inside the house can also be connected to go on.
3. Two ¾-in. empty plastic conduits extending from 2 4-in. boxes in living room wall down to family room and terminating in 4-in. flush boxes. Boxes to be 18 in. AFF and fitted with blank covers. Extend a ¾ in. plastic EC from one 4-in. box in living room to 12-in. speaker backbox recessed in dining room ceiling. Locate in the field. From the second 4-in. box in living room extend a ¾-in. empty plastic conduit to an empty 4-in. box in the master bedroom, 18 in. AFF. Finish with blank cover.
4. Provide television antenna amplifier, recessed in wall box, with hinged ventilated cover, 18 in. AFF. Connections to antenna and to all television outlets by television antenna subcontractor. Provide 120 v outlet at the amplifier, with switch to disconnect.

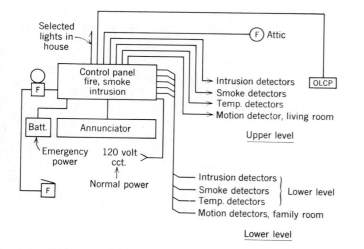

Figure 15.33 Fire, smoke and intrusion alarm system riser.

(b) Show on the annunciator the location of the signal that has operated and what type it is, that is, fire, smoke, or intrusion.

(c) Turn on all the outside lights, via panel OLCP (Outside Lighting Control Panel).

(d) Turn on selected inside lights.

Note also that a battery has been furnished to supply emergency power to the unit. The technologist is responsible for the layout of the devices and the riser diagram. To familiarize him with the symbols normally used in such diagrams a general symbol list of signaling devices is furnished in Figure 15.34. This list is Part VII of the overall symbol list. Photographs of some typical residential signal equipment are shown in Figures 15.35 to 15.39. Figures 15.40 to 15.42 show the telephone equipment for a typical multistory apartment building.

SIGNALLING DEVICES

F 1, A	BELL OR GONG, INSCRIBED LETTER INDICATES SYSTEM (SEE BELOW; AND SUBSCRIPT LETTER OR NUMBER INDICATES TYPE e.g. A – 8", VIBRATING BELL, 12 V. DC; B – 12" WEATHER PROOF SINGLE STROKE GONG, 120 V. D.C., etc.
W A	BUZZER, TYPE A.
F 1	FIRE DETECTOR, TYPE 1. ETC.
I 1	INTRUSION DETECTOR, TYPE 1. ETC.
SD 2	SMOKE DETECTOR, TYPE 2. ETC.
F	MANUAL STATION – WATCHMEN TOUR, FIRE ALARM, ETC. LETTER INDICATES SYSTEM, SEE BELOW.
F 2	ANNUNCIATOR, LETTER AND NUMBER INDICATE SYSTEM AND TYPE.
SD A	CABINET OR CONTROL PANEL, SMOKE DETECTION, USE IDENTIFYING TYPE LETTER IF MORE THAN ONE TYPE IS USED ON THE PROJECT.
BATT	AUXILIARY DEVICE.
• / •• / PB	PUSH BUTTONS
A, S 2	LOUDSPEAKER OR HORN.
B	TELEPHONE OUTLET, TYPE B
IC 2	INTERCOM OUTLET, TYPE 2
A	CLOCK SYSTEM OUTLET, TYPE A
TV	TV ANTENNA OUTLET

SYSTEM TYPES

F, FA	FIRE ALARM
S, SD	SMOKE DETECTION
I, IA	INTRUSION ALARM
W, WF	WATER, WATERFLOW
S, SP	SPRINKLER
W, WT	WATCHMAN'S TOUR
T, TEL	TELEPHONE
TV	TELEVISION
IC	INTERCOM
NC	NURSE CALL

AUXILIARY DEVICES

BATT	BATTERY
CT	CONTROL TRANSFORMER
DH	DOOR HOLDER
DO	DOOR OPENER
S, SP	SPEAKER, LOUDSPEAKER
TC	TELEPHONE CABINET
BT	BELL TRANSFORMER
DB	DOOR BELL
CH	CHIME

Figure 15.34 ARCHITECTURAL-ELECTRICAL PLAN, SYMBOL LIST, PART VII SIGNALLING DEVICES.

Part I, Raceways, Figure 12.6, page 304.
Part II, Outlets, Figure 12.37, page 338.
Part III, Wiring Devices, Figure 12.39, page 341.
Part IV, Abbreviations, Figure 12.43, page 345.
Part V, Single Line Diagrams, Figure 13.4, page 354.
Part VI, Equipment, Figure 13.14, page 369.
Part VIII, Motors and Motor Control, Figure 16.19, page 527.
Part IX, Control and Wiring Diagrams, Figure 16.22, page 530.

Detectors, placed in all
critical spaces, cover up
to 400 sq ft each

Zone indicating panel
is optional, often
containing a test button
and bell silencer

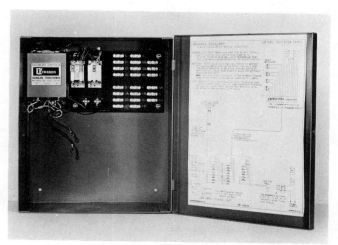

Control unit with integral stand–by power equipment
is available for surface or flush mounting

Audible signals, both indoor
and outdoor type, alert
occupant or neighbors
to actual or incipient fire

Edwards Co., Inc.

Figure 15.35 Typical residential fire alarm
equipment. (Courtesy of Edwards Company, Inc.)

Figure 15.36 One type of smoke detector unit. This unit is very sensitive and will detect a fire long before a temperature type detector. (Courtesy of Statitrol)

Figure 15.37 A modern annunciator unit shows the zone and actual device which operated. On a plan of the house, a light indicates the location of the tripped device. Such an annunciator can show all the devices and distinguish between types by use of different color lights. (Courtesy of Honeywell)

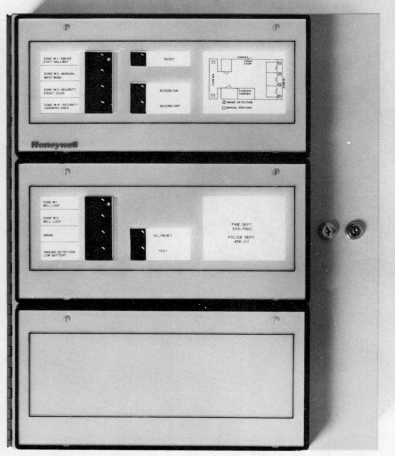

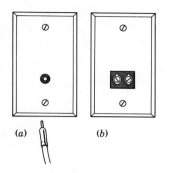

(a) (b)

Figure 15.38 Typical television wall outlets. Outlet *(a)* is a coaxial type that requires professional skill to make up the jack connection shown; *(b)* is of the screw type allowing tenants to make their own installation.

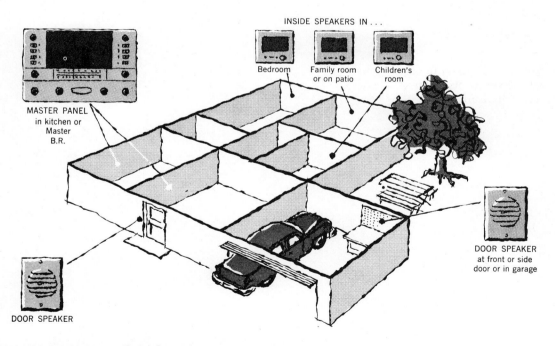

Figure 15.39 Typical residential intercom equipment.

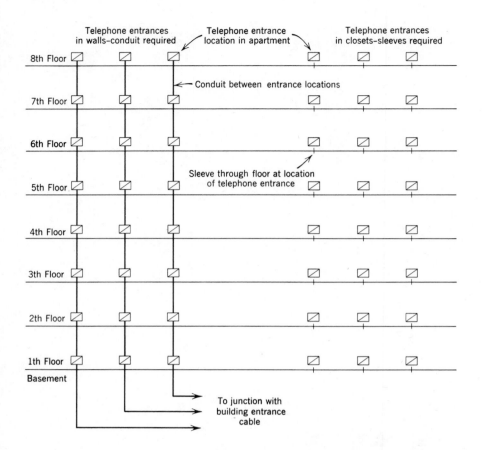

Figure 15.40 Typical telephone riser diagram. Note the need for conduit between apartments when installation is made inaccessible, as in a wall.

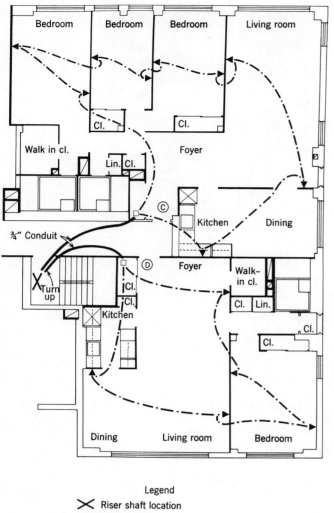

Legend

✕ Riser shaft location
━ ¾" Conduit
-·- Telephone cable
▼ Telephone outlet
▫ Conduit turned up or 4" ✕ 4"
box for concealment

Figure 15.41 Typical telephone floor plan—conduit in riser shaft and connections to apartments. (Courtesy of New York Telephone Company)

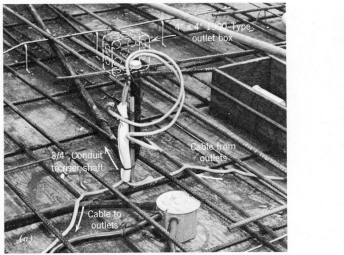

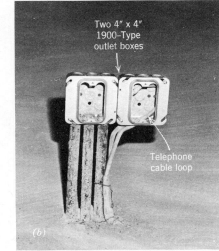

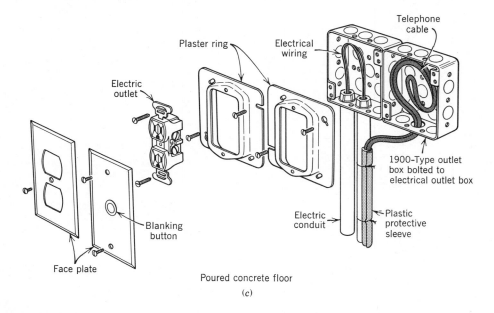

Figure 15.42 In concrete construction, telephone outlets may be prewired by using special telephone company wire that is laid directly into the concrete. The outlets are most often placed adjacent to receptacle outlets. Wiring method is illustrated. The feed is introduced into the apartment via a ¾-in. conduit *(a)* terminating in a 4-in. box. From there, the cable is laid in the concrete, and brought up *(b)* adjacent to receptacles. The finished outlets are assembled as in *(c)*. Prewiring eliminates the requirements for unsightly surface wiring. (Courtesy of New York Telephone Company)

Problems

15.1. Convert the basement of the Basic Plan House into a family room. Show all partitioning. Provide lighting and outlets for children's and adult's use. Revise the electrical plans and schedules for the basement, given in Chapter 13. Draw to scale any necessary details.

15.2. Draw the detail called for in Note 9, Figure 15.1*d*.

15.3. Based on the data shown on panel schedule H, Figure 15.16*c*, compute the difference between the panel load for bus size calculation and the panel load for service calculation. Do not use demand factors.

15.4. Using the load figures for kitchen equipment given in Table 15.2, calculate the safety factor involved when using the recommended wire size and receptacle size. The safety factor is defined as the ratio of spare capacity to load rating. The ratings for wire and receptacles are found in Tables 12.2 and Figure 12.40, respectively. Tabulate the results.

15.5. Using the loads shown in panel schedule P and CC, calculate:
   (a) Volt-amperes per square foot for the building, not counting receptacles.
   (b) Volt-amperes per square foot for the building, counting receptacles as 180 w.
   (c) Volt-amperes per square foot for the building electric heating load (use 1 hp = 1 kva).
   (d) Total volt-amperes per square foot for the building.

15.6. Using the criteria given in Sections 15.3 to 15.6, analyze (a) a private residence, (b) an apartment of, at least, 800 sq ft area, for which you have complete electrical plans. On a plan draw the changes you would make to conform the electrical work to these criteria. When completed, these drawings should be suitable to deliver to an electrical contractor for pricing.

15.7. For the same two residential occupancies, prepare a signal plan providing intercom, security, television outlets and any other devices you think necessary. (Manufacturer's catalogs will be a considerable help.)

Additional Reading

Refer to the Additional Reading of Chapter 16.

16. Nonresidential Electrical Work

In preceding chapters we discuss the principles of electrical layout work. Branch circuit layout and circuiting are emphasized. We also state some general guidelines for electric layout in residential buildings. We then combine these principles and guidelines, and apply them to two residences, complete with circuiting and schedules. Different guidelines apply to layout of lighting and devices in nonresidential buildings. We briefly consider these guidelines here.

In nonresidential buildings, the service entrance, service equipment and interior electric distribution become very important. These items are called the building's electric power system. They include service equipment, switchboards, bus and heavy feeders, distribution panels, and motors and their control. The technologist is responsible for drawing this equipment, and showing how it goes together. To do this, he must know what it looks like, how it mounts, and generally how to handle it. In this chapter, our purpose is to explain these functions. A few additional topics of special interest complete our study.

Study of this chapter will enable you to:

1. Assist in the preparation of power riser diagrams.
2. Draw electric service details, including emergency power provision.
3. Understand electric metering and service arrangements.
4. Calculate required dimensions of pull boxes and draw the related details.
5. Draw a motor wiring diagram.
6. Draw a motor control (ladder) diagram.
7. Show motor wiring on architectural-electrical plans.
8. Utilize layout and circuitry guidelines for commercial and institutional buildings.
9. Draw stair and exit risers.
10. Prepare the schedules relating to electrical power plans.

16.1. General

Refer to Figure 16.1. This is a block diagram of the electric system of a typical commercial building. It corresponds to the pictorial riser shown in Figure 12.14, page 315. The major items are the electric service and related equipment (not shown in Figure 12.14), the building switchboard, the bus duct and cable distribution system, the motors and their control equipment, and the branch circuit equipment (also not shown in Figure 12.14). Also shown in Figure 16.1, are panels, switches, controllers, pull boxes, and small transformers. Not shown on either diagram are auxiliary items such as grounding systems, lightning protection equipment, lighting risers and signal equipment. Obviously not all of these items are found in all nonresidential buildings. You will, however, come across these items in your work experience, and the competent technologist knows how to handle them.

16.2. Electric Service

The service from the utility can be overhead or underground, high voltage (primary) or low voltage (secondary), and any power rating required.

a. Primary Service

This is generally run underground from the utility line to the building. These runs are in rigid steel conduit or in concrete encased nonmetallic duct. This nonmetallic duct is lighter in weight than the Type II of Figure 15.27 and is called Type I. Such underground power cable runs are often combined with UG telephone cable runs, for economy. See typical detail of Figure 16.2. Al-

though service cables are generally run without splices, occasionally splices are required. Splicing and pulling underground cables is done in manholes or handholes, depending on the size and voltage of the cable. These manholes and handholes are large concrete boxes set into the earth. Although many are field-poured, precast units are readily available to fill most requirements. Figure 16.3 shows a typical double manhole that would accommodate the ducts of Figure 16.2. Figure 16.4 shows typical duct termination details at the manhole and the building. A handhole is simply a small manhole. The difference is that a man climbs into a manhole but only reaches into a handhole. A typical handhole detail is given in Appendix A.22. A competent electrical draftsman can develop details like these from engineer's sketches. For this reason these illustrations should be carefully studied and fully understood. Where the UG cable reaches the building, it is connected to a transformer which changes primary (high) voltage to secondary (low) voltage. See the last step of Figure 11.25, page 293. Since the transformer inside the building can create maintenance problems, many designers place it outside on a concrete pad. Appendix A.23 shows typical transformer and pad dimensions and other data related to pad-mount transformers.

b. Secondary (Low Voltage) Service

This is taken either underground or overhead, as explained and illustrated in Sections 15.10 to 15.14. A typical plot plan of a secondary underground service to a nonresidential building is illustrated in Figure 16.5 with service details in Figure 16.6. Note that, as with the Merker residence, the service entrance cable is run in con-

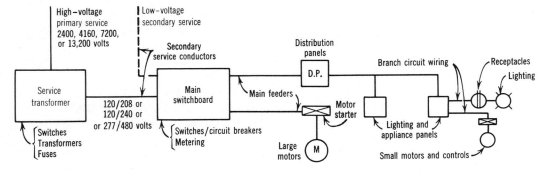

Figure 16.1 Block diagram of the electrical system of a typical commercial building. Note that service can be taken at either high or low voltage. If taken primary (high voltage), it is changed to the usable low voltage by the service transformer.

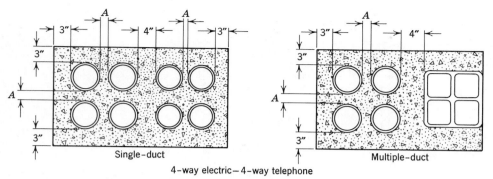

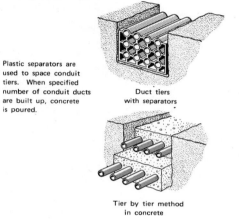

4-way electric—4-way telephone

Notes: $A = 1\frac{1}{2}''$ for clay and soapstone duct and $2''$ for fiber or asbestos cement.
All dimensions are minimum from outside surface of ducts.
Concrete not required between horizontal faces of clay duct.

Installing Type I Nonmetallic
Underground Duct

Plastic separators are
used to space conduit
tiers. When specified
number of conduit ducts
are built up, concrete
is poured.

Duct tiers
with separators

Tier by tier method
in concrete

Figure 16.2 Typical underground duct-bank section, and details of installation. Although nonmetallic duct is illustrated, steel duct is used where high physical strength is required, such as in filled earth. Alternatively reinforced concrete can be used to provide this required physical duct-bank strength.

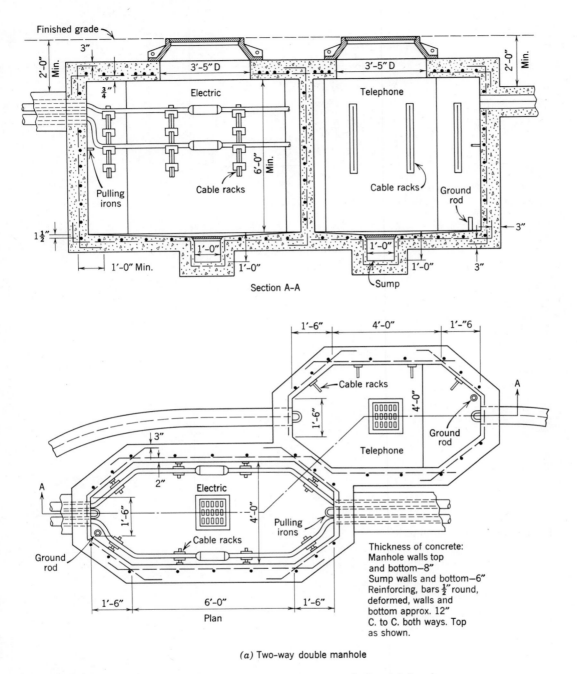

Section A-A

(a) Two-way double manhole

Figure 16.3 Typical details of double power/telephone manhole with hardware.

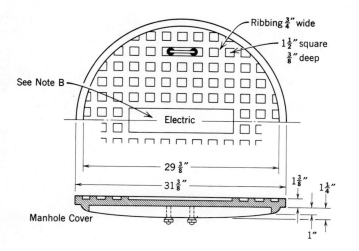

Ribbing $\frac{3}{4}''$ wide

$1\frac{1}{2}''$ square
$\frac{3}{8}''$ deep

See Note B

Electric

$29\frac{3}{8}''$

$31\frac{3}{8}''$

$1\frac{3}{8}''$

$1\frac{1}{4}''$

$1''$

Manhole Cover

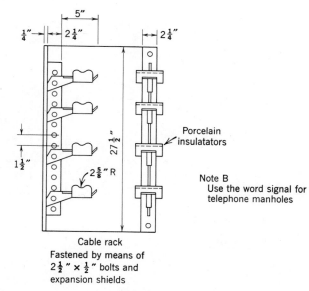

5"

$2\frac{1}{4}''$

$\frac{1}{4}''$

$2\frac{1}{4}''$

$27\frac{1}{2}''$

Porcelain
insulatators

$1\frac{1}{2}''$

$2\frac{5}{8}''$ R

Note B
Use the word signal for
telephone manholes

Cable rack
Fastened by means of
$2\frac{1}{2}'' \times \frac{1}{2}''$ bolts and
expansion shields

(b) Manhole Hardware

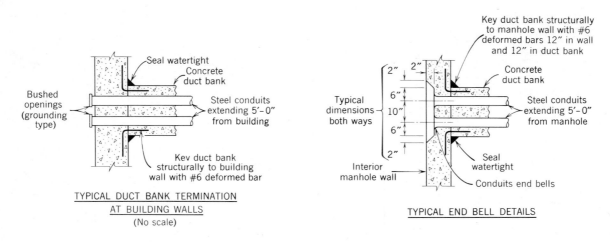

TYPICAL DUCT BANK TERMINATION
AT BUILDING WALLS
(No scale)

TYPICAL END BELL DETAILS

Figure 16.4 Typical underground construction details.

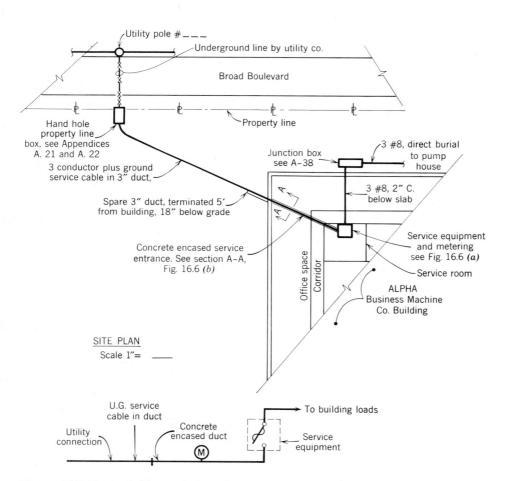

Figure 16.5 Typical electrical plan of underground service to an industrial building, plus the same information shown diagrammatically.

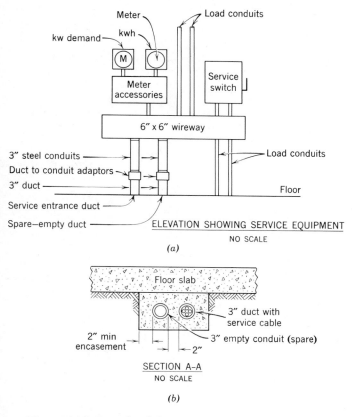

Figure 16.6 Details of electric service pertaining to service illustrated in Figure 16.5.

crete encased duct. This is to keep it "outside the building" by NEC definition, because the service switch is not at the service entrance point. The technologist should remember that *all* structural design is the responsibility of the architect and structural engineer. Therefore, all concrete work and reinforcing details must either come from them or be approved by them.

A brief survey of the available secondary service voltages and their application is helpful. Refer back to Figures 11.26 and 11.27, pages 294-295, and to Tables 16.1 and 16.2 while you read the following service descriptions.

1. 120 v, single phase, 2 wire, up to 100 amperes: used for small residences, farmhouses, outbuildings, barns and the like. Capacity of a 100-amp service of this type is

$$kva = \frac{100 \text{ amperes} \times 120 \text{ volts}}{1000} = 12 \text{ kva max}$$

2. 120/240 v, single phase, 3 wire, up to 400 amperes: this is the usual residential and small commercial service. Maximum power is

$$kva = \frac{400 \times 240}{1000} = 96 \text{ kva}$$

3. 120/208 v, 3 phase, 4 wire, usually not in excess of 2500 amperes: this is the normal urban 3-phase service taken by most commercial buildings. Maximum power is

$$kva = \frac{\sqrt{3} \times 208 \times 2500}{1000} = 900 \text{ kva}$$

Table 16.1 Nominal Service Size in Amperes[a]

Facility	Area in Square Feet				Remarks
	1000	*2000*	*6000*	*10000*	
Single Phase, 120/240 v #3 wire					
Residence	100A	100A	200A	—	Minimum 100A
Store[b]	100A	150A	—	—	
School	100A	100A	—	—	
Church[b]	100A	150A	200A	—	
3 Phase, 120/240 v #4 wire					
Apartment House	—	—	150A	150A	
Hospital[b]	—	—	200A	400A	
Office[b]	—	—	400A	600A	
Store[b]	—	100A	400A	600A	
School	—	100A	150A	200A	

[a]Nominal service sizes are 100A, 150A, 200A, 400A, 800A, 1200A, 1600A, and 2000A.
[b]Fully air conditioned using electric driven compressors.

Table 16.2 Current and Wattage Relationships[a]

Load Watts	120 v Single phase	120/240 v 3 Wire	120/208 v Single Phase, 3 wire	120/208 v 3 Phase	277/480 3 Phase	277 v Single Phase
100	0.83	0.41	—	—	—	0.362
200	1.6	0.8	—	—	—	0.72
500	4.2	2.1	—	—	—	1.8
1,000	8.3	4.2	4.8	2.77	1.2	3.6
2,000	16.6	8.3	9.6	5.5	2.4	7.2
5,000	41.7	20.8	24.0	13.9	6.0	18.0
10,000	83.2	41.6	48.0	27.7	12.0	36.0
20,000	—	—	96.0	55.6	24.0	72.0
50,000	—	—	240.0	139.0	60.0	181.0
100,000	—	—	480.0	277.0	120.0	362.0
	$I = \dfrac{W}{120}$	$I = \dfrac{W}{240}$	$I = \dfrac{W}{208}$	$I = \dfrac{W}{360}$	$I = \dfrac{W}{830}$	$I = \dfrac{W}{277}$

[a]Assuming 100% power factor.

4. 277/480 v, 3 phase, 4 wire, usually not in excess of 2500 amperes: this service is taken by commercial and industrial buildings with large loads and heavy motors. Maximum power is

$$kva = \frac{\sqrt{3} \times 480 \times 2500}{1000} = 2000 \, kva$$

Loads larger than those available on a single service take multiple services. To clear up any confusion that may exist between system and utilization voltage, refer to Table 16.3. The *system* voltage is what the power company supplies, or what the transformer produces. The *utilization* voltage is the one utilized, after some voltage drop. Motors are rated at utilization voltage. Therefore, a 115 v motor is used on a 120 v line, a 200 v motor is used on a 208 v system, a 230 v motor on a 240 v line (system), a 460 v motor on a 480 v line, and so on. *All* drawing notations of voltage should recognize this difference. This means that a transformer should be shown as 240/480 v, a motor should be shown at 230/460 v, and so on. Showing a new motor as being rated 480 v is incorrect, and this should be avoided.

16.3. Emergency Electric Service

In buildings that require electric service when normal power fails, an emergency service is frequently installed. See Figure 16.7. There are many ways to

Table 16.3 System and Utilization Voltages[a]

	Standard Voltages[b]		
System Voltage (Transformers)		Utilization Voltage (Motors)	
Nominal	With 4% Drop[c]	New Standard	Old Standard
120	115.2	115	110
208[d]	199.7	200	208
240[d]	230.4	230	220
480	460.8	460	440
600	576.0	575	550

[a]To eliminate any confusion between system and utilization voltages, the current NEMA standards are tabulated above.

[b]When specifying transformers, use system voltages; for motors use utilization voltage.

[c]Note that utilization voltage corresponds to a 4% drop from system voltage; well within the normal motor tolerance.

[d]Motors for 208 v systems are rated 200 v. Motors for 240 v systems are rated 230 v. They cannot be used interchangeably without seriously affecting motor performance.

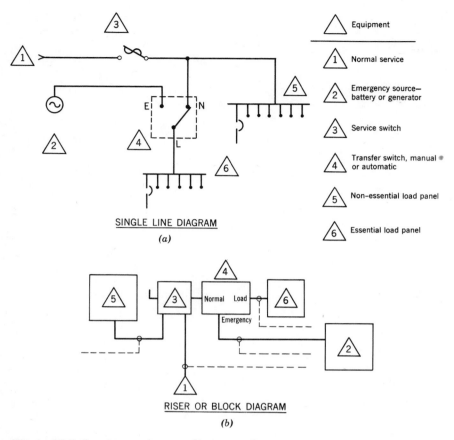

Figure 16.7 Service equipment diagrams showing normal and emergency electric service. *(a)* Single-line diagram. *(b)* Riser or block diagram.

supply emergency power. If all the loads will operate on d-c as well as a-c, and the total load is very light, a battery emergency source can be used. If a-c is required, a generator is usually furnished. Transfer to the emergency source can be done either manually or automatically by using either a manual or automatic transfer switch respectively. This latter item senses voltage loss and automatically transfers to the emergency source.

16.4. Building Main Electric Service Equipment

The NEC requires some means by which the incoming electric service can be completely disconnected. This disconnecting arrangement can consist of up to

six switches or circuit breakers, mounted at the point where the electric service feeders enter the building. The service disconnect(s) can be combined with the metering in a separate enclosure (Figure 16.8), mounted in a switchboard or panelboard (Figs. 16.9 and 16.10) or mounted entirely separate from all other equipment, as in Figure 12.14, page 315, and Figure 16.7. The main disconnect of the building panel or switchboard frequently acts as the service disconnect. When we refer to the service switch, we mean all the service disconnects, whether it is one or six.

The building switchboard or main panelboard controls and protects the feeders running through the building. Many types of switchboards are available. Depending on the type of devices in them, their sizes vary greatly. The engineer on the job will normally select this equipment, and the technologist will assist him in layout. Some typical building switchboard dimensions are given in Appendix A.25, to give the reader a "feel" for size. Accurate data are available in manufacturers' catalogs.

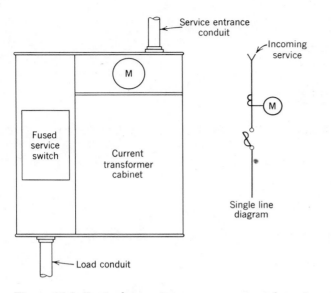

Figure 16.8 Typical metering arrangement and service disconnect for large (above 400 amp) service, including current transformers, meter and service switch.

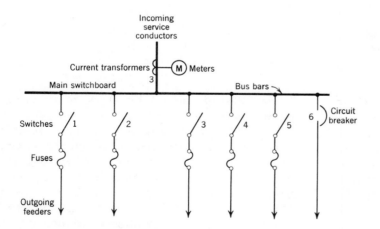

Typical switchboard. Switches are normally shown in the open position. The NEC allows up to 6 switches in parallel as service entrance equipment. Switches must be on the line (supply) side of fuses. Metering is normally placed on the service conductors and the metering equipment built into the main switchboard. Each line in a single line diagram represents a 3-phase circuit. If circuit breakers were used, the entire board would consist of units as illustrated in circuit 6.

Figure 16.9 Typical switchboard. Switches are normally shown in the open position. The NEC allows up to six switches in parallel as service entrance equipment. Switches must be on the line (supply) side of fuses. Metering is normally placed on the service conductors and the metering equipment is built into the main switchboard. Each line in a single-line diagram represents a 3-phase circuit. If circuit breakers were used, the entire board would consist of units as illustrated in circuit 6.

16.5. Electric Power Distribution

By this, we mean the system of conductors and devices that carry power out from the building service equipment. Refer again to Figure 12.14, page 315. The system of bus ducts and feeder cables that is shown constitutes the power distribution system. The system is "power" down to the last panel. After that it is branch circuitry, which we have studied in detail. A few definitions should be given here. In a multistory building, the service equipment is normally in the basement, in a separate room called the switchboard room. The feeders *rise* from the equipment in this room to feed the panels on each floor. For this reason the diagram showing this is called a *riser* diagram. A system on one level, such as in Figure 16.7, is also called a riser diagram for want of another term. A typical riser diagram is shown in Figure 16.11. Note the panel designations. Each riser shaft is given an identifying letter: A, B, or C. Panels are then identified by floor and shaft. Panel 4A is fourth floor, shaft A. This particular building has metering that is separate from the switchboard. The panels feeding fire alarm and stair and exit lighting are connected *ahead* of the four service switches. This is standard practice, so that they will remain energized even if power in the building is turned off. (Power is normally turned off in case of a fire, to prevent electrical faults and injury to firemen.)

We mentioned above the term "bus duct." This item of equipment is illustrated in Figure 16.12. It consists of heavy bars of copper or aluminum which are insulated and put together in a metal enclosure. A bus duct serves to carry heavy current instead of using parallel sets of cables. For instance, to carry 3500 amperes, 10 sets of 500 MCM copper cables in parallel are required. On the other hand, a single bus duct could be substituted measuring approximately 6 in. × 24 in. in cross section. Certain types of bus duct are made with plug-in points, to allow power to be picked off easily. The illustrated unit in Figure 16.12a is of this type. The illustrated unit will carry about 1000 amperes. The plug-in points are like giant receptacles at which up to 200 amperes can be picked off.

Now refer again to the riser diagram of Figure 16.11. We have stated that the individual risers go up in shafts. These *riser shafts* are vertically stacked spaces specifically designed for electrical conduits, cables and possibly bus duct. On each floor the shaft enters a small closet. In this closet a tap is made to

Figure 16.10 Low voltage switchboards are available in various designs, to meet all requirements. Illustrated is a NEMA I, panel-type free-standing, front accessible unit with large air main circuit breaker and molded case branch circuit breakers. The compartment above the main breaker can be used for metering, current transformers, and the like. The section across the top is for wiring. Units are from 14 to 20 in. deep. (Courtesy of Square D Company)

the risers, to feed the floor panel. The floor panel in turn feeds the branch circuits on each floor. Logically, this closet is called an *electric closet*. Two typical electric closets with their equipment are illustrated in Figure 16.13. Figure 16.14 shows a bus duct run through a closet and the metering installation that is tapped off it. In a multifloor building it is very desirable that closets be stacked one above the other. If this is not possible, an offset must be made. Such offsets generally require splicing the riser cables and are best avoided. Offsets in bus duct can be made with special right-angle fittings. See Figure 16.12a. One-story buildings also use electric closets to contain the electric panels. In large buildings or where the local utility companies require them, separate closets are installed for electric and for telephone equipment. The technologist is responsible for laying out the equipment in the closets, to scale. He must make sure that everything fits and

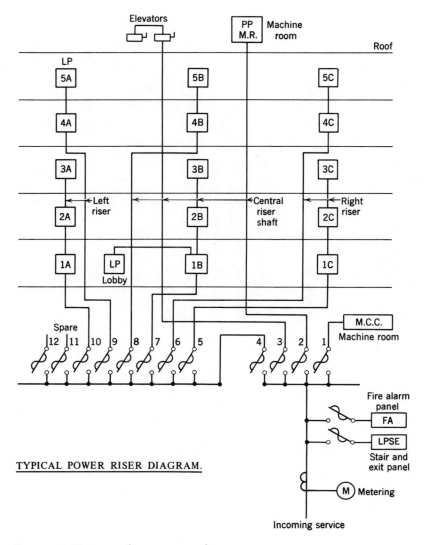

TYPICAL POWER RISER DIAGRAM.

Figure 16.11 Typical power riser diagram.

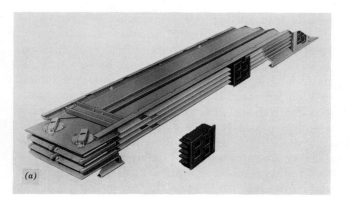

Figure 16.12 (a) Construction of one type of plug-in bus duct. Plug-ins are spaced every 12 in. on alternate sides to facilitate connection of plug-in breakers, switches, transformers or cable taps. Notice that bars are insulated over their entire length and are clamped rigidly at plug-ins with spacer blocks of insulating material. Housing is of sheet steel with openings for ventilation. Cover plate is not shown. (Courtesy of Square D Company)

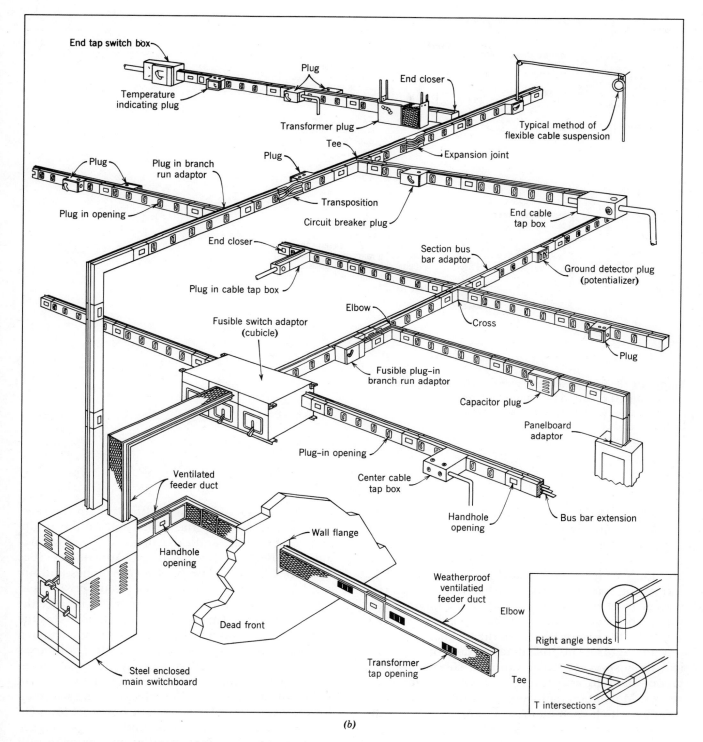

(b)

Figure 16.12 *(b)* Typical bus duct system. (Courtesy of General Electric Company)

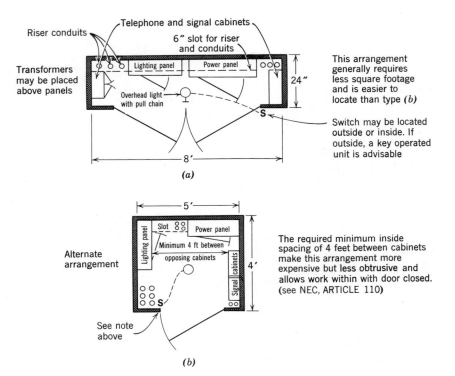

Figure 16.13 Typical electric closets with some usual equipment. If required by amount of equipment, separate closets may be used for signal and telephone conduits and cabinets.

that adequate clearances are maintained for safety and for maintenance. The NEC specifies minimum clearances and working spaces in its article, "Requirements for Electrical Installations." In some buildings a decision is made to install panels in furred-out spaces in walls rather than in electric closets, or even to have panels surface mounted on walls. The factors leading to these decisions are not generally the responsibility of the technologist.

Manufacturers publish detailed data on the physical dimensions of their equipment. This permits the layout man to accurately size all the electrical spaces. Most manufacturers use a modular sizing code by which each element in an assembly is dimensioned modularly. Refer to Figure 16.15, for instance. This extract from a Westinghouse catalog shows how to arrive at the size of a panelboard once the contents are known. Each item—the branch breakers, the main lugs or main breaker and the neutral space is dimensioned in modular "X" units. When these are totaled, the box can be sized. If additional gutter space is required, that too can be

provided in the height or width. To help the technologist get a feel for the dimensions of the equipment that he will handle day in and day out, we provide detailed information in the Appendices. (See Appendix A.26 for data on safety switches and Appendix A.27 for data on enclosed circuit breakers.)

We mentioned above that riser offsets require the use of a pull box. The purpose of a pull box is, as the name indicates, to provide a pulling point. When a raceway run is long or has many angles, pull boxes enable the pulls to be made easily. The exact location and need for a pull box is decided after an examination of the run. The rules covering size, construction and installation of these boxes are found in the NEC in the article, "Boxes and Fittings." Figure 16.16a shows a typical installation requiring pull boxes. The size of the pull box can be arrived at in the following way.

Draw a sketch of the box as in Figure 16.16b. The Code gives two methods for calculating the minimum pull box size. Both methods should be tried and the larger size should be used.

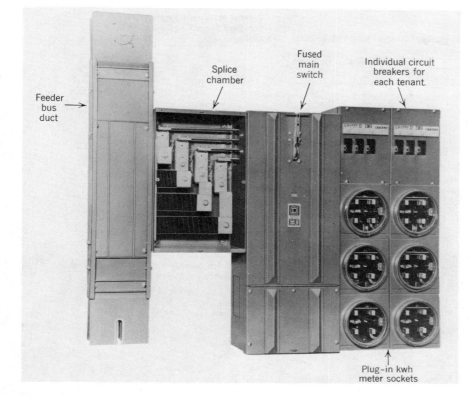

Labels: Feeder bus duct; Splice chamber; Fused main switch; Individual circuit breakers for each tenant; Plug-in kwh meter sockets

Figure 16.14 Typical metering installation. Bank of sockets for kilowatt-hour meters is connected to a feeder bus duct through a fused safety switch. Cover has been removed from the connecting box to show method of splicing bus work. Additional meter sockets are readily installed. Such assemblies are mounted in electric closets and meter rooms with access to utility, building personnel and tenants. (Courtesy of Square D Company)

Method 1: For a right-angle turn box as illustrated, the minimum width of the box is equal to six times the trade diameter of the largest conduit, plus the trade diameters of the other entering conduits. Thus minimum box size is

six times largest conduit	$6 \times 4 =$	24 in.
adding other conduits	$1 \times 4 =$	4 in.
	$2 \times 3 =$	6 in.
	$2 \times 2 =$	4 in.
width (length) of box		38 in.

Method 2: This method states that the minimum distance between raceways containing the same conductor is six trade diameters. Assuming that the arrangement stays the same entering and leaving, the 2-in. conduits should be 12 in. apart. The triangle in the corner (see Figure 16.16b), therefore, is an isosceles triangle with a 12-in. hypotenuse. The sides are therefore 8½ in. To this we add the

minimum conduit spacings obtained from Table 12.7, page 321 as follows:

Corner Distance	8½ in.
4 in. to 4 in.	5⅞ in.
4 in. to 3 in.	5¼ in.
3 in. to 3 in.	4 9/16 in.
3 in. to 2 in.	3¾ in.
2 in. to 2 in.	3⅛ in.
Total	31 1/16 in.

Since this is smaller than the figure obtained by Method 1, we use the larger figure of 38 in. The sketch of Figure 16.16b is then revised, increasing the corner distance by the difference between the two methods.

Corner distance =
$$= (38 - 31\tfrac{1}{16}) + 8\tfrac{1}{2} = 15\tfrac{7}{16}$$

We now have the actual box layout, which is then shown on the drawings.

Bolt-on Breakers in Branches

Panel and Cabinet Layout

Layout panel using X units as directed below to determine box height.

1. Obtain total X units for branch breakers from layout below⑥.
2. Obtain X units for main lugs or main breaker from Table A.
3. Obtain X units for neutral (if required) from layout below.
4. Total of 1, 2 and 3 is panel height in X units. Refer to Table B for box height and Cat. No.

Maximum panel height is 38X. When total X exceeds 38, lay out as two or more single panelboards, adding sub-feed or through-feed lugs (on main lugs only panels②) to all but one panel, and neutral bar to all panels where required. Cross tie cables will not be supplied.

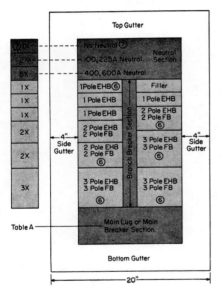

Table A —

Table A: Main Lugs, Main Breakers

Ampere Rating	X Height Dimension		
	Main Lugs Only	Main Breaker 2 Pole	3 Pole
50	2X	(EHB) 3X③	(EHB) 4X③
100	2X	(EHB) 3X③	(EHB) 4X③
150	..	(FB) 3X③	(FB) 4X③
225	2X	(JB, KB) 9X④	(JB, KB)9X④
400	4X	(LAB) 11X④	(LAB) 11X④
400	4X	(LBB) 11X④	(LBB) 11X④
600	5X	(LA) 12X④	(LA) 12X④

Table B: Box Selection
(Boxes 20" Wide, 5¾" Deep I.D.)

Total X Panel Height	Box Height, Inches
8.X	23
10 X	26
12 X	29
14 X	32
16 X	35
18 X	38
20 X	41
22 X	44
24 X	47
26 X	50
30 X	53
32 X	56
34 X	59
36 X	62
38 X	65

Box Knockouts, Top and Bottom

2", 2½" (2)
Space for 3½" Knockout
½", ¾" (12)
¾", 1"(4)

Terminal Wire Ranges, Cu/Al
Standard Main Lug Terminals

Ampere Capacity	Wire Size Range
100 Amperes	#12 – #1/0
225 Amperes	#6 – 300 MCM
400 Amperes	#4 – 600 MCM or 2(#4 – 250 MCM)
600 Amperes	2(#2 – 500 MCM)

Standard Main Breaker and Branch Breaker Terminals, Cu/Al⑤

Type	Ampere Rating	Wire Size Ranges
EHB	15-100	#14 – #1/0
FB	125-150	#6 – #3/0
JB, KB	70-225	#4 – 350 MCM
LBB	250-400	2(3/0 – 250 MCM)
LAB	250-400	1(3/0 – 600 MCM), and 1(#4 – 250 MCM)
LA⑧	500-600	2(250 – 500 MCM)

Minimum Top and Bottom Gutter

For reference only; do not include in box height calculation. Main lugs or main breakers are furnished at bottom of panelboard unless otherwise specified.

50, 100 Amp Mains: 4 In.
225 Amp Mains: 6 In.
400, 600 Amp Mains: 8 In.

② For two section panels with main breaker, sub-feed lugs, or through-feed lugs, refer to Westinghouse for dimensions.
③ Horizontally mounted 15 through 150 ampere EHB and FB main breakers will be furnished as branch breaker construction. Branch breakers 1, 2 or 3 pole as required, may be located opposite these main breakers. The adjacent sections normally occupied by main lugs becomes a blank filler 1X in height.
④ Vertically mounted.
⑤ For terminals providing wire sizes other than shown, refer to Westinghouse.
⑥ The sum of branch breaker ratings connected to any one connector must not exceed 200 amperes.
⑦ This dimension becomes 2X when 15 through 150 amp horizontally mounted main breakers are used, and no neutrals are required.
⑧ Top feed only.

Figure 16.15 Typical dimensional data for panelboards; lighting and distribution. (Courtesy of Westinghouse Company)

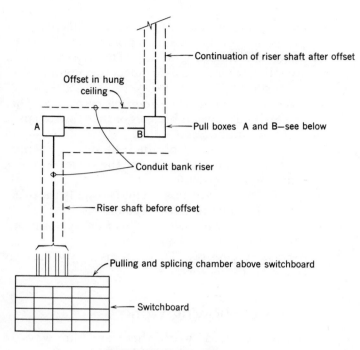

Continuation of riser shaft after offset

Offset in hung ceiling

A

B — Pull boxes A and B—see below

Conduit bank riser

Riser shaft before offset

Pulling and splicing chamber above switchboard

Switchboard

(a) OFFSET RISER WITH PULL BOXES

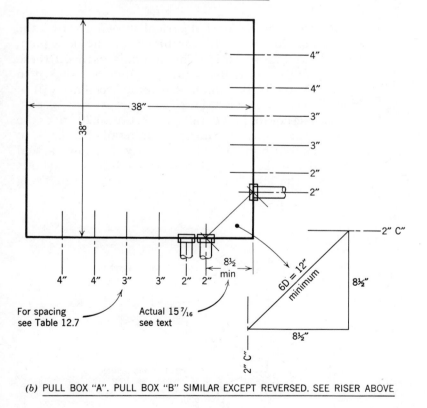

4"

4"

3"

3"

2"

2"

2" C"

38"

38"

6D = 12" minimum

8½"

8½ min

8½"

4" 4" 3" 3" 2" 2"

For spacing see Table 12.7

Actual 15 7/16 see text

2" C"

(b) PULL BOX "A". PULL BOX "B" SIMILAR EXCEPT REVERSED. SEE RISER ABOVE

Not to scale

Figure 16.16 Pull boxes (see chapter discussion). Pull boxes must be properly sized to prevent overcrowding the wires inside.

16.6. Motors and Motor Control

An extremely important part of electric power work is the layout of electric motors and their control. This subject is complicated and is covered in great detail in the NEC. The technologist is responsible for drawing the equipment. He must therefore know the sizes and functions of the items normally used.

a. Equipment

The NEC shows all the items of equipment normally associated with a motor. Let us review them briefly, with Figure 16.17.

1. *The motor feeder.* This connects the motor and its equipment to the power supply, usually at the panel.

2. *Motor feeder overcurrent protection.* This provides protection for the feeder and the controller.

3. *Motor disconnecting means.* This normally is a switch that serves to disconnect the motor *and* its controller.

4. *Motor branch circuit overcurrent protection.* This is usually the circuit breaker or fused switch part of a combination controller.

5. *Motor controller*—also called the starter. Its purpose is to connect the motor to the power. Usually a magnetic contactor.

6. *Motor running protection.* This is usually the overload (OL) contacts in the starter.

7. *Motor inherent protection.* This item is built into the motor and opens on overheat.

8. *Secondary controls.* These items apply only to wound rotor motors.

b. Ways of Showing Work

When doing motor work, there are four ways of showing the work, each of which has a special purpose.

1. The architectural-electrical drawing shows the position of all the equipment on the floor plan. See Figure 16.18. The symbols used on these architectural-electrical drawings are shown in Figure 16.19, which also comprises Part VIII of the master symbol list.

2. The equipment detail (see Figure 16.20) gives the contractor the construction details he needs, to enable him to build correctly. It shows actual detailed construction fittings—the nuts and bolts of the job. Details are required when anything

NEC DIAGRAM 430-1

General		Part *A*
Requirements for over 600 volts		Part *J*
Protection of live parts all voltages		Part *K*
Grounding		Part *L*
Tables		Part *M*

		To supply	
1.	Motor feeder		Part *B* Sec. 430–23 and 430–24 430–25 and 430–26
2.	Motor feeder		Part *E*
	Overcurrent protection		Part *E*
3.	Motor disconnecting means		Part *H*
4.	Motor branch–circuit Overcurrent protection		Part *D*
	Motor circuit conductor		Part *B*
5.	Motor controller		Part *G*
	Motor control circuits		Part *F*
6.	Motor running Overcurrent protection		Part *C*
7.	Motor		Part *A*
	Inherent protection		Part *C*
8.	Secondary controller Secondary conductors		Part *B* Sec. 430–23
	Secondary resistor		Sec. 430–23 and Art 470

Figure 16.17 NEC Chart of code sections relevant to motor circuitry. Reference is to the 1975 edition, Article 430.

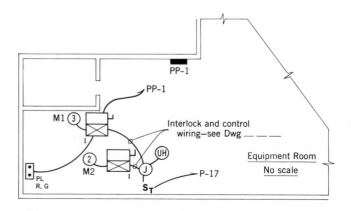

Figure 16.18 Architectural-electrical drawing for motor wiring of motor shown in Figure 16.21.

MOTORS AND MOTOR CONTROL

⊠ I MOTOR CONTROLLER, 3 POLE ACROSS—THE—LINE (ATL) UON, NEMA SIZE I, SEE SCHED., DWG. — — —

⊠ II COMBINATION TYPE MOTOR CONTROLLER; ATL STARTER PLUS FUSED DISCONNECT SWITCH, NEMA SIZE II, SEE SCHEDULE DWG. — — —

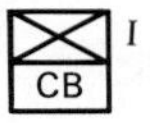

⊠ I CB COMBINATION TYPE MOTOR CONTROLLER; ATL STARTER PLUS CIRCUIT BREAKER, NEMA SIZE I, SEE SCHEDULE DWG. — — —

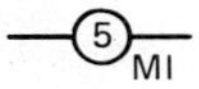

—(5)MI— MOTOR, 3φ SQUIRREL CAGE UON, MOTOR #1, 5 HP.

Ⓣ OR ⊡T DEVICE 'T', SEE LIST OF ABBREVIATIONS, SYMBOLS Part IX

S_T MANUAL MOTOR CONTROLLER WITH THERMAL ELEMENT.

PUSH BUTTON STATION — MOMENTARY CONTACT

PUSH BUTTON STATION — MAINTAINED CONTACT

△,◻,⦶ SYMBOL INDICATING LOCATION OF AN EQUIPMENT ITEM

ABBREVIATIONS

ATL ACROSS THE LINE STARTER — MAGNETIC
CATL COMBINATION ACROSS—THE—LINE—MAGNETIC STARTER
FS FUSED SWITCH
CB CIRCUIT BREAKER
RV REDUCED VOLTAGE
FV FULL VOLTAGE
SR STARTER RACK
MCC MOTOR CONTROL CENTER
S START BUTTON — MOMENTARY CONTACT
ST STOP BUTTON — MOMENTARY CONTACT
S/S PB START—STOP PUSH BUTTON
PL PILOT LIGHT; COLOR INDICATED BY LETTER; A — AMBER, G — GREEN
   B — BLUE R — RED Y — YELLOW
MER MECHANICAL EQUIP. ROOM
NO NORMALLY OPEN
NC NORMALLY CLOSED
LO LOCKOUT
R RELAY
UV UNDERVOLTAGE
OC OVERCURRENT
REV REVERSING

Figure 16.19 Architectural–electrical plan symbols, Part VIII; motors, and motor control. For other parts of the symbol list, see:

Part I, Raceways, Figure 12.6, page 304.
Part II, Outlets, Figure 12.37, page 338.
Part III, Wiring Devices, Figure 12.39, page 341.

Part IV, Abbreviations, Figure 12.43, page 345.
Part V, Single Line Diagrams, Figure 13.4, page 354.
Part VI, Equipment, Figure 13.14, page 369.
Part VII, Signalling Devices, Figure 15.34, page 500.
Part IX, Control and Wiring Diagrams, Figure 16.22, page 530.

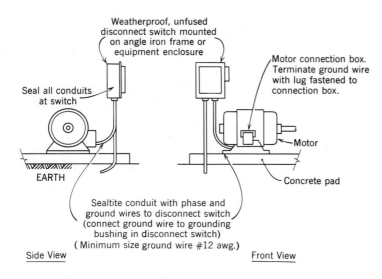

Weatherproof, unfused disconnect switch mounted on angle iron frame or equipment enclosure

Motor connection box. Terminate ground wire with lug fastened to connection box.

Seal all conduits at switch

Motor

EARTH

Concrete pad

Sealtite conduit with phase and ground wires to disconnect switch (connect ground wire to grounding bushing in disconnect switch) (Minimum size ground wire #12 awg.)

Side View

Front View

Figure 16.20 Detail of typical outdoor motor installation.

special or complex is required. Showing a detail will save much time on the job, and will avoid arguments.

3. Motor wiring diagram as in Figure 16.21*a* shows only the motor and the devices controlling it. Motor feeder and motor feeder overcurrent protection (items **a1** and **a2** above) are not shown. Everything is shown with full wiring, and all terminals are numbered. Power wiring is drawn heavy, control light. Symbols for this type of diagram are given in Figure 16.22, which also makes up Part IX of the master symbol list. This type of diagram is used in the field to do actual wiring.

4. The control or *ladder* diagram shows graphically the *control wiring only*. See Figure 16.21*b*. This diagram corresponds to the wiring diagram above except that power wiring is not shown. Terminals are numbered. This diagram, called a ladder because it looks like one, is used by the engineer to develop the control scheme. It is shown on the drawing because it is simple to read. The ladder diagram does not require tracing out of wires as does the wiring diagram. Symbols for it are also shown in Figure 16.22. On-the-job practice will give the technologist experience with all these forms.

c. Controllers

Motor controllers come in different sizes, depending on the horsepower and voltage of the motor. These controllers are available in a number of different type enclosures. Table 16.4 and Appendix A.29 provide data on enclosure types and some typical di-

Table 16.4 NEMA Enclosure—Designations

Type	Description	Application
1A	General Purpose dust resistant	Dry indoor locations.
2	Drip-proof	Indoor, subject to dripping
3R	Raintight, weatherproof	Exterior, vertical rain, sleet, and snow
4	Watertight	Driving rain and sleet
5	Dust tight	Nonhazardous dust-filled areas.
6	Submersible	Self-explanatory
7-11	Hazardous	Hazardous and corrosive areas
12	Industrial	Indoor dust, lint, oil, and moisture resistant, used in place of Type 1 in industrial interiors.

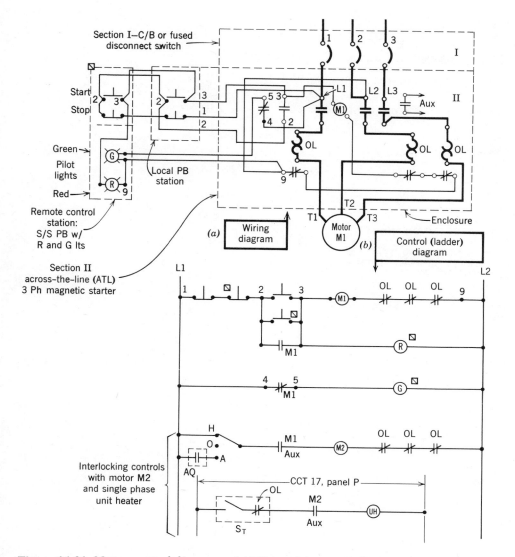

Figure 16.21 Motor control diagrams. *(a)* Wiring diagram showing equipment enclosures and actual connections. Shown are a combination circuit breaker type across-the-line magnetic starter, 3-phase, with integral start-stop momentary contact push-button station. Also shown are a remote s/s push button with red and green pilot lights. The actual "remote" location of this push button station is shown on the corresponding floor plan, Figure 16.18.

(b) Showing a control or "ladder" diagram. The upper section shows the same equipment as in *(a)* above. Note that terminals are numbered, for ease in reading the diagram and tracing the circuit. The lower portion of the ladder diagram shows interlocking (interconnection) with motors M2 and UH. Note that UH is a single-phase motor, connected to circuit 17, panel P, via motor control switch S$_T$, and is interlocked with M2. Refer to Figure 16.22 for symbols, and Figure 16.18 for the same equipment shown on an architectural-electrical plan.

CONTROL DIAGRAMS & WIRING DIAGRAMS

MOMENTARY CONTACT PUSH BUTTON – N.O. – ('START')

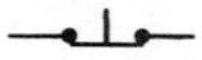

MOMENTARY CONTACT PUSH BUTTON – N.C. – (STOP)

MAINTAINED CONTACT START–STOP PUSH BUTTON ONE N.C. AND ONE N.O. CONTACT.

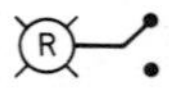

PILOT LIGHT, R–RED, G–GREEN, Y–YELLOW (SWITCH INDICATES PUSH–TO–TEST).

THERMAL OL ELEMENT WITH N.C., OL CONTACT

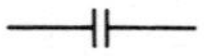

NORMALLY OPEN CONTACT – N O

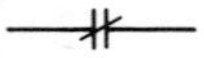

NORMALLY CLOSED CONTACT – N C

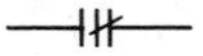

DOUBLE ACTION CONTACT; ONE N O AND ONE N C

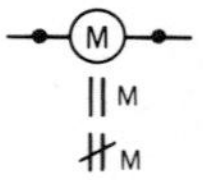

OPERATING COIL FOR RELAY OR OTHER MAGNETIC CONTROL DEVICE. WITH ONE N O AND ONE N C CONTACT. LETTERS NORMALLY USED ARE M, C, R FOR MOTOR, CONTROL COIL AND RELAY.

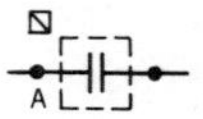

PILOT CONTROL DEVICE TYPE A, SEE LIST OF ABBREVIATIONS. ⊠ INDICATES REMOTE LOCATION.

POWER WIRING

CONTROL WIRING

WIRES CROSSING

WIRES CONNECTED

LIST OF ABBREVIATIONS

T	THERMOSTAT	MOM	MOMENTARY CONTACT
H	HUMIDISTAT	EP	ELECTRO–PNEUMATIC
SD	SMOKE DETECTOR	PE	PNEUMATIC–ELECTRIC
A,AQ	AQUASTAT	BG	BREAK–GLASS
R	RELAY	F, FL	FLOAT SWITCH
M	MOTOR	PS	PRESSURE SWITCH
MD	MOTORIZED DAMPER	H–O–A	HAND–OFF–AUTOMATIC SWITCH
PB	PUSH–BUTTON	LS, HS	LOW SPEED, HIGH SPEED
OL	OVERLOAD		

Figure 16.22 Architectural-electrical plan symbols, Part IX; control diagrams and wiring diagrams. For other of the symbol list, see:

Part I, Raceways, Figure 12.6, page 304.
Part II, Outlets, Figure 12.37, page 338.
Part III, Wiring Devices, Figure 12.39, page 341.

Part IV, Abbreviations, Figure 12.43, page 345.
Part V, Single Line Diagrams, Figure 13.4, page 354.
Part VI, Equipment, Figure 13.14, page 369.
Part VII, Signalling Devices, Figure 15.34, page 500.
Part VIII, Motors and Motor Control, Figure 16.19, page 527.

mensions. Note that the enclosures are called by NEMA numbers after the National Electrical Manufacturers Association, which sets these standards.

In spaces using a large number of motors, it is often good design to group the motor controllers in one location, instead of placing one next to each motor. It saves space, money and wiring and gives a central control point. Such an assembly is called a motor control center, often abbreviated MCC. A typical unit is illustrated in Figure 16.23. Each manufacturer publishes physical data on his unit. These data allow the technologist to assist in sizing and arranging an MCC once he gets a list of the required contents. An MCC can contain, in addition to controllers—switches, circuit breakers, and even whole panelboards. To assist in MCC layout some manufacturers make available blank work sheets. One is shown in Appendix A.14. Also shown in the Appendixes are a typical form for an MCC schedule (Appendix A.15) and a form for a job motor schedule (Appendix A.16). This last form is used to keep track

of all the motors on a job and the related control items. As such, it serves a very valuable purpose.

16.7. Guidelines for Layout and Circuitry

Sections 15.2 to 15.6 and Section 13.9 are devoted to a detailed discussion of the layout of spaces in residences. For nonresidential buildings the number of different types of rooms is so great that discussion of that kind is not practical. Instead, we include the following helpful suggestions.

a. Schools

Since schools contain many different types of room use, including those of classroom, lab, shop, office

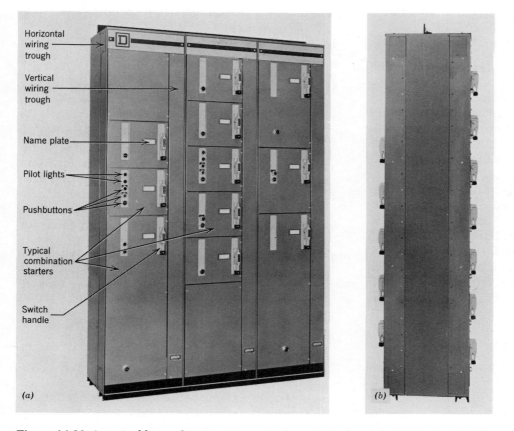

Figure 16.23 A typical low voltage motor control center is shown in (a). Back-to-back construction (b) is space saving, adding only 8 in. to the basic 14-in. depth. (Units with large starters are 20 in. deep for a one-sided unit.) All units are normally 90 in. high and 20 in. wide per section. (Courtesy of Square D Company)

and gym, we cannot generalize on guidelines, except for the following:

1. To handle the projectors often used in classrooms, provide a 20-amp outlet in front and back of each classroom. Wire these two to a 20-amp circuit. Other receptacles in classrooms can be 15 or 20 amp, wired no more than 6 on a 15-amp circuit or 8 on a 20-amp circuit. Each wall should have at least one outlet.

2. Light switching should provide:
   (a) High-low levels, to permit low-level lighting for film viewing. This can be done by the type of alternate ballast wiring and switching shown in Figure 15.1c.
   (b) Separate switching of the lights on the window side of the room. This side is often lighted sufficiently by daylight. Energy conservation demands that lights in daylighted areas be switched off.

3. Special outlets are supplied for all special equipment in labs, shops, cooking rooms, and the like.

4. Use heavy duty devices, key operated switches where the switches are exposed to students, and plastic instead of glass in fixtures. Also use vandal-proof equipment wherever possible. All panels *must* be locked and should be in locked closets.

5. Lighting should be fluorescent or high intensity discharge (HID). Incandescent, which is wasteful of energy, is acceptable where use is infrequent, such as in closets and inactive storage. Wherever possible use high-low switching, for energy conservation.

6. The NEC requires sufficient branch circuitry to provide a minimum of 3 w/sq ft for general lighting in schools. Refer to the NEC Article No. 220, the table entitled "General Lighting Load by Occupancies." This is a *minimum* figure. Remember that unlike residential occupancy this figure does *not* include receptacles. Receptacles are figured separately at 180 w each for ordinary convenience outlets.

7. Keep lighting and receptacles completely separate when circuiting.

b. Office Spaces

By office space we mean generally large open spaces with many desks. The smaller private office is special and must be treated individually. General suggestions for layout and circuitry follow.

1. In office spaces of less than 400 sq ft, provide either 1 receptacle outlet for every 40 sq ft or 1 receptacle outlet for every 10 linear ft of wall space, whichever is greater. In offices larger than 400 sq ft, provide 1 receptacle for every 100 to 125 sq ft, above the original 400 sq ft (10 receptacles).

2. Outlets must be provided in usable locations. A wall outlet is no use for an electric typewriter on a desk 10 ft from the wall. The outlet can be brought to the desk with a service pole (see Figures 12.29 to 12.31, page 330), or by using some sort of floor outlet. This decision is made by the job designer and may involve some sort of underfloor raceway system. These are discussed in Section 16.10a.

3. In view of the heavy business machine load found in offices, no more than six receptacles (frequently less) should be wired to a 20-amp circuit. In general, 20-amp branch circuits should be used. Devices should be 20-amp specification grade. This grade is high quality, and is intended for commercial use. In comparison, standard grade is used frequently for residences. This grade is lower quality than specification grade and is, obviously, cheaper.

4. Corridors should have a 20-amp, 120-v outlet every 50 ft, to supply cleaning and waxing machines.

5. Lighting is generally fluorescent. As in schools, switching should allow shutting off lights in spaces that receive enough daylight, for even part of the day. Lighting must be suitable for the work involved. In areas between work stations, lighting can be reduced to the level required for circulation, which is much lower. Combination low-level general lighting plus high-level supplementary lighting is good energy-conscious design.

6. As with all nonresidential buildings, convenience receptacles are figured at 180 w each.

7. Only specification grade equipment should be used.

c. Industrial Spaces

These areas are so specialized that no meaningful guidelines can be given. The technologist will receive from the engineers the material he needs to do the required layout. Familiarity with the motor, panel, feeder and transformer schedules found in the chapter discussions and Appendices will be a considerable help.

16.8. Classroom Layout

Our previous discussion and detailed analysis has concentrated on residential buildings. We have done this, as explained, because these buildings contain many different design and layout problems in a single, relatively small area. At this point, using the guidelines in the above section, we lay out the lighting and outlets in part of a school building. We chose classroom spaces because the reader is somewhat familiar with them from the discussion above and, in particular, from illustrative problems, 14.3 and 14.4. Also, once the concepts are clear, layout of other types of areas will be relatively straightforward. This layout opportunity will be provided in working out the problems at the end of the chapter.

Refer to Figure 16.24a. This is the basic, stripped-down architectural plan of a portion of a classroom building. Construction is concrete slab, and painted masonry block walls. The building has a clerestory, which produced two ceiling heights—a lower one of 11 ft 8 in. and an upper one of 15 ft 0 in. All these data are given on the stripped architectural. The problem is to lay out the lighting and receptacles, making reasonable choices of equipment. As you will see, we present several solutions to this problem, plus our recommendation. This is exactly the type of work a technologist is called on to do, and obviously should be capable of doing well.

Refer to Figure 16.24b. We will lay out the large room in the upper left first. Present recommendations call for a work area lighting level of about 50 fc, a well-lighted blackboard, low glare, and nonwasteful energy use. In a classroom, work is done throughout the area. Therefore, an overall, even lighting level should be provided. The first solution to be tried is one using fluorescent lighting. A direct-indirect, wrap-around plastic, pendant fixture provides low glare, good visibility and low shadowing. Fluorescent lamps are long-life, efficient, and economical. A mounting height of 9 ft AFF is high enough to minimize direct glare, yet low enough below the ceiling (2 ft 8 in.) to take full advantage of the up-light from the fixture.

Typical characteristics for this type of fluorescent fixture in such a room would be:

CU = 0.35, 0.33, 0.31 for two-, three-, or
          four-lamp fixtures, respectively
MF = 0.6
S/MH = 1.3 max.

We would select a 4 ft cool white RS lamp (3150 lm

from Table 14.4) and calculate for two-, three-, and four-lamp fixtures.

$$\text{number of fixtures} = \frac{\text{illumination} \times \text{area}}{\text{lumens per fixture}}$$

(two-lamp unit)

$$= \frac{50\,(20.67)\,(26.33)}{2\,(3150)\,(0.6)\,(0.35)} = 20.56 \text{ fixtures}$$

(three-lamp unit)

$$= \frac{50\,(20.67)\,(26.33)}{3\,(3150)\,(0.6)\,(0.33)} = 14.54 \text{ fixtures}$$

(four-lamp unit)

$$= \frac{50\,(20.67)\,(26.33)}{4\,(3150)\,(0.6)\,(0.31)} = 11.61 \text{ fixtures}$$

Note that the CU is highest for a two-lamp unit and drops for three- and four-lamp units. Depending on fixture design, the three-lamp may be more efficient (higher CU) than the four-lamp, as here, or vice versa. As an exercise, try to determine why this is true.

The next step is layout. The question to be answered immediately is whether we want the lights parallel to, or at right angles to, the blackboard. Most lighting experts suggest that the units be placed at right angles to the chalkboard to minimize both direct and reflected glare. This, however, has the disadvantage of reducing chalkboard lighting, frequently requiring supplementary lighting for this purpose.

Returning to our calculations, we have either

20 two-lamp units, or
14 three-lamp units, or
12 four-lamp units.

If we elect to run the fixtures parallel to the board (contrary to most recommendations), we have a maximum row length of 24 ft, or six fixtures, each 4 ft long. We would therefore use four-lamp units, in two rows of six fixtures maximum per row. Actually, six fixtures in a row is excessive, since lockers and the aisle at one end, and windows, radiation and the aisle at the other end, reduce the effective working area of the room. We would therefore use no more than five fixtures in a row, centered in the room. Rows 1 and 2 are this solution. Layout is based on the principle of row spacing, S, and side spacing, S/2, as shown in Figure 15.6, page 460.

Fixture spacing is

$$S = \frac{20 \text{ ft } 8 \text{ in.}}{2} = 10 \text{ ft } 4 \text{ in.}$$

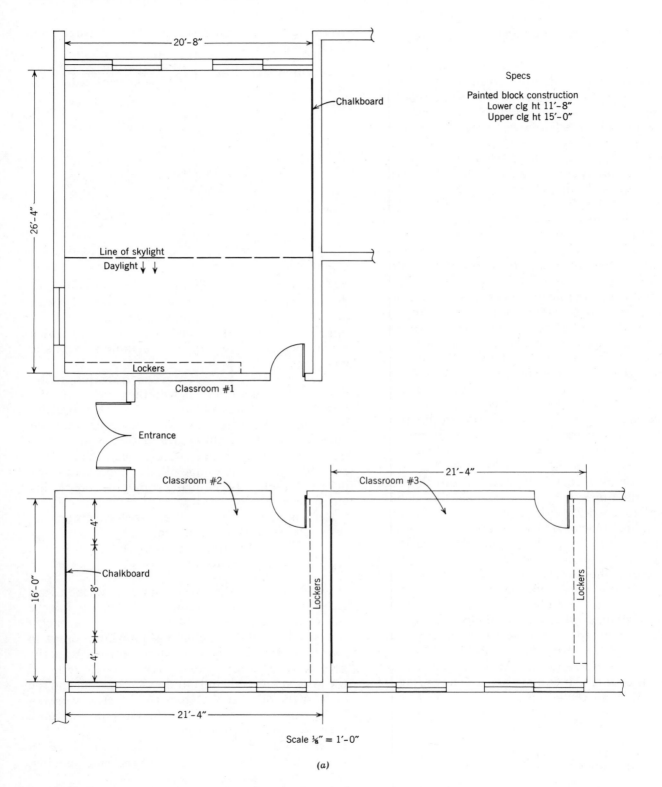

Specs

Painted block construction
Lower clg ht 11'–8"
Upper clg ht 15'–0"

Scale ⅛" = 1'–0"

(a)

Figure 16.24 Typical electrical layout for a portion of a school building. *(a)* shows the basic architectural plan, with the construction and dimensional details required. *(b)* shows the electrical layout, including several variations in classroom No. 1. For a full discussion, see Section 16.8.

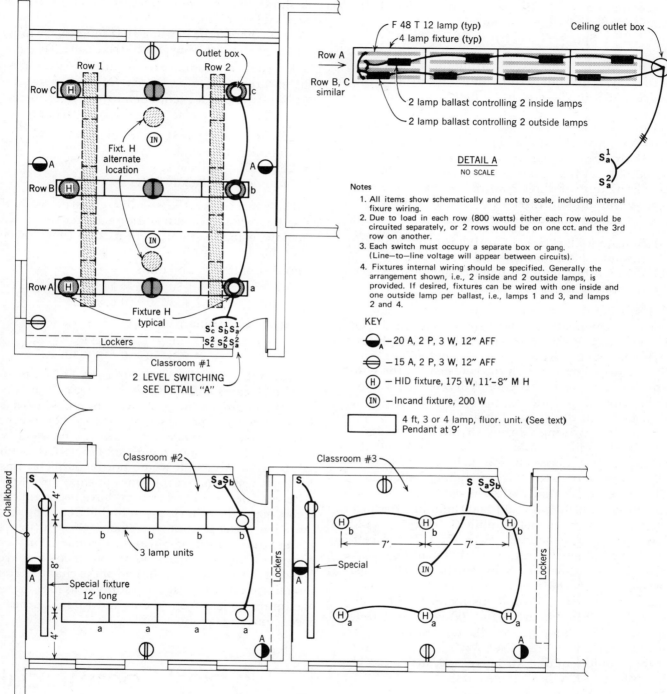

Row A
Row B, C similar

F 48 T 12 lamp (typ)
4 lamp fixture (typ)
Ceiling outlet box

2 lamp ballast controlling 2 inside lamps
2 lamp ballast controlling 2 outside lamps

S_a^1
S_a^2

DETAIL A
NO SCALE

Notes

1. All items show schematically and not to scale, including internal fixture wiring.
2. Due to load in each row (800 watts) either each row would be circuited separately, or 2 rows would be on one cct. and the 3rd row on another.
3. Each switch must occupy a separate box or gang.
   (Line—to—line voltage will appear between circuits).
4. Fixtures internal wiring should be specified. Generally the arrangement shown, i.e., 2 inside and 2 outside lamps, is provided. If desired, fixtures can be wired with one inside and one outside lamp per ballast, i.e., lamps 1 and 3, and lamps 2 and 4.

KEY

— 20 A, 2 P, 3 W, 12" AFF

— 15 A, 2 P, 3 W, 12" AFF

H — HID fixture, 175 W, 11'–8" M H

IN — Incand fixture, 200 W

4 ft, 3 or 4 lamp, fluor. unit. (See text)
Pendant at 9'

Row 1
Row 2
Outlet box

Row C H
Row B H
Row A H

Fixt. H alternate location
IN
IN

A
A

a
b
c

Fixture H typical

$S_c^1 S_b^1 S_a^1$
$S_c^2 S_b^2 S_a^2$

Lockers

Classroom #1
2 LEVEL SWITCHING
SEE DETAIL "A"

Chalkboard

Classroom #2

S
$S_a S_b$

4'
8'
4'

A

b   b   b   b
3 lamp units

Special fixture
12' long

a   a   a   a

A

Lockers

Classroom #3

S
S $S_a S_b$

A
Special

b   b   b
7'   7'

H   H   H
IN
H   H   H
a   a   a

Lockers

A

(b)

Since maximum S/MH is 1.3, and MH is 9 ft, maximum permissible spacing is

$$S_{max} = 1.3 \, (9 \, ft) = 11.7 \, ft$$

Spacing of 10 ft 4 in. is well within this total and is acceptable. The front fixture row is 5 ft 2 in. from the board, and provides adequate chalkboard illumination.

Running the fixture rows in the recommended direction, at right angles to the board, we would need three rows, since two rows would have an unacceptable 13 ft± spacing. We could use either 12 four-lamp units in three rows of four fixtures, or 15 three-lamp units in three rows of five fixtures. We chose the four-lamp units because

(a) Sufficient blackboard lighting is provided with the units 2 ft from the board.
(b) Two-level switching is much easier with an even number of lamps in a fixture.

This solution is shown as Rows A, B and C, and is our recommendation. Switching should be shown immediately. We have provided here both high-low switching and individual row switching. Ordinarily, we would only switch the row nearest the window. Here, however, because of the clerestory, daylight enters *both* sides of the room. Switching is provided accordingly (see Detail A).

One more lighting solution should be considered for an HID source. Note from Table 14.4 that the efficacy of the fluorescent source used is 68 lumens per watt (abbreviated lpw), including ballast loss. The table in Section 14.19 shows an efficacy for metal-halide of 70 lpw and 10,000 hour life. Because of glare problems, a metal-halide HID source would have to be ceiling mounted. Here, the 11 ft 8 in. ceiling is just about the minimum that could be used. Since the metal-halides 70 lpw efficacy is not much better than the 68 lpw of our fluorescent, only a much lower first cost would justify its use. Furthermore, the 70 lpw figure is for 400 w metal halide units. A 175 w lamp has an efficacy nearer 60 lpw, which is below that of a F40CWRS fluorescent, and its use would require a strong reason, such as much lower first cost. Assume these data:
Metal halide lamp, 175 w, 13,000 lm.

Fixture data:
MF = 0.7, CU = 0.4   S/MH = 1.5 max
Number of fixtures

$$= \frac{50 \times 20.67 \times 26.33}{12500 \times 0.7 \times 0.4} = 7.8 \quad or \; 8$$

Although a symmetrical plan requires 9 fixtures, an acceptable layout with 8 is possible. Both are shown on the same plan, as circles with an inscribed letter H.

As a further factor in using HID sources, we must remember slow starting and restrike time. To avoid blacking out the room, we would add two incandescent fixtures, shown as fixture IN. Summarizing, then, we would not recommend the use of metal-halide HID over fluorescent in this case, because

(a) Efficacy and life are both lower.
(b) Color is not as good.
(c) Low ceiling will cause glare.
(d) Incandescent backup units are required.

Now considering the smaller classrooms, and using the experience gained in the larger room, we calculate, using the fluorescent fixtures (note lower CU in smaller room):

number of three-lamp units

$$= \frac{50 \times 16 \times 21.33}{3 \times 3150 \times 0.6 \times 0.31} = 9.7$$

number of four-lamp units

$$= \frac{50 \times 16 \times 21.33}{4 \times 3150 \times 0.6 \times 0.30} = 7.52$$

By reducing the work area size to 16 by 17.5± due to the back lockers and the front aisle, we can readily use 8 three-lamp units, as shown. The chalkboard is illuminated by a special over-the-board fluorescent unit. Consult manufacturers' catalogs for details. Classroom No. 3 is shown with six 175 w HID metal halide units. Although we do not recommend this design, it is shown here for your information.

Receptacles in all three rooms have been laid out according to the criteria given in the preceding section. We suggest that you check these criteria against the layout to confirm its adequacy. Further exercises in layout of classrooms and office space are found in the problems at the end of the chapter. You should work them through carefully so as to gain the maximum benefit from them, and to gain experience in making layouts.

16.9. Typical Commercial Building

To illustrate some of the topics we have already discussed, and to show actual contract drawings, we have chosen a combination office—light industry building. We do *not* show bare floor plans, since that material should be clear by this point. Instead, we concentrate on the power aspects of the building.

Refer to the light and power riser diagram in Figure 16.25a. Follow the discussion with the drawing in front of you.

1. The power riser shows the *entire* electrical system, from the utility through the panels. That is, everything except branch circuits. All elements are shown as boxes, and each is labeled or identified. If a repetitive item occurs such as the 200/90A, 3PSN switch, it is shown once and labeled "Typical." Pull boxes are identified by number such as PB-1, PB-2, and the like. A schedule gives the sizes. (The schedule is not shown here.)

2. Notice how panels are identified. There are basically four types—lighting and power on normal service and lighting and power on emergency service. The code is simple:

Normal Service
L1–2	Lighting panel, first floor, panel no. 2
P2–3	Power panel, second floor, panel no. 3

Emergency Service
L3–EM	Lighting panel, third floor
LB–EM	Lighting panel, basement
PB–EM	Power panel, basement

Therefore, at a glance, we see that the third floor has two lighting panels in normal service (L3–1 and L3–2), one lighting panel on emergency power (L3–EM), and nine power panels on normal service (P3–1 to 9). Other items easily identified are the power panel feeding the elevator equipment P–EL on the roof level, and the roof level motor control center MCC–R.

In the basement we have:

Normal Service
Lighting panels	LB–1, 2, 3
Exterior lighting	LB–4, 5
Power panels	PB–1, 2, 3
Kitchen power	PB–K

Emergency Service
Lighting panel	LB–EM
Power panel	PB–EM

In addition, the basement contains two motor control centers (MCC–B1, B2) other special control panels (LC–1, LC–2) and the emergency generator plus its accessories.

3. The two main switchboards (MS–1 and MS–2) are shown with all their feeders. The switchboard contents are detailed in separate schedules. See Figure 16.25b, where part of the MS–1 schedule is reproduced. Notice that the schedule contains details on the switchboard mains, each individual circuit, what each circuit feeds and the wiring. This schedule makes it unnecessary to show feeder sizes on the riser. This keeps the riser from becoming cluttered.

4. Part of the data shown in block form on the riser are also shown in the system single-line diagram and notes (see Figure 16.25c and d). The one-line is shown complete to allow comparison with the riser. The notes are only given in part to show the technique of presentation.

An interesting fact appears in item 3 of the legend in Figure 16.24d. The designer has given the contractor the choice of using a 4000-amp bus duct or 11 sets of parallel 500 MCM cables. Compare this with our remarks about bus duct in Section 16.5.

Notice that the single-line diagram goes only as far as the switchboards and does not include the panels. There is no reason to repeat. On the contrary, it is bad practice to show the same information in two places since, when we make changes, one location can easily be overlooked. The single-line shows the service entrance, the metering (not shown on the riser) and the electrical layout of the switchboards. The amount of information to be shown on the single-line diagram is up to the designer. It might have been helpful to show the emergency generator arrangement here, but he chose not to. He relied on the riser for this information. The important thing is that the single line diagram shows the arrangement at a glance. The riser is more difficult to follow, and does not show the internal connections in its boxes.

5. Referring back to the riser diagram, Figure 16.25a, note that the utility company supply is 277/480 v. This accounts for the large number of transformers (T–1 to T–18) shown on the riser. They supply 120/208 v. A transformer schedule (Figure 16.25e) gives a partial listing of these transformers. The form of the schedule is useful to the technologist as a guide in future work.

6. Another item of interest on the riser is the building interior lighting control panel shown as located in Room B-24. This panel centrally controls all of the building's lighting for convenience and security. Duplicate local control is provided in some areas. The control wiring to permit this central control is shown (by note) running from each lighting panel (LP). The control is accomplished with large contactors inside each lighting panel that energize the entire lighting panel. The action is similar to the low voltage remote

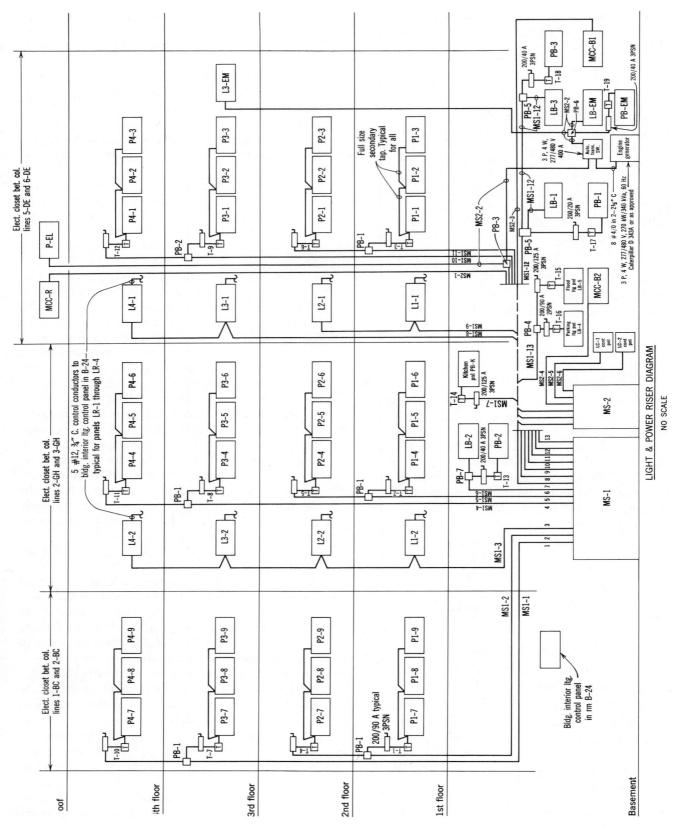

Figure 16.25 (a) Light and power riser diagram (no scale).

Main Switchboard Design	Feeder No.	Description of Loads	Load (kVA)	Circuit Breaker				Wiring				Remarks
		Item Served		Sym. Int. Cap. Min. (kiloamp)	Poles	Frame	Trip	Quan	AWG	Ground	Cond. Size	
	1	T-7 (RP3-7, 8, 9) T-10 (RP4-7, 8, 9)	150	100	3	600	200	4	3/0	6	2½ in.	
	2	T-1 (RP1-7, 8, 9) T-4 (RP2-7, 8, 9)	150	100	3	600	200	4	3/0	6	2½ in.	
MS1 3φ, 4 W. 277/480 v. 400 amp mains and neutral bus, 1350 amp Gnd. bus 4000/3500 amp Main c/b 100 kA I.C. min.	3	L1-2, L2-2, L3-2 L4-2	280	100	3	600	400	4	500 MCM	—	3½ in.	
	4	T-8 (RP3-4, 5, 6) T-11 (RP4-4, 5, 6)	150	100	3	600	200	4	3/0	6	2½ in.	
	5	T-2 (RP1-4, 5, 6) T-5 (RP2-4, 5, 6)	150	100	3	600	200	4	3/0	6	2½ in.	
	6	LPB-2, T-13 (RPB-B)	100	100	3	600	150	4	1/0	6	2½ in.	
	7	T-14 (P-KP)	112	100	3	600	150	4	1/0	6	2½ in.	
	8	L3-1, L4-1	250	100	3	600	300	4	350 MCM	—	3 in.	
	9	L1-1, L2-1	250	100	3	600	300	4	350 MCM	—	3 in.	
	10	T-9 (RP3-1, 2, 3) T-12 (RP4-1, 2, 3)	225	100	3	600	300	4	350 MCM	2	3 in.	

Figure 16.25 (b) Schedule of main distribution switchboards.

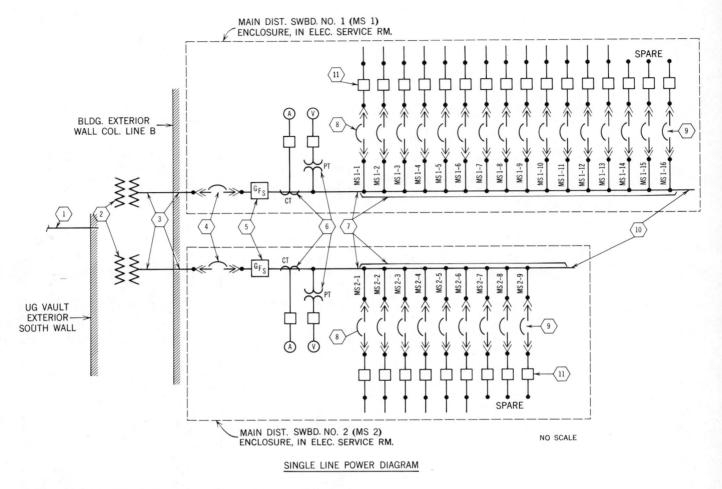

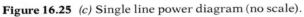

SINGLE LINE POWER DIAGRAM

Figure 16.25 *(c)* Single line power diagram (no scale).

(PARTIAL) LEGEND OF ELECTRICAL WORK AND EQUIPMENT REQUIREMENTS		
ITEM	DESCRIPTION	REMARKS
1	PRIMARY UG SERVICE; 2–5″ (1 ACTIVE, 1 SPARE) FIBER OR ASBESTOS CEMENT CONDUITS IN 3″ CONC ENVELOPE, (INCLUDING MANHOLE) PROVIDED UNDER ELEC SECTION OF WORK; HIGH VOLTAGE CONDUCTORS (INCLUDING CONNECTIONS) PROVIDED BY UTILITY CO.	
2	SERVICE TRANSFORMER WITH 3φ 4W 277/480V WYE GNDED NEUTRAL SECONDARY PROVIDED BY UTILITY CO.	
3	SERVICE FEEDER 3φ 4W 277/480V WYE FULL SIZE GNDED NEUTRAL, EACH CONSISTING OF A 4000 AMP FULLY RATED LOW IMPEDANCE COPPER VENTILATED BUS DUCT ASSEMBLY OR 15–3 1/2″ STEEL CONDUITS (11 ACTIVE, 4 SPARE) CONTAINING 11 SETS OF 4–500 MCM RHW INSUL. COPPER CONDUCTORS. CONDUCTORS AND CONNECTIONS PROVIDED UNDER ELEC. SECTION OF WORK.	
4	SERVICE MAIN CIRCUIT BREAKER 3P, 4000/3500A WITH SPECIFIED AUXILIARY EQUIPMENT.	SEE SCHEDULES FOR MAIN SWBDS MS1 AND MS2.
5	MAIN BUS GROUND FAULT SENSING EQUIPMENT	
6		

(d)

Figure 16.25 *(d)* (Partial) legend of electrical work and equipment requirements.

(PARTIAL) SCHEDULE OF DRY TYPE TRANSFORMERS							
XFMR NO.	KVA	PHASE	INPUT VOLTS	OUTPUT VOLTS	OUTPUT FEEDER	MTG. HT. BOTTOM AFF	REMARKS
T–1	75	3	480V Δ	120/208V Y	4 #4/0 + 1 #2 GND IN 2½″C	7′–6″	TAPS, PRIMARY WINDING, 4 OF 2–2½% ABOVE & 2–2½% BELOW RATED VOLTAGE.
T–2	75	3	480V Δ	120/208V Y	4 #4/0 + 1 #2 GND IN 2½″C	7′–6″	TAPS, PRIMARY WINDING, 4 OF 2–2½% ABOVE & 2–2½% BELOW RATED VOLTAGE.
T–3	112½	3	480V Δ	120/208V Y	4 #350 MCM + 1 #1 GND IN 3″C	7′–6″	TAPS, PRIMARY WINDING, 4 OF 2–2½% ABOVE & 2–2½% BELOW RATED VOLTAGE.
T–4	75	3	480V Δ	120/208V Y	4 #4/0 + 1 #2 GND IN 2½″C	7′–6″	TAPS, PRIMARY WINDING, 4 OF 2–2½% ABOVE & 2–2½% BELOW RATED VOLTAGE.
T–5	75	3	480V Δ	120/208V Y	4 #4/0 + 1 #2 GND IN 2½″C	7′–6″	TAPS, PRIMARY WINDING, 4 OF 2–2½% ABOVE & 2–2½% BELOW RATED VOLTAGE.
T–6	112½	3	480V Δ	120/208V Y	4 #350 MCM + 1 #1 GND IN 3″C	7′–6″	TAPS, PRIMARY WINDING, 4 OF 2–2½% ABOVE & 2–2½% BELOW RATED VOLTAGE.
T–9							

(e)

Figure 16.25 *(e)* (Partial) schedule of dry type transformers.

MECHANICAL SPACE LIGHTING AND RECEPTACLE REQUIREMENTS (PARTIAL SCHEDULE)								
SPACE NAME OR NUMBER	LTG. FIXT.		PANEL AND CKT. NO.	WIRING AND CONDUIT	LOCAL SW.	DUPLEX RECEPT.	PANEL AND CKT. NO.	WIRING AND COND.	REMARKS
	QUAN.	TYPE							
ELECTRIC SWITCHGEAR ROOM B–7	12	W	LB–EM	SEE PNL SCHEDULE	2 3–WAY	6	PB–1	SEE PNL SCHEDULE	
GENERATOR ROOM B–59	8	W	LB–EM	SEE PNL SCHEDULE	1	6	PB–3	SEE PNL SCHEDULE	
H.V.A.C. RM. B–6	12	W	LB–3	SEE PNL SCHEDULE	2 3–WAY	6	PB–1	SEE PNL SCHEDULE	3 FIXTS TO LPB–EM

(g)

Figure 16.25 *(g)* Mechanical space lighting and receptacle requirements. (partial schedule)

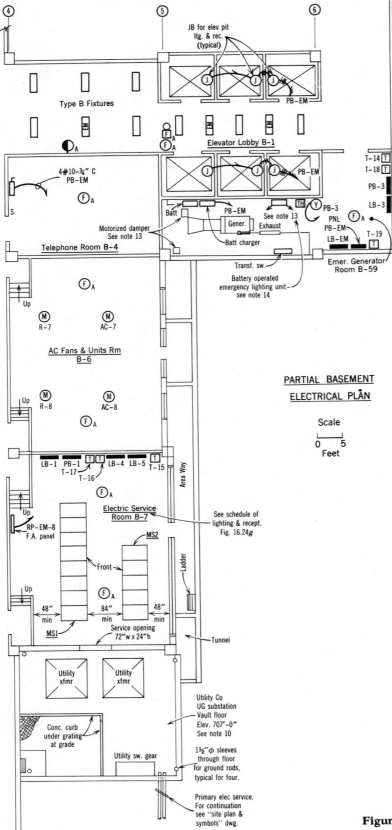

PARTIAL BASEMENT
ELECTRICAL PLAN

Scale

0   5
Feet

Figure 16.25 (f) Partial basement electrical plan.

control discussed in Section 15.4, only bigger. These contactors are normally called RC switches, an abbreviation for Remote Control switches.

7. A partial architectural-electrical floor plan of the basement is given in Figure 16.25f. This now familiar type plan shows the basement equipment in its physical arrangement. Note especially the following:

(a) The vault given to the utility company for its equipment is clearly shown. The work to be done by the building contractor is called out. All remaining work in the vault is by the utility.

(b) In the electric service room the minimum clearances between switchgear are shown. Also, the *front* of any freestanding item must be specified. Otherwise it may be put in backwards.

(c) The room containing the air conditioning fans and units simply shows the motor locations. All wiring, including control, is shown on the MCC schedule. (Although the room looks empty, a glance at the HVAC drawings would show that the space is filled with ductwork.)

(d) Lighting and receptacles for the mechanical and electrical spaces, B–7, B–6 and B–59, are not shown on the drawings. This is because the heavy ductwork, piping and conduit work on the ceilings make it extremely difficult to avoid space conflicts. Such conflicts end up as field changes and often mean extra cost. To avoid this the designer here has prepared a schedule of the lighting and receptacle requirements, to be *field* located. These are given in Figure 16.25g.

(e) Telephone room B–4 is reserved for telephone equipment. The 3-pole, 30 amp unfused switch supplies power to this equipment. Such an arrangement is standard.

(f) Junction boxes, complete with wiring, connected to an emergency service panel, are provided in each elevator pit. This, too, is standard procedure. In addition, it is normal practice to provide a similar junction box at the midpoint (in elevation) of each elevator shaft.

(g) The emergency lighting units throughout the building are easily identified by the letters "EM" on the fixture symbol. Two of these units are shown in the elevator lobby. The lighting in the switchgear and generator rooms is connected to the emergency lighting

panels. See Figure 16.25g. Also, a battery operated emergency lighting unit is provided in the generator room. This is done so that if a problem develops with the generator there will be some lighting in the room to permit servicing of the unit.

(h) It is customary to wire the stair lighting and exit lights on vertical risers in a multi-story building. This is not done for economy but rather to centralize control of these lights. They are fed from a basement stair-and-exit panel (SE panel). This panel is fed from the emergency service. Figure 16.26 shows the SE riser for the above building. Note that these risers are fed from emergency lighting panel LB–EM.

16.10. Special Topics

Some items that are involved in preparing electric drawings have not been considered up to this point. They are not of the same primary importance as the material already discussed. The technologist will find them in his work, and will learn to handle them properly through on-the-job learning. It helps, though, to be familiar with the terms involved. The following discussion is intended to do just that, since a complete coverage of these topics is not possible in the limited size and scope of the book.

a. Underfloor Duct and Other Special Raceway Systems

There are many types of underfloor raceway systems. They all have the same purpose—to make power and telephone wiring available at any location in the room. We discuss in detail in Section 12.8 another system for doing the same thing—Wiremold's Tele-Power Poles. Underfloor duct is different because the wiring is in a floor duct system. Two of these systems are illustrated in Figures 16.27 and 16.28 with captions that explain them. Layout of these systems is highly specialized because of the many different fittings and types of connections. As with so much other electrical equipment, the manufacturers publish complete data on the layout of their particular products.

A special case of underfloor raceway uses a cellular metal structure for ducts rather than burying ducts in the concrete. This is illustrated in Figure 16.29 and is commonly known as Q-Floor. Finally,

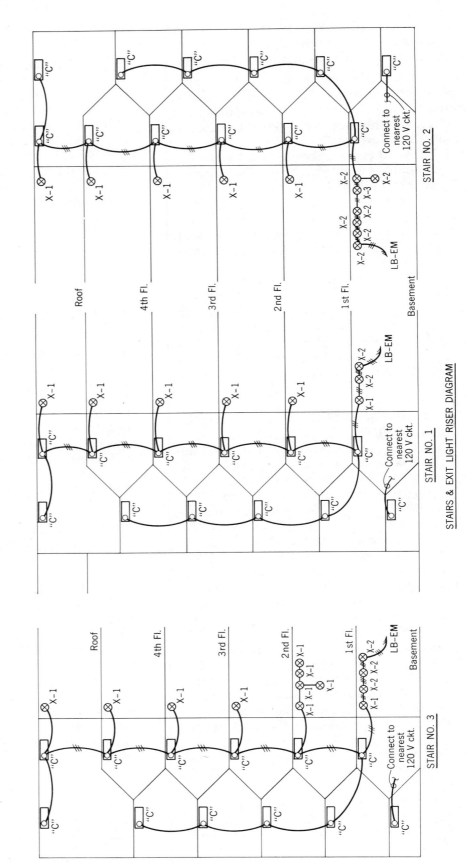

Figure 16.26 Stairs and exit light riser diagram.

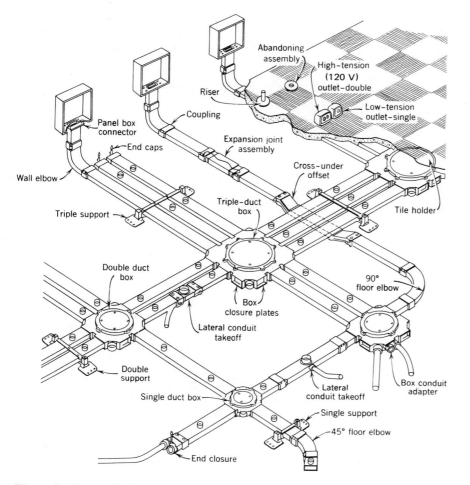

Figure 16.27 Single level underfloor duct systems require complex junction boxes to permit power and telephone wiring to cross without touching.

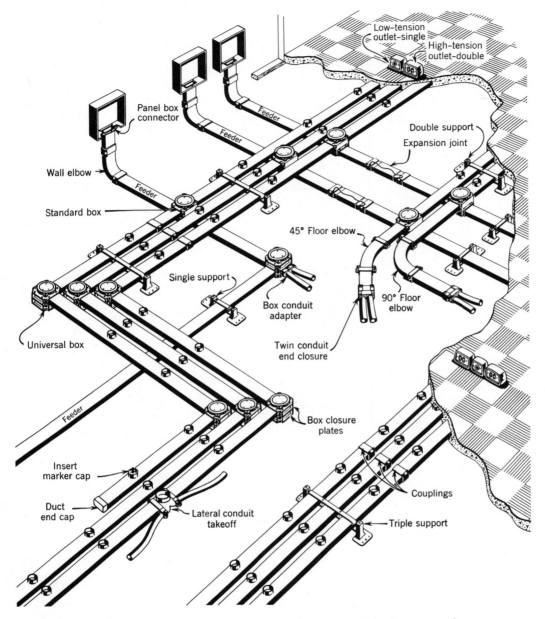

Figure 16.28 Two-level duct systems permit complete separation of power and telephone wiring. They never enter the same enclosure. Feeder ducts are placed below distribution ducts. (Courtesy of General Electric Company)

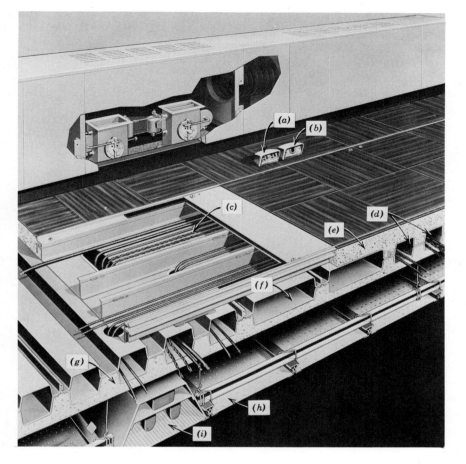

Figure 16.29 Section through a typical Q floor construction: *(a)* High tension (120 v) receptacle. *(b)* Low tension (signal, telephone) receptacle. *(c)* Trench header for wiring between cells and for main feeds. *(d)* Wiring cells. *(e)* The 2½ in. concrete fill. *(f)* Air cell. *(g)* Sprayed-on fireproofing. *(h)* Suspended ceiling. *(i)* Lighting fixture on ceiling below. (Courtesy of H. H. Robertson Company)

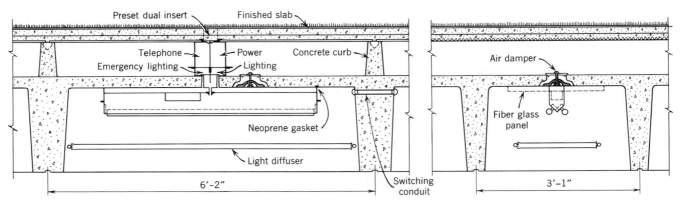

Floor/Ceiling Section above rescrambles the structural and mechanical systems in a new integrated design.

Figure 16.30 This outstanding design was developed for L'Enfant Plaza, Washington, D. C. It provides a 3 ft 1 in. by 6 ft 2 in. grid of floor outlets above, and eliminates duct work and hung ceiling below, while providing integrated lighting. (Courtesy of I.M. Pei and Partners, Architects, and Architectural and Engineering News)

architects favor the combining of the ceiling below with the floor, to form an integrated floor-ceiling system. One such design is shown in Figure 16.30.

b. Forms, Schedules and Details

In our discussion throughout the book we repeatedly refer the reader to detail drawings. Details of this kind are an essential part of the electrical drafting job. The job engineer relies heavily on the technologist to draw these details accurately and to scale. The details are as necessary as the floor plans because without these details the exact intent of the designer would not be clear. In addition to such details, we also provide the forms and schedules needed on most jobs. In addition we also include in the Appendixes the following useful information.

1. Appendixes A.9 and A.10, Dimensional Data for Recessed Fluorescent Luminaires.
2. Appendix A.11, Detail of Grounding System. This detail shows typical grounding and bonding requirements. The information shown is applica-

ble to most steel conduit systems.

3. Appendix A.12, Conduit and Cable Schedule Form. This is useful in large jobs where scheduling of conduit and cable is common.
4. Appendix A.13, Drawing Changes. This form is a way of keeping an accurate record of the time spent on making drawing changes because of "outside" causes. The usefulness of this form will become obvious on the first job on which it is used.

16.11. Conclusion

Our discussion of electrical work is concluded with the above. The material is presented in as direct and useful manner as possible. The technologist who has carefully followed all the preceding discussions and explanations is now in a position to fulfill a major role in the preparation of the HVAC, plumbing and electric drawings.

Problems

16.1. A residence has a load of 12 kw. What size service is required at
   (a) 120 v, single phase.
   (b) 120/240 v, single phase.
   (c) 120/208 v, 3 phase.
   Which service would you recommend? Why?

16.2. The above residence has added central air conditioning, including a 5 hp 240 v single phase compressor. What size service do you now recommend? What voltage?

16.3. In an office space of 2000 sq ft what is the minimum number of receptacles to be provided? How many 20-amp circuits would this require? How much load does this represent?

16.4. The owner of the residence in Problem 16.1 has converted his detached garage into a workshop. After renovation, his electric service comprises the following:
   (a) UG feeder: 3 No. 1/0 AWG, 120/240 v.
   (b) Two service switches in the house: a 150-amp unit for the house and a 100-amp for the garage
   (c) A feeder of 3 No. 4 and 1 No. 8 gnd to the garage, run UG in Type II fiber duct. The garage is 50 ft away from the house.
   (d) A 100 amp disconnect switch in the garage.

Draw a plot plan to scale and a riser diagram. Show all these elements and any other data you feel are necessary.

16.5. **A conduit run consists of these conduits:**
   2–2 in., 2–2½ in., 2–3 in., 2–3½ in.
   The conduits enter a pull box and make a right-angle turn in the same plane but in two directions. One half of the conduits (one of each size) turn left and the other half turn right. Size the pull box, and show how you would lay out the conduits to keep the pull box to a minimum size, while avoiding tangling of cables.

16.6. Select a public building in your neighborhood to which you can get access. The building should have at least three floors. Survey the building and from the data obtained prepare a power riser diagram, The riser should show the service equipment, distribution, switchboard (panel), all building panels and all riser feeders. Diagram should be as complete as possible with sizes, designations, and equipment location.

16.7. For the same building as in Problem 16.6 above, prepare a stair riser and an exit light riser.

Additional Reading

Heat Pumps and Electric Heating, E. R. Ambrose, John Wiley. This is a technical book on heat pumps and electric heating containing fundamentals, theory of operation, and application data. Intended for the reader who wants deeper level study.

Architectural Interior Systems: Lighting, Air Conditioning and Acoustics. J. E. Flynn and A. W. Segil, Van Nostrand Reinhold. This volume is a non-technical, architectural viewpoint study of the above subjects. It provides an understanding of principles.

The Architects Guide to Mechanical Systems, F. T. Andrews, Van Nostrand Reinhold. This book is somewhat more technical than the one above and is limited to HVAC and plumbing systems. One chapter, devoted to costs, may be of special usefulness. Contains many drawings and illustrations which will be of interest to the technologist.

The glossary at the back is also useful.

Mechanical and Electrical Equipment for Buildings
Steel Electric Raceways Design Manual
American Electricians' Handbook
National Electric Code
   All of these publications are discussed in Additional Reading Sections of preceding chapters.

Appendix Contents

A.1 Prepare for Metrication 553

A.2 Typical Porcelain Lampholder 562

A.3 Adjustable Lampholder 562

A.4 Open Prismatic Reflector Fixture 562

A.5 Pendant or Surface-Mounted Industrial Fluorescent Unit—One Lamp 563

A.6 Architectural Element Wall Lighting 564

A.7 Incandescent Wall-Lighting Fixture 565

A.8 Detail of Mounting Method for Building Face Floodlight 565

A.9 Dimensions and Details of Recessed Fluorescent Fixtures 566

A.10 Standard Troffer Details 567

A.11 Typical Requirements for Grounding of System Neutrals and Equipment 568

A.12 Conduit and Cable Schedule 568

A.13 Drawing Changes Record Form 569

A.14 Typical MCC Layout Sheet 570

A.15 Typical MCC Schedule 571

A.16 Typical Job Motor and Control Schedule 571

A.17 Typical Format for Single Phase, Circuit-Breaker Type, Lighting and Appliance Panel 572

A.18 Typical Format for 3-Phase, Circuit-Breaker Type, Lighting and Appliance Panel 573

A.19 Typical Format for 3-Phase, Switch and Fuse Type, Lighting and Appliance Panel 574

A.20 Panelboard Installation Methods 575

A.21 Typical Underground Service Details 576

A.22 Typical Handhole Detail, Giving Table of Dimensions Plus Appropriate Cover Number 578

A.23 Typical Pad-Mount Transformer Data 579

A.24 Typical Exterior Pull Box Detail 580

A.25 Typical Building Switchboard Dimensional Data 581

A.26 Typical Dimensional Data for Heavy Duty Safety Switches 582

A.27 Typical Dimensional Data for Circuit Breakers, Enclosed Type 583

A.28 Dimensional Data on Motor Controllers 584

Metrication is Coming: Prepare for the Transition

by Frank J. Versagi

It is inevitable that the United States will officially adopt the metric system of measurements. This series of articles is being published to help *News* readers prepare for the transition.

The International System of Units, abbreviated SI, for *Systeme Internationale,* is a modified version of the metric system, using modern fundamental discoveries to establish base units of measurement.

There are six, some sources list seven, fundamental physical quantities in SI:

length . . meter . . m
mass . . kilogram . . kg
time . . second . . s
electric current . . ampere . . A
thermodynamic temperature . . Kelvin . . K
luminous intensity . . candela . . cd
amount of substance . . mole . . mol

From these base units and a couple of supplementary units, are developed derived units, some with special names and several of specific interest to the construction industry and mechanical trades, to which we will get in a while.

Discussing things metric with persons unfamiliar with them, one realizes that an obstacle in the way of acceptance of metric measurement is that absence of a mental image and what one writer calls "recognition points." There are those, for example, who have no notion whether one meter is about as long as an inch or a mile.

It may be helpful, therefore, to pause long enough to attempt to establish a few mental images.

• A meter is equivalent to a yard (39 in. compared with 36 in.)

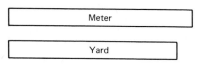

• A kilogram is a little over two pounds.

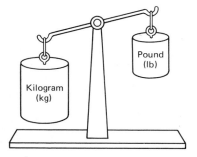

• A liter is just a bit larger than a quart (by the way, a kilogram is the mass of one liter of water).

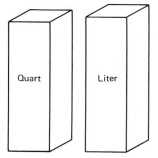

• A square meter is roughly equivalent to a square yard (11 sq ft compared with 9 sq ft).

An Australian publication suggests recognition points called the "10-11-12-13 relationship."

10 meters = 11 yards
10 sq meters = 12 square yards
10 cubic meters = 13 cubic yards

Before we talk about conversion factors and dimensions specifically for our industry, let's touch on a couple of attitudinal and philosophical matters.

Is there anything sacred about, say, a specification for an 8-ft ceiling? Not really. There is no scientific or technological reason for that exact dimension. In most instances, such figures themselves were chosen for numerical simplicity.

That being so, it makes little sense when thinking about going metric to convert 8 ft to 2,438.4 millimeters or even to round it off to 2,440 millimeters. In practice, as the construction industry converts to SI, it is likely that a ceiling height of 2,400 millimeters will substitute for 8 ft. "There are very few instances in the building industry where a high degree of precision is required, or indeed achievable," says an overseas writer.

The experience of other nations has shown that the fear that there will be a period when errors are frequent and serious proved unfounded.

If U.S. law follows the example of other countries, it will also develop that (1) the faster the conversion, the less the cost and inconvenience and (2) many of the feared costs and problems simply do not materialize. Countries which planned 10-year conversion programs have accomplished the conversion in six years. The Australian construction industry has set a 5-yr timetable.

Design professions will quickly experience some added costs:

• Procurement of new scales, instruments, reference publications, and the like.

• Some temporary reduction in productivity caused by unfamiliarity, double-checking, etc.

• Conversion of necessary data where metric data are not available.

• Revision of computer programs and substitution of conventional data with metric data.

Further, according to Australia's Metric Conversion Board, the construction industry may see:

• Some manufacturing equipment will require modification and in some cases complete replacement to produce materials in metric sizes. Plant and equipment which have a weighing function will need to be recalibrated, in some cases to the original units for which they were designed. "Modification or recalibration may be combined with normal maintenance."

• Building regulations, codes of practice, design manuals, drawing scales, will have to be replaced or modified to give metric dimensions or properties.

• Data or records for estimating, costing, marketing, and other purposes will have to be rewritten in metric terms. Computer programs will have to be amended or rewritten in metric terms.

• Technical and marketing support literature will need to be stated in metric terms. "At the same time, advantage may be taken of this opportunity to update and review such literature."

SI, metrication, is inevitable.

An advantage claimed for the metric system is that all dimensions can be expressed as decimal multipliers or submultipliers of the basic unit. "Decimal" means proceeding by 10's.

Compare, for example, the English:

12 inches to one foot
3 feet to one yard
5,280 feet to one mile
1,760 yards to one mile

In metric:

10 millimeters to a centimeter
100 centimeters to a meter
1,000 meters to a kilometer

Multiplying or dividing by 10, 100, 1,000, 10,000 calls for merely moving the decimal point. Thus:

1.75 kilometers = 1,750 meters
300 millimeters = 0.300 meter

Again, in English:

16 ounces to one pound
2,000 pounds to one ton

And in SI:

1,000 grams to a kilogram
1,000 kilograms to a tonne

Even the words used are simpler, at least more uniform, when using metric. Using standard prefixes with the base unit makes all dimensions easily comparable. Again, compare the English and metric terms for length:

inch . . . foot . . . yard . . . mile
millimeter . . . centimeter . . . meter . . . kilometer

Those prefixes—milli, centi, and kilo—apply to all metric units: millileter and kilogram are examples. Table 1 shows the standard multiples and submultiples with their magnitudes and symbols. It is important to use the symbols correctly, capitalized or not, because there are different and specific meanings for "M", and "m", "G", and "g", sometimes used as prefixes, sometimes as the symbol for a base unit or a derived unit.

These prefixes may be applied to all SI units

Multiples and submultiples	Prefixes	Symbols
1 000 000 000 000 = 10^{12}	tera	T
1 000 000 000 = 10^9	giga	G
1 000 000 = 10^6	mega	M*
1 000 = 10^3	kilo	k*
100 = 10^2	hecto	h
10 = 10	deka	da
0.1 = 10^{-1}	deci	d
0.01 = 10^{-2}	centi	c*
0.001 = 10^{-3}	milli	m*
0.000 001 = 10^{-6}	micro	μ*
0.000 000 001 = 10^{-9}	nano	n
0.000 000 000 001 = 10^{-12}	pico	p
0.000 000 000 000 001 = 10^{-15}	femto	f
0.000 000 000 000 000 001 = 10^{-18}	atto	a

*Most commonly used

Table 1

Special names for some derived units are already familiar: hertz for cycles-per-second and watt for joules-per-second are examples. For other derived units, unfamiliar special names will have to be learned: the newton as the unit of force (mass times length divided by time) is an example.

There are several general rules which apply to the use of metric units.

1. Unit symbols do not change in the plural. The designation for one meter is 1 m, for 570 meters is 570 m. For one millimeter is 1 mm, for 750 millimeters is 750 mm.
2. With one exception, all units and prefixes start with a small letter when written in full; the exception is "degree Celsius."
3. Where the prefix is combined with the unit name, the combination is written as one word: kilowatt, meganewton, milligram.
4. Prefix symbols should be written without spacing between the prefix symbol and the unit symbol: mm for millimeter, kW for kilowatt, MN for meganewton.

Here are the metric units with which the industries served by this newspaper will become involved (symbols in parentheses).

Linear measure . . . meter (m), millimeter (mm)
Area . . . square meter (m²), hectare (ha)
Weight . . . gram (g), kilogram (kg)
Volume . . . cubic meter (m³)
Capacity . . . liter (l)
Airflow . . . meter per second (m/s)
Volume flow . . . cubic meter per second (m³/s), liter per second (l/s)
Temperature . . . degree Celsius (°C)
Force . . . newton (N), kilonewton (kN)
Pressure . . . kilopascal (kPa)
Energy, work . . . kilojoule (kJ), megajoule (MJ)
Frequency . . . hertz (Hz)
Power . . . watt (W), kilowatt (kW)
Electric current . . . ampere (A)
Electric potential . . . volt (V), kilovolt (kV)
Electrical resistance . . . ohm (Ω)

A meter is slightly longer than a yard:
meters × 1.094 = yards
yards × 0.914 = meters

Using rounded conversion factors of 1.1 and 0.9, here are some mental images.

A 100-yard football field is 90 m long.

A 1,000 meter, or kilometer, run (1,000 m or 1 km) = 1,100 yards. (The reason that sprints in track competition are 110, 220, and 440 yards is that those English distances are the nearest to 100 meters, 200 meters, and 400 meters.)

A foot is about one-third of a meter.
feet × 0.305 = meters

An 8-ft ceiling, therefore, is roughly 2.4 m high. (Instead of speaking of "7-ft, 11¾-in.," one might refer to "2,440 millimeters," for greater precision.)

A residential lot 75 feet wide and 200 feet deep would, in metric, be 22.5 m wide and 60 m deep.

The nearest equivalent to 1 inch is 25 millimeters (25 mm); the nearest equivalent to 1 foot is 300 mm.
inches × 25.4 = millimeters
inches × 2.54 = centimeters

An 8-in. diameter duct, is, in metric, a 203 mm diameter duct; a 6-in. duct = 152 mm.

A 36-in belt is expressed in metric as 91 cm or 914 mm.

On working drawings, preference is to use only meters and millimeters, not centimeters. Then, dimensions can be used without needing a unit symbol (m, cm, or mm), so long as all millimeter dimensions are expressed in whole numbers, like 3,600, and all meter dimensions are carried to three decimal places, like 3.600.

(Not all countries have agreed on (1) the use or not of a puncutation mark in large numbers—like 3600 or 3,600—and (2) whether the comma or period will be used to express decimal fraction—like 0.462 km or 0,462 km.)

A square yard contains 9 square feet; a square meter (m²) contains about 11 ft².
m² × 10.77 = ft²
m² × 1.196 = yd²
yd² × 0.836 = m²
ft² × 0.093 = m²
in.² × 645.16 = mm²

A 100,000 ft² warehouse has an area of 9,300 m².

A 5,000 yd² field contains 4,180 m².

A 300 ft² window equals 28 m².

A television screen with a 200 in.² area has 129,000 mm², or 0.129 m².

An ounce, avoirdupois, contains about 28 grams and a pound contains about 450 grams.
Ounces × 28.35 = grams (g)
Pounds × 0.454 = kilograms (kg)
g × 0.035 = ounces
kg × 2.204 = pounds

An English ton, 2,000 pounds, is equivalent to 907 kilograms; the metric tonne contains 1,000 kg or 2,205 pounds. The seldom-used "long" ton contains 2,240 pounds, 1,016 kg.

Becoming accustomed to metric weights won't require as much of an adjustment as becoming familiar with some other metric units for two reasons: (1) most persons don't think much about

Common Equivalents and Conversions

Approximate common equivalents		*Conversions accurate to parts per million*	
1 inch	= 25 millimeters	inches × 25.4*	= millimeters
1 foot	= 0.3 meter	feet × 0.3048*	= meters
1 yard	= 0.9 meter	yards × 0.9144*	= meters
1 mile	= 1.6 kilometers	miles × 1.609 34	= kilometers
1 square inch	= 6.5 sq centimeters	square inches × 6.4516*	= sq centimeters
1 square foot	= 0.09 sq meter	square feet × 0.092 903 0	= sq meters
1 square yard	= 0.8 sq meter	square yards × 0.836 127	= sq meters
1 acre	= 0.4 hectare†	acres × 0.404 686	= hectares
1 cubic inch	= 16 cu centimeters	cubic inches × 16.3871	= cu centimeters
1 cubic foot	= 0.03 cubic meter	cubic feet × 0.028 316 8	= cu meters
1 cubic yard	= 0.8 cubic meter	cubic yards × 0.764 555	= cu meters
1 quart	= 1 liter†	quarts (liquid) × 0.946 353	= liters
1 gallon	= 0.004 cubic meter	gallons × 0.003 785 41	= cu meters
1 ounce (avdp)	= 28 grams	ounces (avdp) × 28.3495	= grams
1 pound (avdp)	= 0.45 kilogram	pounds (avdp) × 0.453 592	= kilograms
1 horsepower	= 0.75 kilowatt	horsepower × 0.745 700	= kilowatts
1 millimeter	= 0.04 inch	millimeters × 0.039 370 1	= inches
1 meter	= 3.3 feet	meters × 3.280 84	= feet
1 meter	= 1.1 yards	meters × 1.093 61	= yards
1 kilometer	= 0.6 mile	kilometers × 0.621 371	= miles
1 sq centimeter	= 0.16 sq inch	sq centimeters × 0.155 000	= sq inches
1 sq meter	= 11 sq feet	sq meters × 10.7639	= sq ft
1 sq meter	= 1.2 sq yards	sq meters × 1.195 99	= sq yards
1 hectare†	= 2.5 acres	hectares × 2.471 05	= sq acres
1 cu centimeter	= 0.06 cu inch	cu centimeters × 0.061 023 7	= cu inches
1 cu meter	= 35 cu feet	cu meters × 35.3147	= cu ft
1 cu meter	= 1.3 cu yards	cu meters × 1.307 95	= cu yards
1 liter†	= 1 quart	liters × 1.057	= quarts (liquid)
1 cu meter	= 250 gallons	cu meters × 264.172	= gallons
1 gram	= 0.035 ounces (avdp)	grams × 0.035 274 0	= ounces (avdp)
1 kilogram	= 2.2 pounds (avdp)	kilograms × 2.204 62	= pounds (avdp)
1 kilowatt	= 1.3 horsepower	kilowatts × 1.341 02	= horsepower

† *common term not used in SI*   * *exact*

fractional pounds when they read "12 oz" or "5 oz"; (2) many commodities, especially foods, already contain the weight in grams as well as in ounces.

The cubic meter is to the metric system what the cubic foot is to the English measurements of volume. A cubic meter (m³) contains over 35 ft³ and about one and one-third cubic yards.

$$m^3 \times 1.307 = \text{cubic yards}$$
$$m^3 \times 35.31 = \text{cubic feet}$$
$$yd^3 \times 0.765 = m^3$$
$$ft^3 \times 0.283 = m^3$$

A 10-by-12 room, 13 feet high, contains 10 × 12 × 13 = 1,560 cubic feet. 1,560 × 0.283 = 441 m³.

A concrete floor calling for 5,500 cubic yards of cement: 5,500 × 0.765 = 4,208 m³.

For measuring capacity, as distinct from volume, the liter (l) is the customary unit. (The liter is slightly larger than a quart.) By international agreement the liter is considered to be one-thousandth of a cubic meter. The liter is subdivided into milliliters (ml) but is also spoken of as containing 1,000 cubic centimeters (cc). For practical purposes,

ml = cc
liters × 0.220 = gallons
l × 1.760 = pints
ml × 0.035 = fluid ounces

Because a liter is one-thousandth of a cubic meter, and cubic meter-times-35.31 equals cubic feet, multiply liters by 0.03531 to get cubic feet. And milliliters × 0.610 = cubic inches.

When we work with units of volume and capacity to derive units for **volume rate of flow,** we encounter the phenomenon that the unit of time in the metric system is the second, followed by the hour and the day. **The minute is almost never used.**

Thus, flow rates are expressed in cubic meters per

second (m³/s), liters per second (l/s), or cubic meters per hour (m³/h), instead of cubic meters per minute or liters per minute.

cubic feet per minute $\times$ 0.472 = liters per second
gallons per minute $\times$ 0.0758 = l/s
gallons per hour $\times$ 0.0013 = l/s
m³/s $\times$ 35.31 = ft³/s
l/s $\times$ 13.20 = gal/min
l/s $\times$ 791.9 = gal/h

A design condition that calls for 600 cfm on cooling and 450 cfm on heating:
600 $\times$ 0.472 = 280 l/s
450 $\times$ 0.472 = 212 l/s

That comes from conversion. When the design is performed in metric, it is likely that the airflow would be specified at 300 l/s on cooling and 200 l/s on heating.

Simple velocity of air is expressed as meters per second (m/s) compared to the English feet per minute (fpm).
fpm $\times$ 0.00508 = m/s

So a terminal velocity of 75 fpm equals 0.4 m/s.

Although road speed is expressed in kilometers per hour (km/h), the correct SI unit is m/s.
miles per hour $\times$ 1.609 = km/h
km/h $\times$ 0.621 = miles per hour

So 70 mph = 113 km/h.
And 70 km/h = 43 mph.

The extent of American involvement with temperature conversions, till now, has been from Fahrenheit (°F) to Centigrade (°C). That stays the same in converting to metric, except that the temperature scale is called "degrees Celsius." The term "centigrade" means "by hundreds," just as the term "decimal" means "by tens," and refers to the arbitrary establishment by Celsius of a 100° differential between the freezing and boiling points of water.

Water freezes at 0° C and 32° F; boils at 100° C, 212° F. Conversion is normally accomplished with a chart, but should calculations be required:

$$°C = \frac{(°F - 32)}{1.8}$$

Example:

$$\frac{(212° F - 32)}{1.8} = \frac{180}{1.8} = 100° C$$

°F = 1.8 $\times$ °C + 32

Example:

100° C = 1.8 $\times$ 100 + 32 = 180 + 32 = 212° F

An interesting sidelight is that the two scales cross at −40°.

°F = 1.8 $\times$ °C + 32
= 1.8 $\times$ (−40°) + 32
= −72 + 32 = −40° F

In science, though not often in engineering, one will encounter the "absolute" temperature scale, called "degrees Kelvin." Absolute zero, Kelvin, is −273.15° C or −459.67° F.

Water freezes at:

32° F
0° C
270° K

Water boils at:

212° F
100° C
373° K

When **temperature interval** is mentioned, one degree Celsius equals 1.8 degree Fahrenheit. When one says, for example, that a 14° C differential exists between two objects, the Fahrenheit temperature difference is 1.8 $\times$ 14, or 25°; NOT 1.8 $\times$ 14 + 32, or 57°.

When **heat flow rate** is discussed, there is still controversy because some want to retain traditional metric dimensions and not convert to SI standards. Purists would have all present dimensions, English and traditional metric, dropped in favor of watts (W) whether talking heating or cooling.

Gone would be not only British thermal units (Btu) and therms (100,000 Btu), but also kilocalories (kcal).

watts $\times$ 3.412 = Btuh
kW $\times$ 1.341 = horsepower
Btuh $\times$ 0.293 = watts
horsepower $\times$ 0.746 = kW

So, an 80,000-Btuh furnace would be rated: 80,000 $\times$ 0.293 = 23,440 W, or 23.44 kW.

And a 3-ton, 36,000-Btuh air conditioner: 36,000 $\times$ 0.293 = 10,548 W or 10.55 kW.

The relationships among work, energy, and heat are expressed in the following units:
megajoule (MJ) $\times$ 0.278 = kilowatthours (kWh)
kilojoules (kJ) $\times$ 0.948 = Btu
joules (J) $\times$ 0.7376 = foot-pounds.

There is argument, too, over **whether the newton (N) and/or kilopascal (kPa) will replace more familiar units** in expressions of force, inertia, and pressure. The pertinent conversions:

pounds per square inch (psi) $\times$ 0.068 = atmospheres

psi $\times$ 0.073 = kilograms per square centimeter (kg/cm²)

psi $\times$ 703 = kg/m²
psi $\times$ 6.895 = kPa
pounds per square foot $\times$ 47.88 = pascal (Pa)
kPa $\times$ 20.89 = pounds per square foot

Metric measurement can't be, needn't be, swallowed at one gulp. Most of us have need only for a few of the scores of units and learning those few isn't that difficult.

For reference, use the accompanying alphabetical list of most of the conversions which will be encountered in the refrigeration and hvac industries.

Ultimately, remember, we won't be converting to metric; we'll be designing in metric and we'll have forgotten what a Btu was.

Many companies, agencies, educational institutions, and associations are issuing metric-English conversion charts or booklets for specialized purposes, and many more will be doing so. Unfortunately, some early attempts in the United States contained errors, either in mathematics or in terminology. The best U.S. document we have so far encountered is the "ASTM Standard Metric Practice Guide," published by the American Society for Testing and Materials.

The International Organization for Standardization (ISO) has published "ISO Recommendation R786, Units and Symbols for Refrigeration."

The District Heating Association, of London, England, has published specialized metric information in its 1970/71 handbook.

Heating & Ventilating Publications Ltd., also of London, England, has published "Processed Calculations for Heating System Designs in SI Metric."

The Standards Association of Australia has published a metric handbook, "Metric Conversion in Building and Construction."

Scientific Notation

A majority of the public does not know the meaning of a term like 9.463×10^{-4}, so we have avoided using that format. Instead, the number is listed as 0.0009463.

That's the clue. The exponent (superior number) above the 10 tells one how many places to move the decimal point. If the exponent is preceded by a minus-sign, the decimal point is moved to the left. Examples:

3.62×10^{3} = 3,620
3.62×10^{-3} = 0.00362

Typical abbreviations

acre	no authorized abbreviation
atmospheres	atm
British thermal units	Btu
British thermal units per hour	Btuh
cubic feet	ft³
cubic feet per minute	ft³/min
cubic feet per second	ft³/s
cubic inches	in³
cubic meters	m³
cubic millimeters	mm³
cubic yards	yd³
feet	ft
feet of water	ft H₂O
feet per second	ft/s
foot-pounds of force	lbf/ft
gallons	gal
gallons per hour	gal/h
gallons per minute	gal/min
grams	g
grams per square meter	g/m²
hectares	ha
horsepower	hp
inches	in
inches of mercury	in Hg
inches of water	in H₂O
joules	J
kilocalories	kcal
kilograms	kg
kilograms per cubic meter	kg/m³
kilograms per second	kg/s
kilograms per square meter	kg/m²
kilojoules	kJ
kilojoules per cubic meter	kJ/m³
kilojoules per kilogram	kJ/kg
kilometers	km
kilometers per hour	km/h
kilonewtons	kN
kilopascals	kPa
kilowatts	kW
kilowatt-hours	kWh
liters	l
liters per second	l/s
liters per minute	l/min
megajoules	MJ
meganewtons	MN
megapascals	MPa
meters	m
meters per second	m/s
miles	no abbreviation in metric
miles per hour	mile/h
millimeters	mm

millimeters of mercury .... mm Hg
newtons .... N
ounces .... oz
ounces per square foot .... oz/ft²
pounds .... lb
pounds of force .... lbf
pounds of force per square foot .... lbf/ft²
pounds per cubic foot .... lb/ft³
pounds per second .... lb/s

square feet .... ft²
square inches .... in²
square kilometers .... km²
square meters .... m²
square miles .... mile²
square millimeters .... mm²
watts .... W
watts per square meter .... W/m²
yards .... yd

Useful conversion factors: alphabetized

Multiply	*by*	*to get*
acres	0.4047	hectares
acres	4,047	square meters
atmospheres	33.93	feet of water
atmospheres	29.92	inches of mercury
atmospheres	760.0	millimeters of mercury
atmospheres	1.058	tons per square foot
British thermal units	1,055	joules
British thermal units	0.2520	kilocalories
British thermal units	1.055	kilojoules
British thermal units per hour	0.2929	watts
British thermal units per pound	2.326	kilojoules per kilogram
cubic feet	0.02832	cubic meters
cubic feet	7.481	gallons
cubic feet	28.32	liters
cubic feet	29.92	quarts
cubic feet per minute	0.4719	liters per second
cubic feet per second	0.02832	cubic meters per second
cubic inches	16.39	cubic centimeters
cubic inches	16,387	cubic millimeters
cubic meters	35.32	cubic feet
cubic meters	1.308	cubic yards
cubic millimeters	0.00006102 or (6.102×10^{-5})	cubic inches
cubic yards	0.7646	cubic meters
feet	0.3048	meters
feet	304.8	millimeters
feet per second	0.3048	meters per second
foot-pounds of force	1.356	joules
foot-pounds of force per second	1.356	watts
gallons (liquid)	0.003785	cubic meters
gallons	3.785	liters
gallons per hour	0.001052	liters per second
gallons per minute	0.002228	cubic feet per second
gallons per minute	0.06308	liters per second
grams	0.03527	ounces (avoirdupois)
grams per square meter	0.003278	ounces per square foot
grams per square meter	0.02949	ounces per square yard
hectares	2.471	acres
horsepower	0.7460	kilowatts

Multiply	by	to get
horsepower	746	watts
inches	25.4	millimeters
inches of mercury	0.03342	atmospheres
inches of mercury	1.133	feet of water
inches or mercury	345.3	kilograms per square meter
inches of mercury (60° F)	3,377	newtons per square meter
inches of mercury	0.4912	pounds per square inch
inches of water	0.002458	atmospheres
inches of water	0.07355	inches of mercury
inches of water	25.40	kilograms per square meter
inches of water	0.03613	pounds per square inch
inches of water (60° F)	248.8	newtons per square meter
joules	0.7376	foot-pounds of force
kilocalories	3.968	British thermal units
kilocalories	4,190	joules
kilograms	2.205	pounds
kilograms per cubic meter	0.06243	pounds per cubic foot
kilograms per cubic meter	1.686	pounds per cubic yard
kilograms per second	2.205	pounds per second
kilograms per square meter	0.00009678	atmospheres
kilograms per square meter	0.003281	feet of water
kilograms per square meter	0.002896	inches of mercury
kilograms per square meter	0.2048	pounds per square foot
kilograms per square meter	0.001422	pounds per square inch
kilojoules	0.9478	British thermal units
kilojoules per cubic meter	0.02684	British thermal units per cubic foot
kilojoules per kilogram	0.4299	British thermal units per pound
kilometers	0.6214	miles
kilometers per hour	0.6214	miles per hour
kilonewtons	0.10036	tons of force
kilonewtons	224.8	pounds of force
kilopascals	20.89	pounds of force per square foot
kilowatts	1.341	horsepower
kilowatt-hours	3.6	megajoules
liters	0.03532	cubic feet
liters	61.02	cubic inches
liters	0.2642	gallons
liters	2.113	pints
liters	1.057	quarts
liters per minute	0.0005886	cubic feet per second
liters per second	2.119	cubic feet per minute
liters per second	951.0	gallons per hour
liters per second	15.85	gallons per minute
megajoules	0.2778	kilowatt-hours
meganewtons	100.36	tons of force
megapascals	145.04	pounds of force per square inch
megapascals	9.324	tons of force per square foot
megapascals	0.06475	tons of force per square inch
meters	3.281	feet
meters	1.094	yards
meters per second	2.237	miles per hour
miles	1.609	kilometers

Multiply	*by*	*to get*
miles per hour	1.609	kilometers per hour
miles per hour	0.4470	meters per second
milliliters	0.06102	cubic inches
milliliters	0.03520	fluid ounces
millimeters	0.0394	inches
millimeters of mercury	133.3	newtons per square meter
million gallons per day	0.005262	cubic meters per second
newtons	0.2248	pounds of force
ounces (avoirdupois)	28.35	grams
ounces (fluid)	28.41	milliliters
ounces per square foot	305.15	grams per square meter
ounces per square yard	33.91	grams per square meter
pounds	0.4535	kilograms
pounds of force	4.448	newtons
pounds of force per square foot	47.88	pascals
pounds of force per square inch	6.895	kilopascals
pounds per cubic foot	16.02	kilograms per cubic meter
pounds per cubic yard	0.5933	kilograms per cubic meter
pounds per second	0.4535	kilograms per second
pounds per square foot	4.882	kilograms per square meter
quarts	0.0009463	cubic meters
square feet	0.0929	square meters
square inches	645.2	square millimeters
square kilometers	0.3861	square miles
square meters	10.76	square feet
square meters	1.196	square yards
square miles	2.590	square kilometers
square millimeters	0.00155	square inches
square yards	0.8361	square meters
tons of force	9.964	kilonewtons
tons of force per square foot	107.25	kilopascals
tons of force per square inch	15.44	megapascals
torr (millimeters of mercury at 0° C)	133.3	newtons per square meter
watts	3.412	British thermal units per hour
watts	0.7376	foot-pounds of force per second
watts per square meter	0.3170	British thermal units per square foot
yards	0.9144	meters

Add your own conversion factors here.

Reprinted from

AIR CONDITIONING
HEATING & REFRIGERATION *NEWS*

P.O. Box 6000, Birmingham, Mich. 48012

Copyright 1973

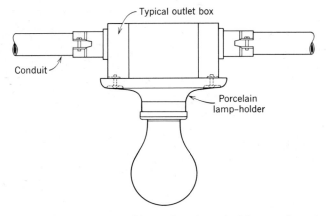

Appendix A.2 Typical porcelain lampholder, surface mounting. Outlet box keyless lamp receptacle to be of porcelain, rated not less than 660 w, and provided with a shadeholder groove. The device should be suitable in size for the outlet box on which it is to be mounted.

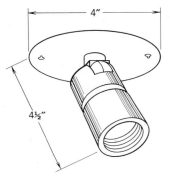

Appendix A.3 Adjustable lampholder. Medium base porcelain keyless socket attached to metal hood with ball socket and tension spring mounted on steel outlet box plates. Socket to be wired with heat-resisting asbestos covered wire 16 gauge to take up to a 300 w R-40 lamp.

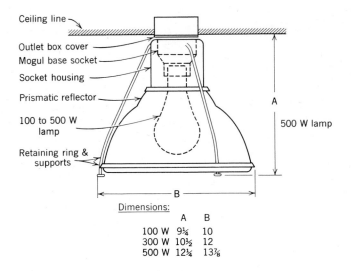

Dimensions:

	A	B
100 W	9¾	10
300 W	10½	12
500 W	12¼	13⅞

Appendix A.4 Open prismatic reflector fixture, 100 to 500 w. Fixture shall be surface mount, stem or box mount. Construction shall be outer aluminum housing, inner prismatic reflector, medium or mogul base, for wattage noted. Optical characteristics of prismatic reflector to be chosen according to use.

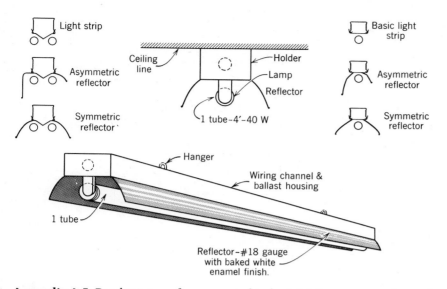

Appendix A.5 Pendant or surface mounted industrial fluorescent unit one-lamp. *Typical specification:* Fixture to be single or two-lamp fluorescent strip suitable for individual or continuous mounting. Body of fixture shall be 18 gauge steel channel. Cover and reflector shall be 20 gauge. Entire unit to be finished in high gloss baked white enamel. Conduit knockouts of ½ and ¾-in. size shall be provided on top, and ½ in. on ends. Accessories shall be suitable for lamp specified. Ballast shall be high power factor class P, sound rated A or B. Each section shall be completely wired with No. 16 AF wire. Removal of ballast without demounting of fixture shall be possible. Provide end caps for each fixture and reflector with same finish as fixture.

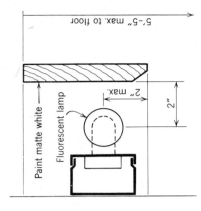

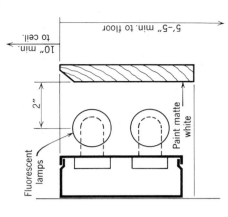

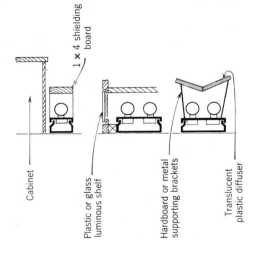

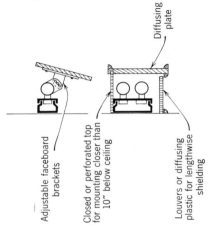

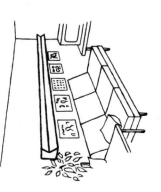

LIGHTED LOW WALL BRACKETS

Low brackets are used for special wall emphasis or for lighting specific tasks such as sink, range, reading in bed, etc. Mounting height is determined by eye height of users, from both seated and standing positions. Length should relate to nearby furniture groupings and room scale.

(a)

LIGHTED HIGH WALL BRACKETS

High wall brackets provide both up and down light for general room lighting. Used on interior walls to balance window valance both architecturally and in lighting distribution. Mounting height determined by window or door height.

(b)

Appendix A.6 Architectural element wall lighting. (a) Lighted low bracket. (b) Lighted high bracket. (Portions of figure courtesy of Architectural Lighting Graphics by Flynn and Mills, 1968, Litton Educ. Publ. Inc., reprinted by permission of Van Nostrand Reinhold Co.)

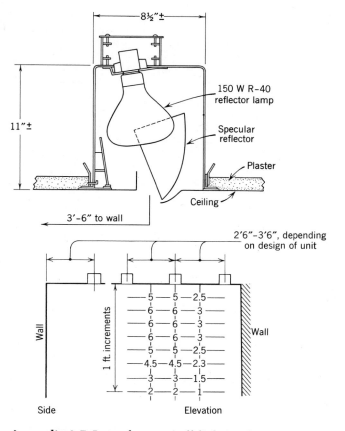

Appendix A.7 Incandescent wall-lighting fixture.

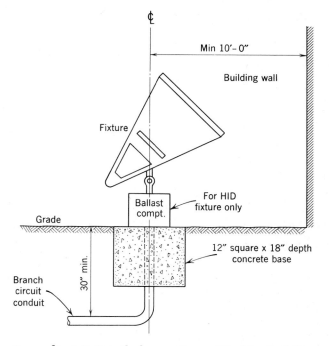
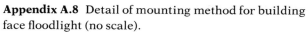

Appendix A.8 Detail of mounting method for building face floodlight (no scale).

Standardized Dimensions for Recessed Fluorescent Luminaires

Type	Width, Inches			Length, Inches		
	Nominal	Minimum	Maximum	Nominal	Minimum	Maximum
Type F	12	—	11¹⁵/₁₆[a]	48	47³¹/₃₂[a]	48[a,b]
	24	—	23¹⁵/₁₆[a]	48	47³¹/₃₂[a]	48[a,b]
Type M	12	11³¹/₃₂	12	48	47³¹/₃₂	48
	24	23³¹/₃₂	24	48	47³¹/₃₂	48
Type G	12	11¹¹/₁₆	11¾	48	47³¹/₃₂[c]	48[c]
	24	23¹¹/₁₆	23¾	48	47³¹/₃₂[c]	48[c]
Type S	12	11⅝	11⅞	48	47³¹/₃₂	48
	24	23⅝	23⅞	48	47³¹/₃₂	48
Type HS	12	11⅝	11⅞	48	47³¹/₃₂	48
	24	23⅝	23⅞	48	47³¹/₃₂	48
Type HF	12	—	11¹⁵/₁₆[a]	48	47³¹/₃₂[a,d]	48[a,b,d]
	24	—	23¹⁵/₁₆[a]	48	47³¹/₃₂[a,d]	48[a,b,d]

[a] Excluding horizontal flange.
[b] Maximum length may be 48⅛ in. for individual fixtures (as contrasted with fixtures for continuous rows).
[c] 47¾ in. maximum up to 1½ in. height to clear web of inverted T.
[d] Not including length of end bracket, hook, etc.
Extracted from NEMA Standard LE-1.

Recommended Ceiling Openings for Recessed Luminaires

To accommodate recessed fluorescent luminaires having the dimensions given above it is recommended that architect-engineers, ceiling manufacturers and installing contractors provide ceiling openings in accordance with the following. The dimensions shown refer to actual openings or to the center lines of suspension ceiling rails.

Type	Width, Inches			Length, Inches		
	Nominal	Minimum	Maximum	Nominal	Minimum	Maximum
Type F	12	12	12¼	48	48¼	48½[a]
	24	24	24¼	48	48¼	48½[a]
Type M	12	11⁶³/₆₄	12¹/₆₄	48	[b]	[b]
	24	23⁶³/₆₄	24¹/₆₄	48	[b]	[b]
Type G[c]	12	11¹⁵/₁₆	12¹/₁₆	48	47¹⁵/₁₆	48¹/₁₆
	24	23¹⁵/₁₆	24¹/₁₆	48	47¹⁵/₁₆	48¹/₁₆
Type S[c]	12	11¹⁵/₁₆	12¹/₁₆	48	[b]	[b]
	24	23¹⁵/₁₆	24¹/₁₆	48	[b]	[b]
Type HS	12	11¹⁵/₁₆	12¹/₁₆	48		
	24	23¹⁵/₁₆	24¹/₁₆	48	[b]	[b]
Type HF	12	12	12¼	48	48¼	48½[a]
	24	24	24¼	48	48¼	48½[a]

[a] When luminaires are installed in continuous rows, the total length of the ceiling opening is assumed to be the nominal length for the number of fixtures in the row, + ½ in.
[b] Since transverse suspension rails may be located at almost any point, the total length in continuous rows is assumed to be the nominal length for the number of fixtures in the row, ± ¹/₁₆ in.
[c] The minimum width between the edges of ceiling rail flanges is assumed to be 1 in. less than the nominal 12, 24, or 48 in.
Extracted from NEMA Standard LE-1.

Appendix A.9 Dimensions and details of recessed fluorescent fixtures.

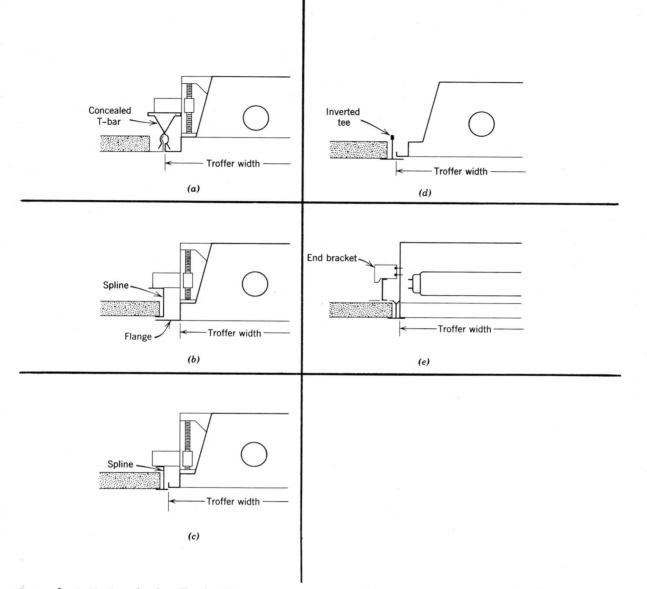

Appendix A.10 Standard troffer details.

(a) A *Type M luminaire* is one having vertical turned-up edges which are parallel to the lamp direction and intended to "snap-in" or otherwise align the luminaire with a concealed T-bar suspension system, the center openings of the Tees being located on modular or other symmetrical dimensional lines.

(b) A *Type F luminaire* is one having horizontal flanges which are parallel to the lamp direction and designed to conceal the edges of the ceiling opening above which the luminaire is supported by concealed mechanical suspension.

(c) A *Type S luminaire* is one which is designed for mechanical suspension from exposed splines and dependent on splines parallel to the lamp direction for concealment of the edges of the luminaire.

(d) A *Type G luminaire* is one having edges which are designed to rest on or "lay-in" the exposed inverted T of a suspension system (customarily described as a grid ceiling system) with the webs of the tees being located on modular or other symmetrical dimensional lines.

(e) A *Type H luminaire* is one having end brackets, hooks or other attachments and designed to be supported at the ends by "hooking-on" to some member of the ceiling suspension system. A *Type HS luminaire* is a Type H luminaire having edges parallel to the lamp direction and dependent on splines of the ceiling suspensions system for concealment of the edges of the luminaire. A *Type HF luminaire* is a Type H luminaire having edges parallel to the lamp direction and designed to conceal the edges of the ceiling opening in which the luminaire is recessed.
(Adapted from NEMA Publication LE-1)

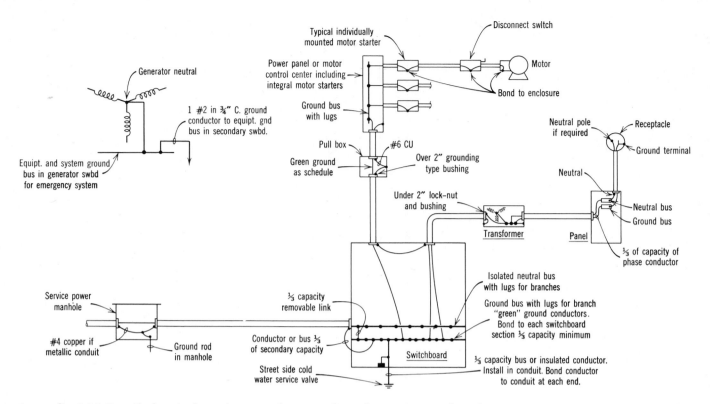

Appendix A.11 Detail of typical requirements for grounding of system neutrals and equipment.

CONDUIT		WIRE				FROM	TO	REMARKS
SIZE	MAT.	NO.	SIZE	TYPE	INS.			
TYPICAL DATA:								
F-16	*4"*	*S*	*4*	*500*	*RHW* *600V*	*SWBD. D-1*	*MCC-14*	
C-12	*1"*	*S*	*16*	*22*	*Tel* *300V*	*MCC #14*	*Annunc. Pnl A-6*	*Pilot Lt. Control Cable*
		S-Steel		*T-Transite*				
		E-Emt		*F-Fibre*				
		A-Aluminum		*Pvc-Plastic*				
		FL-Flex.						

Appendix A.12 Conduit and cable schedule.

ELECTRICAL DWGS CHANGE RECORD

Project   XYZ Building   No.   142 A

Project Manager   B. STEIN

DATE	DESCRIPTION	MOD	MAN-HRS	TOTAL M/HRS
	TYPICAL DATA :			
3/5/75	DWG E-1 Architectural changes - Arch dwgs A-2, A-3, 3/1/75		6	6
3/8/75	DWG E-6 CHANGES TO Mcc #1 due to revised HVAC dwgs of 3/6/75		9	15
3/15/75	DWG E-2,3,4 Arch changes - moved stairs, partition changes, revised elect. closet; changed door swings - Conf with Arch of 3/14/75		4	19

Note: A Record is to be kept of drawing changes on each job, caused by changes in other trades. All changes must be reported to, and approved by the job captain.

Appendix A.13 Drawing changes record form.

SQUARE D COMPANY
INDUSTRIAL CONTROLLER DIVISION
MILWAUKEE 12. WISCONSIN. U.S.A.

CONTROL CENTER INFORMATION SHEET

WK
GO

DATE	PAGE	OF	BRANCH ORDER

JOB TITLE (USER)

LOCATION

ARCHITECT

ENGINEER

ELECTRICAL CONTRACTOR

DISTRIBUTOR

VOLTS	WIRE	MODEL NUMBER	NEMA WIRING TYPE ☐ TYPE A ☐ TYPE B ☐ TYPE C

PHASE	CYCLES	ENCLOSURE NEMA TYPE	UNIT MOUNTING ARRANGEMENT ☐ FRONT-OF-BOARD ☐ BACK-TO-BACK

CUSTOMER ORDER NO.

PREPARED BY

FACT. SALES APPROVAL	FACT. ENG. APPROVAL

WILL BOTH FRONT AND REAR BE ACCESSIBLE ☐ YES ☐ NO

FEEDER CONDUCTORS WILL ENTER

☐ TOP
☐ BOTTOM   SECTION _____

CONDUCTOR SIZE

CONDUCTORS PER PHASE

91½″

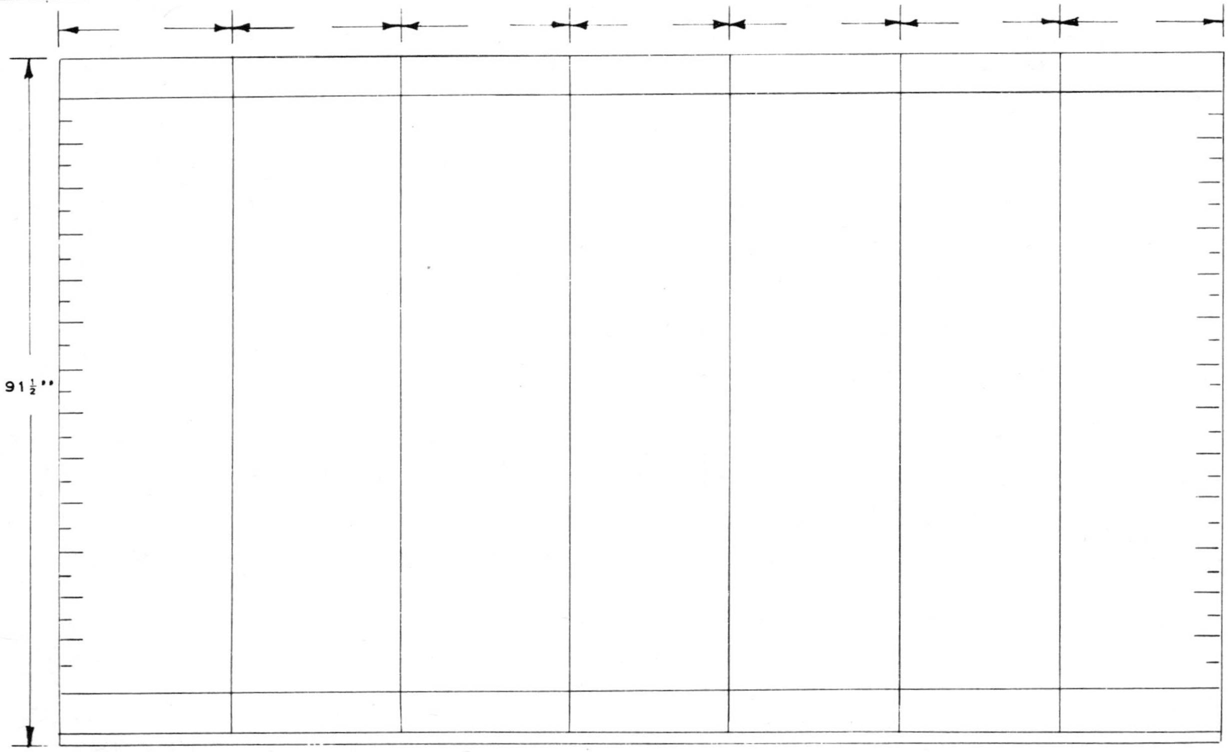

SECTIONS 20″ DEEP UNLESS OTHERWISE INDICATED   OVERALL LENGTH _____

SPECIAL REQUIREMENTS

M-6618

Appendix A.14 Typical MCC layout sheet. (Courtesy of Square D Company)

570

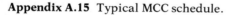

Appendix A.15 Typical MCC schedule.

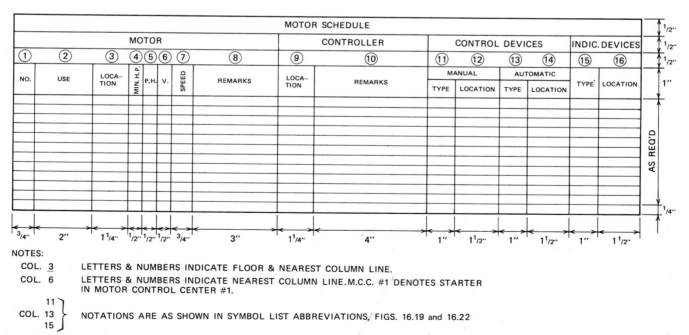

NOTES:

COL. 3   LETTERS & NUMBERS INDICATE FLOOR & NEAREST COLUMN LINE.

COL. 6   LETTERS & NUMBERS INDICATE NEAREST COLUMN LINE. M.C.C. #1 DENOTES STARTER IN MOTOR CONTROL CENTER #1.

COL. 13 } 11 ... 15   NOTATIONS ARE AS SHOWN IN SYMBOL LIST ABBREVIATIONS, FIGS. 16.19 and 16.22

DIMENSIONS SHOWN ARE TYPICAL FOR AN ACTUAL, FULL SIZE FORM.

Appendix A.16 Typical job motor and control schedule.

		PANEL DESIGNATION		
CIRC. NO.	DESCRIPTION	CIRCUIT BREAKERS	DESCRIPTION	CIRC. NO.
		1 | 2		
		3 | 4		
		5 | 6		
		7 | 8		
		9 | 10		
		11 | 12		
		13 | 14		
		15 | 16		
		17 | 18		
		19 | 20		
		21 | 22		
		23 | 24		
		25 | 26		
		27 | 28		
		29 | 30		
		31 | 32		
		33 | 34		
		35 | 36		
		37 | 38		
		39 | 40		
		41 | 42		

PANEL DATA

MAINS AND GND BUS _____

MAIN C/B OR SW/FUSE _____

BRANCH C/B INT. CAP. _____

SURF—RECESSED

REMARKS

VOLTS

Appendix A.17 Typical format for a single phase, circuit breaker type, lighting and appliance panel.

CIRC. NO.	DESCRIPTION	CIRCUIT BREAKERS	DESCRIPTION	CIRC. NO.
		PANEL DESIGNATION		

CIRC. NO.	DESCRIPTION	CIRCUIT BREAKERS	DESCRIPTION	CIRC. NO.
		1 2		
		3 4		
		5 6		
		7 8		
		9 10		
		11 12		
		13 14		
		15 16		
		17 18		
		19 20		
		21 22		
		23 24		
		25 26		
		27 28		
		29 30		
		31 32		
		33 34		
		35 36		
		37 38		
		39 40		
		41 42		

PANEL DATA

MAINS, GND. BUS _____   VOLTAGE
MAIN C/B OR SW/F _____
BRANCH C/B INT. CAP_____
SURF/RECESS
REMARKS:

Appendix A.18 Typical format for 3 phase, circuit breaker type, lighting and appliance panel.

CIRC. NO.			SW	FUSE		FUSE	SW		CIRC. NO.
					PANEL DESIGNATION				

Circuit numbers (center column): 1 2, 3 4, 5 6, 7 8, 9 10, 11 12, 13 14, 15 16, 17 18, 19 20, 21 22, 23 24, 25 26, 27 28, 29 30, 31 32, 33 34, 35 36, 37 38, 39 40, 41 42

PANEL DATA
MAINS, GROUND BUS
SURF/RECESSED
REMARKS:

MAIN C/B OR SW/F
BRANCH FUSE TYPE
VOLTAGE

Appendix A.19 Typical format for 3 phase, switch-and-fuse type, lighting and appliance panel.

Appendix A.20 Panel board installation methods. Preferred mounting heights, panelboard boxes: Mounting heights of panelboard and miscellaneous distribution boxes shall be as follows unless otherwise noted on electrical drawings.

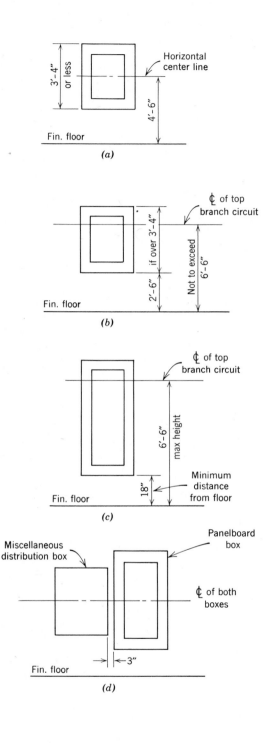

(a) Panelboards having boxes 3'-4" high or less shall be located 4'-6" from the floor to the horizontal centerline of the box.

(b) Panelboards having boxes over 3'-4" high shall be located 2'-6" from the floor to the bottom of the box, except that the highest branch circuit unit shall be not more than 6'-6" from the floor.

(c) When necessary the box may be lowered to a distance not less than 18" from the floor to the bottom of the box. Where a maximum height of 6'-6" above the floor to the upper circuit or a minimum distance of 18" above the floor cannot be adhered to, the panel shall be divided into two sections.

(d) Where two or more boxes are adjacent on the same wall, they shall be installed with the horizontal center line of each equidistant from the floor. The centerline distance of the higher box controlling the boxes shall be installed with a minimum spacing of 3" apart.

Note: Main lugs or main disconnect unit shall be located at the bottom of the panelboard except where the main feeder conduit enters at the top of the box, in which case, main lugs or main disconnect unit shall be located at the top of the panelboard. (Courtesy of the New York Telephone Company)

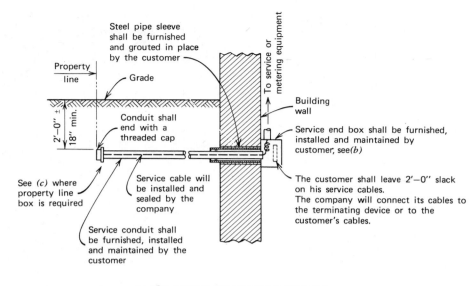

(a) BUILDING SERVICE CABLE DETAILS

SIZE OF SERVICE END BOX	MAXIMUM NO. OF CABLES PER PHASE	
INSIDE DIMENSIONS	COPPER	
	IN	OUT
24″ × 12″ × 8″	1—#4/0	2—#4/0
30″ × 18″ × 12″	1—500MCM	2—500MCM

Appendix A.21 *(a)* Typical underground service details.

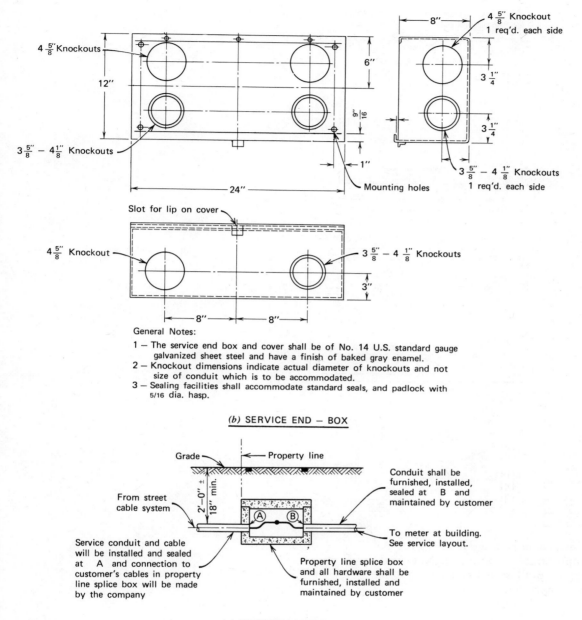

General Notes:

1 — The service end box and cover shall be of No. 14 U.S. standard gauge galvanized sheet steel and have a finish of baked gray enamel.
2 — Knockout dimensions indicate actual diameter of knockouts and not size of conduit which is to be accommodated.
3 — Sealing facilities shall accommodate standard seals, and padlock with 5/16 dia. hasp.

(b) SERVICE END — BOX

(c) PROPERTY LINE BOX

PROPERTY LINE SPLICE BOX(PRECAST)	NO. OF CABLES PER PHASE	
INSIDE DIMENSIONS	COPPER	
	IN	OUT
24" × 12" × 12"	1–#2	2–#2
	1–500MCM	1–500MCM
	1–#4/0	2–#4/0

Appendix A.21 *(b)* and *(c)*.

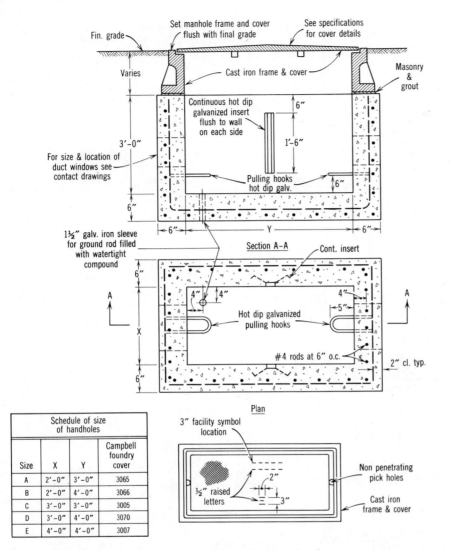

Schedule of size of handholes			
Size	X	Y	Campbell foundry cover
A	2'-0"	3'-0"	3065
B	2'-0"	4'-0"	3066
C	3'-0"	3'-0"	3005
D	3'-0"	4'-0"	3070
E	4'-0"	4'-0"	3007

Appendix A.22 Typical handhole detail, giving table of dimensions plus appropriate cover number. Note wall insert for cable support, pulling hooks for cable pulling, and ground rod inside handhole.

Transformers
Pad Mounted, Three Phase

Liquid Immersed, 75-500 Kva, 18 Kv and
Below, 65°C Rise, Type CTP

Typical Specifications
Electrical

Standard High Voltage Rating	Standard Low Voltage Rating
4160 Grd Y/2400	208Y/120
4160 Grd Y/2400	480Y/277
12470 Grd Y/7200	208Y/120
12470 Grd Y/7200	480Y/277
13200 Grd Y/7620	208Y/120
13200 Grd Y/7620	480Y/277
13800 Grd Y/8000	208Y/120
13800 Grd Y/8000	480Y/277

Kva Continuous 65°C	Losses in Watts	
	No Load	Total
75	360	1350
112½	530	1800
150	560	2250
225	880	3300
300	1050	4300
500	1600	6800

Dimensions and Weights

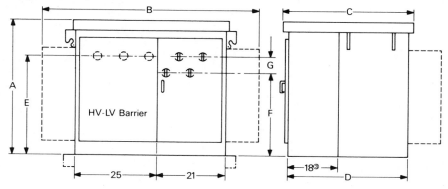

Transformer Dimensions

Kva Cont. 65°C	Dimensions in Inches							Net Weight, Lbs.
	A	B②	C	D③	E	F	G	
75-112½	40	53	42	38	24	22	6	2300-2600
150	40	53	52①	38	24	22	6	2900
225-300	44	53	58①	44	30	27	8	3700-4200
500	52	67①	58①	44	30	27	8	5100

① Dimensions include coolers.
② For loop feed switched units, refer to Westinghouse.
③ Certain combinations require an additional 6 inches, refer to Westinghouse.

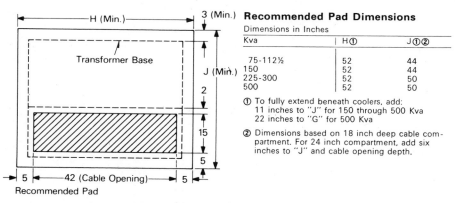

Recommended Pad

Recommended Pad Dimensions

Dimensions in Inches

Kva	H①	J①②
75-112½	52	44
150	52	44
225-300	52	50
500	52	50

① To fully extend beneath coolers, add:
11 inches to "J" for 150 through 500 Kva
22 inches to "G" for 500 Kva

② Dimensions based on 18 inch deep cable compartment. For 24 inch compartment, add six inches to "J" and cable opening depth.

Appendix A.23 Typical pad-mount transformer data. (Courtesy of Westinghouse Company)

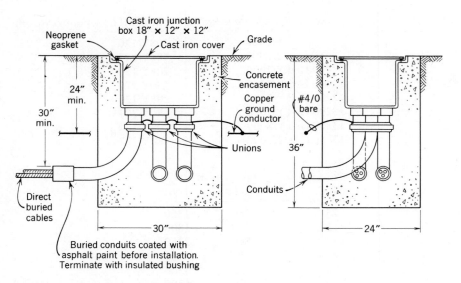

Appendix A.24 Typical exterior pull box detail (no scale).

Type WF – All Front Accessible, Meet NEMA PB-2, 1972 Standards
Incoming or Service Sections – One Main (or tie) Device Only, Individually
Mounted, Front Accessible
Metering C. T. Compartment and Cable Arrangements

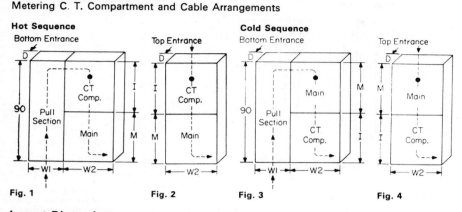

Hot Sequence Bottom Entrance / Top Entrance
Cold Sequence Bottom Entrance / Top Entrance

Fig. 1   Fig. 2   Fig. 3   Fig. 4

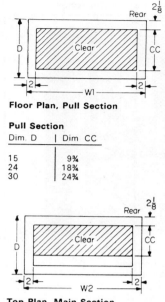

Floor Plan, Pull Section

Pull Section

Dim. D	Dim. CC
15	9¾
24	18¾
30	24¾

Top Plan, Main Section

Layout Dimensions

Main Device	Max Ampere Rating	Dimension in Inches					CC Fig 2, 4
		W1	W2	D	I	M	
Fixed Devices							
Power Circuit Breakers							
DBE	800-1600	②	38	30	45	45	24
DBE	2000	②	38	30	30	60	24
System Circuit Breakers							
SCB-600	600	②	38	15	45	45	9
SCB-1200	1200	②	38	15	45	45	9
SCB-2000	2000	②	38	24	45	45	18
Molded Case Breakers							
LA. HLA	600	②	38	15	45	45	9
MA, HMA	800	②	38	15	45	45	9
NB, HNB	1200	②	38	15	45	45	9
PB	2000	②	38	15	45	45	9
Auto. Transf. Sw.③	100-1200	②	38	15	45	45	9
TRI-PAC Circuit Breakers							
LA	400	②	38	15	45	45	9
NB	800	②	38	15	45	45	9
PB	1600	②	38	15	45	45	9
Bolted Pressure Contact Switches							
BPS	800-2000	②	38	15	45	45	9
Quick-Make, Quick-Break Fusible Switches							
FDP	800, 1200	②	38	15	45	45	9
Combination Sections – Main Plus Distribution – Single Section Only							
Bolt Pr. Sw. w/CDP or FDP	800-2000	②	38	15	45	45	9
Mold Case Brkr. w/CDP or FDP	2000	②	38	15	45	45	9
CDP or FDP 6 Circ. only	2000	②	38	15	45	45	9
Drawout Devices							
System Circuit Breaker							
SCB-600	600	②	38	24	45	45	18
SCB-1200	1200	②	38	24	45	45	18
SCB-2000	2000	②	38	30	45	45	24

Appendix A.25 Typical building switchboard dimensional data. (Courtesy of Westinghouse Company)

General Dimensional Information

The standard arrangements and layout dimensions shown in this section reflect the basic physical, electrical and mechanical considerations governing switchboard and power assembly layout.

These basic considerations are: (1) Accessibility of internal devices and connections. (2) Physical alignment of sections, and (3) electrical safety and integrity.

The three switchboard types offered by Westinghouse to best meet the variable requirements of switchboard specifications, i.e., Type WF, Type WR and Type WPA, may be laid out for close estimating purposes from the dimensional information given in this section.

Heavy Duty Safety Switches

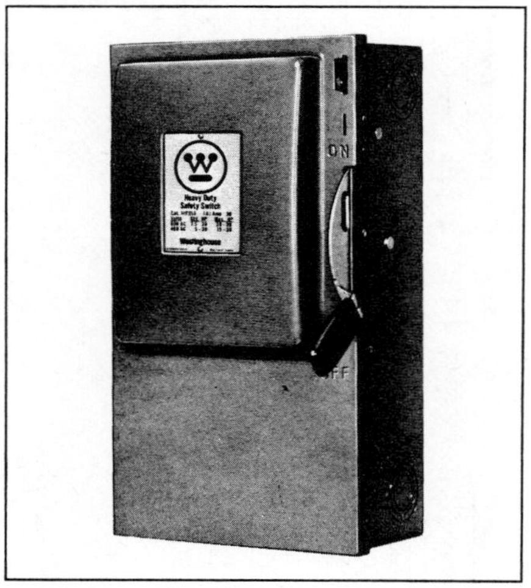

30 to 1200 Amperes
250 Volts Ac, Dc;
2, 3 and 4 Poles
Fusible and Non-Fusible
Copper-Aluminum Terminals

Enclosures

NEMA 1 General Purpose
NEMA 3R Raintight
NEMA 4 Watertight Stainless Steel
NEMA 12 Dust-tight Special Industry

Dimensions, Inches

NEMA 1 Heavy Duty

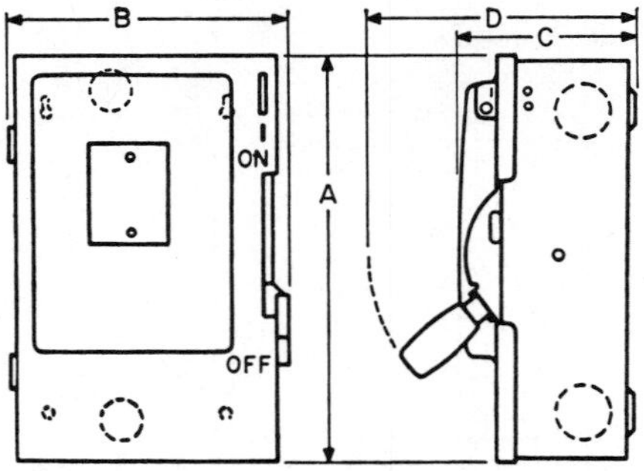

Ampere Rating	No. of Poles	Dimension A	B	C	D	Approx. Wt. Lbs.	Conduit Sizes, Inches (NEMA 1 Only)
250 Volt HF, Fusible Switches							
30	2, 3	9¼	7½	4⅜	6 13/16	7	½, ¾, 1
60	2, 3	12 9/16	8⅞	5⅜	8⅛	12	½, ¾, 1, 1¼, 1½
100	2, 3	20 9/16	10¾	7¼	9¾	34	1, 1¼, 1½, 2
200	2, 3	26¾	13¾	8 7/16	11 3/16	54	1½, 2, 2½, 3
400	2, 3	39	23⅛	11 15/16	15⅞	150	2, 2½, 3, 3½, 4
600	2, 3	44¾	23⅛	11 15/16	15⅞	180	3½, 4
800	2, 3	51 15/16	40⅛	12 15/16	15 1/32	448	....
1200	2, 3	60 1/16	45⅛	13 7/32	15 5/16	495	....
250 Volt HU Non-Fusible Switches							
30	2, 3	9¼	7½	4⅜	6 13/16	7	½, ¾, 1
60	2, 3	12 9/16	8⅞	5⅜	8⅛	12	½, ¾, 1, 1¼, 1½
100	2, 3	20 9/16	10¾	7¼	9¾	34	1, 1¼, 1½, 2
200	2, 3	26¾	13¾	8 7/16	11 3/16	54	1½, 2, 2½, 3
400	2, 3	39	23⅛	11 15/16	15⅞	150	2, 2½, 3, 3½, 4
600	2, 3	44¾	23⅛	11 15/16	15⅞	150	3½, 4

Appendix A.26 Typical dimensional data for heavy duty safety switches. (Courtesy of Westinghouse Company)

Enclosure Dimensions, Inches ***Not to be used for construction purposes unless approved.***

NEMA 1 Surface Mounted

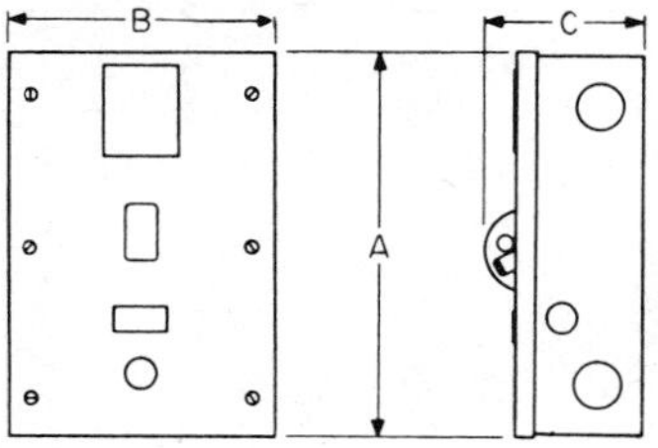

Catalog Number	Maximum Amperes	Dimensions A	B	C	Conduit Sizes
SQC050	50	$8\frac{1}{8}$	$5\frac{5}{8}$	$3\frac{9}{16}$	$\frac{1}{2}, \frac{3}{4}, 1, 1\frac{1}{4}$
SQC	100	$11\frac{1}{4}$	$6\frac{1}{8}$	$4\frac{7}{16}$	$1, 1\frac{1}{4}, 1\frac{1}{2}, 2$
SFB	100	$13\frac{1}{8}$	$7\frac{5}{8}$	$4\frac{13}{16}$	$\frac{1}{2}, \frac{3}{4}, 1, 1\frac{1}{4}, 1\frac{1}{2}, 2$
SCA	225	$17\frac{3}{8}$	$8\frac{7}{8}$	6	$2, 2\frac{1}{2}, 3$
SKA, SKB	225	$20\frac{5}{8}$	$10\frac{3}{4}$	$6\frac{3}{32}$	$2, 2\frac{1}{2}, 3$
SLB	400	$24\frac{5}{8}$	$15\frac{1}{2}$	$7\frac{3}{8}$	$1\frac{1}{2}, 2, 2\frac{1}{2}$
SLA 600	600	$27\frac{5}{32}$	$13\frac{29}{32}$	$8\frac{51}{64}$	
SNB	1200	$42\frac{7}{32}$	$21\frac{23}{32}$	$13\frac{7}{32}$	

NEMA 1 Flush Mounted

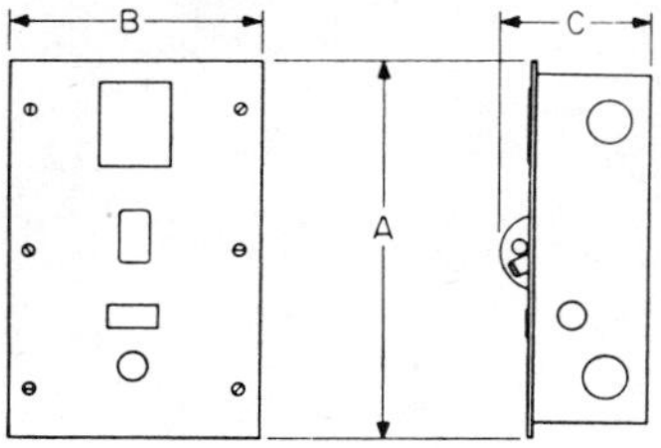

Catalog Number	Maximum Amperes	Dimensions A	B	C	Conduit Sizes
FQC050	50	$8\frac{7}{8}$	$6\frac{3}{8}$	$3\frac{9}{16}$	$\frac{1}{2}, \frac{3}{4}, 1, 1\frac{1}{4}$
FQC	100	12	$6\frac{13}{16}$	$4\frac{7}{16}$	$1, 1\frac{1}{4}, 1\frac{1}{2}, 2$
FFB	100	$14\frac{3}{8}$	$8\frac{1}{2}$	$4\frac{13}{16}$	$\frac{1}{2}, \frac{3}{4}, 1, 1\frac{1}{4}, 1\frac{1}{2}, 2$
FCA	225	$18\frac{1}{8}$	$9\frac{5}{8}$	6	$2, 2\frac{1}{2}, 3$
FKA	225	$21\frac{25}{32}$	$11\frac{29}{32}$	$6\frac{3}{32}$	$2, 2\frac{1}{2}, 3$
FLB	400	$25\frac{25}{32}$	$16\frac{21}{32}$	$7\frac{11}{32}$	$1\frac{1}{2}, 2, 2\frac{1}{2}$
FLA 600	600	$28\frac{29}{32}$	$15\frac{21}{32}$	$8\frac{51}{64}$	

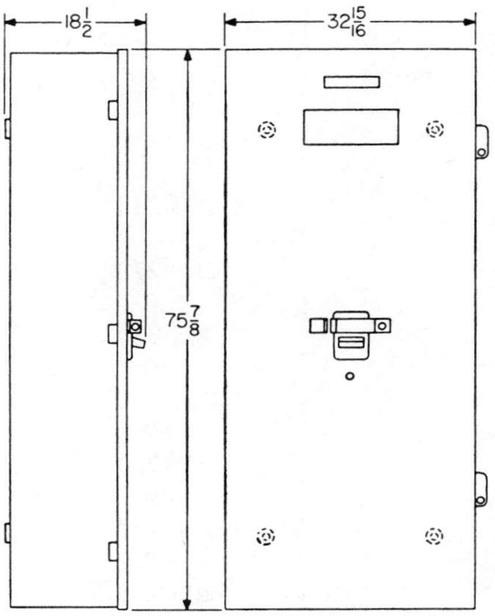

NEMA 1 Semi-Dust-tight

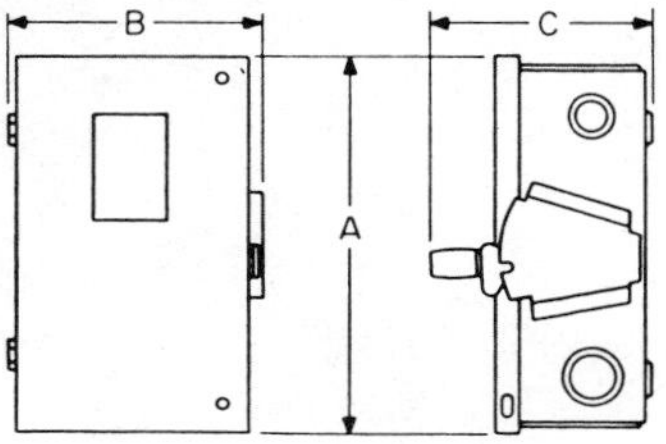

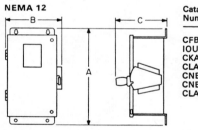

Catalog Number	Maximum Amperes	Dimensions A	B	C	Conduit Sizes
DFB	100	$12\frac{7}{8}$	$8\frac{7}{32}$	$7\frac{13}{16}$	$\frac{3}{4}, 1, 1\frac{1}{4}, 1\frac{1}{2}, 2$
DKA	225	$21\frac{7}{8}$	$12\frac{19}{32}$	$9\frac{13}{32}$	$1\frac{1}{2}, 2, 2\frac{1}{2}, 3$

NEMA 12

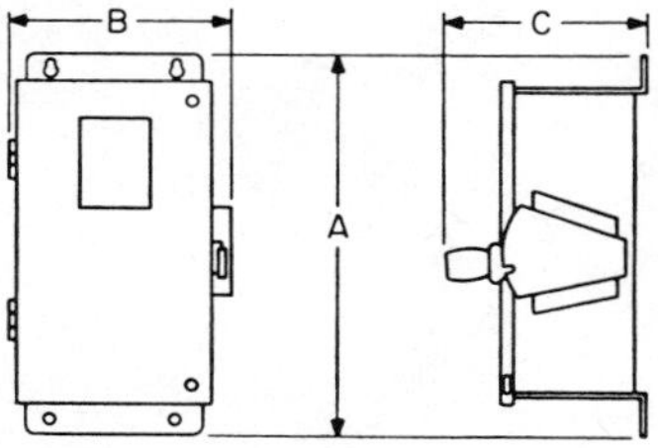

Catalog Number	Maximum Amperes	Dimensions A	B	C
CFB	100	$14\frac{3}{16}$	$8\frac{7}{32}$	$7\frac{13}{16}$
IOU150	150	$23\frac{3}{16}$	$12\frac{19}{32}$	$9\frac{13}{32}$
CKA	225	$23\frac{3}{16}$	$12\frac{19}{32}$	$9\frac{13}{32}$
CLA 600	600	$30\frac{5}{32}$	$15\frac{15}{32}$	$12\frac{15}{64}$
CNB	1200	$45\frac{7}{32}$	$22\frac{17}{32}$	$15\frac{29}{64}$
CNB-P	800	$51\frac{7}{32}$	$22\frac{17}{32}$	$15\frac{29}{32}$
CLA-P	400	$40\frac{5}{32}$	$15\frac{15}{32}$	$12\frac{15}{64}$

NEMA 3R

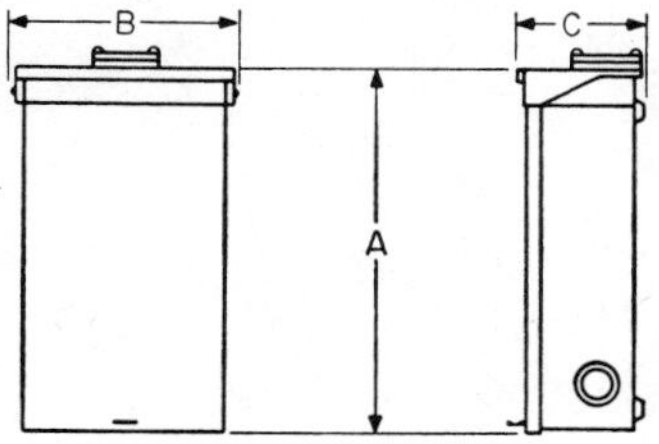

Catalog Number	Maximum Amperes	Dimensions A	B	C	Conduit Sizes
RQC050	50	$8\frac{1}{4}$	$5\frac{3}{4}$	$4\frac{1}{4}$	$\frac{1}{2}, \frac{3}{4}, 1, 1\frac{1}{4}$
RQC	100	$11\frac{3}{32}$	$6\frac{3}{16}$	$5\frac{7}{32}$	$1\frac{1}{4}, 1\frac{1}{2}, 2$
RCA	225	$17\frac{9}{32}$	$9\frac{1}{32}$	$6\frac{7}{16}$	$2, 2\frac{1}{2}, 3$
RLB	400	$25\frac{3}{8}$	$15\frac{9}{16}$	$7\frac{7}{8}$	$2\frac{1}{2}, 3$
RFB	100	$13\frac{1}{4}$	$7\frac{1}{4}$	$4\frac{7}{8}$	$1\frac{1}{4}, 1\frac{1}{2}, 2$
IOU150	150	See NEMA 12 Dimensions			
RKA, RKB	225	$21\frac{9}{16}$	$10\frac{9}{16}$	$5\frac{49}{64}$	$2, 2\frac{1}{2}, 3$
RLA600	600	$30\frac{5}{32}$	$15\frac{15}{32}$	$12\frac{15}{64}$	
RNB	1200	$45\frac{7}{32}$	$22\frac{17}{32}$	$15\frac{29}{64}$	

Note: RLA 600 and RNB enclosures are similar to NEMA 12 construction.

Appendix A.27 Typical dimensional data for circuit breakers, enclosed type. (Courtesy of Westinghouse Company)

Rating and Approximate Dimensions and Weights of a-c Full-Voltage Single-Speed Motor Controllers.

NEMA Size Designation	Voltage	Maximum Horsepower	Weight, Pounds	Width	Height	Depth
					Inches	
0	208-230	2	6	5½	7½	4
	460	2				
1	208-230	5	8	6½	8½	4
	460	7½				
2	208-230	15	20	9	14	5
	460	25				
3	208-230	30	40	12	18	6
	460	50				
4	208-230	50	90	14	25	8
	460	100				
5	208-230	100	200	20	40	12
	460	200				
6[a]	208-230	200	400	30	80	20
	460	400				
7[a]	208-230	300	400	30	80	20
	460	600				

All starters are housed in a NEMA I Indoor ventilated enclosure
[a]Housing is a free-standing unit.

Appendix A.28 Dimensional data on motor controllers.

Glossary

a-c Alternating current, which changes polarity many times per second.

AWG American Wire Gauge, the standard wire size measuring system in the United States.

Access Openings in a house or building through which equipment can be carried in or out, or removable panels in equipment for servicing parts.

Access box Rust-resistant metal box with hinged cover. Set flush with floor and allowing access to a cleanout or other device.

Acidity An acid condition of water that could corrode pipes and equipment.

Air changes (In ventilation) the number of times the air is changed per hour in a room.

Air density The weight of air, pounds per cubic foot.

Airflow pattern Methods by which air is introduced to a space, directed through it and removed.

Air foil vanes Flat blades in a register that can be turned to positions that direct the airsteam.

Airstream Airflow through items such as filters, coils, registers and ducts.

Air vent valve An escape for air at high points in a hot water heating system.

Ampacity A wire's ability to carry current safely, without undue heating. The term formerly used to describe this characteristic was current-carrying-capacity of the wire.

Appliance circuit A branch circuit that supplies outlets specifically intended for appliances. Lighting is not supplied from such circuits and the number of outlets is generally limited.

Appliance outlet An outlet connected to an appliance circuit. It may be a single or duplex receptacle or an outlet box intended for direct connection to an appliance.

Architectural-electrical plan Architectural plan on which electrical work is shown.

Architectural lighting element A light source built into, or onto, the structure. Not a commercial lighting fixture.

Auxiliary resistance heating Electric resistance heaters that supplement the heat from the heat pump.

Average water temperature Average between temperature of water leaving and returning to the boiler.

BX Trade name for NEC type AC Flexible armored cable.

Baseboard Hot water heater or electric heater along the wall.

Blower-coil unit A unit in which a blower moves the airstream across items such as heating coils, cooling coils and filter.

Boiler A unit that produces hot water or steam for heating.

Branch circuit Wiring between the last overcurrent device and the branch circuit outlets.

Breathing wall A method such as the incremental system that has an exterior wall opening for heat and moisture rejection and fresh air supply.

British thermal unit Quantity of heat.

Bus duct An assembly of heavy bars of copper or aluminum that acts as a conductor of large capacity.

Cable An assembly of two or more wires or a single wire larger than No. 8 AWG.

Centralized A system with one heating or cooling source and a ducted distribution network.

Chassis The working parts of an incremental conditioner, shaped to fit into a sleeve.

Check valve A valve that allows fluid to flow in one direction only.

Chilled water The refrigerated water used to cool the air in air systems.

Circuit An electrical arrangement requiring a source of voltage, a closed loop of wiring, an electric load and some means for opening and closing it.

Circuit breaker A switch-type mechanism that opens automatically when it senses an overload (excess current).

Circulating line Piping that permits circulation of domestic hot water for speedy availability at the faucet.

Circulator Centrifugal pump or "booster."

Cleanout A removable plug in a drainage system.

Clearing a fault Eliminating a fault condition by some means. Generally taken to mean operation of the over-circuit device that opens the circuit and clears the fault.

Closet carrier An iron and steel frame to support a water closet that hangs from a wall.

Coefficient of utilization The ratio between "usable" lumens and lamp lumens for a particular combination of fixture and space.

Combined sewer One that carries both storm and sanitary drainage.

Common neutral A neutral conductor that is common to, or serves, more than one circuit.

Compression Produces high pressure in Freon.

Condensing Liquifies high pressure Freon.

Conduit A round cross-section electrical raceway, of metal or plastic.

Connected load The sum of all loads on a circuit.

Continuous circulation Blower runs continuously, evaporator or burner runs intermittently.

Contract documents Legal papers that include the contract, the working drawings and the specifications.

Contrast Difference in brightness between an object and its background.

Control diagram (ladder diagram) A diagram that shows the control scheme only. Power wiring is not shown. The control items are shown between two vertical lines, hence, the name—ladder diagram.

Convector A heating element that warms the air passing over it which, in turn, rises to warm the space by convection.

Convenience outlet A duplex receptacle connected to a general purpose branch circuit and not intended for any specific item of electrical equipment.

Covering Thermal insulation on ducts.

Curb box Access to an underground valve at the street curb. It controls water service to a house or building.

Current (I) The electric flow in an electric circuit, which is expressd in *amperes* (amps).

Curved blade A register blade that directs the airstream.

d-c Direct current, which is unvarying in polarity.

DWV Drainage, waste and vent.

Dead Electrically de-energized, no voltage.

Decentralized Local complete heating and/or cooling system.

Degree day The number of Fahrenheit degrees that the average outdoor temperature over a 24-hr. period is less than 65° F.

Demand The probable maximum rate of water flow as determined by the number of water supply fixture units.

Demand load The actual amount of load on a circuit at any time. The sum of all the loads which are ON. Equal to the connected load minus the loads that are OFF.

Developed length Number of feet of tubing in a water circuit.

Diffuse reflection A type of reflection in which the reflected light is spread out in all directions. The reflection of a matte finish surface.

Diffuseness A measure of the directiveness of the light. Diffuse light comes from many directions.

Diffuser Material placed in front of the lamp for the purpose of controlling the light flux produced.

Domestic hot water Potable hot water as distinguished from hot water used for house heating.

Downstream Electrically speaking, going away from the power source and toward the load.

Drain pit A pit to receive nontoxic water for disposal, often to a dry well.

Dry well Same as a seepage pit, except that it disperses water and *not* sewage effluent.

Duct liner Acoustic liner to absorb sound.

Duct turns Curved vanes that reduce friction and turbulence when square corners are used in ducts.

EEA Electric Energy Association.

Edge loss factor Heat loss, slab to earth.

Efficacy Lumens per watt produced by a lamp.

Effluent A fluid flowing *away* from a process; for example, "The *effluent* of a septic tank."

Electric closet A space containing electric service equipment such as panels, switches, and the like.

Electronic air cleaners A filter somewhat more efficient than the usual bag filter, particularly for the removal of small suspended particles.

Emergency source Standby source of electric power; used when normal electric power fails.

Energy This is expressed in *kilowatt-hours* (kwh) or *watt-hours* (wh), and is equal to the product of power and time.

$$\text{energy} = \text{power} \times \text{time}$$
$$\text{kilowatt-hours} = \text{kilowatts} \times \text{hours}$$
$$\text{watt-hours} = \text{watts} \times \text{hours}$$

Engineering layout A drawing of the design that is the basis for shop drawings.

Evaporation Absorbs heat as Freon becomes a gas.

Expansion fitting A device to allow for the expansion of copper tubing.

Facade The face of a building.

Face velocity The speed in feet per minute by which air leaves a register.

Fall per foot The slope of a drainage pipe.

Fault A short circuit—either line to line, or line to ground.

Feed line A pipe that supplies water to items such as a boiler or a domestic hot water tank.

Feet of head Pressure loss in psi divided by the factor 0.433.

Fill and vent Parts of an oil storage system.

Finned-tube Used for heat transfer between water and air.

Fixture clearance Distance between fixtures, distance from a fixture to an obstruction, distance between a fixture and a wall.

Fixture unit An index of the relative rate of flow of water to a fixture (water supply fixture units) and of sewage leaving a fixture (drainage fixture units).

Flow pressure The pressure necessary to supply a fixture adequately.

Flow rate Cubic feet per minute (cfm) of air circulated in an air system or the number of pounds of water per hour circulated through a hot water heating system.

Flue gas Carbon monoxide, carbon dioxide and the like.

Flush tank A tank that refills automatically and stands ready to flush a water closet.

Flush valve A valve that when operated manually delivers a measured amount of water to flush a water closet.

Foot of head The pressure exerted at the bottom of a column of water 1 ft high.

Footcandle (fc) Unit of light flux density, equal to one lumen per square foot.

Fossil fuels Oil, gas and coal.

Four-way switch A four-terminal switch (abcd) that operates ab—cd and ac—bd. Used to control outlets from three or more locations.

Four-way switching Control of (an) outlet(s) from three locations.

Freon Refrigerant gas.

Fresh air inlet A vent that admits air to the house drain before it joins the house trap.

Frequency   The number of cycles per second for a-c.

Furnace   A unit that warms the air in warm air heating systems.

Fuse   An overcurrent device that opens the circuit by melting out. By nature a single-pole device.

GFI, GFCI   Ground fault (circuit) interrupter—a device that senses ground faults and reacts by opening the circuit.

gpm   Gallons per minute.

Gang   One wiring device position in a box.

Ganged switches   A group of switches arranged next to each other in ganged outlet boxes.

General purpose branch circuit   One that supplies a number of outlets for general lighting and convenience receptacles.

Greenfield   Trade name for flexible steel conduit.

Grills   Perforated or slotted frame, usually used for air *return*.

Ground   Zero voltage. Also, any point connected to ground.

Ground bus   A busbar in a panel or elsewhere, deliberately connected to ground.

Ground conductor   Conductor run in an electrical system, which is deliberately connected to the ground electrode. Purpose is to provide a ground point throughout the system. Insulation color—*green*. Also called "green ground."

Ground electrode   A piece of metal physically connected to ground. Can be a rod, mat, pad, structural member or pipe. See the NEC.

Ground fault   An unintentional connection to ground.

Groundwater   The level of water below the earth's surface.

Grounded   Connected to ground, at zero voltage.

Gutter   Rainwater collecting trough at the edge of a roof.

Gutterspace   The empty spaces around the sides, top and bottom of a panelbox, intended as a wiring space.

Handhole   Small exterior concrete box, intended as a pulling or splicing point for underground cables.

Hardness   Calcium compounds in water. They form a rocklike deposit on the inside of piping.

Heat pump   An all-electric heating/cooling device that takes energy for heating from outdoor air (or groundwater).

Hertz (Hz)   The unit of frequency of a-c. It equals the number of complete cycles per second.

Home-run   The wiring run between the panel and the first outlet in the branch circuit. (Looking upstream, it is the wiring run between the *last* outlet and the panel).

Horsepower (hp)   A unit of power which equals electrically, 746 watts, on 1 hp ≈ ¾ kw.

Hose bibb   Connection for supplying water to a garden hose.

Hot, live   Electrically energized.

House trap   A trap between the house drain and the house sewer.

Hydronic   Heating (or cooling) by water.

Humidifier   A device to vaporize water, and to use it to increase the relative humidity of air.

I-B-R   Institute of Boiler and Radiator Manufacturers.

Impedance (Z)   The quantity in an a-c circuit that is equivalent to resistance in a d-c circuit, inasmuch as it relates current and voltage. It is composed of *resistance* plus a purely a-c concept called *reactance* and is expressed, like resistance, in *ohms*.

Incremental   Self-contained through-wall unit for heating, cooling and ventilating.

Individual branch circuit   One that supplies only a single piece of electrical equipment.

Infiltration   Cold air that leaks in.

Junction box   Metal box in which tap to circuit conductors is made. Junction box is not an outlet, since no load is fed from it directly.

Latent heat   Inherent heat in the form of water vapor.

Lavatory   A wash basin. Also, a room with a lavatory basin and a water closet.

Layout   First drawings of a system showing the relation of the system components. Usually followed by dimensions, sizes and notes.

Leader   A pipe that carries storm water down from a gutter or a roof drain fixture.

Line side   The side of a device electrically closest to the source of current.

Line voltage thermostat   A thermostat that is connected directly to the line. Full power is fed *though* it to the controlled heater or air conditioner.

Lighting system   A method of describing in what directions the light is emitted from the fixture. Systems are direct, semidirect, direct-indirect, general diffuse, semi-indirect and indirect.

Load side   The side of a device electrically furthest from the current source.

Low voltage switching   A system of outlet control by low voltage switches and relays.

Lumen   Unit of light flux.

Lux   Metric unit of lighting flux density, or illumination.

Mbh   Thousands of British thermal units (Btu) per hour.

MCM   Thousand circular mil—used to describe large wire sizes.

Maintenance factor   The ratio between maintained and initial footcandles, for a lighting fixture.

Manhole   Same as handhole except much larger. Intended for underground primary power cables, large secondary cables and telephone cables.

Master (central) control Control of all the outlets from one point.

Mean radiant temperature Average temperature, interior surfaces.

Meter pan Device intended to hold one or more kilowatt-hour meters. Usually contains buswork, and may contain overcurrent devices.

Milinch The pressure exerted at the bottom of a column of water 1/1,000 of 1 in. high.

Milinches per foot The head lost in 1 ft of tubing as caused by friction of water flow.

Mirroring The effect that causes ordinary glass to reflect like a mirror.

Motor control center A single metal enclosed assembly containing a number of motor controllers and possibly other devices such as switches and control devices.

Motor controller The device that puts the motor "on the line." Generally, a magnetically operated contactor. Usually called the "starter."

Multipole Connects to more than 1 pole such as a 2-pole circuit breaker.

NEMA National Electrical Manufacturers Association —an American association that establishes standards of manufacture for electrical equipment. This material is published as NEMA standard and is accepted as the standard throughout the industry.

Net output As applied to a boiler, the *effective* value of the Mbh delivered by the boiler to heat the building.

Neutral The circuit conductor that is normally grounded. Insulated *white*, or *grey*.

Nonintegral trap A trap that is not part of the fixture as manufactured.

Ohms law The relationship between current and voltage in a circuit. It states that current is proportional to voltage and inversely proportional to resistance. Expressed algebraically, in d-c circuits $I = V/R$; in a-c circuits $I = V/Z$.

One-way throw A register that delivers air in only one direction.

Opposed blade dampers Controls for the regulation of the flow rate of air in ducts or through registers.

Outdoor design temperature Temperature to which heat is lost.

Outlet A point on a wiring system at which electric current is taken off to supply an electric load.

Outlet box A metal box containing wires from a branch circuit and connection to wires from electric load.

Output The heat delivered to a room by heating units.

Overcurrent device A device such as a fuse or a circuit breaker designed to protect a circuit against excessive current by opening the circuit.

Overload A condition of excess current; more current flowing than the circuit was designed to carry.

Package unit A boiler plus a number of its controls and accessories.

Panel or panelboard A box containing a group of overcurrent devices intended to supply branch circuits.

Panel directory A listing of the panel circuits appearing on the panel door.

Panel schedule A schedule appearing on the electrical drawings detailing the equipment contained in the panel.

Parallel circuit One where all the elements are connected across the voltage source. Therefore, the voltage on each element is the same but the current through each may be different.

Peformance data Ratings for heating, cooling cfm, and the like.

Planting screen Bushes or other planting that hides a refrigerant compressor.

Plug-in bus duct Bus duct with built-in power tap-off points. Tap-off is made with a plug-in switch, circuit breaker, or other fitting.

Polarity The directions of current flow in a d-c circuit. By convention, current flows from + to −. Electron flow is actually in the opposite direction.

Pole An electrical connection point. In a panel, the point of connection. On a device, the terminal that connects to the power.

Potable Water that is safe to drink.

Power (P) Expressed in *watts* (w) or *kilowatts* (kw), and is equal to:

in d-c circuit, $P = VI$ and $P = I^2R$

in a-c circuit, $P = VI \times$ *Power factor*

Power factor (pf) A quantity that relates the volt-amperes of an a-c circuit to the wattage, or power, that is,

power = volt-amperes × power factor

power factor cannot be greater than 1.0, and is frequently expressed as a percentage figure. In purely resistive circuits, pf equals 1.0 or 100%, and wattage equals volt-amperes.

Primary air Heated or cooled air directly from the conditioner.

Primary service High voltage service, above 600 v.

Private sewage treatment Sewage treatment other than in central city treatment plants.

"Process" hot water Hot water needed for manufacturing processes over and above the "domestic hot water" that is for the personal use of industrial workers.

Psi Pounds per square inch pressure.

Public use Fixture use in a public building where toilet room use is greater than in a private residence.

Pull box A metal cabinet inserted into a conduit run for the purpose of providing a cable pulling point. Cable may be spliced in these boxes.

Raceway Any support system, open or closed, for carrying electric wires.

Radiant cables   Electric cables embedded in the ceiling for heating.

Range hood   Hood over a stove to collect odor-laden air that is to be exhausted.

Receptacle poles   Number of hot contacts.

Receptacles wires   Number of connecting wires including the ground wire.

Recessed   A convector cabinet that extends partially or fully into a pocket in the wall.

Recessed unit   A heating element flush with the floor.

Recharge   Putting water back into the ground.

Reflection factor or reflectance   Ratio between light reflected from, and light falling on, an object.

Register   Slotted frame for control of the *direction* of air delivered to the space and its flow-rate.

Regressed lens   An arrangement where the fixture lens is set back into the body of the lighting fixture. A recessed lens.

Remote control (RC) switch   A magnetically operated *mechanically* held switch, normally used for remote switching of blocks of power.

Resistance (R)   The unit in an electric circuit analagous to friction in a hydraulic circuit, expressed in *ohms*.

Reverse return   A return main that does not go directly back to the boiler but is *reversed* to serve all the further convectors.

Riser diagram   Electrical block-type diagram showing connection of major items of equipment. It is also applied to signal equipment connections, as a fire-alarm riser diagram. Generally applied to multistory building.

Riser shaft   A vertical shaft in a building designed to house the electric riser cables.

Romex   One of several trade names for NEC type NM nonmetallic sheathed flexible cable.

Roof drain   A metal water collector flashed into a flat roof. Usually provided with a strainer to exclude debris.

Roof slope   Pitch of a flat roof to direct rainwater to a roof drain.

Roughing dimensions   Locations of water supply and drainage pipes to assure proper fit of a plumbing fixture.

Runout   A branch pipe (supply or return) from a hot water main to a convector cabinet or a convector baseboard.

R-value   Resistance rating of thermal insulation.

Sanitary drainage   Removal of sewage from a building.

Sealtite   Trade name for waterproof flexible steel conduit.

Secondary service   Low voltage service, up to 600 v.

Secondary air   Air from the space that is drawn along with the primary air, resulting in a milder mixture.

Seepage pit   A chamber that receives the effluent of a septic tank and allows it to seep into the surrounding earth.

Sensible heat   Heat that raises the air temperature.

Septic tank   A tank in which sewage is held and partially purified.

Series circuit   One with all the elements connected end to end. The current is the same throughout but the voltage can be different across each element.

Service drop   The overhead service wires that serve a building.

Service sink   Sometimes called a "slop sink"; a low sink for mopping operations by the custodial staff.

Service switch   One to six disconnect switches or circuit breakers. Purpose is to completely disconnect the building from the electric service.

Shielding and cutoff   Terms indicating the action of a lighting fixture to shield the lamp source from the viewer. Cutoff is the point in the field of vision where this shielding begins.

Shop drawings   Contractors' or manufacturers' drawings giving equipment construction details.

Short circuit   An electric circuit with zero load; an electrical fault.

Shutoff valve   A valve near a plumbing fixture for turning off the water.

Single pole   Connects to a single hot line.

Six-foot rule   The NEC rule that no point along a wall area be more than 6 ft from a wall outlet.

Sleeve   A through-the-wall box into which an incremental "chassis" is placed.

Soil   Major pollutants in plumbing.

Spacing to mounting height ratio (S/MH)   Figure provided by fixture manufacturer indicating maximum S/MH for uniform lighting results.

Specific heat of air   The heat (Btu) necessary to raise 1 lb of water 1F°.

Specular reflection   Mirrorlike reflection.

State labor laws   Ordinances that require adequate health and safety planning for workers.

Storm drainage   Removal of rainwater from a roof or other area.

Static head   Frictional resistance to airflow in a system.

Sweat fitting   A soldered connection of a tube to a fitting.

System voltage   Voltage from the power company; transformer voltage.

Tankless heater   A coil in a hot water heating boiler for heating *domestic* hot water.

Temperature difference   Thermal "pressure", indoors to out.

Temperature drop   As applied to water systems, the *difference* in temperature of water leaving and returning to the boiler.

Temperature drop   As applied to air systems, the difference in temperature of the return air and the heated air delivered.

Temperature rise   The difference in temperature of the return air and the cooled air delivered.

Thermal transfer   Moving heat into or out of occupied space or between thermal media.

Three-way switching   An arrangement for controlling (an) outlet(s) from two locations.

Three-way switch   A three-terminal switch that connects c-a or c-b where c is common.

Throw   The distance (in ft) from the register that air is "thrown".

Tic mark   Hatch mark on drawing raceway symbol, showing number of wires.

U Coefficient   Rate of heat transmission.

Upfeed system   Boiler located below convectors.

Upflow, downflow, horizontal   Furnace types classified by direction of airflow.

Upstream   Electrically speaking, in the direction toward the power source.

Utilization voltage   The voltage that is utilized; motor voltage.

Vacuum breaker   A device to prevent a suction in a water pipe.

Valve   A control to restrict or shut off water flow.

Vent   Air-filled piping that prevents siphonage of trap seals or the bubbling of air through trap seals.

Ventilation   Controlled outdoor air for freshness.

Venturi tee   Directs water through the convector branches.

Voltage (V)   The electric pressure in an electric circuit, expressed in *volts*.

Voltage drop   The voltage drop around a circuit including wiring and loads must equal the supply voltage.

Water cooler   An electric, refrigerated drinking fountain.

Water hammer   Banging of pipes caused by the shock of closing faucets.

Water services   Hot and cold water piping and equipment.

Waste   Minor pollutants in plumbing.

Weather barrier   A divider between the exterior and interior working parts of an incremental conditioner.

Wire-nut   Trade name for small, solderless, twist-on branch circuit conductor connector.

Wireway   Term generally used to mean a surface raceway.

Wiring diagram   Diagram showing actual wiring, with numbered terminals. All wiring is shown.

Wiring device   Receptacle, switch, pilot light, small dimmer, or any device that is wired in a branch circuit and fits into a 4-in. outlet box. Normally 30 amperes or smaller.

Working plane   Generally taken as 30 in. above the floor, but can be set at any desired elevation.

Zone   Section of a heating and/or cooling system separately controllable.

Index

Additional reading, 22, 61, 81, 104, 136, 173, 186, 218, 249, 266, 297, 347, 391, 448, 549
Air conditioners, centralized, 109
  complete system, 9, 19, 20, 139
  principles of, 16
  through wall, 17, 18
Air, duct system, 108
  flow rate, 107
    variable, 145
  fresh, 20, 123
  handling, small unit, 19
  heating, perimeter loop, 108
  systems, 107
  velocities, recommendations, 133
Alternating current, 290
  circuit characteristics, 290, 292
  definition, 290
  frequency, 290
  impedance, 290
  power and energy, 290, 291
  single phase, 293, 294
  three phase, 294, 295
  vs d-c, 290
American Society of Heating, Refrigerating and Air Conditioning
    Engineers, ASHRAE, 3
Ampacity, definition, 284
  table, 310
Aquastat, 69
Architects, Mogensen, 64, 122
  Scheiner, 84
Architectural lighting elements, 458
  brackets, Appendix A-6
  cornice, 469
  coves, 458, 459
  valance, 468
  see also Lighting fixtures
Architectural plan, electrical, 370
  Basic plan house, 372–374, 378–379, 388–389
  classroom layout, 533, 536
  commercial building, 537–543
  definition, 300–301
  design procedure, 368
  equipment layout, 370
  how to show motors, 526–530
  Merker residence, 452, 453, 474–477, 482
    plot plan, 490
  signal device plan, 496–498
  telephone floor plan, 505
  see also circuiting; design criteria; and Schedules, electrical
Association of Home Appliance Manufacturers (AHAM), 17

Baseboard, 33
  electric, 49
  finned tube, 34, 69
Basic plan, 38
  section, 39

Batteries, 268
Blower, air, 16, 19, 111, 113
Boiler, electric, 13, 100
  gas fired, 13, 42
  oil fired, 12, 29, 75
Boiler room, 68, 95
Boosters or circulars, see Circulating pumps
Branch circuits, 302
  capacity, electric heat, 484
  definition, 302
  NEC requirements table, 484
  protection, 350, 351
  types, NEC, 375
  wiring methods, 302–308
Branch vent, 197
Brightness, lighting, 398
  brightness ratio, 398
Building drains, sizing, 235
Building, industrial, 84
Burner, gas, 13
  oil, 12
Bus duct, see Electrical materials and installation methods

Cast iron for DWV, 196
Celsius, 3
Central Equipment, 110
Centralized system, discussion, 175
Cesspool, 211
Circuit breakers, 353
  GFCI type, 361–363
  molded case, illustration, 356
  ratings, 357
Circuiting, 376
  basic plan house 385–389
  guidelines, residential, 371, 375
  load calculation, 480
  Merker residence, 479
  residential circuits table, 464–465
  school, 535, 536
  technique, 376–385
  wire capacity-electric heating, 481
Circuit protection, 350. See also Circuit breakers; Fuses; and GFI, GFCI
Circulating pumps, 42, 68, 93, 100
Classrooms, Vincent Smith School, 123
Cleanout, sanitary drainage, 197
Climate control, modern, evolvement of, 7
Codes, National Electrical Code, 300
  National Standard Plumbing Code, 220
Coil, domestic hot water, 192, 193
Coils, cooling evaporator, 9, 19, 109, 112, 113, 116
Color coding, see Voltage systems; Wire and cable
Comfort, indoors, 2
Compressor, 9, 117
Condensate drain, cooling units, 9, 109
Condenser, 117

Conduit, electric, 307
    fittings, tables, 320, 321
    fittings types, 318, 319, 322
    flexible, 324
    installation details, 317
    properties table, 316
    types, 303, 314
    see also Raceways, electric
Connectors, wire and cable, *see* Electrical materials and installation methods
Contractors shop drawing, 121, 123
Contrast, lighting, 398
Control ladder diagrams, 529
Controls, Merker residence, 157, 158
    incremental unit, 181
Convectors, institutional, 84
    underfloor, 36, 70
Conversion, Fahrenheit, Celcius, 3
    feet of head to psi, 40
    heat, steam to water, 60
Cooling, methods, decentralized, 16
    principles, 16
    rule of thumb, 181
    units, 116, 117
Copper, for DWV, 200
    tubing, 32, 97
Critique, hot water versus electric, 77
Cubic feet per minute (cfm), 107
Current-Carrying-Capacity, *see* Ampacity
Current, electric, 272
    alternating (a-c) 290–292
    comparison of a-c and d-c 290–292
    definition, 272
    direct (d-c), 272
    units of, 273

Damper, opposed blade, 120
    splitter, 120
Dampers, air flow, 120
    opposed blade, 120
    and piping, 89
    splitter, 144
Degree days, by location, 57
Design criteria, electrical, 370
    commercial building, 354, 536, 537
    residential, 370
      appliance ratings table, 464
      circuitry, 371
      kitchen, dining areas, 463
      living areas, 450
      sleeping areas, 467
      utility areas, 470
    schools, 531–532
Design procedure, wiring, 368. *See also* Circuiting; Design criteria
Domestic hot water, 256
    circulation, 188
    heating, 228
    tank storage, 188
Down flow furnace, 112
Drainage pit for water, 254
Drawings, shop, 123
    site, 262, 264
Duct banks, 511. *See also* Underground electric service

Duct, connections, flexible, 131
    control, 119, 120
    covering, 121, 122
    dampers, 120
      splitter, 144
    glass fiber, 119
    rectangular, specifications, 124
    round, specifications, 125
    turning vanes, 120
DWV, cast iron for, 196
    copper for, 200
    drainage waste and vent, 196
    plastics for, 199

Edge loss factors, slabs, 6
Electric circuits, 274
    characteristics, 279
    parallel circuits, 275–278, 284
    power and energy, 279–282
    power loss in conductors, 285
    series circuits, 274, 284
    voltage drop, 283–285
      levels, 292–293
      systems, 293–295
    see also Alternating current; Power, electric
Electric closet, 519, 522
Electric Energy Association, recommendations, 55
Electric heating, design, 54, 56, 75, 78, 138, 176
Electric power, *see* Power, electric
Electrical drawings, 300, 301. *See also* Architectural, electrical plan; Control diagram; One line diagram; and Wiring diagrams
Electrical materials and installation methods, batteries, 268
    bus duct, 520, 521, 523
    circuit breakers, 353
      dimensional data Appendix A-27
    conduit, types, 303, 314–324
    connectors, 313
    fuses, 353, 355, 357
    meters, 285. *See also* Motors; Panel boards; Pull boxes; and Symbols
    outlets, 332–329
    power distribution equipment, 519
    Q floor 547
    raceways, exposed, 325–329
    safety switches dimensional, data Appendix, A-26
    service equipment, 517
    switchboards, 519, 539
      typical dimensions Appendix, A-25
    transformers, 292
      pad-mount details, Appendix, A-23
    underfloor duct, 543, 545, 546
    wire and cable, 309
    wiring devices, 339–346
    see also listings for individual items
Electrical plan, *see* Architectural, electrical plan
Electric service, 510
    cable, 486
    calculation of current, power, 516
    description, 485, 486
    details, 487–491, 493
    diagrams, 485
    emergency services, 516
    metering 492–494

one-line diagram, 517
   primary service, 510
   riser diagram, 517
   secondary service, 510
   service equipment, 515–518
   service size-amperes table, 515
   voltages, 293–295, 516
   See also Overhead electric service; Underground electric service
Elevations, light industry building, 85
   Merker residence, 66
Energy, electric, 279
   calculation in dc circuits, 282, 283
   cost calculation, 284
   units, 279
   see also Power, electric
   sources for heating, 8
Energy saving, 54, 55, 109, 137, 140
Engineers layout, 121, 131
Evaporator, 116
Expansion, chambers, 188, 230
   PVC, Copper, steel cast iron, 207

Fahrenheit, 3
Filter, air, 9, 19, 20, 123
   electronic cleaner, 113
Fire alarm equipment, *see* Signal equipment
Fittings, expansion, 33, 36, 68
Fixture units, drainage, 235
   water, 224
Flexible ducts, 122
Fluorescent lamps, 406
   ballast data, table, 412
   circuit diagrams, 413, 417, 418
   description of types, 416
   general characteristics, 406, 415
   interchangeability, 419
   shape, bases, holders, 411
   typical data, table 414
Footcandles, 296
   measurement, 402, 403
   see also Lighting calculations
Footcandles, ESI, 396
Forced air furnaces, 109, 111, 112, 113
Fresh air, 123
Frequency, a-c circuit, 290
Friction, fittings, allowances, 43
Fuel storage, 10, 74
Furnace, down-flow, 112
   types, 20
   up-flow, 113
Furred space, to enclose pipes, 229
Fuses, 353
   ratings, 357
   types, illustration, 355

Gas piping, 203
GFI, GFCI, *see* Ground fault protection
Glare, lighting, 399
   direct, 399
   indirect, veiling reflections 399–401
Grills, 121, 129, 165–169
Ground fault protection, 358–365
   ground fault circuit interrupters (GFGI), 358–363

Grounding, 358–360
   detail of system grounding requirements, Appendix, A-11
   typical service grounding diagram, 359
Guidelines, for electrical layout and circuitry, *see* Design Criteria

Hangers, pipe, 202
   plumbing fixtures, 260
Heat gain, 14
   calculation sheet, 132
   lights, equipment, sensible-latent, 15
   shading, ventilation, occupancy, 15
   solar gain, thermal lag, 14
   temperature range, 15
   time of day, orientation, latitude, 14
   undersizing, swing, 16
Heating, bathroom heaters, 53
   boiler and controls, 24
Heating, cooling, design, 130, 135, 140
   all electric, 138
   distribution, 30
   electric, 47
     units, 50
   floor, inserts, 51
   hot water, 24
     two pipe, 94, 97
   nonresidential, 84
   radiant, 52
   residential, 63
   system components, 24, 27
Heat loss, 2
Heat pump, 138, 141, 147, 151
   performance, 156
Hot water heating, design, 40, 46, 69, 72, 90, 103
House sewer, 197
House trap, 197
Humidifier, 109

I-B-R, ratings, baseboard, 35
Impedance, electric, 290
Incandescent lamp, 402
   construction, 404
   operating characteristics, 409
   relfector, projector types, 408
   shapes and bases, 405, 407
   typical data, tables, 404, 407
Incremental units, 176
   available types, 177
   exploded and cutaway views, 178, 179
   performance data, 183
Industrial building, 84
   design, heating, 90
     heating-cooling, 182
Infiltration, 4
Insulation, effect on U-coefficients, 4
   duct covering, 121, 122

Light industry, building for, 86
Lighting, 393
   brightness, 398
   classroom, 533, 535
   color, 402, 427
   contrast, 398
   control, 425
   diffuseness, 402

fixtures, 428
footcandle, 396
glare, 399
layout technique, 460
measurements, 402, 403
metric units, 442
reflection, 394
spacing to mounting height ratio, 439
systems, 435
transmission, 395
uniformity, 439
vision, 396
Lighting calculations, 396
classrooms, 533
estimates, 445–447
footcandles, lumen method, 396, 442–445
metric units, 442, 444
zonal cavity method, 439
Lighting fixture, 428
coefficient of utilization, 439
details, 428–433, 435–438, 454, 459, 461, Appendix, A-2 to A-10
diffusers, 435
efficiency, 439
fluorescent, 428–433, 438
incandescent, 428, 429, 431–437
schedule, 371
for Basic Plan House, 373
for Merker residence, 455
see also Architectural lighting elements
Lighting systems, 435, 440, 441
Light sources, 403
efficacy, 425
fluorescent lamps, 406
HID, 419
incandescent lamps, 402
mercury lamps, 420
metal-halide lamps, 425
quartz (tungsten-halogen) lamps, 406
sodium lamps, 425
Lumen, 396. See also Lighting calculations

Measurement, 286
current, 285–286
electric power and energy, 286–289
in electric circuits, 285
voltage, 286
Mercury lamps, 420
ballast data, table, 424
characteristics, 420, 425
shapes, 421
typical data, table, 422–423
Merker Residence, 64, 67, 68
Metal-halide lamps, 425
comparative characteristics, 425
Meters, 285
ammeter, 285–287
service metering, 493–495, 518
submetering assemblies, 523
voltmeter, 285–287
watt-hour meter, 288–289
wattmeter, 288–289
Metrication, 553
Milinch, discussion of, 89
Motors, electric, 526
control and wiring diagrams, 528–529
controllers, starters, control, centers (MCC) 528–531
dimensional data Appendix A-28
schedules, Appendix, A-14 to A-16
detail, of indoor installation, 323
of outdoor installation, 528
enclosure types, NEMA, 528
NEC references, 526
shown on drawings, 526
symbols, 527, 530

National Electric Code (NEC), 300
National Standard Plumbing Code, 220

Ohms Law, 273
in a-c circuits, 290
in d-c circuits, 273
One-line diagram, 300
definition 300
power system, commercial building, 541
service equipment, 517–518
symbols, 354
typical, showing device locations and terminology, 352
Outlets, electrical, 332
boxes, details, 334
on electrical drawings, 335
ganged, 335
how to count, 337
junction boxes, 335
outlet box wire capacity, table, 381
symbols, 338
typical details, 333, 336
Overcurrent protection, see Circuit protection
Overhead electric service, 486
service drop detail 487–489
shown on plot plan 490
see also Electric service

Panelboards, 365, 367–368
dimensional data, 524
identification, 537
illustrations, 364
installation detail, Appendix, A-20
schedules, Basic plan house, 379, 389
Merker residence, 478, 483
switch and fuse type, 353
typical schedules, 366, A-17—A-19
Performance, curve, pump, 42, 93, 100
Pipe, sizing, 99
supports 192, 202
Plans, basic house, 38
light industry building, 86
Merker residence, 67
Plastic pipe, available types, 189
Plastics, for DWV, 199
Plumbing, fixtures, 208, 238, 240, 255, 258, 259
clearances, 253
nonresidential, 251
requirements, residential, 223
residential, 219
section, 252
Power, electric, calculation, in a-c circuits, 290–291
in d-c circuits, 282
generation, 270
mechanical analgies, 280–281
units, 279

Preassembly, DWV systems, 204, 205
Pressure at fixtures, water, 225
Pressure loss and velocity, 41, 43
Pull boxes, 522
    application in risers, 525
    diagram, 525
    exterior, detail, Appendix, A-24
    identification, 537
    size calculation, 523
Pumps, systems and friction, 88

Quartz lamps, tungsten-halogen, 406
    construction, 409
    typical data, table, 410

Raceways, electric, exposed, 316
    surface types, tables, 325
    typical application, 332
    vertical-pole type, 329–331
Receptacles, electric, 339
    residential application, table, 464
    see also Wiring devices
Reflection, light, 394
Refrigeration cycle, 147
Registers and grills, 127, 128, 129, 160, 163, 165, 169
Resistance, electric, 273
    Ohms Law, 273
    unit, 273
Riser diagram, 519
    power riser and schedules, 538, 539
    riser shafts, 519
    riser with offsets, 525
    service equipment, 517
    signal equipment, 499
    stair and exit lights, 544
    telephone riser, 504
    typical power riser, 520
Roof, drain fixture, 263
    drainage, 215, 217
    gutters, sizing, 246

Sanitary drainage, 195, 255
    brances and stacks, sizing, 236
    design, 234
    material for, 196
    plan, 231
    section, 232
    systems, 196
    traps and vents, 195
    typical system, 197, 199, 203, 205
Schedules, electrical, 548
    conduit and cable, Appendix, A-12
    drawing changes, Appendix, A-13
    equipment, 541
    main switchboard, 539
    mechanical space circuits, 541
    motor and control, Appendix, A-16
    motor control center (MCC), Appendix, A-14, A-15
    panel, single phase, Appendix, A-17
      three phase, Appendix, A-18, A-19
    transformers, 541
    see also Lighting fixture
School, Vincent Smith, 122, 126
Schools, electrical work, 531. see also Architectrical plan, electri-

cal Circuiting; Design criteria; and Lighting calculations
Seepage pit, 211
    area, sewage effluent, 243
Septic tank, 212, 214, 211
    capacities, 242
Sewage, flow, by building type, 243
    treatment, private, 207
Shop drawings, ducts, 123
Signal equipment, 492
    fire alarm equipment, 501, 502
    intercom equipment, 503
    symbols, 500
    telephone, 506
    television equipment, 503
Single-line diagram, see One-line diagram
Sodium lamps, 425
    high pressure data, 426
Soil stack, 197
Specific heat, air, 107
    water, 45
Static head, 134
Steam, conversion to hot water, 60
    used in large buildings, 58
Storm drainage, 210, 226, 263
    design, 245
Storm leaders, sizing, 246
Street services, 262
Supports, wall-hung fixtures, 260, 261, 265
Switches, see Electrical materials and installation methods Wiring devices
Switching, 380
    hi-low level, 454
    low voltage, 471
    multiple point, 472
    see also Wiring diagrams
Symbols, Part I, raceways, 304
    Part II, outlets, 338
    Part III, wiring devices, 341
    Part IV, abbreviations, 345
    Part V, one line diagrams, 354
    Part VI, equipment, 369
    Part VII, signaling devices, 500
    Part VIII, motors and motor control, 527
    Part IX, control and wiring diagrams, 530

Tankless heater, external, 193
    internal, 192
Tees, special, 33, 37
    venturi, 68
Temperature, Celsius, 3
    drop, hot water, 45
    Fahrenheit, 3
    design, indoors, outdoors, 2
    differences, 3
    rise, 107
Thermal transfer, air, 106
Thermostat, 33, 54, 69
Tile drains, sewage effluent, 211
Tile length, sewage effluent, 243
Ton of refrigeration, 131
Transformers, see Electrical materials
Transmission, light, 394–396
    through materials, 397
Trap sizes, 234

Tubing, copper, 32, 97
  drain, locations, 47

U-coefficients, 6
Underground electric service, 486
  cable terminations, 493
  direct burial cables, 491, 492
  duct bank details, 511
  exterior pull box, Appendix, A-24
  handhold details, Appendix, A-22
  manhole details, 512–513
  riser pole detail, 490
  typical details, 514, Appendix A-21
  typical plot plan, 514
  see also Electric service
Up flow furnace, 113

Vacuum breakers, 188
Valves, air-vent, 36
  check, gate, globe, 191
  shutoff, 71, 73
Veiling reflections, *see* Glare
Velocity, air in ducts, 133
  fluid in tubes, 41
Ventilation, 20, 102, 170
Vents, sanitary sizing, 236
Vent stack, 197
Vincent Smith School, 122
Voltage, 271
  d-c 271
Voltage levels, 292–293
  system voltage 516
  utilization voltage, 516
Voltage systems, 293
  color coding, 293
  service entrances, 293–295
  single phase, 2-wire, 294
    3-wire, 294
  system/utilization voltage, table, 516
  three phase, 4-wire, 295

Warm air furnaces, types, 20
Water, demand, estimating, 225

distribution, plan, 252
domestic, hot, 190, 194
expansion and shock, 190
flow pressure, 225
main pressure, 221
pipe, sizes to fixtures, 224
  tubing, fittings, valves, 189
service pipe, sizing, 221, 226
services, 188, 221, 252
specific heat, 45
treatment and bypass valve, 188
Wire and cable, 304
  color coding, 312
  conductor gauge (AWG), 309
  definitions, 309
  insulation and jacket types, 310–312
  properties, of bare conductors, table, 309
    of insulated conductors, table, 310, 311
  service entrance, 486
  types, 302–303
    AC (BX), 302–303, 304
    MI, 307
    NM, 302–303, 305
    USE, 486
  see also Ampacity
Wiring devices, 339–346
  low voltage switching, 471
  receptacle configuration charts, 342, 343
  receptacles, 339–340
  switches, 340, 345
  symbols, 341
  typical details, 344–345
  see also individual listings
Wiring diagrams, 300
  definition, 300, 301
  motors, 529
  for split receptacle, 376
  switching, 380, 472
    low voltage, 471
Wiring methods, *see* Branch circuits; Conduit; Wire and cable

Zones, 73, 97, 143